272
CHE

AN INTRODUCTION
TO AGRICULTURAL
BIOCHEMISTRY

WITHDRAWN

D1394020

* ५ ५ 5 1 4 6 9 2 *

JOIN US ON THE INTERNET VIA WWW, GOPHER, FTP OR EMAIL:

WWW: http://www.thomson.com
GOPHER: gopher.thomson.com
FTP: ftp.thomson.com
EMAIL: findit@kiosk.thomson.com

A service of I(T)P

AN INTRODUCTION TO AGRICULTURAL BIOCHEMISTRY

J.M. Chesworth

Honorary Lecturer in Agricultural Biochemistry and Nutrition
Department of Agriculture
University of Aberdeen, UK

and

Independent Consultant
Mar Technical Services
Huntly, UK

T. Stuchbury

Lecturer in Agricultural Biochemistry and Plant Physiology
Department of Agriculture
University of Aberdeen, UK

and

J.R. Scaife

Lecturer in Agricultural Biochemistry and Nutrition
Department of Agriculture
University of Aberdeen, UK

CHAPMAN & HALL

London · Weinheim · New York · Tokyo · Melbourne · Madras

Published by Chapman & Hall, 2–6 Boundary Row, London SE1 8HN, UK

Chapman & Hall, 2–6 Boundary Row, London SE1 8HN, UK

Chapman & Hall GmbH, Pappelallee 3, 69469 Weinheim, Germany

Chapman & Hall USA, 115 Fifth Avenue, New York, NY 10003, USA

Chapman & Hall Japan, ITP-Japan, Kyowa Building, 3F, 2-2-1 Hirakawacho, Chiyoda-ku, Tokyo 102, Japan

Chapman & Hall Australia, 102 Dodds Street, South Melbourne, Victoria 3205, Australia

Chapman & Hall India, R. Seshadri, 32 Second Main Road, CIT East, Madras 600 035, India

First edition 1998

© 1998 Chapman & Hall

Typeset in 10/12 Palatino by Saxon Graphics Ltd, Derby

Printed in Great Britain by St Edmundsbury Press, Bury St Edmunds, Suffolk

ISBN   0 412 64390 1

Apart from any fair dealing for the purposes of research or private study, or criticism or review, as permitted under the UK Copyright Designs and Patents Act, 1988, this publication may not be reproduced, stored, or transmitted, in any form or by any means, without the prior permission in writing of the publishers, or in the case of reprographic reproduction only in accordance with the terms of the licences issued by the Copyright Licensing Agency in the UK, or in accordance with the terms of licences issued by the appropriate Reproduction Rights Organization outside the UK. Enquiries concerning reproduction outside the terms stated here should be sent to the publishers at the London address printed on this page.

The publisher makes no representation, express or implied, with regard to the accuracy of the information contained in this book and cannot accept any legal responsibility or liability for any errors or omissions that may be made.

A catalogue record for this book is available from the British Library

Library of Congress Catalog Card Number: 97-68949

WARWICKSHIRE COLLEGE
LIBRARY

Class No:
630.27419

Acc No:
00514692

∞ Printed on permanent acid-free text paper, manufactured in accordance with ANSI/NISO Z39.48-1992 and ANSI/NISO Z39.48-1984 (Permanence of Paper).

CONTENTS

PREFACE

Biochemistry books tend to fall into two distinct categories: specialist volumes written for the researcher that deal in great depth with narrow areas of the subject, and large, comprehensive textbooks that aim to give an overall coverage of the material. Many of the latter emphasize areas of particular interest to students of medicine and related disciplines. By their emphases and their very size they sometimes prove daunting to students from other disciplines.

In writing this book our main aim was to provide an introduction to biochemical principles for students of agriculture, but we hope that it also proves a useful text for students of forestry, veterinary science and other areas of applied biology. These students, for whom biochemistry may not be the primary interest, need to understand how organisms function at a chemical level and how this relates to the more complex biological systems which they are studying. We have attempted to target those areas of biochemistry which are of most relevance and to keep coverage of the subject simple. The book is intended for use principally in first- and second-year courses where biochemistry is first introduced to students. Those students who find the subject interesting and wish to delve deeper are provided with references to more detailed texts.

The book is divided into four parts. Part 1, The Cell and Cellular Constituents, and Part 2, Metabolism, provide a traditional treatment of the structure and function of biological molecules and how they are synthesized and degraded, but have been written with an agricultural flavour, drawing on examples from plants and animals of agricultural importance.

Parts 3 and 4 are more of a departure from most biochemistry books. Part 3 is devoted to plants. It looks at the important stages in plant growth and development and examines the biochemical process which underpin them; it also discusses the regulation and opportunities for manipulation of such processes. Part 4 concentrates on animals and bridges the gap between biochemistry and nutrition. It examines the ways in which animals acquire nutrients through the digestive tract and looks at the biochemical basis of maintenance, growth, lactation and meat production. More specialized texts are available for the student who wants a more in-depth treatment of these areas.

Discussions with many of our colleagues have contributed to the completion of this book and to them we extend our thanks. Greatest thanks must go to our wives who have shown remarkable patience and understanding during the preparation of the manuscript.

ABBREVIATIONS

2,4,5-T (2,4,5-trichlorophenoxy)acetic acid
2,4-D (2,4-dichlorophenoxy)acetic acid
ABA abscisic acid
ACC 1-aminocyclopropane 1-carboxylic acid
ACP acyl carrier protein
ADP adenosine diphosphate
AIDS acquired immune deficiency syndrome
ALA δ-aminolevulinic acid
α-LA α-lactalbumin
AMP adenosine monophosphate
AOA aminooxyacetic acid
APS adenosine-5′-phosphosulphate
ATP adenosine triphosphate
ATPase enzyme that hydrolyses ATP
AVG aminoethoxyvinylglycine
BA benzyladenine
BAT brown adipose tissue
BCCP biotin carboxyl carrier protein
β-LG β lactoglobulin
BSE bovine spongiform encephalopathy
bST bovine somatotropin
CAM crassulacean acid metabolism
cAMP 3′,5′-cyclic adenosine monophosphate
CCC chlorocholine chloride (cycocel)
CCN cerebrocortical necrosis
cDNA complementary deoxyribonucleic acid
CDP cytidine diphosphate
CMP cytidine monophosphate
CoA coenzyme A
CTP cytidine 5′-triphosphate
Cyt cytochrome
DAG diacylglycerol
DFD dark firm dry (of meat)
ΔG free energy change for reaction
DGDG digalactosyldiacylglycerol
DHA docosahexaenoic acid
DM dry matter
DMAPP dimethylallyl pyrophosphate

DNA deoxyribonucleic acid
DUP digestible undegradable protein
EFA essential fatty acid
EPA eicosapentaenoic acid
EPSP synthase 5-enolpyruvylshikimic acid-3-phosphate synthase
ER endoplasmic reticulum
ES enzyme–substrate complex
F-2,6-P fructose-2,6-bisphosphate
F-6-P fructose-6-phosphate
f-Met N-formyl methionine
FAD flavin adenine dinucleotide (oxidized)
$FADH_2$ flavin adenine dinucleotide (reduced)
Fd/FdH_2 ferredoxin (oxidized and reduced forms)
FIGLU formiminoglutamic acid
FLKS fatty liver and kidney syndrome
FMN flavin mononucleotide (oxidized)
$FMNH_2$ flavin mononucleotide (reduced)
FPP farnesyl pyrophosphate
FW fresh weight
GA gibberellic acid
GDP guanosine diphosphate
GH growth hormone
GMP guanosine-5′-monophosphate
GOGAT glutamate synthase
GTP guanosine triphosphate
GS glutamine synthetase
HMG-CoA hydroxymethylglutaryl coenzyme A
hnRNA heterogeneous nuclear ribonucleic acid
HPLC high performance liquid chromatography
IAA indole-3-acetic acid
IBA indole butyric acid
IGF-I insulin-like growth factor-1

IGF-II   insulin-like growth factor-2
IGFs   insulin-like growth factors
IMP   inosine monophosphate
IPP   isopentenyl pyrophosphate
kDa   kilodaltons
kJ   kilojoule (joule $\times 10^3$)
K_m Michaelis constant
LHC   light-harvesting complex
Met   methionine
MGDG   monogalactosyldiacylglycerol
MJ   megajoule (joule $\times 10^6$)
mRNA   messenger ribonucleic acid
NAA α-naphthalene acetic acid
NAD^+ nicotine adenine dinucleotide (oxidized)
NADH   nicotine adenine dinucleotide (reduced)
NADP   nicotine adenine dinucleotide phosphate (oxidized)
NADPH   nicotine adenine dinucleotide phosphate (reduced)
NEFA   non-esterified fatty acid
NPN   non-protein nitrogen
OxAc   oxaloacetate
P_{680}, P_{700} reaction centres of photosystem II and photosystem I, respectively
PA   phosphatidic acid
PABA   *p*-aminobenzoic acid
PAL   phenylalanine ammonia lyase
PC   phosphatidylcholine
Pc   plastocyanin
PCR   polymerase chain reaction
PDH   pyruvate dehydrogenase
PE   phosphatidylethanolamine
PEM   polioencephalomalacia
PEP   phosphoenolpyruvate
PES   prostaglandin endoperoxide synthetase
PFK1   phosphofructokinase 1
PFK2   phosphofructokinase 2
PG   phosphatidylglycerol
PGE_2 prostaglandin E_2

PGF_2 prostaglandin F_2
PGR   plant growth regulator
P_i orthophosphate
PI   phosphatidylinositol
PME   pectin methylesterase
PMF   proton motive force
PP_i pyrophosphate
Pq   plastoquinone
Pr/Pfr   phytochrome (red and far-red absorbing forms)
PRPP   phosphoribosyl pyrophosphate
PS   phosphatidylserine
PSE   pale soft exudative (of pig meat)
PSI   photosystem 1
PSII   photosystem 2
PSS   porcine stress syndrome
PUFA   polyunsaturated fatty acid
PV   peroxide value
RET   resonance energy transfer
RFLP   restriction-fragment length polymorphism
RNA   ribonucleic acid
rRNA   ribosomal ribonucleic acid
Rubisco   ribulose-1,5-bisphosphate carboxylase/oxygenase
S   Svedberg
SAM   S-adenosylmethionine
SMCO   S-methyl cysteine sulphoxide
snRNP   small nuclear ribonuclear particle
SRP   signal recognition particle
TCA   cycle tricarboxylic acid cycle
TPP   thiamin pyrophosphate
tRNA   transfer ribonucleic acid
UDP   uridine diphosphate
UDPG   uridine diphosphate glucose
UMP   uridine monophosphate
UTP   uridine triphosphate
UV   ultraviolet
VFA   volatile fatty acid
VLDL   very low density lipoprotein

PART ONE

THE CELL AND CELLULAR CONSTITUENTS

CELL STRUCTURE AND FUNCTION 1

1.1 INTRODUCTION

The cell is the unit from which living organisms are built. Despite the wide diversity of organisms, the cells of which they are composed have many common features and most carry out very similar biochemical processes.

The cell consists of a plasma membrane surrounding the cytoplasm, in which a variety of structures may be present. Some of these structures may themselves be bounded by membranes. In certain cells the plasma membrane may be enclosed by a cell wall. Depending on the complexity of the internal structure, cells have been classified into two basic types – prokaryotic and eukaryotic cells.

Cells that have no internal, membrane-bounded structures and no clearly defined nucleus are defined as prokaryotic cells. Bacteria and blue-green algae are examples of prokaryotic cells. Although such cells have no intracellular architecture, they contain all of the metabolic machinery necessary to allow them to grow and multiply.

Eukaryotic cells are much more complex than those of prokaryotes. Cells that contain a nucleus surrounded by a membrane, the nuclear envelope, are defined as eukaryotic. Brown, red and green algae, protozoa, fungi and multicellular plants and animals consist of eukaryotic cells. They have developed an internal system of membranes that separates the cell into distinct areas, called organelles, which have specific biochemical functions (Table 1.1) and allow more ordered and directed metabolism to occur. In addition, multicellular eukaryotic organisms have evolved cells with very specialized functions and structures which are often associated in large numbers to form clearly identifiable tissues. The structure of typical animal and plant cells is shown in Figure 1.1.

1.2 COMPONENTS OF CELLS

1.2.1 PLASMA MEMBRANE

The boundary of the cell is the plasma membrane. The composition, structure and function of plasma membranes are described in Chapter 22. The plasma membrane isolates the internal contents of the cell from its environment, and thus the cell is able to maintain a relatively ordered and constant environment

Table 1.1 Function of cell structures

Cell structure	Function/pathway
Cell membrane	Transport of nutrients; hormone/receptor interactions; cell recognition; permeability barrier
Cytoplasm	Glycolysis; gluconeogenesis; pentose phosphate pathway; polysaccharide breakdown; complex lipid breakdown; fatty acid synthesis; protein breakdown; amino acid synthesis
Smooth endoplasmic reticulum	Fatty acid elongation and desaturation; complex lipid synthesis; detoxification reactions
Ribosomes	Protein synthesis (proteins destined for storage or secretion are synthesized by ribosomes attached to the endoplasmic reticulum)
Chloroplast	Photosynthesis; fatty acid synthesis; complex lipid synthesis; synthesis of some amino acids; synthesis of organelle proteins; Calvin cycle (stroma); light reactions (thylakoids); reduction of nitrate and sulphate, part of photorespiration
Mitochondria – matrix	TCA cycle; fatty acid oxidation; amino acid oxidation; gluconeogenesis; synthesis of organelle proteins
Inner mitochondrial membrane	Electron transport chain; oxidative phosphorylation; metabolite transport
Nucleus	DNA synthesis; RNA synthesis; processing of RNA
Golgi bodies	Carbohydrate synthesis (e.g. lactose); glycoprotein synthesis (addition of sugar residues to existing proteins); packaging of cell products
Peroxisomes	fatty acid oxidation; amino acid oxidation; photorespiration
Glyoxysomes	Glyoxylate pathway; fatty acid oxidation; amino acid oxidation
Lysosomes	Lipoprotein breakdown; recycling of cellular constituents; destruction of foreign bodies
Plant vacuoles	Maintenance of turgor; storage of toxic and waste products
Cytoskeleton	Cytoplasmic streaming; movement of subcellular organelles; maintenance of cell shape
Flagella and cilia	Movement

despite large changes in the composition of the medium in which it lives and grows. This is possible because the plasma membrane of all organisms is a selectively permeable barrier that controls the movement of molecules into and out of the cell. Essential nutrients required for growth and metabolism are allowed to cross the plasma membrane into the cell, often by tightly controlled, energy-dependent transport processes. Waste products produced by the cell, which if allowed to accumulate would be toxic, are excreted by similar mechanisms. Although the plasma membranes of cells have adapted with time to have specialized functions, their essential feature is the presence of a lipid bilayer, composed mainly of phospholipid and protein. This bilayer forms a hydrophobic barrier that prevents the passage of most polar molecules such as inorganic ions, sugars and proteins. Some of the proteins within the membranes have specific functions. They may act as transport proteins, enzymes, or recognition proteins such as receptors. These are discussed in more detail in later chapters.

1.2.2 CYTOPLASM

Inside the plasma membrane is the cytoplasm. This is composed of the cytosol, and the structures contained within it. The cytosol is an aqueous solution in which are dissolved many organic compounds such as sugars, amino acids, proteins and inorganic materials. In

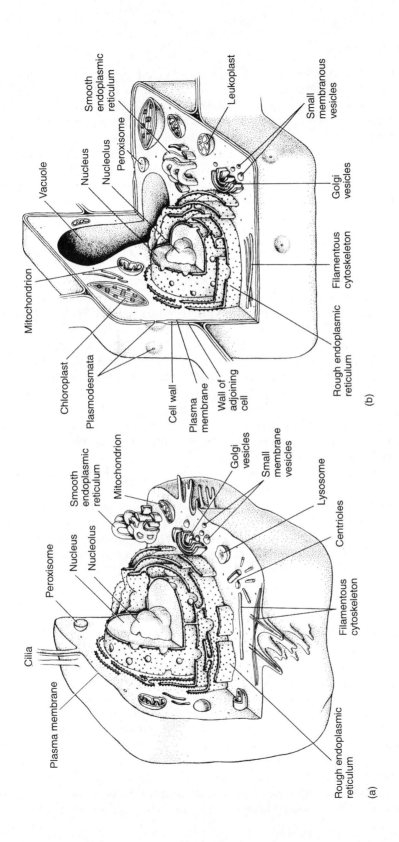

Figure 1.1 Structure of (a) an animal cell and (b) a plant cell. Not all of the intracellular organelles and plasma membrane-associated structures are found in every cell: the numbers of some organelles and the complexity of the intracellular structure vary in cells according to their function. (Reproduced with permission from Darnell, J.E., Lodish, H.F. and Baltimore, D. (1990) *Molecular Cell Biology*, Scientific American Books Inc.)

addition, the cytoplasm may contain insoluble particles and organelles. In the cells of higher organisms these structures can be numerous and complex, but in bacteria the intracellular structure is quite simple.

1.2.3 THE NUCLEOID AND NUCLEUS

The DNA in bacteria is in the form of a single circular strand, tightly folded with protein to form the nucleoid. Some bacteria also contain circular DNA in the cytoplasm, referred to as plasmids.

The most noticeable feature of eukaryotic cells is the presence of a nucleus surrounded by a double membrane known as the nuclear envelope. This envelope has holes or pores that allow communication and controlled exchange of material between the cytoplasm and the nuclear contents. For example, the messenger and ribosomal RNA synthesized in the nucleus is transferred to the cytoplasm where protein synthesis takes place. DNA contained in the nucleus forms a complex with proteins to give chromatin. The part of the DNA required for the production of ribosomes is located in a dense area within the nucleus known as the nucleolus.

1.2.4 CELL WALLS

The cells of plants and most prokaryotes are surrounded by a cell wall which is outside the cell membrane.

There are essentially two different types of bacterial cell wall. One type consists of a thin layer of peptidoglycan, a complex polymer of sugars and amino acids which is surrounded by an outer membrane, similar in structure to the plasma membrane, but with a somewhat different composition. This type of cell wall prevents the uptake of the dye Gentian violet, first used by Gram as a means of identifying bacteria (bacteria with this type of wall structure are referred to as Gram-negative bacteria). The other type of cell wall is composed of a much thicker layer of peptidoglycan and has

no outer membrane. Bacteria with this type of wall are stained by Gentian violet and are referred to as Gram positive.

Plant cell walls consist mainly of cellulose and other complex sugars. Because plants have no skeleton, the rigidity provided by the wall is important in providing support and protection to the plant and in enabling it to grow upright and to reach a considerable height.

1.2.5 RIBOSOMES

Ribosomes are responsible for the synthesis of proteins. They are composed of two subunits, both of which contain RNA and protein. The subunits of prokaryotic ribosomes are classified as 50S and 30S in size and these combine to form a 70S ribosome. Eukaryotic ribosomes consist of 60S and 40S subunits which combine to form 80S ribosomes. An explanation of these terms is given in Chapter 21. In eukaryotes some ribosomes may be attached to the membranes of the endoplasmic reticulum.

1.2.6 ENDOPLASMIC RETICULUM

In eukaryotes the nuclear envelope is continuous with a system of membranes in the cytoplasm called the endoplasmic reticulum (ER), which is effectively a network of membrane-bounded tubes. There are two types of ER, rough and smooth.

The rough ER is the site of synthesis of proteins destined to be secreted from the cell (such as secretion of digestive enzymes or of casein granules by the mammary tissue), or which will become confined within lysosomes or targeted to the nucleus. These proteins are synthesized on ribosomes attached to the ER membranes and are secreted into the lumen of the ER where they can be directed to their specific destinations. The 'rough' appearance of this part of the ER is due to the presence of ribosomes required for this process.

The smooth ER is devoid of ribosomes but is continuous with the rough ER. It is here that the enzymes involved in complex lipid synthesis are located. In addition, the enzymes involved in detoxification of potentially harmful compounds are also found in the smooth ER.

When cells are disrupted in the laboratory in order to study the biochemical activity of their subcellular components, the structure of the ER is lost. The continuous membrane structure is broken into many small segments which form small vesicles referred to as microsomes. These vesicles contain many of the enzymes associated with the ER; these are often referred to as microsomal enzymes, indicating that they originated from the ER.

Eukaryotic cells contain specialized areas of endoplasmic reticulum called Golgi bodies (also called Golgi apparatus or dictyosomes). Under the microscope the Golgi bodies appear to consist of layers of smooth ER stacked one on top of another. Each layer is referred to as a cisternum. The ends of the cisternae give rise to small vesicles which act as transport vesicles and contain the products of biosynthetic processes which take place in the Golgi bodies. These organelles are particularly numerous in secretory cells and it is clear that they are a major site of packaging, modification and sorting of cellular products. Many proteins are modified in the Golgi bodies, for example by the addition of sugar residues to form glycoproteins.

Secretory vesicles produced by the Golgi bodies move to the cell membrane, where the membrane surrounding the vesicle merges with the cell membrane to discharge its contents outside the cell. This process is called exocytosis or reverse pinocytosis.

1.2.7 VACUOLES AND SPECIALIZED VESICLES

Plant cells contain vacuoles which are bounded by a single membrane known as the tonoplast. In the young growing plant cell, several vacuoles may be found. These merge to form a larger vacuole as the cell matures. The movement of water to and from the vacuole causes changes in the turgor of the cell, altering the rigidity of plant tissues. The vacuole also serves as a site where materials may be stored, separated from the main biochemical processes of the cell. Thus pigments and toxic and waste materials may accumulate in the vacuole. The presence of a vacuole also reduces the amount of cytoplasm required by plant cells and allows them to reach a larger size than most animal cells.

Lysosomes are spherical membrane-bounded vesicles found in the cytoplasm of animal cells. They are specialized types of vesicles produced by the Golgi bodies. Lysosomes contain hydrolytic enzymes such as the cathepsin proteases, lipases, nucleases and carbohydrate-degrading enzymes. The contents of the lysosome are more acidic than the cytoplasm, pH 5.0 as compared to pH 7.0. The function of the lysosome is to act as a focus for the recycling of redundant cellular components and in the digestion of foreign material, e.g. bacteria brought into the cell by phagocytosis. The primary lysosome, produced by the Golgi bodies, fuses with a vesicle containing a foreign body or cell component (phagocytic or autophagic vesicles) allowing the hydrolytic enzymes to degrade the contents to simple compounds such as amino acids, sugars and fatty acids which are released into the cell for re-use.

Several other types of vesicles are found in animal and plant cells. Peroxisomes are the site of specialized amino acid and fatty acid degradation, which involves the production of hydrogen peroxide and free radicals. Both of these chemical species are highly reactive and if released into the cytoplasm could cause extensive damage. To prevent this happening several antioxidant mechanisms are found in peroxisomes. The first is the presence of the enzyme catalase which degrades hydrogen peroxide to water and oxygen. The second is the presence of the antioxidant vitamin E in the peroxisomal membrane which reacts with free radicals to produce stable, less-reactive compounds. Glyoxysomes are found only in the cells of certain plants. They contain the enzymes of the glyoxylate cycle (Chapter 11), a specialized adaptation of the TCA cycle which allows some plants to convert stored lipid into carbohydrate. They are usually found in seeds where significant quantities of fat can be stored. Their numbers increase during seed germination when the stored fat is mobilized for sugar production.

1.2.8 MITOCHONDRIA

Both plant and animal cells contain mitochondria. These subcellular organelles make up the powerhouse of the cell. They are major sites of ATP synthesis and oxidative metabolism (Chapters 11–13). In shape and dimensions they are similar to bacteria, and it is generally agreed that mitochondria have evolved from bacteria which developed a symbiotic relationship with early eukaryotic cells. In support of this view is the observation that mitochondria contain DNA, RNA and ribosomes which resemble those in bacteria. Some mitochondrial proteins are synthesized *in situ* in the mitochondrial matrix, whereas others are produced in the cytoplasm from nuclear DNA (Chapter 21).

Mitochondria have a double membrane structure. The outer membrane is relatively simple and unfolded, whereas the inner membrane is highly folded forming cristae which project into the centre of the mitochondria. The outer membrane has limited biochemical activity and is permeable to most molecular species. Thus, the composition of the cytoplasm and the contents of the mitochondrial intermembrane space is similar. The inner mitochondrial membrane has many important biochemical functions, and this is reflected in its composition which is approximately 75% protein and 25% lipid. It is the site of the mitochondrial electron transport chain and the synthesis of ATP by oxidative phosphorylation (Chapter 12). The central compartment of the mitochondrion is called the matrix. This is a concentrated aqueous solution containing many enzymes, for

example those of the TCA cycle and fatty acid oxidation. The number of mitochondria in cells varies greatly. White adipose tissue, for example, contains few mitochondria, whereas brown adipose tissue owes its colour to the large number of mitochondria it contains. Liver and muscle tissue contain many hundreds of mitochondria per cell.

1.2.9 CHLOROPLASTS

Chloroplasts are found in plant cells and share some common features with mitochondria. They are also organelles with a double membrane structure and are involved in the trapping of energy in the form of ATP. However, they are adapted to converting the energy in sunlight into ATP, in contrast to mitochondria, which use the chemical energy released during the oxidation of organic compounds (Chapter 17). The inner membrane of chloroplasts is highly folded to form numerous stacks of membrane called thylakoids. Each fold, referred to as a thylakoid disc, contains the units which carry out photosynthesis. The ability to trap solar energy is conferred by the presence of specialized pigments, mainly chlorophylls a and b, in the inner membrane of the chloroplast. The aqueous contents of the chloroplast are called the stroma. It is believed that chloroplasts may have evolved from an early form of cyanobacter which developed an endosymbiotic relationship with progenitor eukaryotic cell lines.

Photosynthetic prokaryotes lack chloroplasts but are able to carry out photosynthesis because they have pigments embedded in other specialized membranes.

1.2.10 CYTOSKELETON

Many free-living cells are able to move in their growth medium. In the case of bacteria and protozoa this is due to the presence of specialized protein fibres called flagella which rotate to propel the cell forwards. Other organisms such as amoeba are able to move without flagella and this movement is attributed to cytoplasmic streaming.

Many eukaryotic cells, from simple unicellular organisms to complex multicellular higher plants and animals, contain a cytoskeleton. This is a system of protein filaments and tubules distributed around the cell, which acts not only as a mechanism for cell movement but also as a framework for the organisation, positioning and relocation of subcellular organelles, and provides strengthening to stabilize cell shape. Although the nature of the cytoskeleton of eukaryotic cells is complex and diverse, three principal components have been identified: actin filaments, microtubules and intermediate fibres.

Actin filaments are composed of thousands of units of the protein actin, which can polymerize and depolymerize as the cell's requirements change. Another protein, myosin, can bind to actin, and in the presence of ATP myosin is able to move along the actin filament. Both actin and myosin are also components of muscle cells and are involved in muscle contraction (Chapter 32). In the context of the cytoskeleton it is thought that myosin can attach to subcellular organelles and move them around the cell along the framework of actin filaments. This may be one mechanism by which cytoplasmic streaming is achieved.

Microtubules are hollow tubular structures composed of two protein components, α and β-tubulin. Associated with these microtubules are two other proteins, kinesin and dynein, which are thought to act on actin filaments in a similar way to myosin: they may be anchored to subcellular organelles and, in the presence of ATP, can move along the microtubules. Microtubules become prominent during mitosis when cell division requires that the replicated DNA is separated into the two new cells. They also appear to be important in giving support and movement to cilia and flagella.

Intermediate filaments are composed of keratins, proteins commonly found in hair, hoof and nails. They are thought to provide a framework under the cytoplasmic membrane which confers the strength to hold a particular shape and allows positioning of subcellular organelles.

1.3 CELL SPECIALIZATION AND INTERACTION

The size which individual cells can reach is limited mainly by the ratio of their surface area to volume. As cells become larger their volume increases in relation to their surface area, so that the speed with which nutrients and gases diffuse from the cell membrane into all regions of the cell becomes limiting. Large cells usually have some adaptation which allows them to cope with this situation. In many large plant cells, for example, the cytoplasm is restricted to a narrow layer inside the plasma membrane and they may have specialized structures to increase cytoplasmic streaming. This also reduces the energy-requiring need to synthesize cytoplasm as much of the remainder of the cell is occupied by the vacuole. Another adaptation which increases the surface area-to-volume ratio includes changing shape: for example, neurones of higher animals are very long thin cells, and ganglia and kidney cells have a highly convoluted cell membrane.

Cells in multicellular organisms may become highly specialized in the processes which they carry out. Thus the biochemical processes described in this book do not have equal prominence in all cells, and this is often reflected in their structure. All cells must carry out basic biochemical processes required for growth and division. To do this they must respire to produce energy, synthesize proteins and nucleic acids, and make other cellular components as they grow. In general, each prokaryotic cell carries out these processes independently of other cells. In complex multicellular organisms such as higher animals and green plants, individual cells have become restricted in their functions and may make a specialized contribution to the well-being of the organism as a whole. Thus these organisms contain a variety of specialized cell types which are normally organised into organs. This process of specialization of function in multicellular organisms is called differentiation.

In animals and plants, therefore, individual organs are specialized in carrying out specific functions. Some of these, which will be referred to frequently throughout the book, are presented in Table 1.2.

The efficient operation of a multicellular organism in which each part contributes to the whole requires complex control mechanisms by which the biochemical processes can be regulated, and efficient systems by which cells can communicate with one another over both short and long distances. These control systems present opportunities for the manipulation of productive processes in both animals and plants.

Table 1.2 Examples of tissue and organ specialization

Tissue/organ	Specialization
Animals	
Brain and nervous tissue	Generation and transmission of nerve impulses requiring ion transport across membranes by active transport – the ATP required is generated by oxidation of glucose
Liver	Processing and redirection of tissue metabolites: glycogen synthesis, gluconeogenesis, fatty acid synthesis (main site in birds and some non-ruminants), lipoprotein synthesis, cholesterol synthesis, synthesis of bile salts
Adipose tissue	Fat storage, triacylglycerol synthesis, lipolysis, fatty acid synthesis (main site in ruminants), non-shivering thermogenesis (uncoupled fatty acid oxidation in brown adipose tissue)
Muscle	Glycogen synthesis and breakdown, movement –the ATP required is generated by oxidation of glucose
Kidney	Excretion of waste products and resorption of minerals, gluconeogenesis (approx. 10% of total)
Digestive tract	Secretion of digestive enzymes requires high rates of protein synthesis, absorption of end products of digestion, often by active transport – the ATP required is generated by oxidation of glucose
Mammary tissue	Synthesis of protein, lactose, fatty acids and triacylglycerols; secretion of milk
Bone	Synthesis of collagen and formation of crystalline matrix of bone, provision of the rigid internal framework of the body, reserve of calcium and phosphorus
Plants	
Seeds	Synthesis and breakdown of starch and triacylglycerols, synthesis of sucrose, synthesis and oxidation of fatty acids, protein synthesis
Leaves and other green tissue	Photosynthesis, photorespiration, protein synthesis, transpiration
Roots	Uptake of nutrients and water from the soil, assimilation of nitrate, nitrogen fixation
Tubers and storage organs	Starch synthesis and remobilization
Fruits	Synthesis and storage of sugars, respiration, cell wall degradation, syntheis and degradation of pigments, accumulation of fruit acids

WATER AND SOLUTIONS

2.1 INTRODUCTION

As a chemical compound water is somewhat unusual. It has a molecular weight of just 18. Most chemical compounds of this size are not liquids at normal temperatures and pressures. For instance molecular oxygen, O_2, has a molecular weight of 32 and does not condense into a liquid until the temperature reaches $-183°$ C. Hydrogen sulphide (H_2S) with a molecular weight of 34, nearly twice that of water, is a gas at normal temperatures and pressures. Water obviously has some peculiar properties, sometimes referred to as its anomalous properties. The puzzle becomes even stranger when we start to look at the way in which things dissolve in water, and at the nature of solutions.

The difference between a true gas and a liquid is that in the gas the individual molecules (or atoms in a few cases) are not attracted to one another or are only weakly attracted, and so can freely move throughout all the space that is available to them. In a liquid there is some attraction between individual molecules and the whole quantity sticks together. On the other hand, this process of adhesion is not strong enough to force the liquid to cohere into a definite shape and, for this reason, liquids can flow.

Water is a compound of hydrogen and oxygen, two atoms of hydrogen being joined to each one of oxygen. The bonding in water is covalent: each pair of electrons is not fixed in place, but follows a complex three-dimensional path which takes the electrons for part of their time around the oxygen atom and the rest of the time around the hydrogen. When the electrons are situated near the oxygen atom then all that is 'visible' of the hydrogen is a positively charged nucleus. The part of this orbital around the oxygen atom is much bigger than the part around the hydrogen, so the electrons spend more of their time associated with the oxygen than with the hydrogen. The result is that the hydrogen atoms, in the temporary absence of their electrons, take on a small positive charge and the oxygen atom gains a slight negative one. The charges are quite small, but are big enough to have enormous effects on the properties of the water. A hydrogen atom in one water molecule will have an attraction for the oxygen in another – this pattern of attraction of one molecule for another has the effect of aggregating the molecules. They are not free to move as in a gas. Aggregation of molecules caused by hydrogen atoms with small, unshielded, electrical charges is called hydrogen bonding and is very common in biochemical systems.

2.2 THE IONIZATION OF WATER

Because water carries different amounts of electrical charge in different parts of the molecule it is known as a polar chemical compound. Another interesting property of water is that in a small proportion of water molecules the bond that holds one of the hydrogen atoms breaks, this comes away from the rest of the molecule but minus its electron. The only part of a hydrogen atom that is left is the nucleus, which is a single positively charged particle called a proton or hydrogen ion, H^+. (A tiny proportion of hydrogen atoms also have a neutron in the nucleus but this can be ignored for most purposes.)

Once the hydrogen ion has departed, the part of the water that is left has an extra electron which gives it a negative charge, this is the hydroxyl ion, OH^-. For every proton that is formed there must be a hydroxyl ion, which means that overall the electrical charge in the water is balanced. In pure water only one water molecule in ten million is split in this way. In most liquids any ion that drifted away from the rest of a molecule would soon be recaptured. The difference is that in water, hydrogen bonding again comes into play. As soon as a hydrogen ion comes into existence it is immediately surrounded by a layer of water molecules all arranged with their negative charges (the ones on the oxygen atom) pointing towards the positive charge of the free hydrogen (Figure 2.1). The same thing happens to the free hydroxyl ion, although in this case the water molecules of the surrounding layer are aligned so that their positively charged hydrogen atoms point towards the negative ion (Figure 2.1). These layers stabilize the ions so that they can exist separately.

2.2.1 THE PH OF WATER

Water dissociates into hydrogen and hydroxyl ions according to the equation:

$$H_2O \rightleftharpoons H^+ + OH^-$$

For every chemical reaction that comes to an equilibrium point we can calculate an equilibrium constant; for the equation above this is given as:

$$K = \frac{[H^+][OH^-]}{[H_2O]}$$

Note that the figures in square brackets are the concentrations.

Because the concentration of the water in the form of H_2O is massive in comparison to the amounts of ionized water, we can neglect any changes in its concentration. A new constant can be defined which is called the ionic product of the water:

$$K_w = [H^+][OH^-]$$

At $25°$ C the concentration of H^+ is 1×10^{-7}M. In pure water the concentration of OH^- ions must be exactly the same. We can therefore calculate K_w as:

$$K_w = (1 \times 10^{-7}) \times (1 \times 10^{-7})$$

$$K_w = 1 \times 10^{-14}$$

At any one temperature this figure is a constant, so that if for any reason the concentration of H^+ ions increases then the concentration of OH^- ions will have to decrease.

The concentration of H^+ ions is a measure of the acidity of a solution. Unfortunately a scale that uses figures such as 10^{-14} is not very useful in practice, so acidity is expressed as pH, which is related to the hydrogen ion concentration by the formula:

$$pH = -\log_{10}[H^+]$$

When the hydrogen ion concentration is exactly the same as the hydroxyl ion concentration ($[H^+] = [OH^-] = 10^{-7}$), the pH equals 7,

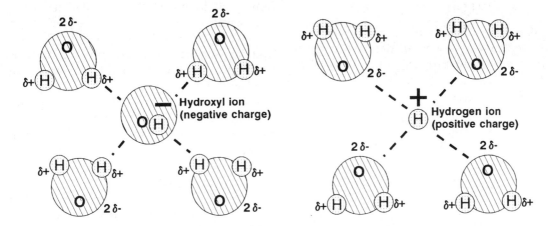

Figure 2.1 Hydroxyl (OH⁻) ions and hydrogen (H⁺) ions are stabilized by being surrounded by a layer of water molecules held in place by electrical charges.

and this is referred to as a neutral solution. Where the pH is lower than 7 there is an excess of hydrogen ions and the solution is said to be acid. If the pH is greater than 7 the solution is alkaline.

Water is a very good solvent for a whole range of chemicals (solutes). It dissolves solids such as common salt and sugar extremely well and it mixes freely with a number of other liquids such as alcohol (ethanol). On the other hand there are many substances that do not dissolve at all well in water: fats and oils are obvious examples. Liquids such as kerosene or petroleum are not miscible with water although they will dissolve fats, oils and grease. In general, materials that dissolve in water are themselves polar compounds, whereas those that are insoluble are non-polar.

Many of the compounds that dissolve in water dissociate into ions in the water. The ions that are formed are stabilized by hydration using hydrogen bonds. Some compounds are completely dissociated when they are in solution. For example common salt, NaCl, breaks down into the positively charged Na⁺ (a cation) and a negatively charged Cl⁻ ion (an anion). Other compounds dissociate only partially.

The organic acid, acetic acid, is a very good example of a compound which is soluble in water but which is only partially dissociated in solution:

$$CH_3.COOH \rightleftharpoons CH_3.COO^- + H^+$$

In a pure solution in water, only about one molecule in 250 would be dissociated. Acetic acid is a weak acid because it does not supply as many H⁺ ions as a strong acid such as HCl, which is completely dissociated. Organic acids such as acetic acid are usually present in biological solutions in the form of their salts and in this case would be completely dissociated. The formula for the acetate anion is CH_3COO^- but in this book the formula CH_3COOH is frequently used for simplicity.

A similar situation exists with alkaline compounds. These supply hydroxyl, OH⁻, ions and consequently reduce the concentration of hydrogen ions in the solution. Some alkalis dissociate completely, good examples being the hydroxides of sodium or potassium.

$$NaOH \rightleftharpoons Na^+ + OH^-$$

These are the strong alkalis or strong bases. On the other hand, weak alkalis or weak bases such as ammonium hydroxide are only partially dissociated in solution.

$$NH_4OH \rightleftharpoons NH_4^+ + OH^-$$

For any compound that is only partially dissociated into ions we can define a constant that describes the proportion that is dissociated.

For acetic acid:

$$K_a = \frac{[CH_3.COO^-][H^+]}{[CH_3.COOH]}$$

For ammonium hydroxide:

$$K_b = \frac{[NH_4^+][OH^-]}{[NH_4OH]}$$

Where K_a and K_b are the dissociation constants.

2.3 WHAT ARE ACIDS AND BASES?

So far we have used the terms acid and base without defining them. As we move into biochemistry, it will be necessary to have a very clear idea of what acids and bases are. One of the simplest and best working definitions is:

● an acid is a hydrogen ion (or proton) donor
● a base is a hydrogen ion (or proton) acceptor.

If we look again at the equation for the dissociation of acetic acid:

$$CH_3.COOH \rightleftharpoons CH_3.COO^- + H^+$$

Acid	Base
(donates a proton)	(can accept a proton)

By this definition; water itself is an acid because it can donate a proton and the hydroxyl ion is a base because it accepts one.

$$H_2O \rightleftharpoons H^+ + OH^-$$

2.4 BIOLOGICAL SYSTEMS, IONIC STRENGTH AND PH

As we shall see in the chapter on enzymes (Chapter 6), it is extremely important that the pH of biological environments is maintained within very close limits. Any large changes in ionic strength or pH will lead to damage of the very sensitive molecules that are responsible for the metabolism of cells. In agriculture, animals and plants manage to survive under conditions which are not always helpful to maintaining constantly favourable conditions. Crop plants often suffer from periodic loss of water, and even under mild drought conditions some still manage to survive. Animals too can suffer from water deprivation. Cattle in the hot tropics may only have access to water every 3 days or so. In the interim they will lose large amounts of water both by excretion and by the evaporation necessary for them to keep cool. Some breeds of cattle are capable of losing up to 25% of their body weight in this way and of making good the losses within a few minutes when drinking water becomes available. The changes in the amounts of water lead to enormous variations in the strength of solutions both inside and outside cells. Despite all these changes, animals and plants manage to survive them on a regular basis and even to thrive. Within biological systems there must be some way in which the properties of solutions are stabilized so as not to damage the other constituents of cells. Much of the stabilization comes from a process of buffering whereby some of the compounds in cells are able to cushion the changes. The pH values for a number of biological materials are shown in Table 2.1.

2.4.1 STABILIZATION OF PH BY BUFFERS

Solutions of weak acids and their salts can act to stabilize pH over a given range and are therefore known as buffers. If small amounts

Table 2.1 pH values for some common biological systems

Material	pH
Blood plasma	7.4
Milk (fresh)	6.9
Egg white	8
Grass silage	3.8–4.8
Tomato juice	4.3
Soils (note extreme variability)	3–11

of either OH⁻ ions (from an alkali) or H⁺ ions (from an acid) are added to these solutions there will be only a small change in pH. In the absence of the weak acid and its salt the change in pH would have been much greater.

If we return to the dissociation of acetic acid:

$$K_a = \frac{[CH_3.COO^-][H^+]}{[CH_3.COOH]}$$

(This is known to have a value of 1.75×10^{-5} at 25° C.)

We can rewrite the equation so that:

$$[H^+] = K_a \times \frac{[CH_3.COO^-]}{[CH_3.COOH]}$$

This can be converted to an equation for pH by taking the negative logarithms (base 10) for both sides:

$$pH = pK_a \times \frac{[CH_3.COO^-]}{[CH_3.COOH]}$$

Note that $pKa = -\log_{10}(K_a)$.

Example: if we have a solution which is half molar with respect to sodium acetate and to acetic acid then in solution we have:

$$CH_3COONa \rightarrow CH_3COO^- + Na^+ \quad (1)$$

$$CH_3COOH \rightleftharpoons CH_3COO^- + H^+ \quad (2)$$

The dissociation of the salt, sodium acetate, will be complete in solution so that we can say that the concentration of acetate (CH_3COO^-) from Equation 1 will be 0.5 M. On the other hand, the extent of dissociation of the acid is very small so that the concentration of CH_3COOH will be very little different from 0.5 M. In other words:

$$\frac{[CH_3.COO^-]}{[CH_3.COOH]} = \frac{0.5}{0.5} = 1$$

and the $\log_{10}$ of this must be zero. Therefore:

$$pH = pK_a$$

$$pK_a = -\log (1.75 \times 10^{-5})$$

$$pH = pK_a = 4.76$$

Acetic acid is a good example to use in the chemical laboratory but it is not commonly used by biochemical systems. In practice, in cells there is a whole range of compounds which act as buffers and are able to stabilize the pH close to neutral. Some of their K_a and pK_a values are shown in Table 2.2.

One place where pH is very important is in the rumen of animals such as the cow. This is the first compartment of the 'stomach' and is extremely important in the digestion of plant material which is the main source of nutrition for these animals. If the pH in the rumen drops too low then digestion of food will be halted and there may be serious effects on the health of the animal. Dairy cows are often fed on diets containing large amounts of grain, and these can lead to very low pH values in the rumen. Work in many countries has shown that the addition of buffers to the diet can maintain pH at normal levels and allow animals to thrive on diets that would otherwise damage their health. The commonest buffer added under these circumstances is sodium bicarbonate.

2.5 COLLIGATIVE PROPERTIES

There are a number of properties of solutions that depend upon the strength of the solution expressed in terms of the number of particles dissolved in a given volume of solution (or solvent). The word particles is used rather than molecules, because many molecules such as salts that dissolve in water will dissociate into two or more particles. On the other hand, sugar molecules which cannot dissociate will contribute just one particle. Two of these properties, the depression of freezing point and osmotic pressure, are extremely important in agriculture.

Table 2.2 Dissociation constants and pK_a values for common organic acids

Acid	Formula	Dissociation constant	pK_a
Formic acid	HCOOH	1.77×10^{-4}	3.75
Acetic acid	CH_3COOH	1.75×10^{-5}	4.76
Propionic acid	CH_3CH_2COOH	1.34×10^{-5}	4.87
Lactic acid	$CH_3CH(OH)COOH$	1.39×10^{-4}	3.86

2.5.1 DEPRESSION OF FREEZING POINT

It is well known that solutions have a lower freezing point than their pure solvents. In cool climates in winter, salt is routinely spread on roads and paths to melt ice and snow, preventing accidents to vehicles and pedestrians. This property has more subtle applications in that many plants can survive without being frozen at temperatures below 0° C. The depression in freezing point is about 1.86° C for every mole of particle dissolved. This is fairly simple for a sugar such as glucose (MW 180). A solution of 180 g glucose per litre will have a freezing point of –1.86° C. For sodium chloride (MW 58.5) a solution of 58.5 g l^{-1} will yield 1 mole of Cl$^-$ ion (particle weight 35.5) and 1 mole of Na$^+$ ion (particle weight 23). The depression of freezing point will therefore be 3.72° C (i.e. 2 × 1.86° C).

A practical application is to be found in the standard testing of milk for adulteration with water. Milk ought to have a freezing point of between –0.54 and –0.59° C; if it freezes at –0.52° C or higher then it is likely that the milk has been diluted with water. Freezing point depressions of other biological fluids are shown in Table 2.3.

2.5.2 OSMOTIC PRESSURE

The osmotic pressure of different parts of plants and animals is essential to maintaining their function. The principle of the phenomenon rests upon the fact that when solutions of different concentrations are put together their concentrations will tend to equalize. Thus, if two solutions of differing strengths are carefully layered in a test tube, there will be a slow intermingling of the layers until eventually all of the contents of the test tube are at the same concentration. If the solutions are of widely different specific gravities the process may take a long time but equilibrium will eventually be achieved.

If the two layers are separated by a solid and impermeable partition then no mixing can take place. A semipermeable barrier will allow some small molecules to pass through it whilst retaining larger ones. Many will only let water through. Most semipermeable materials allow the passage of particles only when they are formed into a very thin barrier. For this reason they are often called semipermeable membranes.

The principles of osmotic pressure are illustrated by Figure 2.2. This shows two compartments, one filled with water, the other filled with a solution, and separated by a semipermeable membrane. Water molecules can pass freely through the membrane in both directions. There is a tendency for the concentrations of solute to equalize across the membrane and so more molecules will pass from the pure water side to the solution than in the opposite direction, so that the solution increases in volume. As the volume grows so the pressure increases, until it becomes so great that more water molecules cannot pass through the membrane and the flow ceases. In the system shown in Figure 2.2, the extra pressure will push water up the left-hand tube until there is a difference in the height (*h*) of the columns of water. The height of this column of water is the osmotic pressure of the solution, and can be expressed in any of the normal units of pressure: atmospheres, mm of water, bar or Pascal.

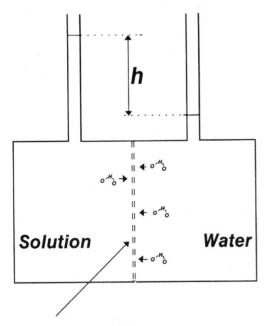

h

Solution   **Water**

Semipermeable membrane

Figure 2.2 Osmosis: water molecules pass through the semipermeable membrane in the direction of the solution in an attempt to balance concentrations on both sides of the membrane. The pressure in the solution compartment increases until it reaches a point (height h) where it prevents further water from passing through the membrane.

The value of the osmotic pressure for an ideal solution (one in which the individual molecules of solute have no effect on one another) is given by the equation:

$$B = MRT$$

Where B (the osmotic pressure in atmospheres) is determined by M (the total molarity of the solution), R (the gas law constant) and T (the absolute temperature in degrees Kelvin). In these units, $R = 0.0821$.

The depression of freezing point of a solution can be used to calculate its osmotic pressure using the equation:

$$B = 0.0441T.\Delta T$$

where ΔT is the depression of freezing point.

Measuring osmotic pressure in this way is a useful concept in chemistry, but in practice the solutions that are met in biological systems are so complex that the equations do not have much meaning. For instance, much of the osmotic pressure of biological fluids comes from proteins, and it is almost impossible to determine how many of the charged groups on the protein will be ionized at any one time and to what extent. This means that we cannot calculate M with any degree of accuracy.

For biological systems we use a unit called osmolarity, which is the molarity of an ideal solution which exerts the same osmotic pressure as the test solution. Osmolarity is measured in Osmoles (osm) or, more commonly, in the smaller milliosmoles (mosm). One interesting finding is that, in most mammalian systems, body fluids exert more or less the same osmotic pressure (Table 2.3).

Cell membranes are semipermeable and the cytoplasm inside has an osmotic pressure. If cells are surrounded by a fluid of a different osmotic pressure there will be a flow of water from the region of low osmotic pressure to that of high pressure. This is easily demonstrated by putting red blood cells into water: so much water flows into the cells that they simply burst. On the other hand, if the cells are placed in strong salt solutions, water flows out of the cells so that they shrivel. Solutions that have the same osmotic pressure as blood are said to be isotonic.

2.5.3 SEMIPERMEABLE MEMBRANES THAT ALLOW SOME SOLUTES TO PASS

In looking at osmotic pressure we have assumed that the only thing that could pass through the membrane was water. In real biological systems, cell membranes do not behave like this – they allow some solutes (usually small molecules or ions) to pass, but others are retained (see Chapter 22). The problem may be further complicated by the fact that the cell itself is able selectively to modify and regulate the permeability of its plasma membrane.

Table 2.3 Colligative properties, depression of freezing point and osmotic pressure in some biological fluids

Fluid	Depression of freezing point (°C)	Osmotic pressure (mosm)
Blood	−0.54	302
Cerebrospinal fluid	−0.57	306
Milk	−0.56	304
Semen	−0.57	296

Consider Figure 2.3a: two compartments of the same volume are separated by a semipermeable membrane that can allow the passage of both K^+ and Cl^-. If water is placed in one side and 0.2 M KCl on the other, there will be a flow of ions from one side to the other so as to equalize the concentration. At equilibrium there will be a 0.1 M solution of KCl on each side. But if we place a solution that consists of 0.1 M KCl on either side but in addition one has 0.05 M KR as well (where R' is a cation too big to pass through the membrane) the system cannot come to equilibrium with the same concentrations on each side of the membrane (Figure 2.3b).

On either side of the membrane, the K^+ ions must balance the total of negative ions. In the system described we will finish up with the conditions shown in Table 2.4.

This means that the concentration of K^+ and Cl^- is different on each side of the membrane; it also means that the total electrical charges are not the same on the two sides (Figure 2.3b). This situation is an example of what is known as the Donnan equilibrium, and it has big implications for the behaviour of cells.

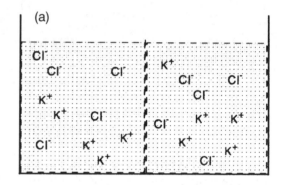

(a)

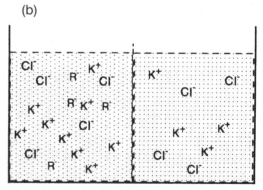

(b)

Table 2.4 Concentrations of solutes after equilibrium is attained

Element	Left-hand side	Right-hand side
K^+	0.139	0.111
Cl^-	0.089	0.111
R^-	0.05	0
$Cl^- + R^-$	0.139	0.111

Figure 2.3 Donnan equilibrium. (a) The membrane is freely permeable to both K^+ and Cl^- ions and the concentration of each will equalize across the membrane. (b) The membrane is not permeable to R^- ions and thus an imbalance of concentration and charge will result.

THE CARBOHYDRATES

3.1 INTRODUCTION

The carbohydrates are a group of organic compounds that includes sugars and related compounds. The sugars are compounds with between 3 and 7 carbon atoms having many hydroxyl (alcohol) groups and either a ketone group or an aldehyde group. Typically, each carbon atom that does not bear an aldehyde or ketone group (collectively called carbonyl groups) will have a hydroxyl group. Simple sugars may be strung together in chains to form long polymers. When they were first carefully investigated, in the 19th century, the simple sugars were all found to have the empirical formula $C_nH_{2n}O_n$, and for this reason they were assumed to have arisen by some form of hydration of carbon (i.e. $C_n + nH_2O$). The name carbohydrates has stayed with them long after their real structures were discovered.

Cells of all types use sugars as a convenient source of energy and as the raw materials for many chemical syntheses. The biochemical pathways for making sugars are relatively simple, and many cells, both in animals and plants, make huge quantities of them. They are water soluble and are therefore easily transported either in blood or phloem.

3.2 STRUCTURES OF SUGARS

The simplest sugars contain just three carbon atoms. Two different formulae are possible (Figure 3.1a) depending upon whether the sugar has a ketone or an aldehyde group. Although these are the simplest of all the sugars, they are not commonly found in these forms but occur as their phosphate esters, as intermediates in metabolic pathways. The structure of the one with the ketone group, dihydroxyacetone, is unique but when we look at the aldehyde sugar, glyceraldehyde, a new level of complexity emerges in the form of optical isomers.

3.2.1 OPTICAL ISOMERS

Chemically the element carbon is known as tetravalent, which means that each carbon atom is joined to neighbouring atoms by four bonds. If all of are these are single bonds then they are arranged around the carbon atom in a three-dimensional tetrahedral shape. A strange thing happens if one particular carbon atom is joined to four atoms or groups which are themselves different: it becomes possible to arrange the surrounding atoms or groups in either of two different ways (Figure 3.2). On paper they look similar but in biological terms they may be quite different. Take for example the simplest of the aldehyde sugars, glyceraldehyde. The two carbons at the end of the chain do not have a chance to show this property; one has two identical hydrogen atoms attached and the other has two of its bonds as a double bond attached to the same oxygen atom. The middle carbon atom (marked with an asterisk) is attached to four different groups; -H, -OH, CH_2OH and -CHO. The two compounds in Figure 3.2 are actually mirror images of one another. Their solutions in water also differ in the effect that they have on polarized light as it passes through: one will deviate the beam of polarized light to the left and the other to the right. They have the same formula so they are isomers, and because they differ in the effect that they have on light they are known as optical isomers. The carbon atom that can be bonded in either of two ways is called the optical centre and is said to be optically active or asymmetric.

Although the structural formulae appear very similar on paper, in real biological systems these two compounds will behave quite differently. Almost all carbohydrates exist as structural and optical isomers, which leads to a fascinating diversity in the properties of these compounds.

3.3 NAMING OF SUGARS

The chemical names of the sugars and many of the more complex carbohydrates end with the suffix -ose. They are also named on a basis of the number of carbon atoms that they contain; tri- for three, and tetra-, pent-, hex-, and hept- for 4, 5, 6, and 7, respectively. The type of carbonyl group is denoted by the prefix of aldo- for an aldehyde and keto- for a ketone. For example, glyceraldehyde is an aldo-triose.

3.4 SUGARS WITH FOUR CARBON ATOMS, THE TETROSES

Four different aldo-tetroses are possible because there are two different asymmetric carbons in the chain (see Figure 3.1b). In the natural world, the tetroses are quite rare in their free form although several do appear as intermediates in metabolic processes.

The carbon at the bottom of each structure is, by convention, the one with the highest number – in this case, 4. This carbon atom is always a CH_2OH group. Note that two of the sugars are referred to as 'L' sugars and two are 'D'. All sugars (with the exception of dihydroxyacetone) are divided into these two groups depending upon the particular optical isomerization at the carbon next to the CH_2OH group. By convention the hydrogen atom on

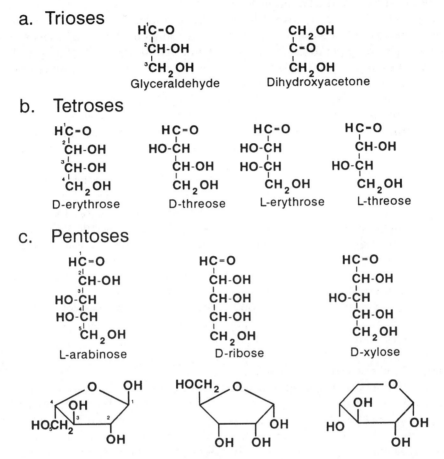

a. Trioses

H'C-O
²CH-OH
³CH₂OH
Glyceraldehyde

CH₂OH
C-O
CH₂OH
Dihydroxyacetone

b. Tetroses

H'C-O
²CH-OH
³CH-OH
⁴CH₂OH
D-erythrose

HC-O
HO-CH
CH-OH
CH₂OH
D-threose

HC-O
HO-CH
HO-CH
CH₂OH
L-erythrose

HC-O
CH-OH
HO-CH
CH₂OH
L-threose

c. Pentoses

HC=O
²CH-OH
³HO-CH
⁴HO-CH
⁵CH₂OH
L-arabinose

HC=O
CH-OH
CH-OH
CH-OH
CH₂OH
D-ribose

HC=O
CH-OH
HO-CH
CH-OH
CH₂OH
D-xylose

Figure 3.1 Structures of some sugars with three, four or five carbon atoms. (a) Trioses: glyceraldehyde exists as two different optical isomers (see Figure 3.2). (b) Tetroses: the tetroses shown are all aldo sugars, the difference between them lies in the optical isomerization at positions 2 and 3. (c) Pentoses: these can exist as straight chain sugars or in rings made by the formation of hemiacetal bonds. These are three of the commonest pentoses found in agricultural materials.

this penultimate carbon is drawn pointing to the left on an L sugar and to the right on a D sugar. Almost all of the natural sugars are D

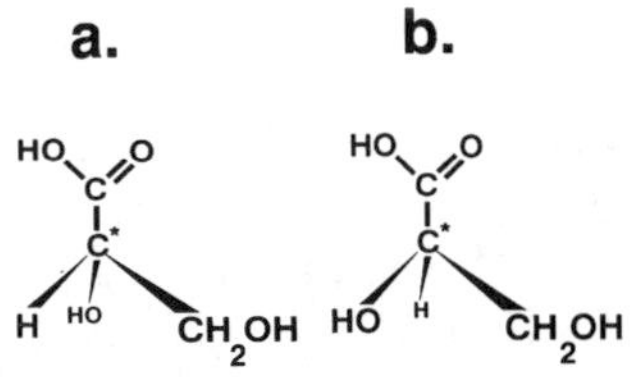

Figure 3.2 The two optical isomers of glyceraldehyde are mirror images of one another. The carbon atom marked ★ is asymmetrical.

isomers, the only L sugar normally encountered is L-arabinose, one of the components of some complex structural carbohydrates in plants.

3.5 SUGARS WITH FIVE CARBON ATOMS, THE PENTOSES

Both ketoses and aldoses can be extended by the addition of a further carbon atom and hydroxyl group to give pentoses. This introduces yet another asymmetric carbon atom into each type of sugar and thus multiplies the number of possible isomers by two. Hence there are four possible keto-pentoses and eight aldo-pentoses. The aldo-pentoses are much more common in nature than the keto ones. Before looking at the individual pentoses it is necessary to understand some of the chemistry of these compounds, because the added chain length brings with it the possibility of the straight chain looping round on itself to form a ring.

3.5.1 RING FORMATION IN SUGARS

As described above, sugars contain both alcohol and carbonyl (aldehyde and ketone) groups. Many of the properties of the carbohydrates depend on addition reactions between these groups. The reactions can take place between any alcohol and any carbonyl group. Figure 3.3 shows an alcohol reacting with the carbonyl carbon atom of an aldehyde to form a new compound called a hemiacetal. This reaction is easily reversed and, if an aldehyde and alcohol are both dissolved in water, the solution will usually contain a mixture of free aldehyde and alcohol molecules together with hemiacetals. A similar reaction takes place between ketones and alcohols to form hemiketals. In the sugars, hemiacetals (or hemiketals) are formed within the same mole-

cule between the carbonyl group and one of the alcohol groups further down the chain. This results in the straight chain of the sugar being bent round on itself to form a ring which also includes an oxygen atom (for instance, see Figure 3.1c).

Once a hemiacetal has been formed, it can react with a further alcohol group to form an acetal. During this reaction, water is given off (it is therefore called a condensation reaction) and the process is not easily reversed. In the sugars the formation of acetals involves a reaction between the hemiacetal group on one sugar molecule and an alcohol from another. In this way two sugar molecules are linked together to form new and more complex carbohydrates.

There is another consequence of the formation of hemiacetals: the carbon of an aldehyde is not asymmetric but that of a hemiacetal is, so that there are two possible optical isomers of each sugar hemiacetal.

3.5.2 FIVE- AND SIX-MEMBERED RINGS

Three examples of aldo pentoses are illustrated in Figure 3.1c. Arabinose, ribose and xylose are quite common constituents of carbohydrates. All have five carbon atoms including the aldehyde group but the orientation (optical isomerization) of the carbon in the middle of the chain is different, which means that the

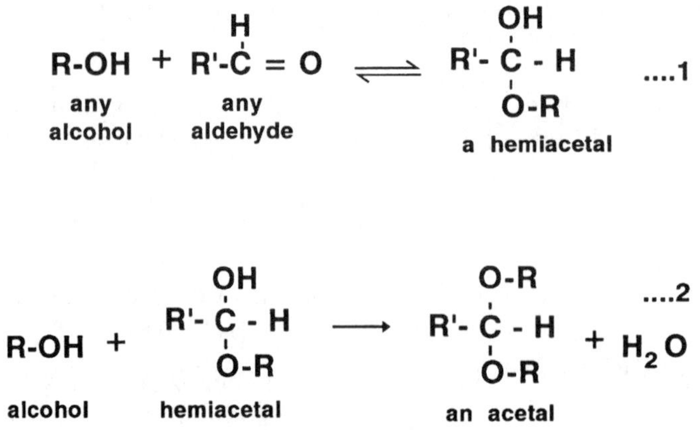

Figure 3.3 The formation of hemiacetals (1) and of acetals (2). These reactions occur between aldehyde and alcohol groups within the same sugar molecule.

hydroxyls on these carbons stick out in opposite directions. These small differences affect the way in which each of these molecules is formed into a ring. In arabinose and ribose the hydroxyl of carbon 4 reacts with carbon 1 (the aldehyde) to form a five-membered ring. Carbon 5 sticks out from the ring. The five-membered ring is called furan. In the case of xylose the hydroxyl on carbon 5 reacts with the aldehyde to give a six-membered, pyran, ring.

3.5.3 RING FORMATION IS NOT PERMANENT

When sugars are in water solution, part will be in the straight-chain form and the rest in the ring form. If some process removes all the straight-chained sugar molecules from the solution then some of the ring structures will open in order to replace the lost straight chains and to restore the balance.

3.6 SUGARS WITH SIX CARBON ATOMS, THE HEXOSES

These sugars are the most common, indeed their polymers probably account for a large proportion of the solid carbon in the earth's biosphere. They have one more asymmetric carbon atom than the pentoses and thus eight keto-hexoses and 16 aldo-hexoses are possible. Luckily there are only a few which are very common in nature, and of these one aldose (glucose) and one ketose (fructose) are particularly abundant. Like the pentoses, they form either internal hemiacetals or hemiketals as appropriate. Theoretically, these can be formed between the carbonyl group and any of the alcohol groups, but in practice they only form between carbons 1 and 5 in the case of the aldo-hexoses, and between carbons 2 and 5 in the case of the keto-hexoses. The formation of an internal hemiacetal or hemiketal makes the whole molecule of the sugar bend into the ring shape. In the case of glucose, the ring which is formed is six-membered (pyran) whereas in the case of fructose it is five-membered (furan).

3.6.1 GLUCOSE

The formation of the ring in glucose links together carbons 1 and 5 of glucose through the atom of oxygen and in doing so makes carbon 1 asymmetric (Figure 3.4). The two different compounds are denoted as α- and β-glucose. In the straight-chain form of glucose, carbon 1 is not asymmetric because it has two of its bonds attached by a double bond to an atom of oxygen. In ring-shaped glucose, this carbon is effectively joined to four different groups. The important difference to notice is that for α-glucose we draw the hydroxyl group on carbon 1 as pointing downwards, whereas in the β form it points up. These two versions of glucose are actually quite different compounds which do not even have the same melting point. When glucose is dissolved in water there is a rapid equilibrium between the three forms:

$$\beta\text{-D-glucose} \rightleftharpoons \text{Straight chain glucose} \rightleftharpoons \alpha\text{-D-glucose}$$

In solution at any one time, most of the glucose is in the α form (65%) with smaller amounts as β-ring (32%) and straight-chain (3%) forms.

3.6.2 FRUCTOSE

This is probably the commonest keto-sugar found (Figure 3.4). The commonest natural source of free fructose is honey, and this accounts for its very sweet flavour. Fructose is quite commonly encountered as a component of more complex carbohydrates, and it accounts for half of the commercial sugar, sucrose.

3.6.3 OTHER HEXOSES

The ring forms for several other common hexoses are shown in Figure 3.5. Many of these do not commonly occur as pure compounds but are to be found in combination with other sugars. Figure 3.5 also shows the structures of the related sugar acids, galacturonic acid and glucuronic acid, which occur in complex carbohydrates.

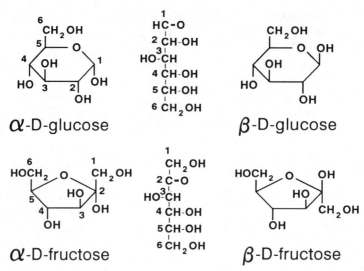

Figure 3.4 The structures of glucose and fructose. In addition to the straight-chain form, each of these sugars can exist in either of two ring structures, differing in the orientation of the carbonyl carbon (carbon 1 in glucose, carbon 2 in fructose) when the ring is formed.

3.7 REDUCING AND NON-REDUCING SUGARS

True sugars all contain a carbonyl group, either an aldehyde or a ketone. Both of these are capable of being oxidized. In the case of aldehyde groups, the product of oxidation is the corresponding carboxylic acid. The ketones are not as reactive and yield a whole series of compounds. These sugars are therefore reducing agents. Many of the chemical methods used for detecting and measuring sugars depend on their reducing properties. In alkaline solution, copper (II) sulphate will oxidize an aldehyde to its corresponding carboxylic acid leading to the formation of copper (I) oxide. The liquid changes from a bright blue solution to a suspension of a dark red precipitate. This provides the standard method for measuring sugars in fruits.

Sugars can act as reducing agents only whilst the carbonyl carbon atom is not firmly linked to any other molecule. In sugars the open-chain form has a free carbonyl group. If this is oxidized then any hemiacetal or hemiketal rings can open to yield more of the open-chain version of the sugar ready to take part in oxidation/reduction reactions.

3.8 FORMATION OF SUGAR ACETALS

Acetals can be formed between any hemiacetal and any alcohol group. Wherever there are sugars there are plenty of 'spare' alcohol groups, but in fact natural sugar acetals tend to be very specific. In the aldo-hexoses such as glucose, the hemiacetal is always formed at the number 1 carbon atom so that this is always involved in acetal formation. The free hydroxyl group used in acetal formation is almost always the one on either carbon 4 or carbon 6 of another sugar molecule.

Once an acetal has been formed it is quite difficult to break it. As long as this acetal bond remains intact then the hemiacetal formed at carbon number 1 cannot be disrupted.

Once an acetal has been formed the ring structure of the sugar is 'locked' – the ring cannot be opened and must remain in either its α or β form. This locking of the carbonyl group of one of the sugars means that the group is

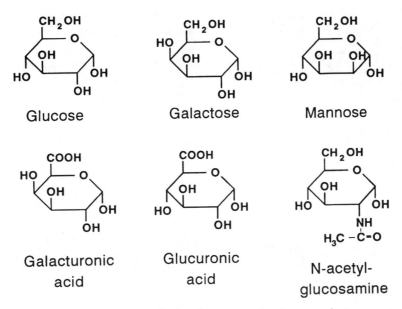

Figure 3.5 Some common hexose sugars and related compounds: glucose, galactose, mannose, galacturonic acid, glucuronic acid, N-acetylglucosamine. The acid and N-acetyl derivatives are found in complex carbohydrates.

unable to take part in any oxidation/reduction reactions.

3.8.1 FORMATION OF DISACCHARIDES

Two molecules of a simple sugar linked together as an acetal are known as a disaccharide. The disaccharide with a bond between the 1 carbon of an α-glucose and the 4 carbon of another α-glucose is called maltose (Figure 3.6a). The bond is called an α-1,4 glycosidic link. If the left-hand sugar had been in the β form before linking then the compound would be a β-linked disaccharide. The compound of this sort which is comparable to maltose is called cellobiose. Lactose, the sugar found in milk, resembles cellobiose but the left-hand sugar is galactose instead of glucose. All these disaccharides are formed between the hemiacetal carbon of the left-hand sugar and the 4 carbon of the right-hand one. Thus, whilst the ring shape of the left-hand sugar is

fixed, the right-hand one is free to open to the straight-chain form.

This means that in maltose, cellobiose and lactose the aldehyde group of the left-hand glucose is unable to take part in any oxidation/reduction reactions. However, the right-hand glucose group can react in this way, so these disaccharides still have reducing properties.

The monosaccharide components of carbohydrates are sometimes distinguished by referring to them as the non-reducing or the reducing end of the chain. Thus in the diagrams all of the left-hand glucose groups are non-reducing, and the right-hand ones reducing.

3.8.2 SUCROSE

The commonest disaccharide of commerce is sucrose, a compound of glucose (in the α orientation) and fructose (in the β form). Unusually, it is an acetal formed between the

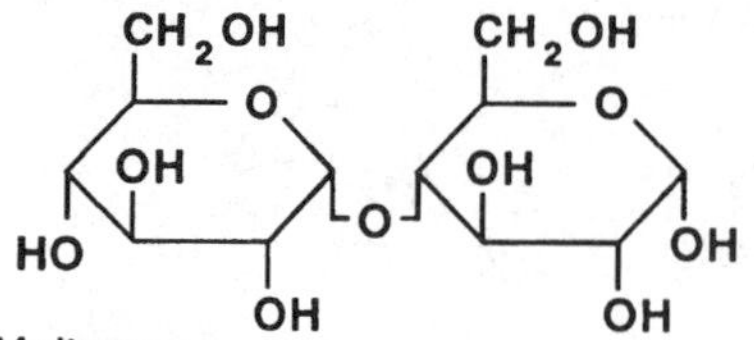

a.   Maltose

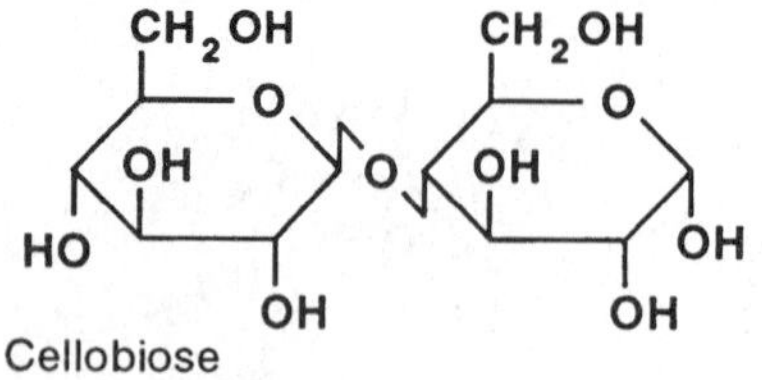

b.   Cellobiose

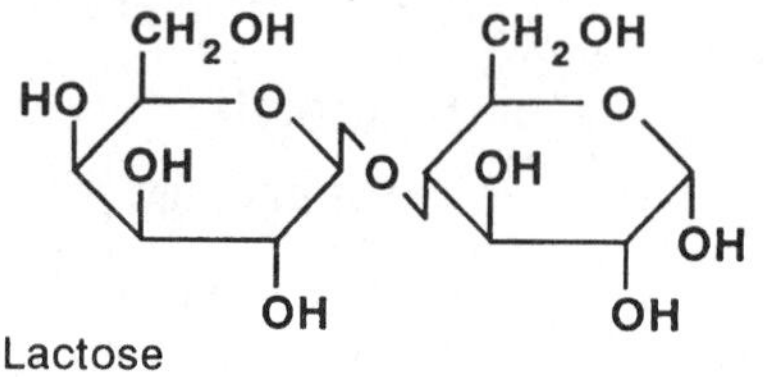

c.   Lactose

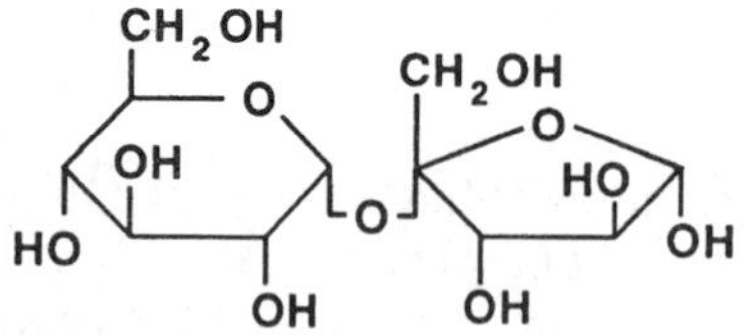

d.   Sucrose

Figure 3.6 Common disaccharides: maltose, cellobiose, lactose, sucrose. With the exception of sucrose, the ring of the right-hand glucose unit can open exposing a free aldehyde group and giving reducing properties to the sugar.

hemiacetal carbon of glucose and the hydroxyl group of the hemiacetal in fructose. This means that neither of the rings in sucrose can open to the straight-chain form and therefore it is a non-reducing sugar. Huge quantities of this sugar are produced either from sugarcane in tropical and semi-tropical areas or from sugar beet in temperate zones.

3.9 POLYSACCHARIDES

Looking at the structures of maltose or cellobiose, it is quite clear that the right-hand glu-

cose unit of this new compound is free to join with a further glucose unit, in which case it would form a trisaccharide. The process can be repeated many times to form long chains of the simple sugars, the polysaccharides. In most polysaccharide chains only the single sugar group at one end of the molecule is free to open and express chemical reducing properties. The chains of a polysaccharide thus carry both reducing and non-reducing ends.

Glucose is the commonest sugar in polysaccharides and it forms two main families of carbohydrates which differ simply in whether the units from which they are made are in the α or β form. In general, those that are made up of β-glucose units are physically much stronger, are less soluble in water and are much more difficult for animals to digest. Many of the common polysaccharides contain only one type of sugar group and are termed homopolysaccharides. Those with two or more different sugars are heteropolysaccharides.

Almost all of the carbohydrates that are commonly found in agriculture come from plants, and here polysaccharides have two main functions: storage and structure.

3.9.1 THE STORAGE CARBOHYDRATES – STARCH AND GLYCOGEN

Plants, and to a lesser extent animals, use carbohydrates as a way of storing nutrients for future need. During the normal processes of growth and regeneration, the plant stores materials in both seeds and roots to cover periods when its ability to supply nutrients from photosynthesis is inadequate. As these materials have to be broken down by the plant they are not physically strong, nor do they have to be water-resistant. They are mainly polysaccharides made up of glucose units in the α form. The commonest examples are the starches and starch-like materials that come from seeds such as corn, and tubers such as potatoes, yams and cassava. Unlike the structural polysaccharides, animals digest them very easily and they form the staple ingredient of the human diet in most parts of the world.

Starch

Starch consists of a mixture of two different types of molecules: amylose, which is a long chain of glucose atoms joined by α-1,4 linkages; and amylopectin, which consists of a mixture of α-1,4 links with occasional α-1,6 branches. The branches occur after about 25 straight α-1,4 bonds (Figure 3.7). Starches from different sources vary in the ratio of amylose and amylopectin, in the size of the individual molecules, and in the degree of cross-linking in the amylopectin molecule. In general, amylopectin accounts for about 70% of starch.

The long chains of amylose roll themselves into a stable helix shape which is held in place by hydrogen bonding (Figure 3.8). The helix is a tube into which other molecules or atoms can fit. One example of this is the fact that iodine can fit inside the helix and form a blue-coloured complex with the amylose, a reaction which is often used to detect the presence of either starch or iodine.

Glycogen

Animals also make use of an α-linked polysaccharide for storage of energy, although the amount of material stored is very small. The compound used is called glycogen and is very similar in structure to amylopectin although the molecules are larger and the cross-linkages

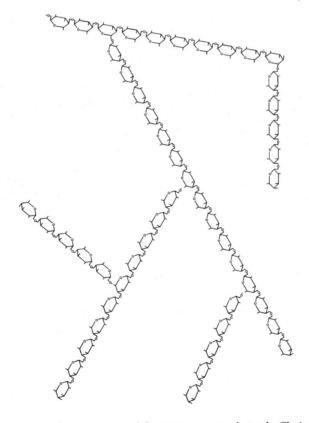

Figure 3.7 The structure of amylopectin, one of the components of starch. Chains of glucose groups, linked α-1,4, contain occasional α-1,6 bonds that provide branching points. In native starches the branching points occur approximately every 25 glucose residues.

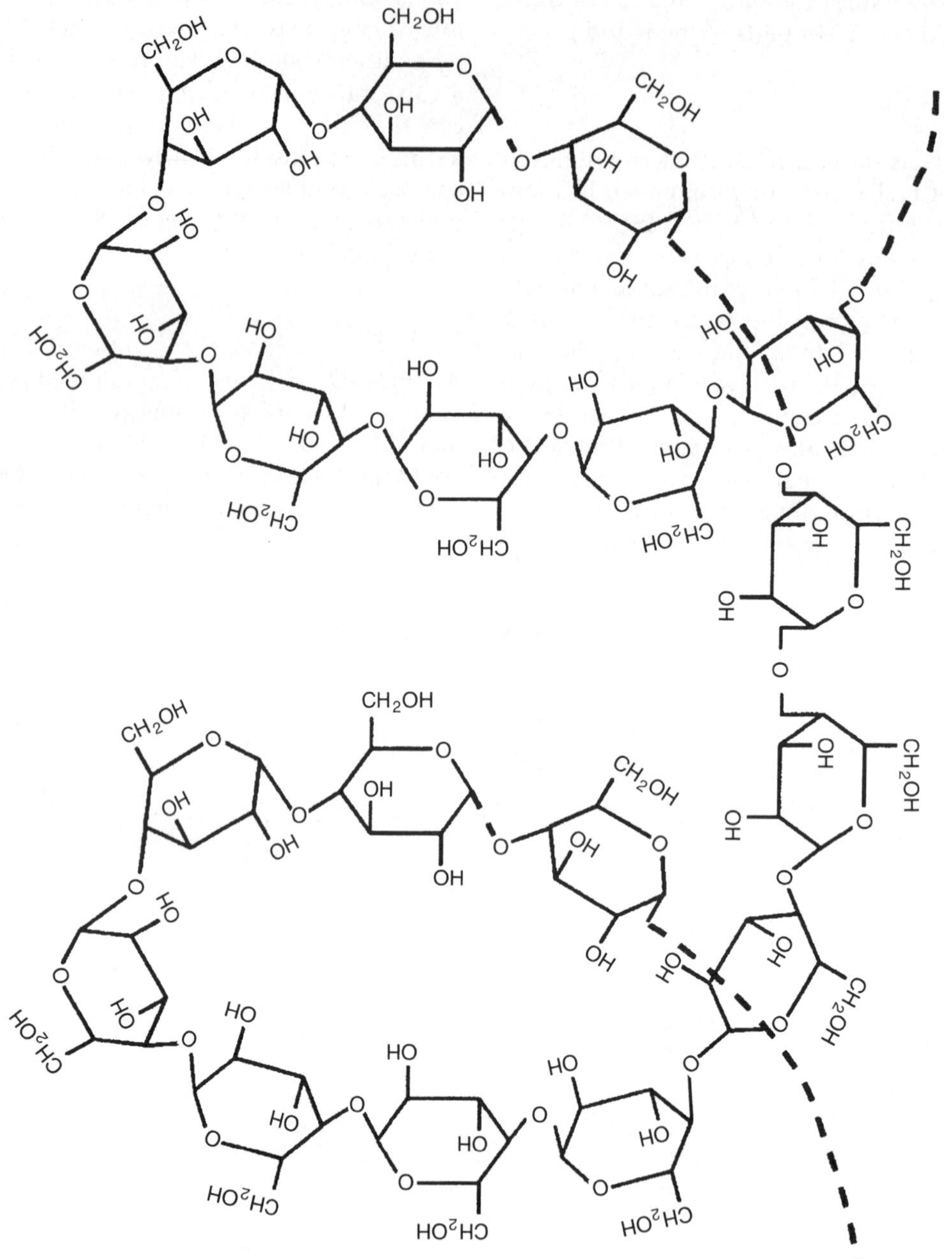

Figure 3.8 The structure of amylose, the second component of starch. Molecules consist of long chains of glucose groups, linked α-1,4. These are held in their helical form by hydrogen bonding to form long tubes.

more frequent (once every 15 or so straight bonds).

3.9.2 STRUCTURAL POLYSACCHARIDES IN PLANTS

In agriculture the non-starch polysaccharides can be viewed in two distinct ways. To the crop scientist they are the materials from which plants fashion much of the physical components of their structures. The basic units of the plant are its cells, and their structural integrity comes from the materials that make up the cell wall. On the other hand, the animal scientist regards non-starch polysaccharides as the materials that form the fibre components of the diets eaten by livestock. This has led to the development of a number of different terms for what are essentially the same materials: fibre carbohydrates, structural carbohydrates and cell-wall carbohydrates.

At least 90% of the structural material of cell walls of all higher plants is polysaccharide. The remaining 10% is made up of protein which, like some structural proteins of animal cells, is rich in hydroxyproline.

The main polysaccharides are cellulose, hemicellulose and pectin. Originally these terms referred to their relative solubilities in strong acids and alkalis, however they are now used to describe their molecular structures as carbohydrates. In addition to the carbohydrates, phenolic materials called lignins are also present in some cell walls.

Cellulose

Cellulose fibres make up the main structure of the cell wall and give the wall much of its strength. The wall consists of cellulose rods or microfibrils embedded in an amorphous matrix of non-cellulose polysaccharides (hemicelluloses and pectins). Chemically the cellulose consists of long, linear chains of glucose residues covalently linked by β-1,4-glycosidic bonds. Long, linear chains of pure glucose are termed glucans (Figure 3.9). Between 2000 and 6000 residues are polymerized in each glucan chain of the primary wall. A single cellulose fibre of 3.5–4 nm in diameter contains 30–40 glucan chains held together by a very large number of hydrogen bonds. Within the fibre, individual glucan chains are very much shorter than the length of the fibre; they overlap at random and probably all have the same polarity, i.e. the reducing groups are all at the same end.

Hemicellulose

In dicotyledonous plants the main hemicelluloses are xyloglucans, mixed polysaccharides principally of glucose and xylose. Some xylan, a pure polymer of xylose, is also present. In addition to the major components of xylose and glucose, xyloglucan also contains fucose, galactose and small amounts of arabinose. The xyloglucan has a β-glucan backbone very like that of cellulose and this can probably interact very strongly with cellulose by hydrogen bonding (Figure 3.10a). Fixed to about 75% of the glucose units of the glucan backbone are side chains of α-xylose. The xylose is α-1,6 linked. A minority of the xylose units may also bear another sugar such as galactose attached by α-1,4 bonds.

In monocotyledons and in legume leaves and stems, the main hemicelluloses are arabinoxylans which consist of β-1,4-linked xylose residues with side chains at various points, but some xyloglucan is also present (see Figure 3.10b). Single arabinose units are the most common but arabino-xylose and arabino-xylogalactose side chains are also found. The polymeric xylan backbone serves the same function as the xyloglucan in dicotyledons, in that it can attach itself to cellulose through hydrogen bonds.

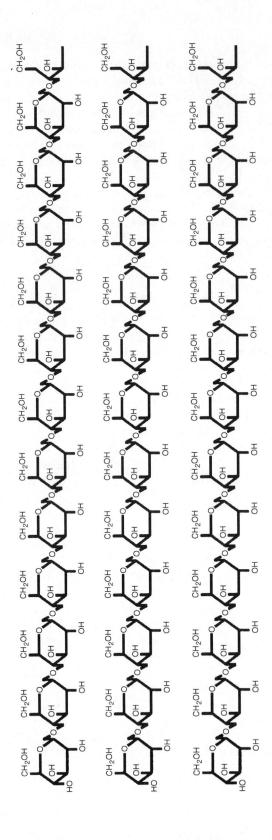

Figure 3.9 The structure of cellulose. Long chains of glucose groups, linked β-1,4, associate by hydrogen bonding to form sheets. Adjacent chains may lie parallel (as here) or may alternate in direction.

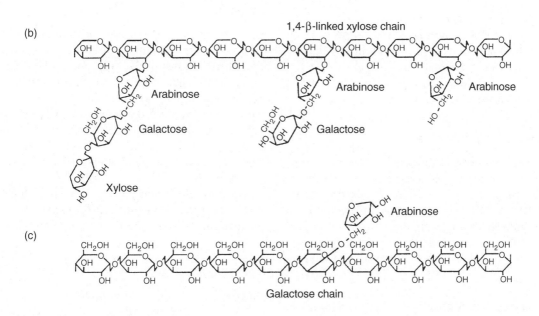

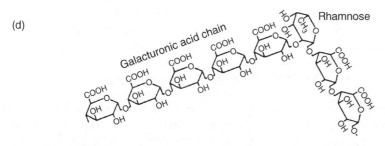

Figure 3.10 Structures of hemicelluloses and pectins. (a) The xyloglucans of dicotyledonous plants have a backbone of glucose units with side chains of xylose. (b) Monocotyledons and legumes possess arabinoxylans which have a principal chain of xylose residues substituted with arabinose and other sugars. (c) Galactosan pectins have a spine of β-1,4-linked galactose units with arabinose branches. (d) The rhamno-galactonuran pectins have chains of galacturonic acid, broken every 10 or so residues by a rhamnose group.

Pectins

The main feature of the pectins is the presence of linear chains of galacturonic acid residues (polygalacturonans), some of which are present in the form of methyl esters. The polygalacturonans also usually contain other sugars, e.g. rhamnose (rhamnogalacturonans). Pectins are generally the most soluble of the cell wall components. They are easily extracted using hot water and they have physical properties which make them important in the food industry.

The polygalacturonans may also be associated by covalent bonds to neutral pectic polysaccharides such as the arabinans, galactans and arabinogalactans.

The most common types of pectic polysaccharides are indicated below.

Galactans and arabinogalactans

These arabinose- and galactose-containing neutral pectic polysaccharides are thought to consist of homo-β-1,4-linked galactose units with few side chains linked to an arabinose polymer with more arabinose branches (Figure 3.10c). Pectin occurs in the primary and secondary cell walls and in the middle lamella.

Rhamnogalacturonans

Pure polygalacturonans are quite rare – usually chains of polygalacturonic acid residues are broken by neutral rhamnose residues (Figure 3.10d). In general, they contain about one rhamnose for each 10 galacturonic acid units. Galacturonic acid chains may be linked to both 2 and 4 positions of the rhamnose unit and this results in a zig-zag chain. The pectin in primary and secondary walls is called protopectin and has more COOH groups esterified than in the middle lamella. In the middle lamella the COO⁻ groups of the polygalacturonan are held together by Ca^{2+} cross links. The degree of cross linking and hence the strength of cell–cell adhesion may be regulated by the degree of methylation of the COOH groups in the pectin.

The commercial importance of pectins results from their ability to form gels. Aqueous solutions of pectin, heated with sugar under acidic conditions (pH 2–3.5) solidifies to a clear gel on cooling and this forms the basis of jam-making.

3.9.3 OTHER POLYSACCHARIDES AND RELATED COMPOUNDS

Saponins

Many forage legumes grown in temperate climates contain saponins. These are glycosides consisting of a non-polar aglycone and a polar sugar group. The non-polar group is either a steroid or a similar polycyclic compound. They therefore have strong detergent properties. Saponins in alfalfa (lucerne) have been studied in detail. They are bitter and reduce intake when alfalfa is added to diets for non-ruminant animals.

Saponins form complexes with cholesterol and can reduce serum cholesterol levels because they prevent readsorption from bile.

Glucosinolates

Oilseed rape (canola) is widely grown as a source of oil and the residue may be used as a feed component. However it contains types of compound which restrict its use: the glucosinolates and erucic acid (see Chapter 4). The glucosinolates (Figure 3.11a) are glycosides of β-D-thioglucose with aglycones which yield toxic isothiocyanate, thiocyanates, nitriles or oxazolidone derivatives such as goitrin under the influence of the enzyme glucosinolase (myrosinase) which is released when the plant is crushed. Swelling of the thyroid (goitre) is a common symptom of glucosinolate poisoning. Modern varieties have much lower levels of glucosinolates than earlier ones and the meal obtained from these varieties is therefore of higher value.

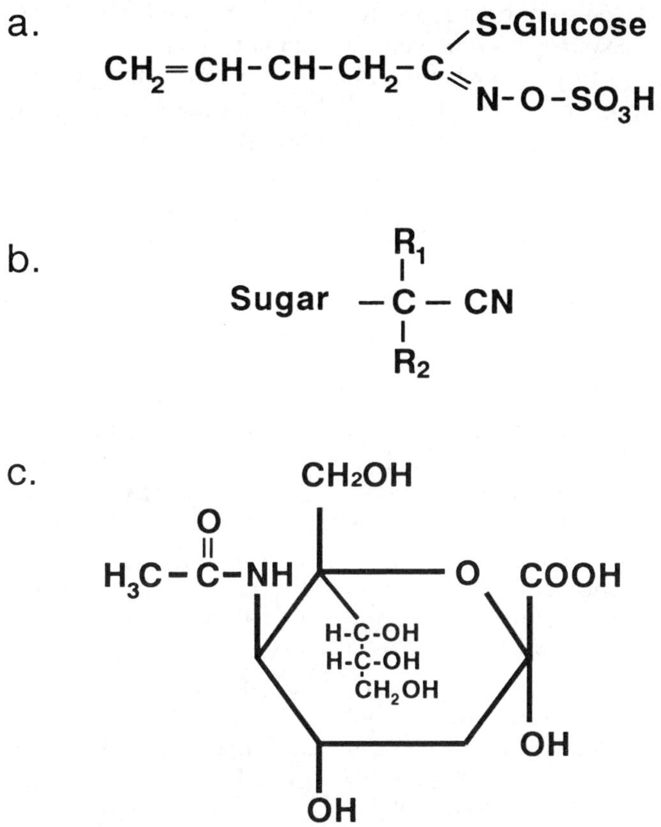

a.

$$CH_2=CH-CH-CH_2-C\underset{N-O-SO_3H}{\overset{S\text{-Glucose}}{\lessgtr}}$$

b.

$$Sugar -\underset{R_2}{\overset{R_1}{\underset{|}{\overset{|}{C}}}}- CN$$

c.

Figure 3.11 Sugars conjugated with other types of compound are quite commonly found in agricultural materials. (a) Glucosinolates are found in some brassicas (e.g. rape) and may limit their usefulness as crops. (b) The cyanogenic glucosides can be hydrolysed to yield toxic quantities of cyanide. (c) Sialic acid is found in some cell membrane carbohydrates.

Cyanogenic glucosides

These are glycosides of a sugar and a cyanide-containing aglycone. The most important cyanogens are amygdalin and prunasin (wild cherries, almond, apricot, peach, apple kernels), dhurrin (sorghum) and linamarin (white clover, cassava, linseed and lima beans) which have the general structure shown in Figure 3.11b.

Cyanogens can be broken down by the action of glucosidases and hydroxynitrile lyases to release HCN, which is extremely toxic because it inhibits cytochrome oxidase which catalyses the final step in the electron transport chain (see Chapter 12).

Normally the glucosides are found in the vacuoles of plants whilst the enzymes which degrade them are located in the cytoplasm, so negligible breakdown occurs. However, wilting, frost or mechanical treatments bring the two together and cause degradation. The enzymes are also produced by rumen bacteria, so ruminants are particularly sensitive to these compounds.

Exposure to these compounds affects livestock which browse on the leaves of plants such as chokecherry (*Prunus virginiana*) and Saskatoon serviceberry (*Amelanchiar alnifolia*). Some varieties of cassava also contain cyanogens which are poisonous to humans who eat it. Traditional methods of preparation minimize the risk of poisoning and involve either grating and soaking to remove the HCN, or cooking to destroy the enzymes and prevent cyanide production.

Lectins and their action

Lectins are proteins which are able to recognise and bind to specific sugar sequences. They may therefore agglutinate cells which bear carbohydrates on the outside of their walls, or precipitate glycoproteins (proteins that contain one or more carbohydrate groups bound to them). Many lectins of plant origin are known. Thus concanavalin A from jack bean (*Canavalia ensiformis*) recognises oligomannosyl *N*-linked sugars, wheatgerm agglutinin binds sialic acid (Figure 3.11c) and *N*-acetylglucosamine, and ricin from castor bean bind galactose. These compounds are of interest to science as probes of carbohydrate structure and function, especially in relation to membrane structure, but at a more practical level many of them are toxic constituents of feeds. Within the organism they have important functions in cellular recognition. One example of this is in development of self-incompatibility in plants: germination of pollen grains is prevented in many types of plants if lectins in the pollen recognise the stigma as being from the same plant. It is also now recognised that animals contain similar proteins which may serve similar functions.

Many mammalian cell membranes have projecting carbohydrate groups on their surfaces, and it is through these carbohydrates that many of the processes of cell recognition take place. For instance, red cells from individuals with A-, B- or O-type blood groups differ in the nature of the carbohydrate groups exposed at the cell surface. The cells of the intestinal wall also bear carbohydrate side chains, many of them mannose polysaccharides (mannans). On their outer cell membranes, the infecting bacteria carry lectins which are able to recognise and bind to the specific sequences of carbohydrates in the tissues that they are about to invade. The extent of bacterial invasion can be reduced in animals fed high levels of dietary mannans. This has the effect of occupying all of the lectin-binding sites so that they are unable to lock onto the intestinal polysaccharides.

Promising results have been obtained in feeding complex polysaccharides of glucose and mannose (glucomannans) extracted from the cell walls of cultivated yeasts. Reductions in the incidence of infectious respiratory and enteric illness have been demonstrated in poultry, pigs and pre-ruminant calves. The advantages for animal health are expected to be confined to non-ruminant animals which do not have the capability to break down glucomannans in the gut.

4.1 INTRODUCTION

The term lipids is used to describe a chemically heterogeneous group of organic compounds which have in common the general property that they are insoluble in water but are soluble in organic solvents such as hydrocarbons (e.g. hexane and toluene), chloroform and alcohols. In its widest sense the term encompasses natural products such as the fat-soluble vitamins, carotenoids, steroids, terpenes, bile salts, fatty acids and their ester and amine derivatives. Commonly it is used more narrowly to include only fatty acids and their derivatives, waxes, steroids and steroid esters.

Several other terms in common use need to be defined more precisely for the purposes of this book. 'Fat' is often used in a very general sense to mean any substance that is fatty in texture. Similarly, in common usage 'oil' is a term used to describe liquids as different as cooking oil and engine oil. In the context of this chapter, and for most nutritional and biochemical applications, fats and oils refer to triacylglycerols (triglycerides in older texts) in which a fatty acid is esterified to each of the hydroxyl groups of the trihydric alcohol glycerol. It is generally accepted that, at room temperature, fats are solids such as lard or dripping whereas oils are liquids such as rapeseed oil or olive oil. These different physical properties are determined by the type of fatty acids incorporated into the triacylglycerols, as described in more detail later in this chapter.

4.2 STRUCTURE AND OCCURRENCE OF LIPIDS

4.2.1 FATTY ACIDS

Fatty acids are a group of aliphatic carboxylic acids which can contain from two to 24 or more carbon atoms. The most abundant types of fatty acids are saturated and unsaturated straight-chain fatty acids. Other types of fatty acids, such as branched-chain fatty acids, hydroxyl-substituted fatty acids and cyclic fatty acids, are usually minor components of lipids.

Saturated and unsaturated straight-chain fatty acids

Saturated and unsaturated fatty acids usually contain an even number of carbon atoms, typically between 10 and 24 carbon atoms in most plant and animal tissues. Small amounts of odd-numbered fatty acids (mainly 15 and 17 carbon atoms) are also found in plant and animal lipids. Unsaturated fatty acids contain one or more double bonds. Fatty acids with one double bond are called monounsaturated fatty acids, those with more than one double bond are referred to as polyunsaturated fatty acids or PUFAs. The double bonds may have one of two configurations, *cis* or *trans* (Figure 4.1). As a general rule the double bonds in most naturally occurring unsaturated fatty acids have the *cis* configuration, although fatty acids with *trans* double bonds are found in bacterial lipids. PUFAs with a combination of both *cis* and *trans* double bonds are produced from *cis* unsaturated fatty acids by chemical hydrogenation of vegetable oils, during the manufacture of margarines and during biohydrogenation of dietary PUFAs in the rumen of ruminant animals.

Fatty acids are named in a number of different ways. For the most commonly occurring fatty acids trivial names are often used, although a systematic naming convention can be used for all fatty acids indicating the number of carbon atoms and the number, type and position of double bonds and substituent groups. Carbon atoms in a fatty acid are normally identified with respect to the carboxyl carbon, which is carbon 1. The position of the double bonds is identified by one of two naming conventions. In the first convention, double bonds are numbered from the carboxyl carbon, and their position indicated by the notation Δ^x (where x is the number of carbon atoms between carbon 1 and the double bond). The second convention uses the number of carbon atoms between the methyl carbon and the nearest double bond and uses the *n-* (*n* minus) notation. This convention is particularly useful for identifying families of fatty acids derived from a common precursor fatty acid, as

described in Chapter 19. When using the Δ notation it is usual to indicate the configuration of the double bonds in the full systematic name. Thus palmitic acid, a saturated fatty acid with 16 carbon atoms, is the trivial name for hexadecanoic acid which is also referred to as C16:0 using a shorthand notation. Linoleic acid is the trivial name for a fatty acid which contains 18 carbon atoms and two *cis* double bonds, one between carbons 9 and 10 and the other between carbons 12 and 13. Its systematic name is *all-cis* $\Delta^{9,12}$ octadecadienoic acid which in this book is represented in the shorthand notation as $\Delta^{9,12}$ C18:2 or C18:2*n*-6.

Prior to the introduction of the numbering system, the Greek alphabet was used to label fatty acid carbon atoms. The lettering system did not label the carboxyl carbon, thus, the α-carbon was equivalent to carbon 2. The methyl carbon was always referred to as the ω-carbon. Although this lettering system is no longer in use in modern texts it is closely linked with the discovery of the pathways for the metabolism of fatty acids, and it is for this reason that the three pathways of fatty acid oxidation are still referred to as α-, β- and ω-oxidation (see Chapter 13).

Table 4.1 contains a list of saturated and unsaturated fatty acids commonly found in plant and animal tissues, with their common name, systematic name, shorthand notation and chemical structure.

The presence of a *cis* double bond markedly reduces the melting point of a fatty acid, as can be seen from Table 4.2. This depression occurs because the *cis* double bond introduces a bend into the otherwise linear structure of a saturated fatty acid, preventing the molecules from stacking closely together into a crystalline structure as the temperature decreases. The greater the number of double bonds in a fatty acid of a given chain length, the lower its melting point. In contrast, a *trans* double bond does not change the linear nature of a fatty acid and has a much smaller effect on melting point.

Lauric and myristic acids are minor components of most animal fats, but occur in signifi-

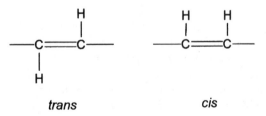

Figure 4.1 The configuration of *trans* and *cis* double bonds.

Table 4.1 Nomenclature and structure of commonly occurring fatty acids

Trivial name	Systematic name	Shorthand notation	Structure
Saturated fatty acids			
Lauric acid	Dodecanoic acid	C12:0	$CH_3(CH_2)_{10}COOH$
Myristic acid	Tetradecanoic acid	C14:0	$CH_3(CH_2)_{12}COOH$
Palmitic acid	Hexadecanoic acid	C16:0	$CH_3(CH_2)_{14}COOH$
Stearic acid	Octadecanoic acid	C18:0	$CH_3(CH_2)_{16}COOH$
Arachidic acid	Eicosanoic acid	C20:0	$CH_3(CH_2)_{18}COOH$
Behenic acid	Docosanoic acid	C22:0	$CH_3(CH_2)_{20}COOH$
Lignoceric acid	Tetracosanoic acid	C24:0	$CH_3(CH_2)_{22}COOH$
Monounsaturated fatty acids			
Palmitoleic acid	cis-Δ^9-hexadecenoic acid	Δ^9 C16:1	$CH_3(CH_2)_5CH=CH(CH_2)_7COOH$
Oleic acid	cis-Δ^9-octadecenoic acid	Δ^9 C18:1	$CH_3(CH_2)_7CH=CH(CH_2)_7COOH$
Gondoic acid	cis-Δ^{11}-eicosenoic acid	Δ^{11} C20:1	$CH_3(CH_2)_7CH=CH(CH_2)_9COOH$
Erucic acid	cis-Δ^{13}-docosanoic acid	Δ^{13} C22:1	$CH_3(CH_2)_7CH=CH(CH_2)_{11}COOH$
Nervonic acid	cis-Δ^{15}-tetracosenoic acid	Δ^{15} C24:1	$CH_3(CH_2)_7CH=CH(CH_2)_{13}COOH$
Polyunsaturated fatty acids			
	all cis-$\Delta^{6,9}$-octadecadienoic acid	$\Delta^{6,9}$ C18:2	$CH_3(CH_2)_7CH=CHCH_2CH=CH(CH_2)_4COOH$
Linoleic acid	all cis-$\Delta^{9,12}$-octadecadienoic acid	$\Delta^{9,12}$ C18:2	$CH_3(CH_2)_4CH=CHCH_2CH=CH(CH_2)_7COOH$
γ-Linolenic acid	all cis-$\Delta^{6,9,12}$-octadecatrienoic acid	$\Delta^{6,9,12}$ C18:3	$CH_3(CH_2)_4CH=CHCH_2CH=CHCH_2CH=CH(CH_2)_4COOH$
α-Linolenic acid	all cis-$\Delta^{9,12,15}$-octadecatrienoic acid	$\Delta^{9,12,15}$ C18:3	$CH_3CH_2CH=CHCH_2CH=CHCH_2CH=CH(CH_2)_7COOH$
Arachidonic acid	all cis-$\Delta^{5,8,11,14}$-eicosatetrenoic acid	$\Delta^{5,8,11,14}$ C20:4	$CH_3(CH_2)_4CH=CHCH_2CH=CHCH_2CH=CHCH_2CH=CH(CH_2)_3COOH$
EPA	all cis-$\Delta^{5,8,11,14,17}$-eicosapentaenoic acid	$\Delta^{5,8,11,14,17}$ C20:5	$CH_3CH_2CH=CHCH_2CH=CHCH_2CH=CHCH_2CH=CHCH_2CH=CH(CH_2)_3COOH$
	all cis-$\Delta^{7,10,13,16,19}$-docosapentaenoic acid	$\Delta^{7,10,13,16,19}$ C22:5	$CH_3CH_2CH=CHCH_2CH=CHCH_2CH=CHCH_2CH=CHCH_2CH=CH(CH_2)_5COOH$
DHA	all cis-$\Delta^{4,7,10,13,16,19}$-docosahexaenoic acid	$\Delta^{4,7,10,13,16,19}$ C22:6	$CH_3CH_2CH=CHCH_2CH=CHCH_2CH=CHCH_2CH=CHCH_2CH=CHCH_2CH=CH(CH_2)_2COOH$

Table 4.2 Melting point of commonly occurring fatty acids

Trivial name	Shorthand notation	Melting point (°C)*
Palmitic acid	C16:0	60.7
Palmitoleic acid	cis-Δ^9-C16:1	1.0
Stearic acid	C18:0	69.6
Oleic acid	cis-Δ^9-C18:1	16.0
Elaidic acid	trans-Δ^9-C18:1	44.0
Linoleic acid	all cis-$\Delta^{9,12}$-C18:2	−5.0
α-Linolenic acid	all cis-$\Delta^{9,12,15}$-C18:3	−11.0
Arachidic acid	C20:0	75.4
Gondoic acid	cis-Δ^{11}-C20:1	24.0
Erucic acid	cis-Δ^{13}-C20:1	24.0
Arachidonic acid	all cis-$\Delta^{5,8,11,14}$-C20:4	−49.5

*From Gurr, M.I. and Harwood, J.L. (1991) *Lipid Biochemistry, An Introduction*, 4th edn, Chapman & Hall, London.

cant quantities in certain vegetable oils such as coconut and palm-kernel oils. The properties of these oils make them useful ingredients in the food industry. Coconut oil is used extensively as an ingredient of calf and lamb milk replacer due to its high digestibility. When included in ruminant feeds the lauric acid oils can adversely affect rumen function. The shorter-chain fatty acids (C4:0; C6:0; C8:0 and C10:0) are characteristic of milk fats. Some species of mammals produce milk which is particularly rich in these acids. For example, in fat from rabbit milk, C6:0 and C8:0 constitute more than 40% of the total fatty acids, whereas C8:0 and C10:0 make up more than 60% of the fatty acids of elephant milk.

Palmitic acid is the major saturated fatty acid found in most vegetable oils and is also found in significant quantities (15–25%) in animal fats. Stearic acid is the predominant saturated fatty acid in ruminant fats, accounting for as much as 40–45% of the total fatty acids. In pig and poultry fats it occurs in lower proportions and it is often only a minor component in fish and vegetable oils. Longer-chain saturated fatty acids (C20:0, C22:0 and C24:0) occur as only trace components in both animal and vegetable lipids

Although many monounsaturated fatty acids have been identified, by far the most abundant is oleic acid, which is found in varying proportions in all animal and plant lipids. It is the major fatty acid of olive and almond oils. Other monoenoic fatty acids are gondoic acid and erucic acid (Δ^{11} C20:1 and Δ^{13} C22:1) which are found in oils from seeds of the genus Cruciferae. In certain varieties of rape and mustard, erucic acid may constitute in excess of 50% of the fatty acids. Because of the toxic properties of this fatty acid, varieties of rape grown for human and animal consumption have been bred that are low in erucic acid (<5.0%).

PUFAs are found in both plant and animal lipids. The C18 PUFAs, linoleic and α-linolenic acids, are found in many vegetable oils. Rich sources of linoleic acid are soyabean oil, maize (corn) oil and safflower oil which contain typically 50–75%. These oils also contain lower proportions of α-linolenic acid. The best-known plant source of α-linolenic acid is linseed (flax) oil. γ-Linolenic acid is usually a trace component of plant and animal fatty acids, but it is found in significant quantities in the seeds of the evening primrose, borage and blackcurrant. Although its medicinal properties are not understood, consumption of oils rich in this fatty acid appears to have beneficial effects on sufferers of multiple sclerosis. In general, plant lipids do not contain significant amounts of PUFAs with more than 18 carbons. However, examination of animal tissue lipids reveals a more complex picture, with a range of PUFAs with chain lengths from 18 to 22 carbon atoms. Fatty acids in fat depots (adipose tissue) have a relatively simple fatty acid composition containing $\Delta^{9,12}$ C18:2 and some $\Delta^{9,12,15}$ C18:3 as the main classes of PUFAs, whereas other tissues such as liver and muscle tissue contain a wider range including $\Delta^{9,12}$ C18:2; $\Delta^{9,12,15}$ C18:3; $\Delta^{5,8,11,14}$ C20:4; $\Delta^{5,8,11,14,17}$ C20:5 and $\Delta^{4,7,10,13,16,19}$ C22:6, plus many other minor components. This complexity reflects the higher proportion of membrane phospholipids found in these tissues. Most marine oils contain substantial amounts of the longer-chain C20 and C22 PUFAs, particularly; $\Delta^{5,8,11,14,17}$ C20:5, $\Delta^{7,10,13,16,19}$ C22:5 and $\Delta^{4,7,10,13,16,19}$ C22:6. Typical fatty acid compositions of vegetable and animal lipids are given in Table 4.3.

Although saturated and unsaturated straight-chain fatty acids make up by far the greatest proportion of those found in plant and animal tissue lipids, there are a number of other important, minor types which can have an impact on the properties of lipids and, in some cases, their nutritional value.

Branched-chain fatty acids

This term is normally used to describe fatty acids which contain one or more methyl (and rarely ethyl) substituents along the carbon chain. Many microorganisms contain branched-chain fatty acids which are mainly of the iso and anteiso type (Figure 4.2).

These fatty acids are typical of most Gram-positive and some Gram-negative organisms.

Trace quantities of iso and anteiso fatty acids (0.1–0.3%) are found in ruminant animal fat depots. They arise as a result of the digestion and absorption of the lipids from rumen microorganisms as they pass through the small intestine. When certain unusual diets are provided the proportion of branched-chain fatty acids in sheep and goat fat can be greatly increased (see Chapter 19).

Hydroxy and cyclic fatty acids

A number of hydroxy fatty acids are found in bacterial lipids. These are mainly saturated in nature, such as 3-hydroxymyristic acid, and are found predominantly in the lipopolysaccharide fraction of the cell membrane. The best known example of a hydroxy fatty acid in plant lipids is the occurrence of ricinoleic acid (12-hydroxy-oleic acid) which constitutes between 80 and 95% of the fatty acids in castor oil.

Cyclopropane and cyclopropene fatty acids are found in a number of plant species, particularly the Malvaceae and Sterculaceae, e.g. malvalic acid and sterculic acid which is a minor component of cottonseed oil (Figure 4.3).

4.2.2 TRIACYLGLYCEROLS AND OTHER ACYLGLYCEROLS

Most purified fats and oils isolated from plant and animal sources and used in human and animal diets are triacylglycerols. The term triacylglycerol encompasses a wide spectrum of molecular species in which each of the three hydroxyl groups of glycerol are esterified to a fatty acid (Figure 4.4).

Because of the wide variety of fatty acids that occur naturally, many thousands of molecular species of triacylglycerols are possible. Those which contain three identical fatty acids

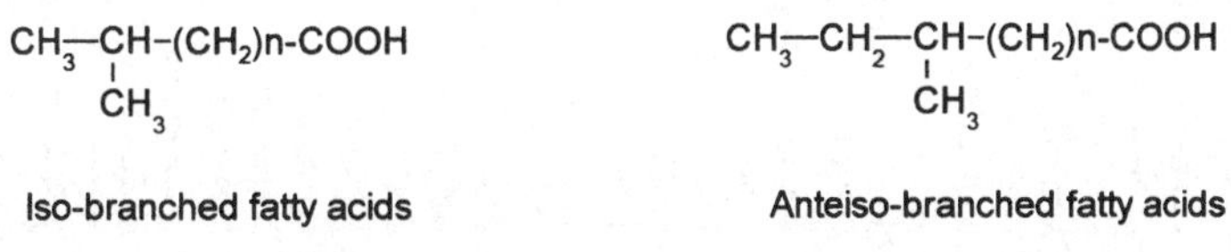

$$CH_3\!-\!CH\!-\!(CH_2)n\!-\!COOH$$
$$|$$
$$CH_3$$

Iso-branched fatty acids

$$CH_3\!-\!CH_2\!-\!CH\!-\!(CH_2)n\!-\!COOH$$
$$|$$
$$CH_3$$

Anteiso-branched fatty acids

Figure 4.2 The structure of iso- and anteiso-branched chain fatty acids.

Table 4.3 Fatty acid composition of vegetable and animal lipids

Lipid	C10:0	C12:0	C14:0	C16:0	C16:1[§]	C18:0	C18:1[§]	C18:2[§]	C18:3[§]	C20:0	C20:1[§]	C20:4[§]	C20:5[§]	C22:1[§]	C22:5[§]	C22:6[§]
Vegetable lipids																
Coconut oil*	7.0	47.0	17.0	8.0		4.0	5.0	2.0								
Palm kernel oil*	4.0	51.0	17.0	8.0		2.0	13.0	2.0								
Palm oil*			1.0	48.0		4.0	38.0	9.0								
Cottonseed oil*			1.1	27.3	1.4	3.1	16.7	50.4								
Maize oil*				12.6	0.8	1.8	30.0	54.3								
Olive oil*				14.0	2.0	2.0	64.0	16.0	0.5							
Rape oil (low erucic acid)†				4.0			56.0	26.0	10.0		2.0					
Rape oil (high erucic acid)†				3.0		1.0	16.0	14.0	10.0	1.0	6.0			49.0		
Safflower oil*				6.6	0.6	3.4	12.2	77.0	0.2							
Soyabean oil*				12.0		3.6	23.7	51.4	8.8							
Sunflower oil*				6.0	3.0	27.0	64.0									
Linseed oil†				6.1	0.1	3.2	16.1	14.2	59.8							
Animal lipids																
Beef tallow*			3.0	26.0	6.0	17.0	43.0	4.0		0.2						
Lard*			2.0	26.0	4.0	14.0	43.0	10.0								
Chicken fat[b]			1.2	20.3	3.9	6.5	33.5	18.5	2.6		0.5	0.3	0.4			
Sheep liver			0.8	17.0	1.1	29.3	15.4	9.5	1.8			6.9	1.4	2.6	5.2	1.9
Chicken liver[b]			0.8	22.9	1.9	14.5	25.5	8.8	0.5		0.4	2.4	1.8		0.3	2.0
Chicken breast muscle[b]			1.4	19.9	3.6	9.0	25.5	17.5	1.6		0.4	2.8	0.7		1.1	2.1
Cod liver oil*			4.0	10.0	12.0		20.0	4.0			15.0		12.0			11.0
Menhaden oil*			6.0	18.0	12.0	5.0	16.0	4.0	4.0				13.0			13.0

* From Allen, J.C. and Hamilton, R.J. (eds) (1989) *Rancidity in Foods*, 2nd edn, Elsevier Applied Science, London and New York.

† From Gunstone, F.D., Harwood, J.L. and Padley, F.B. (1994) *The Lipid Handbook*, 2nd edn, Chapman & Hall, London.

[b] From Scaife, J.R., Moyo, J., Galbraith, H., Michie, W. and Campbell, V. (1994) Effect of different dietary supplemental fats and oils on the growth performance and tissue fatty acid composition of female broilers. *British Poultry Science*, **33**, 107–118.

§ C16:1 = Δ^7 C16:0; C18:1 = Δ^9 C18:1; C18:2 = $\Delta^{9,12}$ C18:2; C18:3 = $\Delta^{9,12,15}$ C18:3; C20:1 = Δ^{11} C20:1; C20:4 = $\Delta^{5,8,11,14}$ C20:4; C20:5 = $\Delta^{5,8,11,14,17}$ C20:5; C22:5 = $\Delta^{7,10,13,16,19}$ C22:5 and C22:6 = $\Delta^{4,7,10,13,16,19}$ C22:6

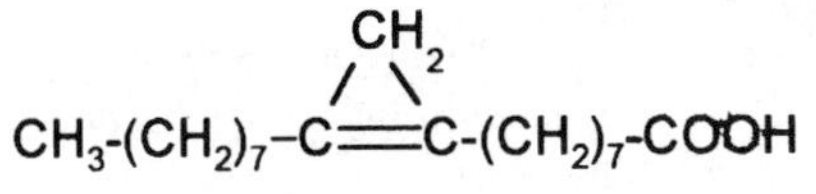

$$CH_3\text{-}(CH_2)_7\text{-}C\overset{\displaystyle CH_2}{=\!=\!=}C\text{-}(CH_2)_7\text{-}COOH$$

Sterculic acid

Figure 4.3 The structure of sterculic acid.

constitute a relatively small proportion of the naturally occurring triacylglycerols; the vast majority contain at least two different fatty acids and are called mixed triacylglycerols.

The great economic importance of triacylglycerols for industrial, pharmaceutical and food use can be seen in Table 4.4.

The physical properties of triacylglycerols are determined by the nature of the fatty acids

they contain. Fats such as beef and pork tallow (lard) contain a higher proportion of saturated fatty acids than plant oils. The degree of unsaturation of fats and oils is often measured as the iodine value (IV), a measure of the reaction of iodine with the double bonds of unsaturated fatty acids. High iodine values indicate a high degree of unsaturation. The saponification value (SV) of triacylglycerols gives comparative information about the chain length of the fatty acids they contain. It represents the yield of fatty acid from one gram of triacylglycerol. High values indicate the presence of significant amounts of short- and medium-chain length fatty acids (see Table 4.4). Saponification is the process in which lipids are hydrolysed by heating in dilute ethanolic KOH. The fatty acids released form water-sol-

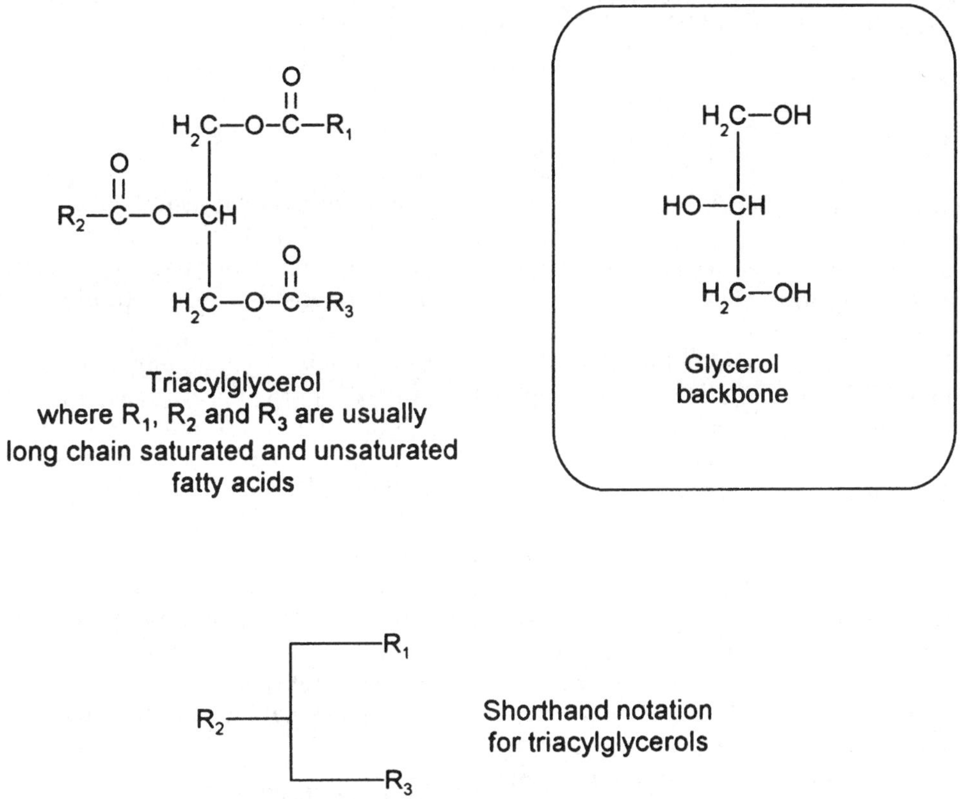

Triacylglycerol
where R₁, R₂ and R₃ are usually
long chain saturated and unsaturated
fatty acids

Glycerol
backbone

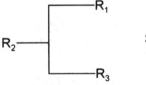

Shorthand notation
for triacylglycerols

Figure 4.4 The outline structure of triacylglycerols (where R_1, R_2 and R_3 are usually long-chain saturated and unsaturated fatty acids), showing the glycerol backbone and the shorthand notation used to represent triacylglycerols.

Table 4.4 World production and physicochemical properties of fats and oils

Fat/oil	Annual production (tonnes)#	Melting point	S/U ratio*	Saponification value	Iodine value
Beef tallow	}	40–50	0.85	190–200	32–47
Pig tallow	}10.6 × 10⁶	28–48	0.72	193–200	46–66
Butter	6.3 × 10⁶	28–33	1.70	216–235	26–45
Cod liver oil	}		0.16	182–193	155–170
Menhaden oil	} 1.4 × 10⁶		0.41	189–193	160–180
Coconut oil	3.4 × 10⁶	23–26	13.29	251–264	7–10
Palm kernel oil	1.9 × 10⁶	25–30	8.09	244–254	14–20
Palm oil	11.0 × 10⁶	38–45	1.13	196–202	48–56
Cottonseed oil	3.3 × 10⁶		0.46	191–196	100–112
Maize oil	0.4 × 10⁶		0.17	187–196	84–102
Olive oil	1.9 × 10⁶		0.19	187–196	117–130
Rapeseed oilᵛ	7.8 × 10⁶			173–181	105–120
Soyabean oil	16.5 × 10⁶		0.18	189–195	124–133
Sunflower oil	8.0 × 10⁶		0.09	186–194	127–136
Linseed oil	0.8 × 10⁶		0.11	188–195	180–185

\# 1990 figures from Agra Europe (1993) No 1537. Agra Europe (London) Ltd, Tunbridge Wells.
* Ratio of saturated to unsaturated fatty acids.
ᵛ Low erucic acid rapeseed oil.

uble potassium salts (soaps), hence the term saponification.

The distribution of fatty acids between the three hydroxyl groups of glycerol is by no means random. During the biosynthesis and later modification of these molecules (discussed in more detail in Chapter 19), the type of fatty acid esterified to each position (saturated or unsaturated, long-chain or short-chain) is determined by the specificity of the enzymes which catalyse this addition. Thus, in general, fatty acids found in position 1 are predominantly saturated and those found in position 2 are unsaturated. The fatty acids in position 3 appear to have a more variable nature, although in mammals they tend to be rich in PUFAs (e.g. $\Delta^{5,8,11,14,17}$ C20:5, $\Delta^{4,7,10,13,16,19}$ C22:6). In fish oils these fatty acids tend to occupy the 2 position. Milk fats which are synthesized in the mammary tissue are characterized by the presence of short- and medium-chain-length fatty acids. These fatty acids are found mainly in position 3.

Other acylglycerols, such as diacylglycerols and monoacylglycerols, usually occur as minor components of tissue lipids and are important intermediates in both the synthesis and breakdown of triacylglycerols.

Diacylglycerols contain only two fatty acids and can exist in two forms, 1,2-diacylglycerols and 1,3-diacylglycerols. The first species is the initial breakdown product of triacylglycerols during lipid digestion in the monogastric animal. Monoacylglycerols, which contain only one fatty acid, can also exist in two forms, 1- or 2-monoacylglycerols. In practice 2-monoacylglycerol occurs most commonly. It is a major end product of monogastric lipid digestion and the substrate for triacylglycerol resynthesis in the intestinal mucosa (see Chapter 19).

4.2.3 GLYCEROPHOSPHOLIPIDS

Glycerophospholipids are found in all living organisms. They are based on the structure of glycerol and are important components of bio-

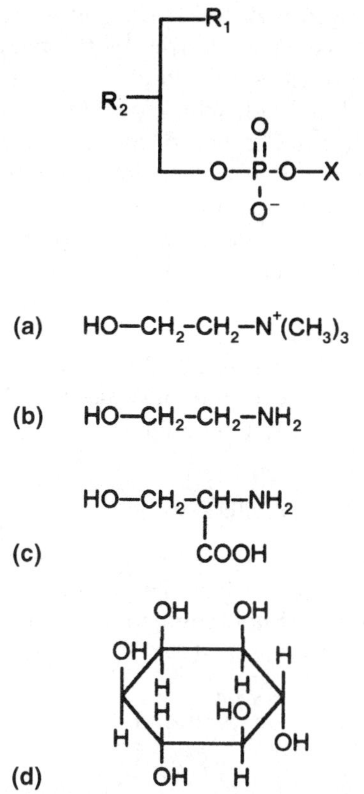

Figure 4.5 The outline structure of glycerophospholipids. R_1 and R_2 are usually long-chain saturated or unsaturated fatty acids. The X group may be one of a number of compounds, the structure of some of which is shown: (a) choline → phosphatidylcholine; (b) ethanolamine → phosphatidylethanolamine; (c) serine → phosphatidylserine; (d) inositol → phosphatidylinositol.

logical membranes. Their structures can be represented as shown in Figure 4.5.

Hydroxyl groups in the 1 and 2 positions of the glycerol backbone are esterified to fatty acids, usually long-chain fatty acids. In bacteria and animals the distribution of fatty acids between the 1 and 2 position is similar to that of triacylglycerols, i.e. predominantly saturated fatty acids in position 1 and unsaturated fatty acids in position 2. Plants, however, do not show the same consistent pattern of fatty acid distribution.

The hydroxyl group in the 3 position is phosphorylated. In the simplest form of glyc-erophospholipid, phosphatidic acid, the phosphate group is not linked to any other sub-stituent group. Phosphatidic acid is normally a minor constituent of tissue phospholipids although it is an important intermediate in glycerophospholipid synthesis.

Most phospholipids contain a substituent group linked to the phosphate in position 3. These groups, a number of which are basic compounds, are usually polar in nature, the most common being choline, ethanolamine, serine and inositol. The choline-containing glycerophospholipid has traditionally been called lecithin, and although this name is still in use it is now more commonly referred to as phosphatidylcholine. Similarly those glyc-erophospholipids containing ethanolamine, serine and inositol are known as phos-phatidylethanolamine (old name cephalin), phosphatidylserine and phosphatidylinositol, respectively.

Because these molecules contain both hydrophobic long-chain fatty acids and hydrophilic phosphate and substituent groups, they have physicochemical properties which make them ideal building blocks for mem-branes. They are often described as amphiphilic molecules because of their ability to act as an interface between a polar aqueous environ-ment and a non-polar lipid environment.

In animals, in addition to fulfilling an impor-tant structural role, membrane glycerophos-pholipids have a role in inter- and intracellular signalling by acting as a reservoir of PUFAs, which are precursors for the synthesis of a family of related compounds such as prostaglandins, leukotrienes and thrombox-anes. These compounds are powerful local reg-ulators involved in processes as varied as the inflammatory response, platelet aggregation, smooth muscle contraction and ovulation.

4.2.4 GLYCOSYLGLYCERIDES

These glycerol-based lipids are characterized by the presence of sugar residues. They are found in small quantities in bacteria and animals, but are most characteristic of photosynthetic tissues in plants, algae and cyanobacteria, particularly as components of chloroplast membranes (Figure 4.6).

In plants the most commonly occurring species are the mono- and digalactosyl diacylglycerols. These contain either a single galactose residue or two galactose residues (linked together by an α-1,6 glycosidic bond) esterified to the 3 position of the glycerol backbone. In plants galactose is almost exclusively the sugar found in glycosylglycerides, but in algae and bacteria diglycosyldiacylglycerols are found which contain other sugars, mainly two glucose or two mannose residues which may be linked α-1,2 or β-1,6.

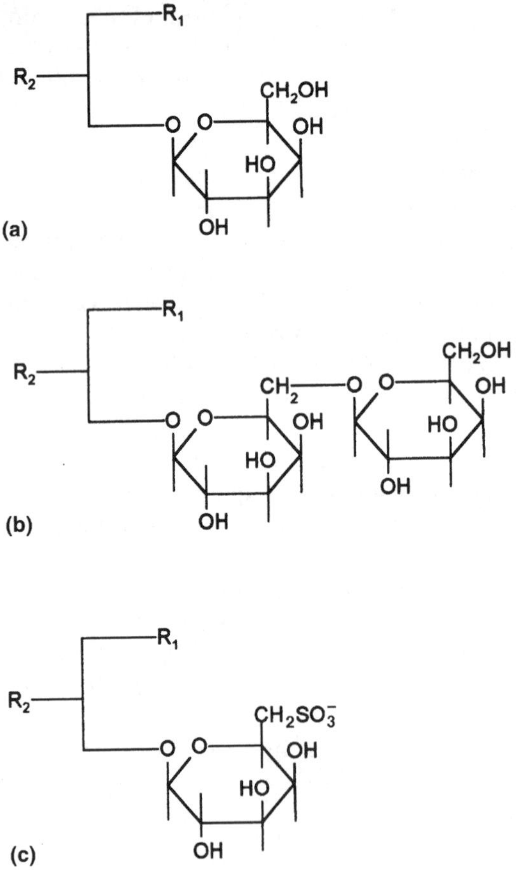

(a)

(b)

(c)

Figure 4.6 The outline structure of (a) monogalactosyldiacylglycerol, (b) digalactosyldiacylglycerol and (c) sulphoquinovosyldiacylglycerol (plant sulpholipid).

The glycosylglyceride fraction from chloroplasts is also characterized by the presence of sulphoquinovosyldiacylglycerol, so-called plant sulpholipid, which contains a sulphate group on the 6 carbon of the sugar residue. A number of other sulpholipids are found as minor constituents of algae and bacteria.

Glycosylglycerides have a high content of polyunsaturated fatty acids, particularly $\Delta^{9,12}$ C18:2 and $\Delta^{9,12,15}$ C18:3 and in some cases, such as spinach chloroplast monogalactosyldiacylglycerol, the unusual fatty acid $\Delta^{7,10,13}$ C16:3. These lipids form the major source of dietary fatty acids in grazing animals.

4.2.5 SPHINGOLIPIDS

Sphingolipids are structural lipids found mainly in membranes. They are based on the structure of the long-chain amino alcohol sphingosine (Figure 4.7).

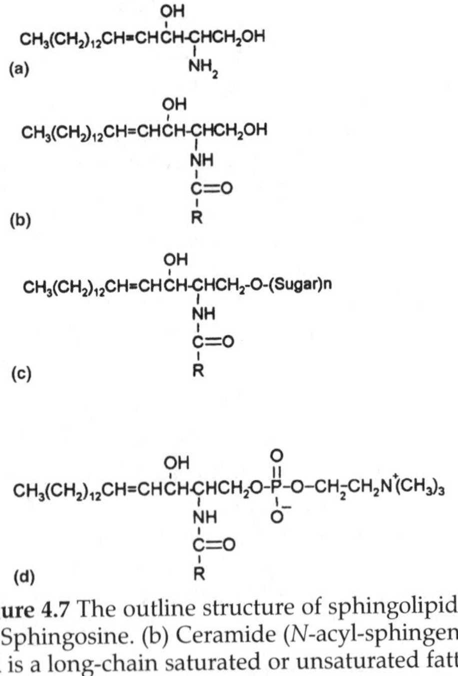

Figure 4.7 The outline structure of sphingolipids. (a) Sphingosine. (b) Ceramide (*N*-acyl-sphingenin – R is a long-chain saturated or unsaturated fatty acid. (c) Cerebrosides and gangliosides – (cerebrosides may contain a number of sugar residues, gangliosides are characterized by the presence on one or more residues of sialic acid (*N*-acetylneuraminic acid). (d) Sphingomyelin.

The attachment of a fatty acid to the amino group of sphingosine produces N-acylsphingosine, or ceramide. Various other substituent groups can be esterified via the primary alcohol group. In sphingomyelin the substituent group is phosphocholine and this lipid may, therefore, also be classified as a phospholipid. Most other types of sphingolipids contain one or more sugar residues linked to N-acylsphingosine. The most commonly occurring sugars are glucose, galactose and N-acetylglucosamine, usually linked by β-1,4 or β-1,3 glycosidic bonds to form mono-, di-, tri- and tetraglycosylceramides. These sphingolipids are classified under the generic name cerebrosides. Gangliosides are a further type of sphingolipid similar to cerebrosides but containing one or more residues of sialic acid (N-acetylneuraminic acid) normally linked β-2,3 to a galactose residue. The names of sphingolipids reflect their association with nervous tissue and particularly brain lipids, however they are widely distributed in other animal tissues and are found in small quantities in plants, although gangliosides appear to be unique to the animal kingdom.

4.2.6 TERPENES AND STEROIDS

Terpenes and sterols are related compounds which are constructed from the five-carbon building block isoprene (Figure 4.8).

Terpenes containing two isoprene units are called monoterpenes, those containing three isoprene units are called sesquiterpenes, and those containing four, six and eight isoprene units are called diterpenes, triterpenes, and tetraterpenes, respectively. Terpenes may be linear or cyclic molecules and some terpenes contain structures of both types.

A large number of terpenes have been discovered in plants and many of these compounds have characteristic smells or flavours, and are major components of the essential oils derived from such plants. The monoterpenes geraniol, limonene, menthol, pinene, camphor and carvone are major components of gerani-um oil, lemon oil, mint oil, turpentine, camphor oil and caraway oil, respectively.

An example of a sesquiterpene commonly found in essential oils is farnesol. Perhaps one of the most commonly occurring diterpenes is the phytol component of chlorophyll. Squalene is probably the most abundant linear triterpene (Figure 4.9).

Other higher terpenes in plants include the carotenoids such as β-carotene (Figure 4.10f), and the xanthophylls such as lutein. These compounds are strongly coloured and are often used as colorants in foods. For example, the synthetic xanthophyll, carophyll yellow, is added to poultry diets to intensify the yellow colour of egg yolk, and astaxanthin is added to commercial salmon diets to reproduce the pink flesh found in wild salmon. Flowers are a rich source of pigments and marigolds are grown in large quantities for the carotenoids and xanthophylls they contain.

Among the most important terpenes in the animal kingdom are three fat-soluble vitamins, A, E and K (Figure 4.10). In addition, the other fat-soluble vitamin, vitamin D, is related to the terpenes as it is synthesized from sterols.

Another important group of terpenoid compounds function as coenzymes in a number of oxidation–reduction reactions in nearly all living organisms. This is the ubiquinone coenzyme Q (CoQ), family of compounds. In higher organisms these are located mainly in the mitochondria and are components of the electron transport chain (see Chapter 12). These compounds contain a substituted quinone ring which can be reduced and reoxidized, and a long isoprenoid side chain. The length of the side chain differs from organism to organism. Similar compounds called plastoquinones occur in plant chloroplasts and are involved in photosynthesis (Figure 4.10).

Yet another important class of terpenes is the polyprenols. These compounds are long-chain linear polyisoprenes with a terminal alcohol group. Certain polyprenols have an important role to play in the synthesis of cell walls in bacteria and plants, and in the synthesis of cell-sur-

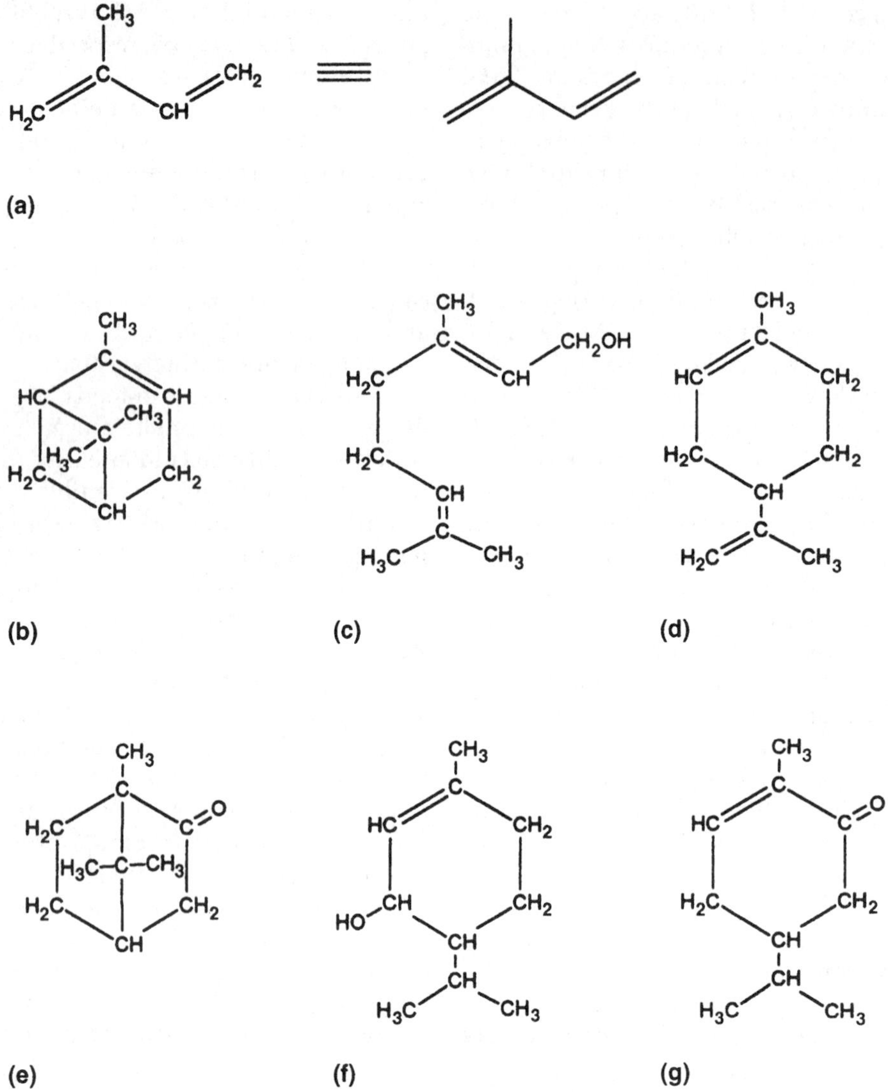

Figure 4.8 The structures of (a) isoprene, the basic five-carbon building block of terpenes, and some naturally occurring monoterpenes in plant oils: (b) α-pinene, (c) geraniol, (d) limonene, (e) camphor, (f) menthol and (g) carvone.

face lipopolysaccharides, peptidoglycans and glycopolysaccharides. Compounds such as bactoprenol are found mainly in bacteria and plant tissues. The dolichols, found mainly in animal plasma membranes, are involved in the transport of glycopeptides across the cell membrane. They are thought to act as carriers for the glycopeptides, transporting them to their site of incorporation into macromolecules such as glycoproteins.

A number of important plant hormones are also terpenoids, for example abscisic acid and the gibberellins. These are discussed in more detail in Chapter 27.

A number of terpenoid compounds occur in insects. For example, compounds known as

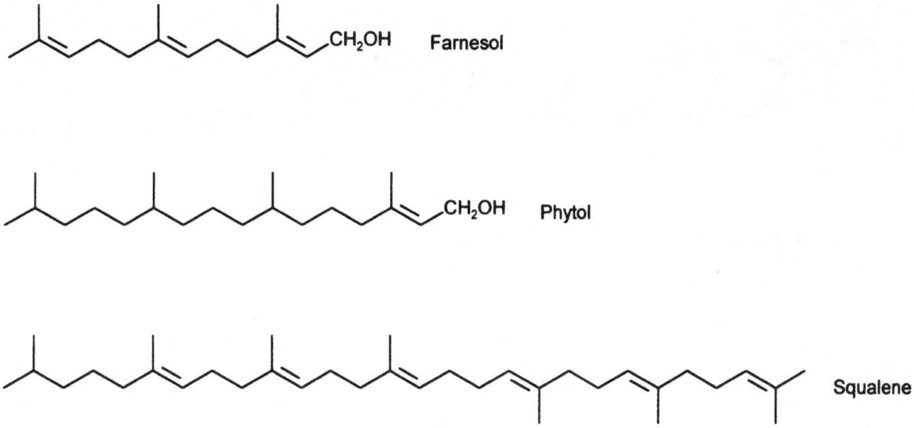

Figure 4.9 The structures of farnesol, phytol and squalene.

the juvenile hormones control the complicated post-embryonic development of insects. In addition, a group of compounds known as the ecdysones, found in insects and arthropods, control moulting. It is interesting to note that ecdysones are also found in plants, and it has been suggested that production of these compounds by plants is a defence mechanism which interferes with the development of insect predators.

A variety of steroids are found in eukaryotic organisms but few are found in bacteria and other prokaryotes. The basic building block of steroids, which are derived from squalene, is a fused, four-membered ring structure, sometimes called the steroid nucleus. The four most commonly occurring steroids in plants and animals are shown in Figure 4.11. These compounds have a hydroxyl group on carbon number 3 and are therefore steroid alcohols or sterols.

Steroid esters can be formed by reaction of a long-chain fatty acid with this hydroxyl group. Alternatively they can form glycosidic links with sugars to give steroid glycosides.

In plants the most abundant sterol is sitosterol. In addition, there are smaller amounts of stigmasterol, campesterol and cholesterol, all of which occur as free sterols, steroid esters and steroid glycosides in varying proportions, depending on the plant species and tissue.

In animals the principal sterol is cholesterol. In its free form and in the ester form it is a component of membranes and appears to play an important role in the modulation of membrane fluidity. Cholesterol is the precursor for the synthesis of a family of steroid hormones found in animal tissues.

4.2.7 WAXES

The strict chemical definition of a wax is an ester formed between a long-chain alcohol and a long-chain fatty acid. However, the term is used more widely for plants and animals to describe the surface lipids of stems and leaves and the sebaceous secretions associated with skin, hair and feathers. The main components of waxes are long-chain alkanes, acids, alcohols and aldehydes and their esters.

In plants these compounds form complex polymers and constitute a significant proportion of suberin and cutin, the structural layers of the cuticle. It has been observed that the alkane composition of plants can be used as a 'fingerprint' to identify plant species. Because these alkanes are not absorbed in the digestive tract, analysis of alkanes in faeces can be used

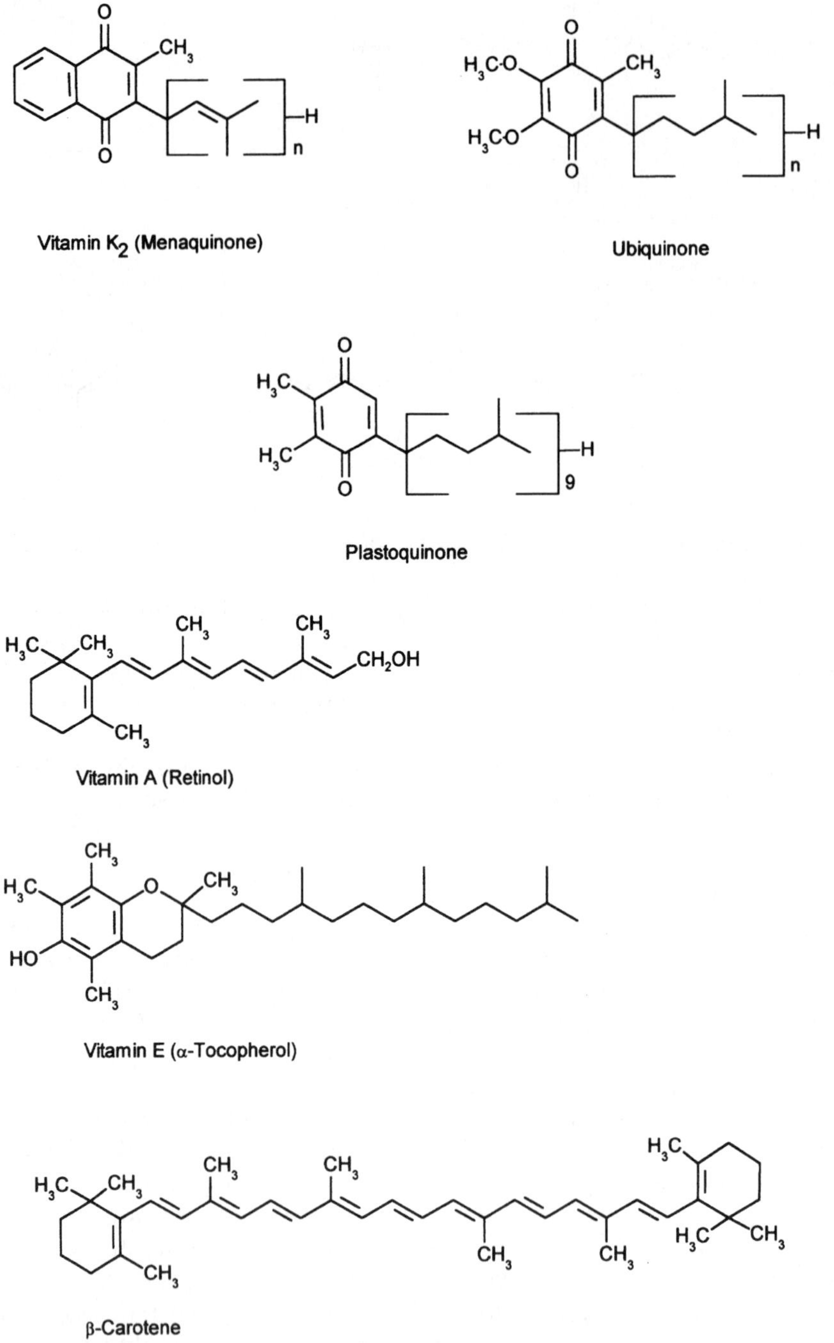

Figure 4.10 The structures of some of the higher terpenes found in plants and animals: vitamin K_2 (menaquinone); ubiquinone; plastoquinone; vitamin A (retinol); vitamin E (α-tocopherol); β-carotene.

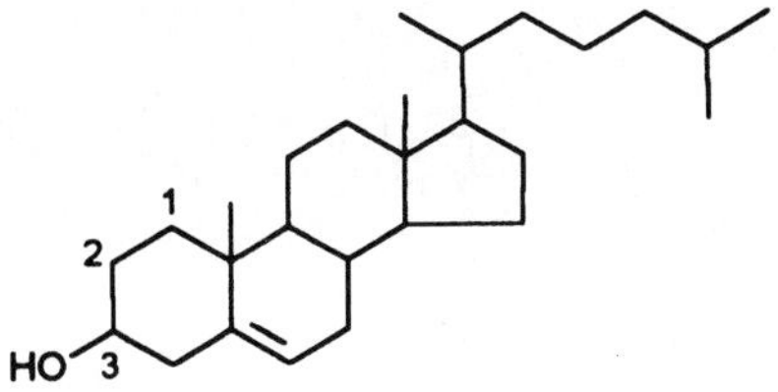

Cholesterol

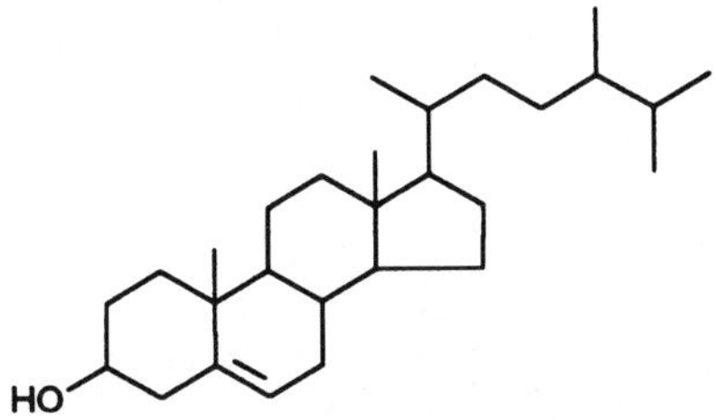

Campesterol

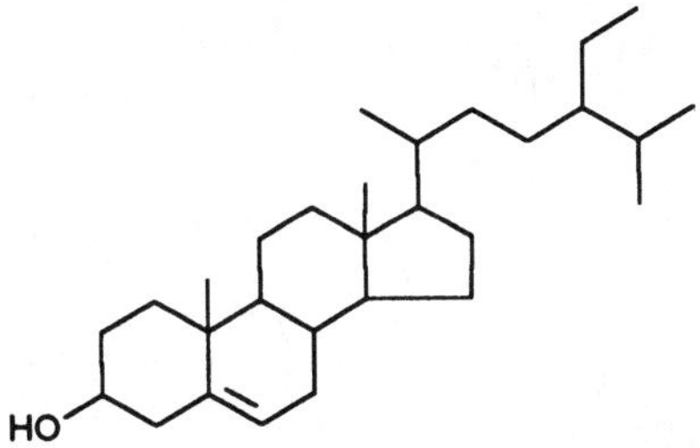

Sitosterol

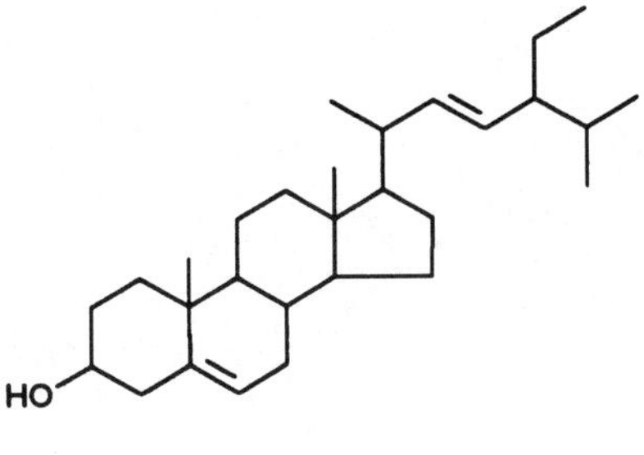

Stigmasterol

Figure 4.11 The structures of the most common sterols found in plants and animals: cholesterol; campesterol; sitosterol; stigmasterol.

to establish the quantity and types of plants eaten by grazing animals.

In certain plant and animal species, waxes may be used as an alternative form of energy store, for example in some marine animals, in zooplankton and in certain oil seeds, jojoba being the most often quoted.

5.1 INTRODUCTION

Amino acids are small molecules containing both $-NH_2$ (amino) and $-COOH$ (carboxylic acid) groups. They are found in all types of cells. Most of the amino acid molecules are found linked together to form proteins. Proteins have vital functions including acting as enzymes, as structural components of the cell and in molecular recognition. It is through the production of specific proteins that the genetic information carried in the DNA is able to control the activities of the cell. Thus the synthesis of proteins is a vital process in all cells, and this requires adequate supplies of all of the amino acids of which they are composed.

Although amino acids occur most commonly as components of proteins, some free amino acids are also found in cells. The concentrations of these are normally relatively low but when plants are subjected to water or salt stress, protein synthesis is slowed down and some free amino acids, especially proline, may accumulate and reach quite high concentrations.

5.2 AMINO ACIDS

5.2.1 STRUCTURE OF AMINO ACIDS

There are approximately 20 different amino acids which occur in proteins. They all have the basic structure shown in Figure 5.1a.

In all of the amino acids except glycine the α-carbon atom is an asymmetric centre, so all amino acids except glycine are optically active. All amino acids found in proteins are the L-isomers. Although they are referred to as amino acids, proline and hydroxyproline do not contain a true amino group. Instead the nitrogen atom forms part of a five-membered ring.

In solution all free amino acids exist in the form of zwitterions (doubly charged ions) as shown in Figure 5.1b.

The amino acids can be divided into several classes. Depending on the nature of the R group amino acids are classified as aliphatic, hydroxy, sulphur-containing, aromatic, basic, acidic and imino acids. The structures of the amino acids commonly found in proteins are shown in Figure 5.2. The side chain (R) can also carry charges and its nature is important in determining whether the amino acid is hydrophobic or hydrophilic, as well as determining the properties of any proteins in which it is found.

5.3 NON-PROTEIN AMINO ACIDS AND RELATED COMPOUNDS

In addition to the amino acids which are found in proteins, there are also a large number of other, non-protein amino acids which exist in the free form, especially in plants. Several hundred non-protein amino acids have been extracted from plants and they are particularly widespread in legumes. Some of them are toxic, or have physiological effects on animals and may be important components of animal feeds. Some examples are given in Figure 5.3.

5.3.1 CANAVANINE

The non-protein amino acid canavanine is present in jack bean (*Canavalia ensiformis*) seeds. This amino acid resembles arginine in structure, and interferes with the metabolism of arginine and its incorporation into proteins in animals which eat the seeds.

5.3.2 SELENIUM-CONTAINING AMINO ACIDS

Some plants growing on soils which are rich in the element selenium may accumulate high levels of selenium-containing amino acids, in which selenium replaces the sulphur atom which normally forms part of their structure. Thus they may contain amino acids such as Se-methylselenomethionine or Se-methylselenocysteine (Figure 5.3). Such plants are toxic to livestock eating them and may cause 'alkali

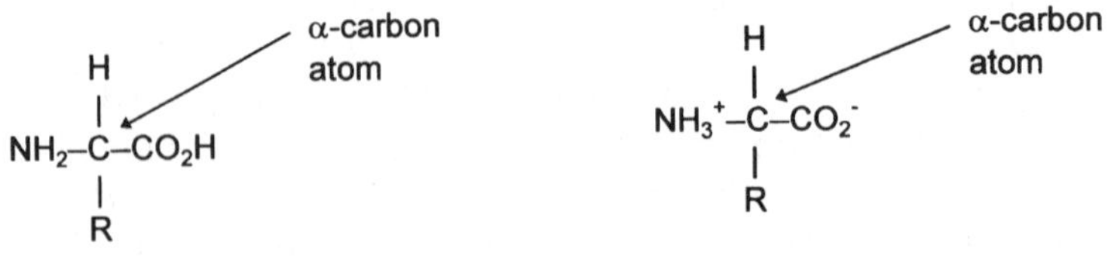

Figure 5.1 The structure of an amino acid in (a) the un-ionized form and (b) the zwitterion form.

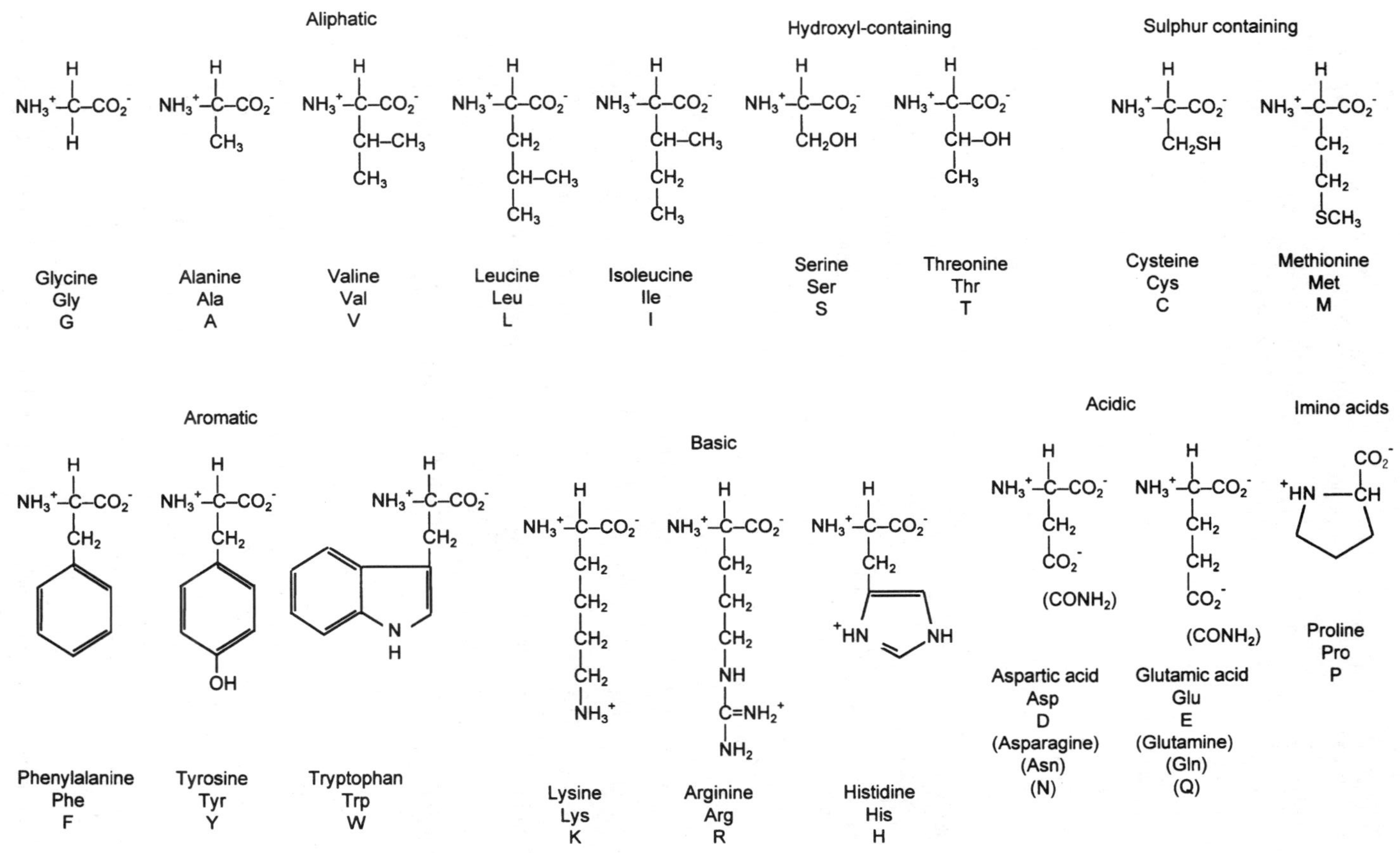

Figure 5.2 The structures of the amino acids commonly found in proteins.

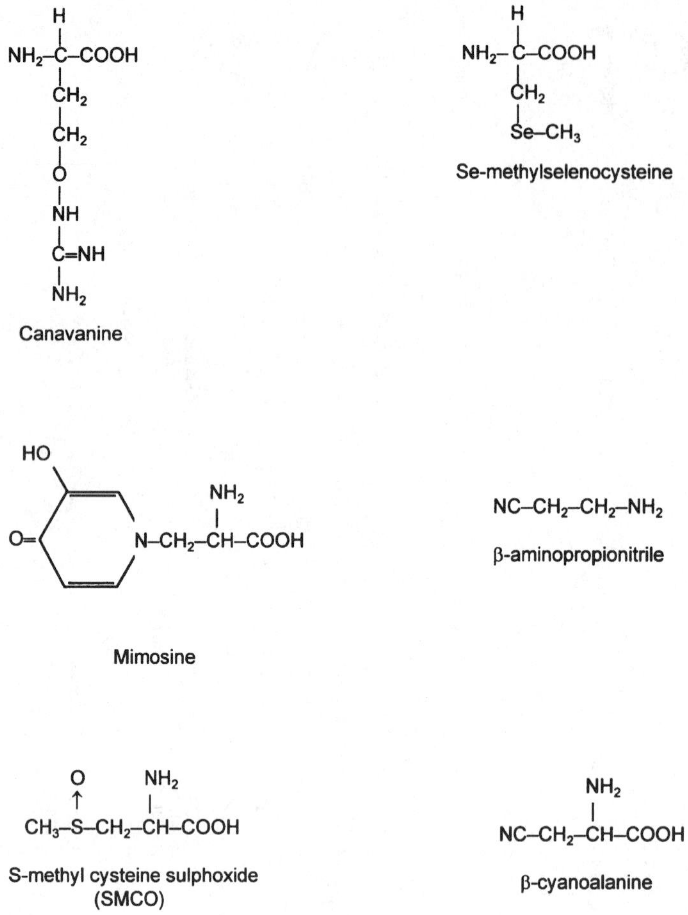

Figure 5.3 The structures of some non-protein amino acids: canavanine; Se-methylselenocysteine; mimosine; β-aminopropionitrile; S-methyl cysteine sulphoxide (SMCO); β-cyanoalanine.

disease' or 'blind staggers'. These amino acids are toxic to animals because they are incorporated into proteins in place of the normal sulphur-containing amino acids, but the proteins which are produced are inactive. The plants which accumulate the amino acids have adapted to avoid these effects. Low levels of selenium-containing amino acids, particularly seleno-methionine, may be used in animal feeds as a selenium supplement for livestock. Sometimes this is achieved by persuading yeasts to form proteins containing selenium amino acids and then incorporating the killed yeasts in commercial feedstuffs.

5.3.3 MIMOSINE

Mimosine is a toxic amino acid present in leaves and seeds of the tropical legume *Leucaena leucocephala* (Figure 5.3). Although the plant is potentially of high nutritional quality, its uses are limited by the presence of this amino acid. Some of the toxic effects may be due to mimosine, but others are caused by a product of its breakdown in the rumen, dihyroxypyridine (DHP). *Leucaena* is not toxic to ruminants from Hawaii because their rumen microorganisms further degrade DHP to nontoxic products. Introduction of Hawaiian rumen bacteria into Australian animals protects them against mimosine poisoning.

5.3.4 LATHYROGENS

Lathyrism is a disease which is caused by eating seeds of the genus *Lathyrus*, including *Lathyrus sativus*, the chickpea and *Lathyrus odoratus*, the annual sweet pea. The chickpea is mainly consumed by humans and causes neurolathyrism, whilst sweet peas cause osteolathyrism in livestock, especially horses.

Symptoms of osteolathyrism are skeletal deformity and rupture of the aorta. It is caused mainly by β-aminopropionitrile (Figure 5.3) which interferes with cross-linking of lysine residues in collagen.

Neurolathyrism is caused by compounds such as β-*N*-oxalyl α,β-diaminopropionic acid which is found in chickpea, and β-cyanoalanine in common vetch (Figure 5.3). These compounds attack nerve cells and lead to weakness and eventual paralysis of the legs. The effects are most common in man and are rarely seen in animals.

5.3.5 S-METHYL CYSTEINE SULPHOXIDE (SMCO)

Brassicas contain SMCO (Figure 5.3) as well as glucosinolates (see Chapter 3). SMCO appears to be the cause of severe anaemia in ruminants. In brassicas, garlic and onion this amino acid makes up 4–6% of dry matter. The most likely cause of the toxicity is not SMCO itself but dimethyl disulphide, which is produced from it in the rumen. This compound may react with reduced glutathione, preventing it from protecting haemoglobin against oxidation.

5.3.6 ALKALOIDS

Alkaloids are nitrogen-containing, basic compounds which are found in many plants. The nitrogen normally forms part of heterocyclic rings. Several thousand alkaloids have been identified but their functions in the plants which make them are uncertain, although they may confer some resistance to animal or insect attack.

Some alkaloids have dramatic physiological effects on man and animals and many, including morphine and nicotine (Figure 5.4), have pharmaceutical and medicinal uses and may be of great commercial and sociological importance. The presence of alkaloids renders some plants toxic to man and to livestock, as described below.

The pyrrolizidine alkaloids (Figure 5.4) occur in *Senecio* species, and are the cause of poisoning by *S. jacobea* (ragwort) and other species. Although ragwort is not very palatable it may be consumed when there is no other forage or when it has been dried, as in hay. Pyrrolizidine alkaloids are toxic as a result of conversion in the liver into toxic metabolites such as pyrroles. Liver function is seriously impaired, the liver is reduced in size and other organs may also be affected. Cattle and horses are susceptible to this type of poisoning but sheep are relatively very resistant, possibly because their liver is less able to convert the alkaloids to pyrroles. Some pyrrolizidine alkaloids have also been shown to be carcinogenic to livestock.

Indole alkaloids include the ergot alkaloids, produced by fungi, which grow on the seed of some grasses and cereals. These alkaloids are based on lysergic acid (Figure 5.4) and may cause convulsions, numbness, gangrene of extremities and reproductive defects in humans and livestock. Ergot is the name commonly used for fungi of *Claviceps* species, and a condition resulting from consumption of ergot-infected grain is known as ergotism. Rye is particularly susceptible to attack by *C. purpurea*, and infected rye caused many epidemics of ergotism in Europe in the Middle Ages.

Quinolizidine alkaloids (Figure 5.4) occur in lupins (*Lupinus* spp.) and laburnum. Cultivated lupins have been selected to have low levels of these compounds, but wild species have caused great losses of sheep as a

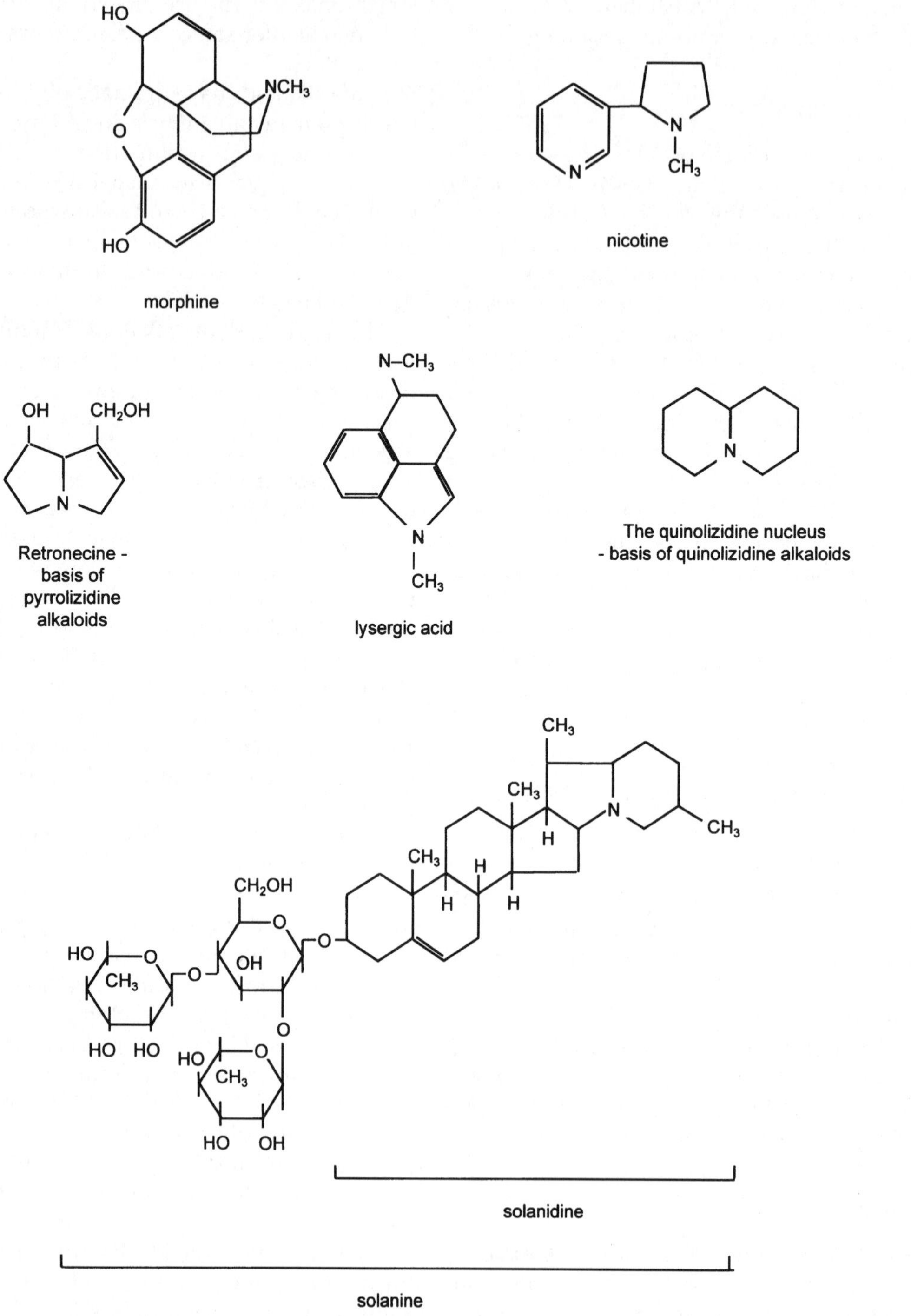

Figure 5.4 The structures of some alkaloids: morphine; nicotine; retronecine (basis of pyrrolizidine alkaloids); lysergic acid; the quinolizidine nucleus (basis of quinolizidine alkaloids); solanine and solanidine.

result of respiratory paralysis, especially in the USA.

Steroid alkaloids are present in plants of the genus *Solanum* such as potatoes, tomatoes and nightshades, and are therefore known as solanum alkaloids. The glycoalkaloid solanine (consisting of the aglycone solanidine and a side chain of sugars – Figure 5.4) is present in potatoes and can cause poisoning of both humans and livestock. Concentrations of alkaloids are highest in green sprouts and green tubers. Solanum alkaloids cause irritation of the gastro-intestinal tract, neurological impairment (they inhibit acetylcholinesterase) and have been suggested to be teratogenic.

Larkspurs (*Delphinium* spp.) have probably been responsible for greater losses of cattle in the USA than any other plant. Their toxicity is due to the presence of polycyclic diterpenoid alkaloids which cause respiratory paralysis.

5.4 PHENOLICS

Phenolics are compounds containing aromatic rings substituted with -OH groups. They are not amino acids, but as they contain phenyl rings which are synthesized in the same way as the phenyl rings of the aromatic amino acids, it is convenient to discuss them here.

Animals are unable to synthesize aromatic compounds but plants can and may contain a wide range of them. The structure of some typical phenolic compounds is shown in Figure 5.5. These are often found as components of more complex molecules such as phytoalexins, coumarins, anthocycanins, tannins or lignins.

The production of a number of phenolic compounds appears to be related to disease resistance in plants. Thus protocatechuic acid is present in onions which are resistant to the smudge fungus *Colletotrichum circinans* but absent from those which are susceptible, and chlorogenic acid may be oxidized to fungistatic quinones in resistant plants. Ferulic acid is found in suberin which protects the underground parts of plants and which is formed in

scar tissue after wounding or abscission of leaves.

5.4.1 LIGNIN

Lignin is a component of the plant cell wall but is found in only small quantities in primary cell walls. Extensive lignification is restricted to tissues such as xylem and phloem and takes place only after growth has stopped.

Lignins are formed by polymerization of coniferyl alcohol, sinapyl alcohol and *p*-hydroxycinnamyl alcohol through a free radical mechanism which results in random formation of bonds and forms a very complex structure (Figure 5.6).

Whilst the gymnosperms (and pteridophytes) have lignin which contains almost exclusively coniferyl alcohol, hardwood trees and dicotyledonous and monocotyledonous crops contain comparable quantities of coniferyl and sinapyl alcohols (Table 5.1). In addition, monocots contain appreciable quantities of *p*-hydroxyphenyl residues.

Lignin is extremely resistant to either chemical or enzymatic attack. Materials with a high lignin content are therefore durable structural materials and have low digestibility when they are part of the diet.

5.4.2 TANNINS

Tannins are polyphenolic materials which are able to precipitate proteins from solution. The name is derived from the use of extracts containing such compounds to tan leather, making it more resistant to microbial attack, heat and abrasion. Tannins contain *o*-dihydroxyphenol groups which allow them to form hydrogen bonds and hydrophobic bonds with proteins such as collagen in animal skins.

Precipitation of proteins by plant extracts is significant to animals. It reduces the digestion of proteins in feeds, both by inhibiting digestive enzymes and by precipitating (and therefore making unavailable) protein in the feed. Precipitation of proteins in the mouth is

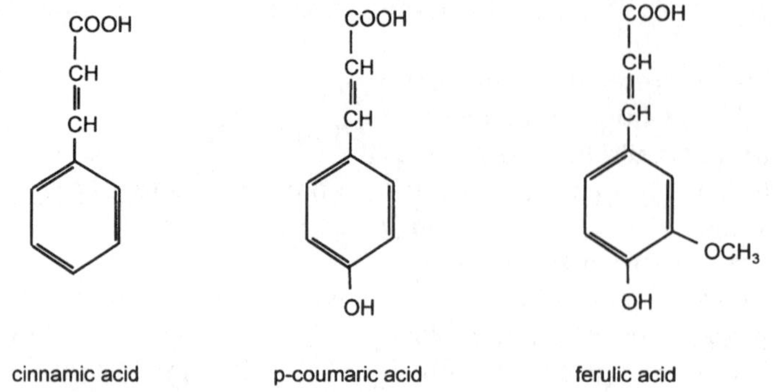

Figure 5.5 The structures of some phenolic compounds commonly found in plants: cinnamic acid; *p*-coumaric acid; ferulic acid.

responsible for the astringent (bitter) taste of fruits such as persimmon in their unripe state. This has a major effect on palatability of feeds and provides protection against herbivores. Inhibition of enzymes also confers resistance to attack by insects, bacteria etc., rendering wood more durable. Tannins in forage legumes reduce the incidence of bloat by precipitating bloat-inducing proteins.

The presence of tannins in crops such as sorghum reduces the palatability and digestibility of the protein and reduces value, especially for feeding of non-ruminants. Sorghum tannins inhibit amylase and trypsin in the digestive tract. High-tannin lines are generally more resistant to birds and normally have pigmented seed coats.

Tannins may be divided into the condensed tannins (proanthocyanidins), which are complex and not readily hydrolysed, and the hydrolysable tannins which consist of simple phenolics, such as gallic acid, condensed with glucose (Figure 5.7).

5.4.3 FLAVONOIDS

Flavonoids are 15-carbon compounds which are widely distributed in plants. Their structure is based on the flavonoid ring system (Figure 5.8).

The most important groups of flavonoids are the anthocyanins and the flavones.

Anthocyanins

These are coloured compounds that commonly occur in flowers and in certain types of fruits, stems, leaves and even roots. In flowers, they facilitate pollination and aid dispersal of seeds in coloured fruits. The red, purple and blue colours of most flowers, and red colours of most fruits and autumn leaves, are due to antho-

Table 5.1 Percentage composition of lignin from different sources

Source	Coniferyl alcohol (guaiacyl group)	Sinapyl alcohol (syringyl group)	p-Hydroxycinnamyl alcohol
Softwoods	80	6	14
Hardwoods	56	40	4
Grasses	44	34	22

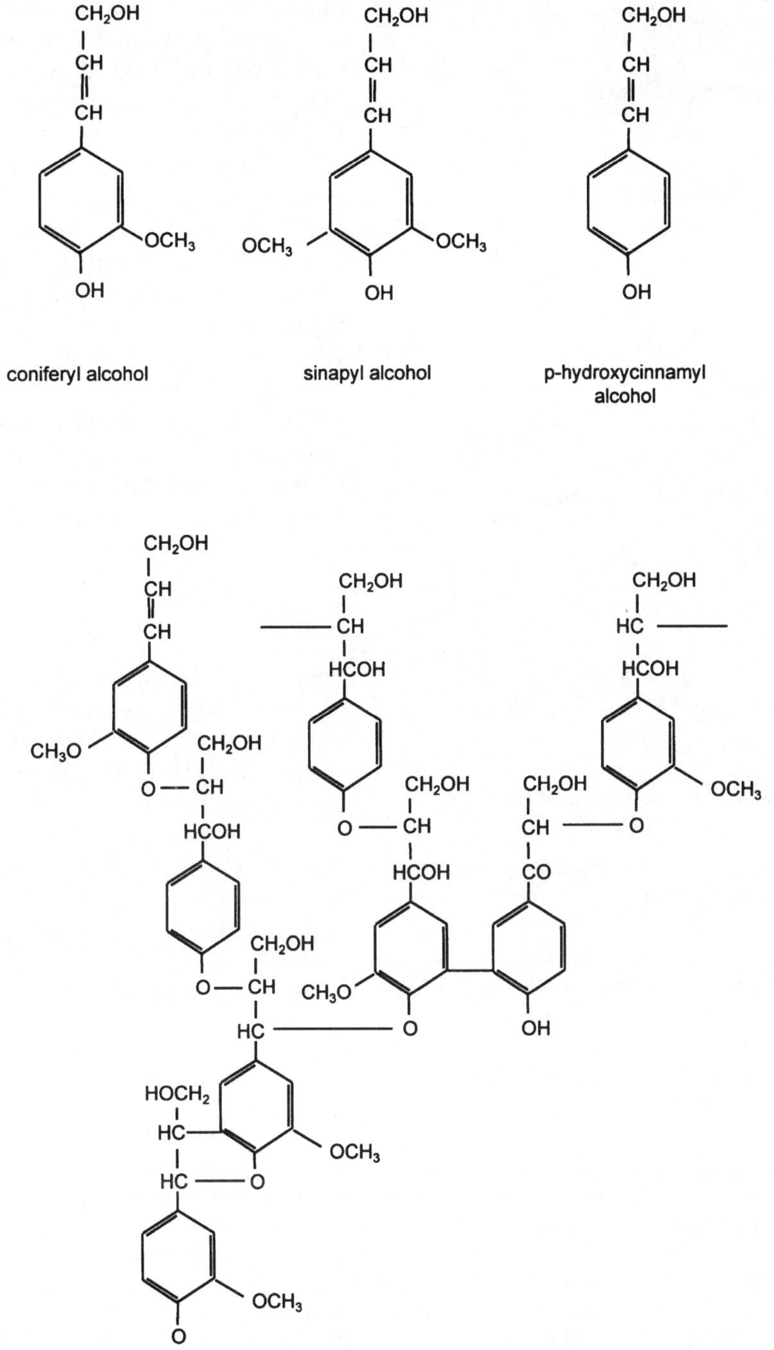

Figure 5.6 The structure of lignin precursors and typical components of lignin: coniferyl alcohol; sinapyl alcohol; *p*-hydroxycinnamyl alcohol.

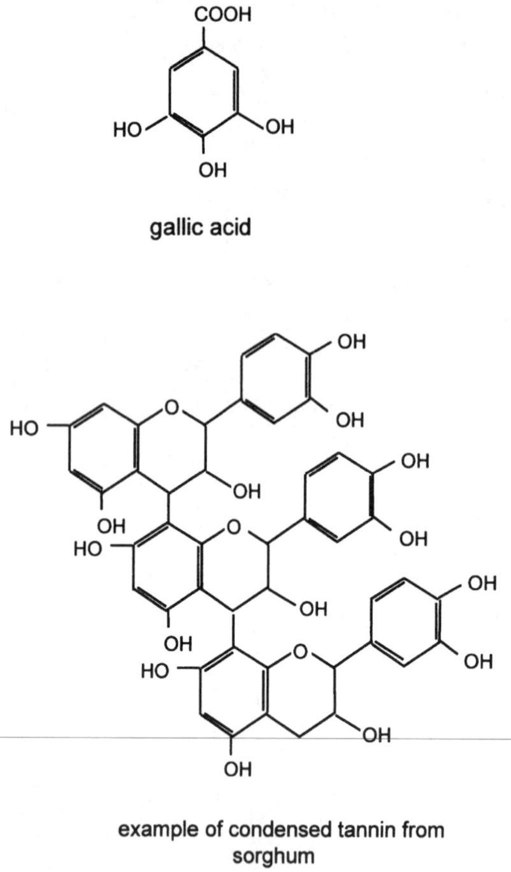

gallic acid

example of condensed tannin from
sorghum

Figure 5.7 The structure of gallic acid and a condensed tannin from sorghum.

are β-glucosides of anthocyanidins. The commonest are indicated in Table 5.2.

A sugar is normally present at the 3 and sometimes the 5 position, and some of these may be di- or trisaccharides. The colour is influenced by the chemical structure – generally methylating OH groups and glycosylation have a reddening effect. pH also has a strong effect on colour: pelargonidin is red in acid solution but blue in alkaline. The colour is also altered by the presence of co-pigments such as flavone-glucosides and hydrolysable tannins, which intensify the colour and shift it towards the blue. The presence of metal ions such as Fe^{2+} or Al^{3+} also influence colour. Thus the colour of both blue cornflowers and red roses is due to cyanidin derivatives, but in cornflower these are complexed with iron and flavones.

Flavones

These exist as glycosides of flavones and flavonols which are mostly yellow and cream. The basic structure of flavones and flavonols is shown in Figure 5.8. They may contribute body to white or cream flower colours and act as co-pigments. They absorb light strongly in the UV which may help them attract insects and prevents damage to the photosynthetic apparatus from excess light.

5.5 PEPTIDE BONDS

During the formation of proteins, amino acids are joined together by condensation reactions to form long chains (polymers).

cyanins. (Yellow and orange colours as in tomatoes and some yellow flowers tend to be due to carotenoids.) In the cell anthocyanins are concentrated in the vacuoles (not in the plastids as carotenoids are). Chemically the anthocyanins

Table 5.2 Properties of some anthocyanins

Anthocyanin	Colour	Typical source	R_1	R_2
Pelargonidin	Red	Geranium	H	H
Delphinidin	Blue	Delphinium	OH	OH
Cyanidin	Red or blue	Ripe apples, cornflower	OH	H
Petunidin	Purple	Petunia	OCH_3	OH
Peonidin	Reddish	Peony	OCH_3	H
Malvidin	Mauve	Mallows	OCH_3	OCH_3

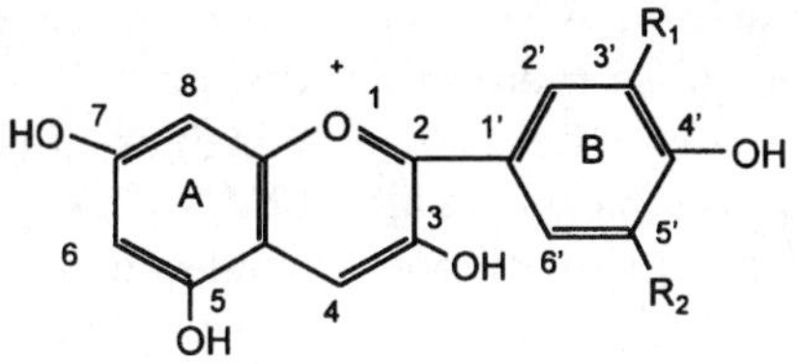

Anthocyanin ring structure

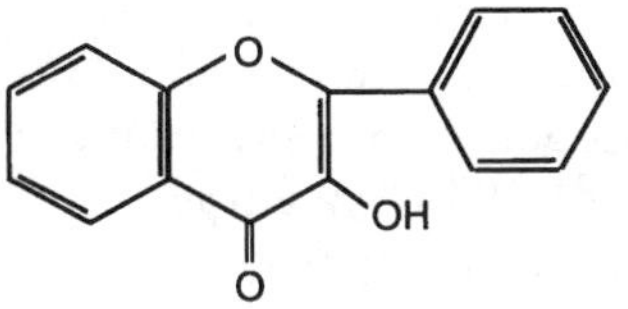

a flavonol

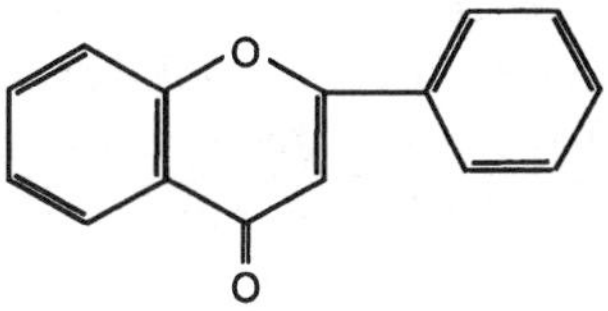

a flavone

Figure 5.8 The structure of some flavonoids: anthocyanin ring structure; a flavonol; a flavone. Substitution of various groups at R_1 and R_2 in the anthocyanin ring gives rise to pigments with a wide range of colours (see Table 5.2).

These chains of amino acids are known as peptides and the bonds that join them are peptide bonds. The structure of a peptide bond is shown in Figure 5.9.

The groups marked R_1 and R_2 are the side chains of the amino acids. This compound is an example of a dipeptide: it consists of two amino acids, but other amino acids can be added to extend the chain. Some small peptides are very important compounds in their own right and have strong biological activity.

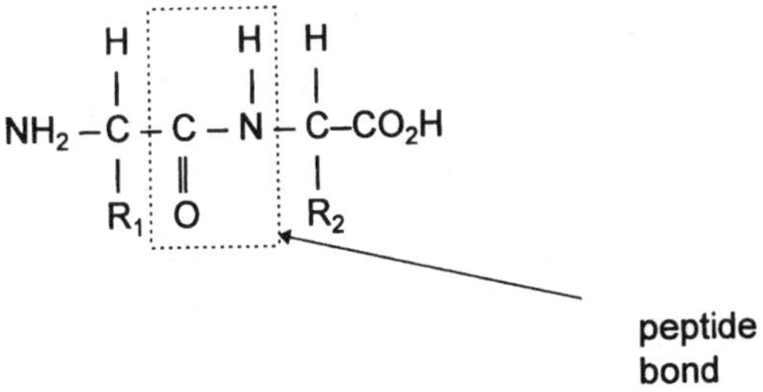

peptide bond

Figure 5.9 The structure of a peptide bond between two amino acids.

An example is the pentapeptide enkephalin which is produced and used in the brain as a natural pain-killer.

5.6 PROTEIN FUNCTION AND STRUCTURE

Proteins are polymers of amino acids linked together in straight chains to form polypeptides. They consist of between approximately 100 and 3000 amino acid residues. As the average molecular weight of an amino acid is about 110, protein molecular weights vary between about 10 000 and 300 000.

The functions of proteins are very varied. They may act as enzymes, as means of storing nitrogen in a biologically accessible form, as structural components of cells, etc. The unique property that makes them particularly important is their ability to fold into well-defined three-dimensional shapes or conformations so that they can bind very strongly to other molecules. This gives them the ability to 'recognise' other molecules in an extremely specific way. Thus enzymes can recognise their substrates, antibodies can recognise the corresponding antigen, etc. Even very small changes in the shape of proteins usually prevent them from carrying out their normal functions correctly.

The structure of proteins can be described at several levels, as summarized in Table 5.3.

5.6.1 PRIMARY PROTEIN STRUCTURE

This is the sequence in which amino acids are linked to one another by peptide bonds. As we shall see when we come to look at the way in which a protein is synthesized, the order of the amino acids is 'written down' in the genetic material of the cell.

At one end of the chain there is a free amino group (N-terminal) and at the other a carboxyl group (C-terminal). These groups normally carry a positive and negative charge, respectively. Some of the side chains (or R groups) also carry positive or negative charges. These features of the primary structure of a typical protein are shown in Figure 5.10.

There is a further type of covalent bond that occurs in proteins. This is the disulphide bridge formed between cysteine side chains on the same or neighbouring chains. The cross-linked, double amino acid formed in this way is called cystine. Figure 5.11 shows the structure of a very simple protein, bovine insulin. This has two polypeptide chains which are linked together by three disulphide bridges. Of these bridges one (A) is within a single chain whereas the others, (B) and (C), link the two chains.

5.6.2 SECONDARY PROTEIN STRUCTURE

Each of the peptide bonds in a peptide chain can rotate so that a long chain would have a very flexible and bendable structure. It achieves a fairly rigid structure because of the formation of hydrogen bonds between atoms within the chains. In a hydrogen bond, a hydrogen atom essentially becomes shared between two other atoms, e.g. the H of an NH_2 can be shared between its own N atom and the O atom of a C=O group. Although individual hydrogen bonds are really quite weak, there are so many of them that they hold the polypeptide chain very firmly in shape. Examples of the formation of such hydrogen bonds are shown in Figure 5.12.

A number of distinct arrangements of the polypeptide chain occur commonly in proteins. In some cases hydrogen bonding takes place between groups which are close together in the same polypeptide chain, resulting in the chain being twisted into a helix, which is usually called the α-helix. In this structure there are 3.6 amino acids for each turn of the

Table 5.3 Levels of organization in the structure of a protein

Protein structure	Organization
Primary	The sequence of amino acids in the polypeptide chain or chains. The chemical bonds which maintain the primary structure are covalent ones (peptide bonds).
Secondary	Repeating structures recognizable within the 3-D structure of the polypeptide chains. There are several types of secondary structure: α-helix; β-pleated sheet; triple helix, etc. Secondary structure is maintained largely by hydrogen bonds formed between the atoms of different peptide bonds.
Tertiary	The overall 3-D shape of the polypeptide chain describing its bending, twisting and meandering in space. This is maintained by large numbers of weak bonds, e.g. hydrogen bonds, hydrophobic bonds, ionic bonds, together with covalent disulphide bridges.
Quaternary	The way in which a number of polypeptide chains, each with its own tertiary structure, come together, perhaps with other molecules, to form the final, active protein.

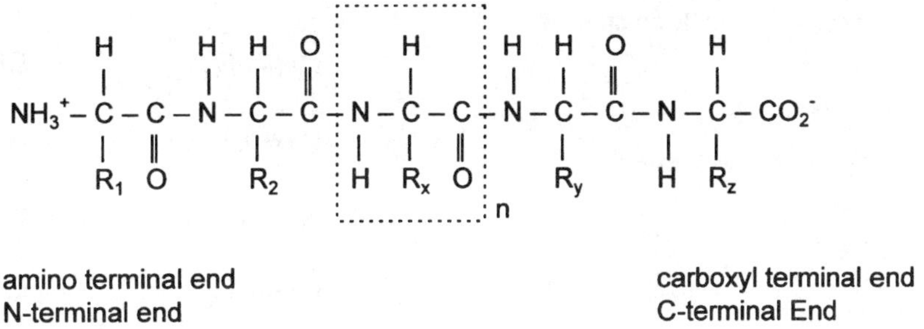

$$\text{NH}_3^+ - \underset{\underset{R_1}{|}}{\overset{\overset{H}{|}}{C}} - \underset{\underset{O}{\|}}{\overset{\overset{H}{|}}{C}} - \underset{H}{\overset{H}{N}} - \underset{\underset{H}{|}}{\overset{\overset{H}{|}}{C}} - \underset{\underset{R_2}{|}}{\overset{\overset{O}{\|}}{C}} \!\!\vdash\!\! \underset{\underset{H}{|}}{\overset{H}{N}} - \underset{\underset{R_x}{|}}{\overset{\overset{H}{|}}{C}} - \underset{\underset{O}{\|}}{\overset{\overset{H}{|}}{C}} \!\!\vdash\!\! \underset{H}{\overset{H}{N}} - \underset{\underset{R_y}{|}}{\overset{\overset{H}{|}}{C}} - \underset{\underset{H}{|}}{\overset{\overset{O}{\|}}{C}} - \underset{H}{\overset{H}{N}} - \underset{\underset{R_z}{|}}{\overset{\overset{H}{|}}{C}} - \text{CO}_2^-$$

amino terminal end   carboxyl terminal end
N-terminal end   C-terminal End

Figure 5.10 The primary structure of a protein. The chain has an amino (N-terminal) end and a carboxyl (C-terminal) end, and normally contains 100 or more amino acids.

helix and each turn occupies about 0.54 nm along the length of the chain. The hydrogen bonds between peptide groups lie parallel to the peptide chain (Figure 5.13).

Some proteins such as α-keratin, found in hair, wool and hooves, consist almost entirely of amino acids arranged in the form of an α-helix. Helical chains may aggregate to form long, twisted fibres which give the structures their great strength. In most globular proteins, however, only parts of the peptide chain exist as α-helix whilst others have different conformations.

Another type of secondary protein structure is called the β-pleated sheet. In this structure several individual peptide chains, laid side-by-side, are held together by hydrogen bonds between the peptide groups. This results in a very strong structure, rather like corrugated cardboard. The peptide chains involved may run in the same direction (parallel pleated sheet) or in opposite directions (anti-parallel pleated sheet) and they may be part of the same, or different, polypeptides (Figure 5.13). Silk is an example of a material consisting of proteins composed mainly of antiparallel β-pleated sheets. Most globular proteins contain some regions where the peptide chains form a β-pleated sheet as well as some which form an α-helix. The difference between α-helix and β-pleated sheet may be of great biological significance in prions, which are described in section 5.8.

5.6.3 TERTIARY STRUCTURE

The tertiary structure describes how the polypeptide backbone of the protein is bent and twisted. Much of this is determined by the nature of the amino-acid side chains and whether they are hydrophilic or hydrophobic. In biological systems most proteins exist in solution or suspension in water, and the presence of this water has a great influence on the shape of the protein. The protein chain will arrange itself so that as many hydrophobic groups as possible point towards the middle of the protein structure, whilst the hydrophilic ones point towards the surrounding water. This is a stable structure as in the middle of the molecules the non-polar groups can interact with one another and are kept away from the water, whilst the polar groups on the outside of the molecule are stabilized by interaction with water molecules. Any change in the nature of the charges on the side chains (for instance by changing the pH of the solution) will alter the relationship between the water and the protein. As its environment changes the protein will change its shape to restore the most favourable interactions between the water and the hydrophilic and hydrophobic groups.

The tertiary structure is maintained by:

- hydrogen bonds formed between amino acid side chains;
- ionic bonds formed between groups with opposite charges;

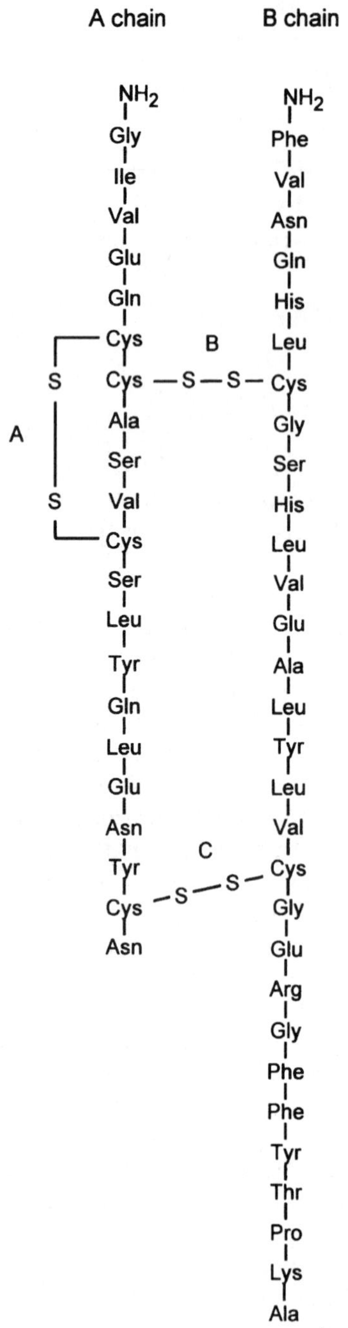

A chain   B chain

A chain	B chain
NH$_2$	NH$_2$
Gly	Phe
Ile	Val
Val	Asn
Glu	Gln
Gln	His
Cys	Leu
Cys —S—S— Cys (B)	
Ala	Gly
Ser	Ser
Val	His
Cys	Leu
Ser	Val
Leu	Glu
Tyr	Ala
Gln	Leu
Leu	Tyr
Glu	Leu
Asn	Val
Tyr	Cys
Cys —S—S— Cys (C)	Gly
Asn	Glu
	Arg
	Gly
	Phe
	Phe
	Tyr
	Thr
	Pro
	Lys
	Ala

(A: the A chain inner Cys residues joined by S—S bond)

Figure 5.11 The primary structure of bovine insulin. The protein consists of the A chain (21 amino acids) and the B chain (30 amino acids). The chains are linked by two disulphide bridges, and two cysteine residues in the A chain are joined by a further disulphide bond.

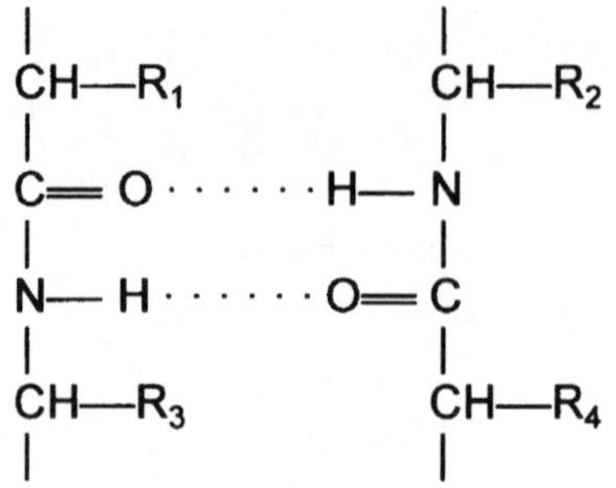

$$CH—R_1 \qquad CH—R_2$$
$$C{=}O \cdots\cdots H—N$$
$$N—H \cdots\cdots O{=}C$$
$$CH—R_3 \qquad CH—R_4$$

Figure 5.12 Hydrogen bonds formed between components of peptide bonds. Such bonds help to stabilize the secondary structure of the protein.

- hydrophobic interactions between hydrophobic amino-acid side chains;
- covalent disulphide (-S-S-) bridges between cysteine residues.

Although many of these bonds are individually quite weak, their large number stabilizes the structure of the protein.

5.6.4 QUATERNARY STRUCTURE

Many proteins do not consist of a single polypeptide chain; instead they may have several chains or subunits, which may be identical to or different from one another. In addition, many proteins have prosthetic groups which are not made of amino acids but are essential for the activity of the protein. Prosthetic groups may be complex carbohydrates, metal ions or complex polycyclic compounds such as haem in haemoglobin, myoglobin or cytochromes. In the molecule of haemoglobin (Figure 5.14) one haem group is associated with each of the four subunits and can be identified from its flat ring structure.

Also regions of the polypeptide where the chain has taken up α-helix or β-pleated sheet structures can be seen. These regions are interspersed with other regions where no regular structures can be seen.

Interactions between subunits are important in maintaining protein structure. If the relationship between the subunits is changed then the protein may alter its biological activity.

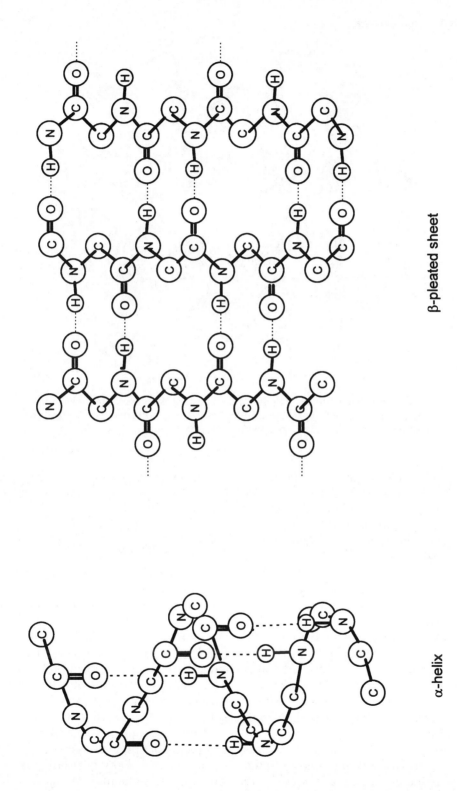

β-pleated sheet

α-helix

Figure 5.13 Outline structures of α-helix and β-pleated sheets showing how the structures are stabilized by hydrogen bonds between components of the peptide bonds.

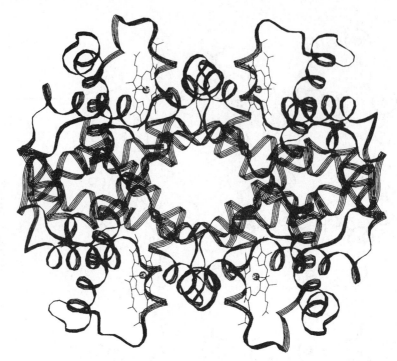

Figure 5.14 The structure of haemoglobin. The four haem groups, each associated with one of the subunits, can be seen. (Reproduced with permission from Smith, C.A. and Wood, E.J. (1991) *Biological Molecules*, 1st edn, Chapman & Hall, London, p. 54.)

Most commonly, biological activity is lost when the subunits dissociate. Some enzymes consist of several subunits, and changes in the interactions between them are of great importance in controlling the activity of the enzyme according to the needs of the cell (see Chapter 6).

After their synthesis, many proteins are further modified by the action of enzymes. The amino acid hydroxyproline, for example, is found in collagen and is involved in formation of cross-links between adjacent molecules. The amino acid is not incorporated into the protein in this form. Instead proline is inserted and is subsequently hydroxylated by the action of the enzyme proline hydroxylase which requires the presence of ascorbic acid (vitamin C) for activity.

Carbohydrates are also added to many proteins by covalent attachment, normally to the O atom of serine or threonine residues or to the N atom of asparagine side chains, and these may play an important part in determining the protein's functions.

5.7 PROPERTIES OF PROTEINS

Most proteins are soluble in water or in dilute salt solutions but are denatured by organic solvents. However, some proteins do dissolve in organic solvents, and such differences in solubility can be used to divide proteins into classes which show broadly similar behaviour. A system of classification of plant proteins based on their solubility in different solvents has been used for almost 100 years. In spite of its age this system is still often used for the classification of seed proteins and represents a useful way of grouping other proteins as well (Table 5.4).

This classification is useful because the glutelins and prolamins are deficient in some

Table 5.4 Classification of proteins on the basis of their solubility

Type of protein	Properties	Functions
Albumins	Soluble in water and in dilute salt solutions	Mainly globular, metabolic proteins; includes most of the enzymes, etc.
Globulins	Insoluble in water but soluble in dilute salt solutions	Similar functions to the albumins, also main type of storage protein in legumes and oats
Glutelins	Insoluble in water or dilute salt but soluble in dilute acids and bases	Insoluble enzymes, ribosomal and membrane proteins etc., main storage protein in rice seeds
Prolamins	Insoluble in water, dilute salt or dilute acids but soluble in 70–90% ethanol	Storage proteins in most cereal seeds except for rice and oats

of the essential amino acids, which limits their usefulness for feeding non-ruminant animals. This point will be discussed in more detail in Chapter 23.

The properties of albumins and globulins are rather similar and the distinction between them is often not very clear-cut. These types of proteins have the widest distribution and they have the properties which are normally thought of as being typical. The remaining parts of this section discuss the properties of these groups. The solubility of most proteins is influenced by a number of factors which are discussed below.

5.7.1 IONIC STRENGTH AND PRESENCE OF SPECIFIC IONS

Most proteins are not very soluble in pure water but their solubility is usually increased by the addition of small amounts of some inorganic salts. However if larger amounts of these salts are added then the solubility of the proteins may be decreased again. These processes are known respectively as 'salting in' and 'salting out' (Figure 5.15a). The important factor which determines the solubility of the protein is the ionic strength of the solution rather than the concentration of the salt.

Ionic strength is defined as:

$$\mu = \sum c_i z_i$$

where c_i = the concentration of the ith ion and z_i = the charge of the ith ion.

Thus salts containing multivalent ions have a greater influence on the solubility of proteins than those containing only monovalent ones.

Because of the effect of ionic strength on proteins, they are usually studied when dissolved in dilute salt solutions under conditions similar to those which exist in the cell. Most enzymes usually work best under these conditions.

Salting out is a very mild process and is often employed to precipitate proteins from solution without damaging them. This may be used as a means of concentrating the protein or when separating one protein from another. Ammonium sulphate is often used as the salt in these processes because of its high solubility in water and its ability to dissociate into highly charged ions.

Proteins carry charges in solution and, at alkaline pHs, they have a net negative charge and therefore behave as anions. In the presence of many metal ions, particularly those with valencies greater than one, proteins will coagulate. Heavy metal ions such as Cd^+, Cu^{2+},

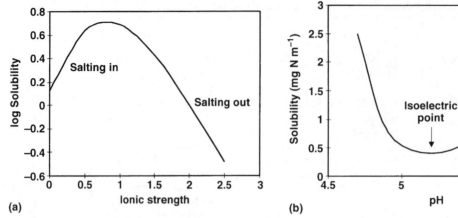

Figure 5.15 (a) The effect of ionic strength on solubility of a typical protein. At low concentrations salts normally increase the solubility (salting-in) but at higher concentrations they cause precipitation (salting-out). (b) The effect of pH on the solubility of typical proteins. Minimum solubility occurs at the isoelectric point.

Fe^{3+}, Hg^{2+}, Pb^{2+} and Zn^{2+} all cause proteins to precipitate. This accounts for the toxic nature of most of these metals if they enter the food chain or are accidentally included in the diets for animals. Some (e.g. Hg^{2+} and Pb^{2+}) combine specifically with -SH groups in the proteins, which may be essential for enzyme activity.

At acidic pH levels, protein molecules carry a net positive charge. Under these conditions many anions will cause proteins to precipitate. The anions from some acids are very effective at this, for example trichloroacetic acid (TCA) is frequently used in analysis to precipitate all proteins before determination of some other analyte. Under natural conditions tannins, found in many plant materials, will precipitate proteins. This phenomenon occurs naturally in animal feeds which contain high levels of tannins and it may reduce the availability of proteins in the diet.

5.7.2 EFFECT OF pH

As the pH is altered the number of charges on a protein molecule in solution changes. This results from changes in the extent of ionization of the N- and C-terminal groups at the ends of the polypeptide and of groups in the side chains. As the solution is made more acidic the net charge becomes positive, and as it is made more alkaline the net charge becomes more negative. At some point the number of negative charges on the protein due to ionized acid groups (-COO⁻) and the number of positive charges on the basic amino acid groups (-NH_3^+ or =NH_2^+) are equal. When this point, called the isoelectric point (pI), is reached, the solubility of the protein is at its lowest and many proteins precipitate from solution under these conditions (Figure 5.15b). This is because repulsion between individual protein molecules is minimized at this pH so that they can aggregate and precipitate. The effect is demonstrated very easily by adding lemon juice to milk. The proteins will precipitate as the pH falls to about 4 which is the isoelectric point of casein, the most abundant protein in milk. This process can be used for the commercial preparation of casein and occurs during the preparation of some milk products such as cottage cheese.

5.7.3 DENATURATION

The addition of ammonium sulphate, or adjusting the pH to the isoelectric point, usually results in precipitation of proteins. However, removing the salt or adjusting the pH again often allows the proteins to redissolve. Under some circumstances precipitation is irreversible and the proteins become denatured. Denaturation occurs when a protein uncoils and loses its three-dimensional shape. When this happens, the non-polar groups, which are usually folded into the centre of the protein molecule, are exposed to the water surrounding it. This is an unstable arrangement and it is usual for protein molecules to coalesce so that the non-polar groups of one can interact with those of another. Heating, extremes of pH, addition of organic solvents or addition of detergents may cause denaturation.

5.7.4 EFFECT OF HEAT

Heating proteins usually has drastic effects on their biological activity. As they are heated the atoms vibrate more rapidly and the carefully arranged secondary, tertiary and quaternary structures are lost. The chains usually assume more randomly arranged structures with fewer charged groups to the outside of the molecules, which greatly reduces their solubility. The effects of such denaturation can readily be seen during cooking of protein-rich foods such as eggs.

Denaturation is not always undesirable. Cooking proteins may make them more susceptible to attack by enzymes and thus hasten their digestion in animals. It can also be useful to denature proteins which are potentially harmful such as some enzyme inhibitors discussed in Chapter 28.

5.8 PRIONS

Evidence is accumulating for an important role of some proteins in the determination of their own three-dimensional structure. The disease scrapie has been known for many years to cause severe nervous disorders in sheep and, more rarely, goats. It is a progressive disease marked in the early stages by intense itching and trembling. As it develops over a period of 1.5–6 months, walking is impaired and in the final stages the animal is unable to rise. Microscopic examination of the brain tissue showed a sponge-like structure with large plaques of protein. Related conditions were described in humans (Creutzfeld–Jacob Syndrome and Gerstmann–Strussler Syndrome) and in mink, deer and cats. The conditions were however quite rare. The diseases are known in general as transmissible spongiform encephalopathies (TSEs).

In the 1980s, particularly in Britain, there appeared a bovine version of the disease (bovine spongiform encephalopathy – BSE, sometimes called mad cow disease). From a few sporadic cases in the earlier part of the 1980s, the incidence had risen to hundreds of thousands of cases by the early 1990s. The blame was laid on the practice of feeding meat and bone meal to cattle and to a change in the conditions used for its manufacture. Until the early 1980s solvents were used to extract as much fat as possible from the meal because this was a high-value product. The last traces of organic solvent were removed by using superheated steam. With a drop in the world price of oils and fats there was no longer any incentive to remove it all from the meat and bone meal, and this step was dropped from the processing.

It has been known for many years that the causative agent of scrapie was neither a bacterium nor a virus, and a suggestion was made in the 1960s that it was caused by a protein free of nucleic acids. The protein was extremely resistant to denaturation, surviving all but the most extreme physical conditions. Similar proteins were then shown to be associated with the related degenerative conditions in man, mink and deer. The name prions has been given to these proteins. Further research has shown them to be a modified form of a glycoprotein that occurs as a normal component of nerve membranes. It is rooted in the membrane through a phosphoinositol group. These proteins are known as prion proteins (PrP), the one which causes scrapie is known as PrP^{Sc} whereas the normal cellular one is PrP^C. The two proteins have the same polypeptide sequence coded by the same DNA sequence of the gene. In sheep the immune system fails to give any protection against scrapie. This is explained by the fact that the causative agent is actually a variant of a quite normal protein that is not in any way foreign to the animal. Their molecular weights are identical at around 34 kDa. In both cases there is a considerable degree of post-translational modification of the polypeptide by the addition of seemingly identical carbohydrate side chains fixed to an asparagine group and a glycolipid which contains phosphoinositol near the C-terminal end of the chain.

However, the two proteins differ from one another in their sensitivity to proteolytic enzymes, in the conformation of their polypeptide chains (PrP^C has 42% α-helix and 3% β-pleated sheet, but PrP^{Sc} has 30% helix and 43% pleated sheet) and in their stability in the cell as well as in their capacity to transmit spongiform encephalopathy.

The exact way in which the disease is spread is not clear, however a suggested way is shown in Figure 5.16. In this theory the infective PrP^{Sc} protein reacts with a normal molecule of PrP^C to form a dimer. This induces the PrP^C to change shape so it is in the same conformation as PrP^{Sc}. The dimer now separates to release two molecules of PrP^{Sc} which are free to transform more PrP^C.

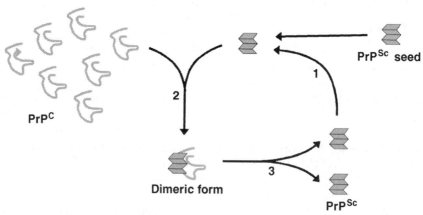

Figure 5.16 Proposed means of multiplication of prion proteins. Interaction of infective PrP^{Sc} (mainly β-pleated sheet) causes normal PrP^c (mainly α-helix) to change its conformation to that of the infective form.

ENYMES

6.1 INTRODUCTION

Enzymes are molecules which catalyse biochemical reactions. They act on their substrates and convert them into products. Although a few examples of catalytic RNA molecules are known these are quite unusual, and the vast majority of biological catalysts are proteins.

Enzymes have a number of particularly interesting and important properties.

- They bring about changes in their substrates without being changed themselves, i.e. they are true catalysts.
- They increase the rates of reactions many thousands or even millions of times.

- They change the speed of reactions, but cannot make reactions occur which would not otherwise take place. Thus they change the rates but not the equilibrium constants of reactions.
- They are highly specific. This means that they will act only on substrates which have a particular structure. They convert substrates into products with almost 100% efficiency.

6.2 TYPES OF REACTIONS CATALYSED BY ENZYMES

Although thousands of different reactions are catalysed by enzymes they can be classified into six basic types (Table 6.1). As we study metabolism we shall see examples of all of these.

6.3 MODE OF ACTION OF ENZYMES

Enzymes function by lowering the activation energy of reactions. They do this by providing an easy route for the reaction. This usually occurs because they allow a reaction to proceed via several small steps instead of one large one.

When molecules react together they do so by forming an unstable intermediate called a transition state, which has a higher energy content than the original reactants. The formation of this intermediate therefore represents an energy barrier through which the molecules must pass, and the rate of the reaction is determined by the number of molecules with sufficient energy to pass over the barrier. An enzyme may function by lowering this activation energy barrier, as shown in Figure 6.1. Notice that the overall free energy change for the reaction (difference between the energy of the reactants and products) is unaltered by the presence of the enzyme.

The first step in any enzyme-catalysed reaction is for the enzyme to bind its substrate or substrates to form an enzyme–substrate complex (ES). Groups of atoms on the surface of the enzyme molecule interact with specific parts of the substrate molecule. The part of the enzyme at which the substrates are bound is called the active site (Figure 6.2). Binding of the substrate to the enzyme is accomplished through a combination of many of the same forces which are used to maintain the structure of the proteins themselves – ionic bonding, hydrophobic bonding, hydrogen bonding, etc. (see Chapter 5). The binding of substrate to the enzyme is a reversible process and the substrate is free to dissociate from the enzyme–substrate complex again.

Many enzymes do not use a single substrate, but instead bring about a reaction between two or more substrates, all of which have, at some time, to bind to the enzyme. At the active site reactions take place between the substrates and they are converted into products which then dissociate from the enzyme, freeing the enzyme to bind more substrate.

This may be represented in the following way:

$$E + S \underset{k_2}{\overset{k_1}{\rightleftharpoons}} ES \overset{k_3}{\rightarrow} E + P \qquad \text{Equation 6.1}$$

The active site usually takes the form of a cleft in the surface of the enzyme. This cleft is lined by amino acids from different parts of the polypeptide chain which has been folded to form the site. It thus involves amino acids which are distant from one another in the primary amino-acid sequence but which come close together in space. For this reason enzymes are very susceptible to conditions which cause even small changes in their three-dimensional structure, as this distorts the critical structure of the active site.

There are usually many points of attachment between the substrate and the surface of the enzyme protein. The substrate must fit the active site very precisely if it is to interact with all of these attachment sites. This is one of the

Table 6.1 Types of enzyme-catalysed reactions

Enzyme	Type of reaction
Oxidoreductases	Catalyse oxidation and reduction reactions
Transferases	Transfer groups from one molecule to another
Hydrolases	Hydrolyse bonds by the 'addition' of water
Lyases	Catalyse the removal of groups from substrate molecules without hydrolysis
Isomerases	Catalyse the interchange of optical or structural isomers
Ligases	Catalyse the linking together of two or more molecules

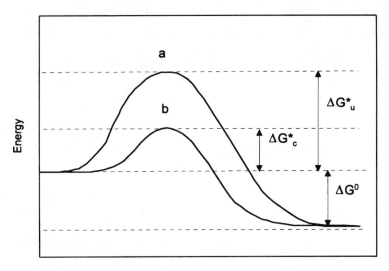

Reaction co-ordinate

Figure 6.1 Reaction profile for (a) uncatalysed and (b) catalysed reactions. The activation energy for the catalysed reaction (ΔG^*_c) is less that that for the uncatalysed reaction(ΔG^*_u). Hence more molecules can cross the energy barrier and the reaction is accelerated. Note that the overall free energy change for the reaction (ΔG^0) is unchanged.

factors which allows enzymes to be so specific in binding their substrates and even to distinguish between optical isomers of the same compound.

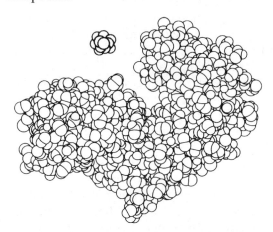

Figure 6.2 The enzyme hexokinase and its substrate glucose. The substrate is small in comparison with the enzyme, and binds to the enzyme at the active site which lies in a cleft in the enzyme surface. (Reproduced with permission from Smith, C.A. and Wood, E.J. (1991) *Biological Molecules*, 1st edn, Chapman & Hall, London, p. 109.)

The idea that the shape of an enzyme's active site is complementary to that of the substrate is called the lock and key model and it is a useful way of looking at the relationship between enzymes and substrates. However in many cases it seems that enzymes may actually fold themselves around their substrates making an even better fit - this is called the induced fit model.

6.4 FACTORS CONTRIBUTING TO ENZYME ACTIVITY

Although each individual enzyme has its own particular mechanism, the catalytic efficiency of most enzymes is influenced by a number of common factors.

6.4.1 PROXIMITY OF SUBSTRATES AT THE ACTIVE SITE

Although it is common to talk about an enzyme and its substrate, most enzymes actually catalyse the reaction between two or more substrates. One of the functions of an enzyme is to bind the substrates in such a way that the groups which will react are close together and have the correct orientation. This greatly increases the rates at which they react and is called the proximity effect.

6.4.2 ENVIRONMENT

The environment in the immediate vicinity of the active site of many enzymes is very different from that found in the aqueous solutions surrounding the enzyme. For example the active site is often lined by the side chains of non-polar amino acids which may greatly influence the rates at which many chemical processes take place.

6.4.3 ACID–BASE CATALYSIS

Enzymes can make some atoms or functional groups in their substrates more reactive by adding a proton or removing a proton from them. This is known as acid–base catalysis. Usually this is the function of the side chain of one specific amino acid at the active site. Groups such as the side-chain carboxyl groups of aspartic or glutamic acids, or the imidazole ring of histidine, can function in this way. Depending on the pH they may be able to function as either acids or bases.

6.4.4 EFFECTS ON THE STABILITY OF SUBSTRATES AND REACTION INTERMEDIATES

Binding of substrates to enzymes may create strains in the bonds between atoms of the substrates. This occurs because some distortion of the substrate may take place as it binds to the active site. In other enzymes, reaction intermediates or transition states may be stabilized by spreading the charges on them as a result of their interaction with the enzyme. Destabilizing the substrate and stabilizing reaction intermediates both decrease the activation energy barrier and accelerate conversion of substrates into products.

6.4.5 FORMATION OF COVALENT ENZYME–SUBSTRATE INTERMEDIATES

The side chains of some amino acids at the active sites of enzymes can form covalent bonds with groups within the bound substrates. The CH_2–OH group of serine is one such group which is present at the active site of a number of proteolytic (protein hydrolysing) enzymes such as trypsin and chymotrypsin. The CH_2–OH group attacks the peptide bond and the acyl group becomes covalently attached to the CH_2–O– group (Figure 6.3). This intermediate then breaks down to release the acid and regenerate enzyme.

Another series of enzymes, which includes proteolytic plant enzymes such as papain and bromelain, use the side chain of cysteine (–CH_2–SH) in a similar way. These enzymes are called the cysteine proteases. In yet another group of enzymes, using ATP as a substrate, the side-chain carboxyl groups of aspartate or glutamate take part in similar reactions.

Metal ions contained within enzymes may destabilize substrates. In carboxypeptidase a zinc atom polarizes the carbonyl group of the peptide bond of the substrate and thereby renders the peptide bond more easily broken.

In many enzymes, several of these effects may operate at once. For example in chymotrypsin and papain, not only does the serine or cysteine group attack the substrate molecule, but a nearby histidine residue also acts as an acid–base catalyst. The presence of this histidine results in the serine or cysteine residue at the active site of these enzymes being very much more reactive than similar residues found in other parts of the protein.

The net result of all these processes taking place at the active site is that substrates are converted into products which then dissociate from the enzyme, leaving it free to act on further substrate molecules. The speed at which products are formed varies enormously from enzyme to enzyme. It can be represented by the turnover number of the enzyme, which is the maximum number of substrate molecules

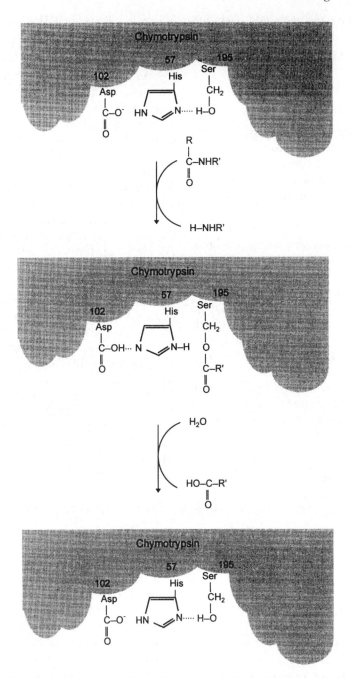

Figure 6.3 Mechanism of chymotrypsin action. Chymotrypsin hydrolyses peptide bonds. The reaction involves transfer of the acyl (R–CO) group of the substrate to a serine residue (at position 195 in the primary sequence) at the active site to form an acyl-enzyme intermediate which then breaks down to release the acid and regenerate free enzyme. The aspartic acid residue (position 102) and histidine residue (position 57), also at the active site, activate the serine and speed up both formation and breakdown of the acyl-enzyme by acid–base catalysis.

acted on by one enzyme molecule each second. This varies from about $0.5\ s^{-1}$ for lysozyme to about $4 \times 10^7\ s^{-1}$ for catalase.

6.5 FACTORS AFFECTING THE RATES OF ENZYME-CATALYSED REACTIONS

The rates at which enzymes convert their substrates into products depend on many factors. Their study is called enzyme kinetics and is important because enzymes determine the rates at which most vital processes occur in all living organisms.

When a substrate is added to an enzyme, the enzyme begins to convert it into products. Enzymes also catalyse conversion of the products back into substrates. Eventually the reaction will reach equilibrium when products are converted back into substrates at the same rate as they are formed. The position of the equilibrium in these enzyme-catalysed reactions is no different from that of the uncatalysed reaction.

The rate at which product is formed when an enzyme and substrate are first mixed together is called the initial rate of the reaction (v_o) and it depends on many factors including:

- enzyme concentration
- substrate concentration
- temperature
- pH
- presence of inhibitors
- presence of co-factors.

These will be considered in turn.

6.5.1 ENZYME CONCENTRATION

Under normal circumstances the initial rate of reaction (v_o) is directly proportional to the concentration of enzyme present. This allows enzyme activity to be determined by measuring v_o. This parameter is often used in the diagnosis of human and animal diseases as a measurement of enzyme activity in tissue or blood samples.

6.5.2 SUBSTRATE CONCENTRATION

The dependence of v_o on substrate concentration [S] for most enzymes is shown in Figure 6.4. As the substrate concentration is raised, v_o increases but the rate of increase becomes less until increasing substrate concentration has no further effect on v_o. At low substrate concentrations v_o is proportional to [S] but at high substrate concentrations it is independent of [S].

The shape of this curve can be understood by referring to Equation 6.1. At low substrate concentrations, increasing [S] causes more enzyme–substrate complex (ES) to form, and the rate of formation of product (P) increases. At high substrate concentrations the active sites of the enzyme become saturated by substrate molecules. This means that substrate is bound to the active site of every enzyme molecule (in terms of Equation 6.1 all E is converted to ES). Under these conditions, adding more substrate cannot increase the rate of the reaction any further as there is no free enzyme to which it can bind.

The shape of the plot of v_o against [S] is a hyperbola and it can be described mathematically by the following equation. This is the Michaelis–Menten equation and K_M is the Michaelis constant.

$$v_0 = \frac{V_{max}}{\left(K_M / S + 1\right)} \qquad \text{Equation 6.2}$$

The terms V_{max} and K_M determine the shape of the curve. V_{max} is the maximum rate that the reaction can reach at high substrate concentrations. Under these conditions the enzyme is said to be saturated with substrate and the active sites of all of the enzyme molecules are occupied by substrate. Adding more substrate cannot increase the rate of the reaction any further. V_{max} therefore tells us something about k_3 (see Equation 6.1), i.e. about how efficiently the enzyme converts ES into P; the larger the value of V_{max} the more rapidly the enzyme will convert substrates to products.

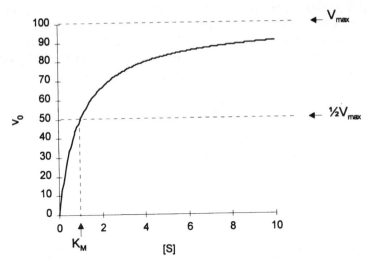

Figure 6.4 Dependence of rate of enzyme-catalysed reaction (v_0) on substrate concentration [S]. At low substrate concentrations v_0 is proportional to [S] but at high substrate concentrations v_0 becomes independent of [S]. Under these conditions the enzyme is 'saturated' by substrate. The shape of the plot is given by the Michaelis–Menten equation: $v_o = V_{max} / (K_M/S+1)$.

K_M is defined as the substrate concentration needed to achieve half of the maximum rate of reaction and it thus has the same units as substrate concentration. It can be shown to be given by Equation 6.3.

$$K_M = \frac{(k_2 + k_3)}{k_1} \qquad \text{Equation 6.3}$$

where k_1, k_2 and k_3 are rate constants for each step as shown in Equation 6.1.

The binding of substrates to enzymes is a reversible process. Once the substrate is bound to the enzyme it may dissociate again or it may be converted into products by the enzyme. The attachment of substrate to the active site and its dissociation again are usually very much faster than the rate at which ES is converted into P. For many enzymes, therefore, k_2 and k_1 are much greater than k_3. Under these circumstances k_3 is very much smaller than k_2 and can be ignored, thus K_M approximates to k_2/k_1. This is the dissociation constant for the enzyme–substrate complex and indicates how tightly the substrate is bound to the enzyme. A low value for K_M generally means a

strongly bound substrate, whilst a high value denotes a weakly bound one.

It should be noted that not all enzymes obey Michaelis–Menten kinetics. For some complex, allosteric enzymes, the plot of v_0 against [S] is not hyperbolic but sigmoid.

To determine K_M and V_{max} in practice it is necessary to measure the initial rates of an enzyme-catalysed reaction (v_0) at differing substrate concentrations [S]. These should be chosen so that they range from about 0.5–10 K_M. It is also necessary to decide the way in which the initial rates of reaction will be measured. One way is to mix enzyme and substrate together and, after fixed times, to stop the reaction and determine the quantities of substrates or products present. Any suitable method (high-performance liquid chromatography (HPLC), radiochemical methods, etc.) can be used to analyse the mixture. However this approach is slow and laborious, and it is much more convenient to monitor the course of the reaction continuously. For some reactions this is very easy as conversion of substrate into product may result in a change in a property, such as the absorption of light, which can be measured easily and non-destructively. A particularly useful application of this type of method follows the conversion of the coenzyme NAD$^+$ to NADH or *vice versa*, which is described in more detail in Section 6.5.6.

Once the values of v_0 at different values of [S] have been measured they can be analysed to determine K_M and V_{max}. This can be done using commercial computer programs or by transforming the data into reciprocal form and analysing it in the form of Lineweaver–Burk or Eadie–Hofstee plots (Figure 6.5).

Measurement of enzyme activity has many direct, practical applications, for example in diagnosis of disease in humans and farm animals, or in identification of plants for particular uses. Some enzymes are restricted to particular locations in normal healthy animals, and only appear in other body parts in diseases. Thus the transaminase enzymes are normally restricted to liver and muscle tissues and creatine kinase to muscle, and the appearance of high levels of these in blood is an indication of some disorder. As blood analysis is

(a)

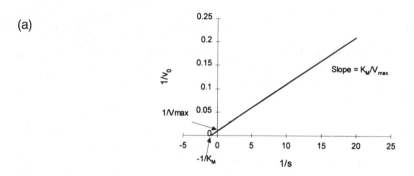

(b)

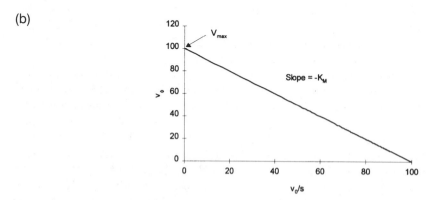

Figure 6.5 Reciprocal plots are used to examine the dependence of v_0 on [S]. (a) Lineweaver–Burk plot: $1/v_0$ is plotted against $1/S$. The intercept on the x axis is $1/V_{max}$ and that on the y axis is $-1/K_M$. (b) Eadie–Hofstee plot: v_0 is plotted against v_0/S. The intercept on the y axis is V_{max} and the line has a slope of $-K_M$ The Lineweaver–Burk plot gives a poor representation of experimental errors but is a convenient way of examining the nature of enzyme inhibitors.

easily performed it provides a convenient and sensitive diagnostic tool in both human and veterinary medicine. Other examples are given in Table 6.2.

Measurement of enzyme activity is also often required in the food industry. As examples, alkaline phosphatase activity has been used for many years as a measure of the completeness of pasteurization of milk, and determination of the activity of specific enzymes in feeds can be used as a measure of their quality. Soyabeans contain a protein which inhibits the activity of digestive proteolytic enzymes, and also contain the enzyme urease. Heating the beans denatures both of these proteins. Measurement of urease activity can therefore be used to determine the effectiveness of heating in preparation of soyabean meal.

6.5.3 TEMPERATURE

The rate of chemical reactions increases as temperature increases. The way in which this takes place is described by the Arrhenius equation:

$$\ln(k) = \text{constant} - \frac{E}{RT} \qquad \text{Equation 6.4}$$

where: k = rate constant; E = energy of activation; R = gas constant; T = temperature (°K).

The rate of a chemical reaction approximately doubles for each 10°C rise in temperature and those catalysed by enzymes behave just like any other. However, because they are proteins, enzymes are denatured by high temperatures. As the temperature of an enzyme is raised, two opposing effects take place. The rate is increased by the greater energy and more frequent collisions of reactant molecules; but the rate is also decreased due to enzyme denaturation. In general, at low temperatures the former is more important, so that the rate increases until denaturation becomes more important, when the rate decreases often in quite an abrupt way. A temperature is seen at which the enzyme appears to be working at its fastest. This is sometimes called the optimum temperature of the enzyme. However it does not have a unique value: increases in rate occur almost instantaneously as the temperature is raised but denaturation may be a relatively slow process. Thus the optimum temperature measured depends on the length of time that the enzyme is held at the temperature before its activity is measured. This is illustrated in Figure 6.6a.

In many instances it is found that enzymes work best at the temperature of the environment in which the organism is usually found. Thus in many warm-blooded animals the optimum temperature of many enzymes is close to the body temperature, whilst thermophilic bacteria found in hot springs have enzymes which retain activity above 90°.

6.5.4 pH

Most enzymes function best at a particular pH, so that the plot of their activity against pH is a bell-shaped curve. Some work well over only very narrow ranges whilst others work over a wide range of pHs. Enzymes also differ very widely in the pHs at which they function best, e.g. pepsin has a pH optimum of about 2, and alkaline phosphatase about pH 10. Most enzymes work most rapidly between pH 5 and 7 as shown in Figure 6.6b.

Table 6.2 Diagnostic uses for some enzymes

Enzyme	Tissue for assay	Disorder
α-amylase	Serum and urine	Pancreatitis
Creatine kinase	Serum	Disorders of skeletal and cardiac muscle
Alkaline phosphatase	Serum	Bone diseases, obstructive jaundice
Acid phosphatase	Serum	Prostatic tumours
Transaminases	Serum	Liver and muscle diseases

(a)

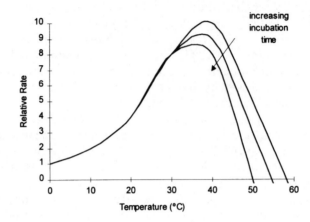

(b)

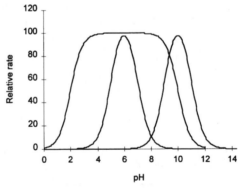

Figure 6.6 (a) Dependence of the rates of enzyme-catalysed reactions on temperature. The optimum temperature is lower when the enzyme activity is measured after being held at the temperature for long enough for significant denaturation to take place. (b) Dependence of the rates of enzyme-catalysed reactions on pH. Both the optimum pH, and the pH range over which an enzyme is active, vary widely.

Changes in pH alter the number of charges on both the substrate and the enzyme, which alters the ability of groups on the enzyme and substrate to interact with one another by ionic and hydrogen bonding. Thus there is a pH at which interaction of substrate with the active site is most favourable. On either side of this pH, binding of substrate is less favoured. The ability of the enzyme to convert enzyme–substrate complex to product may also change with pH for the same reason. Within a certain range of pH, on either side of the pH optimum, the changes which take place are reversible so that activity can be restored by adjusting the pH back to the optimum. However, towards extremes of pH the enzyme may become denatured, resulting in permanent loss of catalytic activity.

6.5.5 PRESENCE OF INHIBITORS

Inhibitors are compounds which reduce the activity of enzymes. Because the interaction of substrates with the active site of enzymes is very specific, it can be upset in a variety of ways.

Enzyme inhibitors are divided into classes depending on the way in which they work. The main division is into reversible and irreversible inhibitors. In the former case removal of the inhibitor from the enzyme restores activity. In the latter case the inhibitor is so tightly bound, sometimes by formation of covalent bonds, that it cannot be removed.

Reversible inhibitors

These inhibitors become reversibly bound to their enzymes as follows.

$$E + I \underset{}{\overset{K_i}{\rightleftharpoons}} EI \qquad \text{Equation 6.5}$$

Reversible inhibitors are divided into two types.

Competitive inhibitors

Here the inhibitor binds to the active site of the enzyme and, in doing this, prevents the substrate from binding. The binding of these inhibitors, however, like that of the substrate, is reversible so that an equilibrium is established. Either substrate or inhibitor, but not both, may be bound at the active site at any one time. The proportion of enzyme molecules with substrate or inhibitor bound will depend on the affinities of the enzyme for substrate and inhibitor and on their relative concentrations in the solution. Adding more substrate will displace the inhibitor and overcome the inhibition. The structure of these inhibitors usually resembles very closely that of the substrate because they bind to the same part of the enzyme by similar interactions. Addition of an excess of substrate can overcome the inhibition, so that V_{max} is unaltered but more substrate is needed to achieve the maximum rate than in the absence of inhibitor. Thus K_M is increased.

Non-competitive inhibitors

Non-competitive inhibitors bind, not to the active site of the enzyme but at other locations, and their binding is unaffected by the presence of substrate. Binding of the inhibitor to the enzyme distorts the active site so that substrate is transformed less effectively by the enzyme. Increasing the substrate concentration has no effect on binding of the inhibitor to the enzyme. The structure of these inhibitors may bear no resemblance to that of the substrate as they bind to different parts of the enzyme. Their presence can be detected by measuring the rate of reaction in the presence of increasing concentrations of inhibitor. V_{max} is decreased in the presence of this type of inhibitor but K_M is unchanged.

Irreversible inhibitors

These become very tightly bound to the enzyme: usually covalent bonds are formed with an amino acid in the protein or with a tightly bound coenzyme. The point of attachment is often, but not always, the active site of the enzyme. They may prevent the substrate from binding to the enzyme or prevent the enzyme from converting enzyme–substrate complex into product. Increasing the substrate concentration cannot overcome the inhibition because there is no competition between the substrate and inhibitor for binding to the enzyme.

Practical applications of enzyme inhibitors

Many enzyme inhibitors have very potent effects on animals or plants. Some are highly toxic and have undesirable effects on a wide range of organisms. Others are more selective in their action and may be used very effectively in the control of pests and diseases.

Insecticides

Insecticides have allowed many important diseases of animals, plants and man to be controlled. Some of the commonest interfere with the function of insect nerves, in which they selectively inhibit a particular enzyme involved in the transmission of messages. Nerves consist of long fibrous cells (axons) along which electrical impulses are transmitted by changes in the concentration of ions in the cell. At the end of each axon the 'message' has to be passed on to the next cell and this is done at synaptic junctions. The synapse at the end of the nerve fibre produces a compound called acetylcholine. This passes to the next nerve cell, in which it induces changes in the internal ionic concentration and initiates another nerve impulse. The acetylcholine produced does not enter the second cell, but binds to a protein on the cell surface (Figure 6.7). Once the second cell has responded the acetylcholine must be rapidly destroyed by the initiating cell to stop the second cell sending further impulses. The enzyme that destroys acetylcholine is called acetylcholine esterase. If it is inhibited then acetylcholine levels increase and nerve impulses flow in an uncontrolled way.

Organophosphorus insecticides function by reacting with the active site of acetylcholine esterase. They covalently attach a phosphate group to the side chain of a serine residue at the active site. This has the effect of permanently and irreversibly inhibiting the enzyme (Figure 6.8).

The carbamate insecticides have structures which are quite similar to those of acetylcholine itself and they thus compete for the binding sites and act as competitive inhibitors of the esterase. Both these groups of compounds stop the nervous system from exerting control over muscles, and the symptoms of insecticide poisoning of insects (tremors,

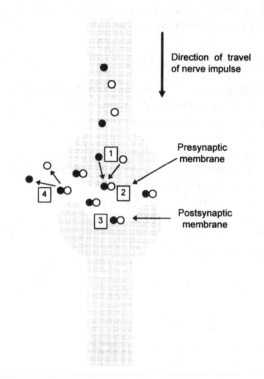

Figure 6.7 Transfer of a nerve impulse across a synaptic junction using acetylcholine. Acetylcholine (●○) is formed from acetyl-CoA (●) and choline (○), and is released from the presynaptic membrane in response to a nerve impulse. Acetyl-CoA stimulates the postsynaptic membrane and initiates an impulse before it is broken down by acetylcholinesterase. Inhibition of this enzyme causes nerve impulses to flow in an uncontrolled way.

hyperactivity, convulsions, paralysis and death) reflect this fact.

Organophosphorus and carbamate compounds are toxic to all types of animals, but by careful selection examples can be found that are much more harmful to insects than they are to man. One use of these has been as components of sheep dip. However, long-term exposure of farm workers to these compounds has resulted in serious health problems.

Some organophosphorus compounds which are exceptionally toxic to man have been made, and used, as nerve gases in chemical warfare.

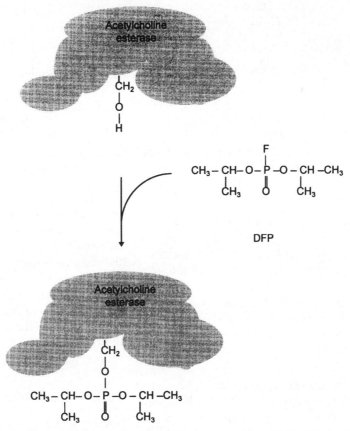

Figure 6.8 Inhibition of acetylcholine esterase by diisopropyl fluorophosphate (DFP). DFP reacts with the –OH group of a serine residue at the active site and becomes covalently attached to the enzyme, irreversibly inhibiting its activity.

Herbicides

A number of herbicides are enzyme inhibitors. One of these is glyphosate, which acts on enzymes of a biochemical pathway (the shikimic acid pathway) that animals do not have (Chapter 20). In plants and bacteria the aromatic amino acids (tryptophan, phenylalanine and tyrosine) are synthesized by this route. Without these amino acids plants cannot synthesize proteins and thus they die. Glyphosate has a structure that resembles phosphoenolpyruvate and competes with it for binding sites on the EPSP synthetase enzyme (Figure 6.9).

A number of other herbicides also act as enzyme inhibitors. For example, several groups of herbicides inhibit the enzyme acetoxy acid synthetase, which is required for the synthesis of the branched-chain amino acids (Chapter 20). Herbicides which act on this enzyme include the sulphonylureas, imidazolinones and the triazolopyrimidines.

Like the aromatic amino acids, branched-chain amino acids cannot be synthesized in animals because they lack the enzymes required to catalyse the steps in the biosynthetic pathways. The targets on which the herbicides act are therefore not present in animals and, unless they have other separate effects, they are normally of low toxicity to man and to animals.

COOH
|
CH$_2$
|
NH
|
CH$_2$
|
HO – P = O
|
OH

Glyphosate

COOH
|
C = CH$_2$
|
O
|
HO – P = O
|
OH

**Phosphoenol
pyruvate**

Figure 6.9 Some herbicides inhibit amino acid biosynthesis. Glyphosate resembles phosphoenolpyruvate and inhibits the enzyme EPSP synthetase which uses phosphoenolpyruvate as a substrate.

Pharmaceuticals

Many very useful pharmaceuticals function by inhibiting, often competitively, specific enzymes in man and animals. Because competitive inhibitors resemble the substrates for an enzyme, once the specificity of an enzyme is known it is possible to design and synthesize potential competitive inhibitors. It is therefore possible to carry out rational design of potential drugs.

PABA and the sulphonamides The compound *p*-aminobenzoic acid (PABA) is a component of the coenzyme folic acid. This is required for the transfer of groups containing single carbon atoms. Reactions of this type are common and include one of the steps by which the purine nucleotide IMP is synthesized. The structure of the sulphonamide drugs closely resembles that of PABA (Figure 6.10).

The sulphonamides competitively inhibit the biosynthesis of folic acid at the step where *p*-aminobenzoic acid is incorporated. Higher animals cannot synthesize folic acid and require it in their diet as a vitamin. Their metabolism is therefore unaffected by the sulphonamides. Bacteria however synthesize their own folic acid and growth of these organisms is selectively inhibited by sulphonamides.

Allopurinol A further example of a commonly used enzyme inhibitor is allopurinol. This compound is used in the treatment of gout: a condition which occurs in man and poultry and results from excessive accumulation of uric acid from the degradation of purines. The uric acid crystallizes in joints, which become extremely painful. It seems to occur because the enzyme (xanthine oxidase) which converts hypoxanthine into uric acid is particularly active. Allopurinol closely resembles hypoxanthine and is used as a substrate by the enzyme (Figure 6.10). The product however remains tightly bound to the enzyme, inhibiting its action and decreasing uric acid production.

6.5.6 PRESENCE OF COENZYMES

Some enzymes require the presence of small molecules called coenzymes for activity. In animals many of these are synthesized from vitamins in the diet (see Chapter 8).

Redox coenzymes (NAD$^+$, NADP$^+$, FMN and FAD) and biological oxidations

The redox coenzymes have important functions in biological oxidation and reduction reactions. Biochemical oxidations usually take place by removal of pairs of hydrogen atoms from the molecules to be oxidized. A typical example is the oxidation of malate in the TCA cycle shown in Figure 6.11.

Although the reactions involving these coenzymes are oxidations they do not in themselves require oxygen. However the reduced coenzymes may later be re-oxidized in the electron transport chain, which does require oxygen.

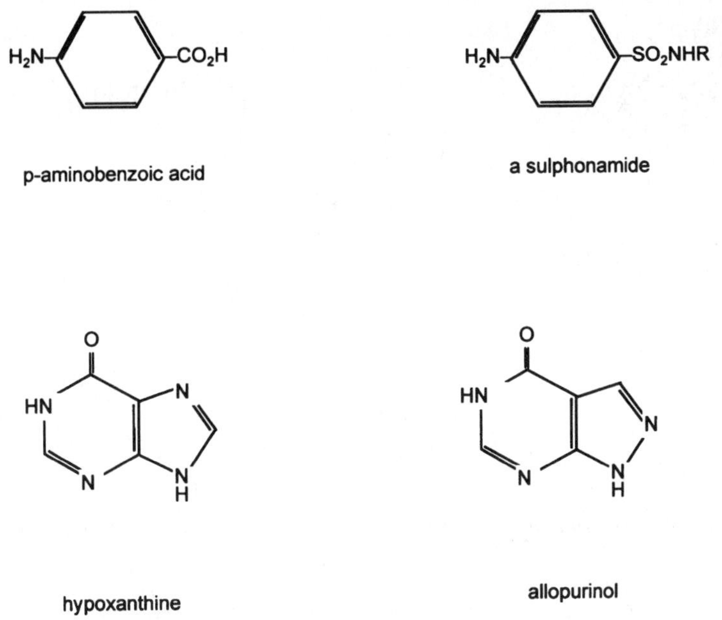

p-aminobenzoic acid

a sulphonamide

hypoxanthine

allopurinol

Figure 6.10 Many pharmaceuticals act as competitive inhibitors of enzymes because they resemble their substrates. Sulphonamides resemble *p*-aminobenzoic acid and inhibit bacterial synthesis of folic acid. Allopurinol is an analogue of hypoxanthine which is used in the treatment of gout.

When the pairs of hydrogen atoms are removed they are transferred to a hydrogen acceptor such as NAD^+, $NADP^+$, FMN or FAD.

NAD^+ (nicotinamide adenine dinucleotide) and $NADP^+$ (nicotinamide adenine dinucleotide phosphate) differ chemically only in the fact that $NADP^+$ has an extra phosphate group. Their structures are shown in Figure 6.12. Despite their chemical similarity, the functions of NADH and NADPH are usually quite different. NADH is almost invariably reoxidized in the electron transport chain, during which ATP is synthesized. It is therefore mainly involved in energy conservation. On the other hand, NADPH is used as a reducing agent and serves as a source of hydrogen atoms to be used in biosynthetic reactions.

NAD^+ and NADH act as normal substrates for their enzymes and so are not attached permanently to any one enzyme molecule, but are temporarily bound to the active site during the reaction.

NADH strongly absorbs UV light at a wavelength of 340 nm whereas NAD^+ absorbs only

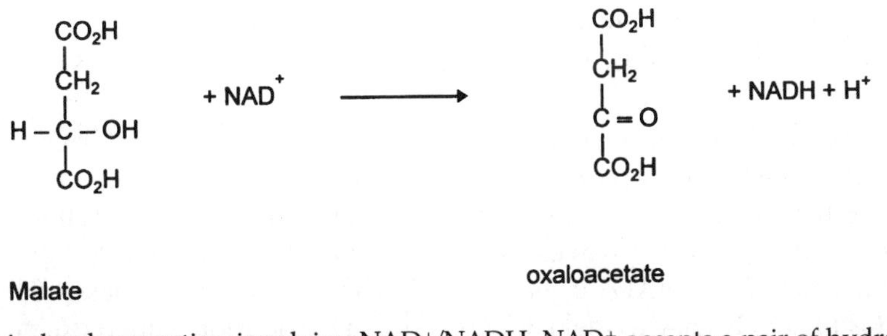

Malate

oxaloacetate

Figure 6.11 A typical redox reaction involving NAD^+/NADH. NAD^+ accepts a pair of hydrogen atoms from malate as it is oxidized to oxaloacetate.

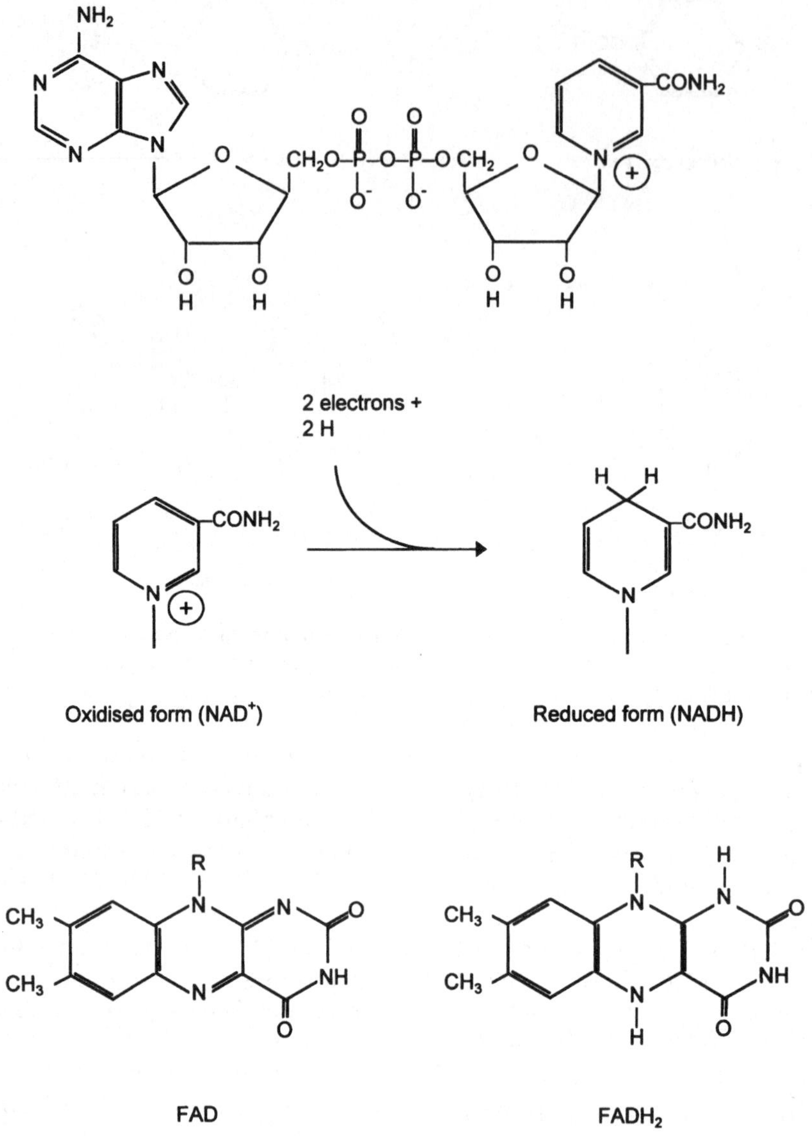

Figure 6.12 NAD$^+$/NADH and FAD/FADH$_2$ are important redox coenzymes. Each may be converted from oxidized to reduced form by accepting hydrogen atoms from a substrate which is oxidized.

weakly at this wavelength. The formation or disappearance of NADH can therefore be followed continuously in a spectrophotometer by measuring the absorbance at this wavelength. As many enzymes use either NAD$^+$ or NADH as substrates, it is relatively easy to make kinetic measurements of their reactions. In addition,

it is possible to couple reactions which do not involve these coenzymes to others which do, so that they can also be followed in this way.

Like NAD$^+$, the flavin coenzymes (FAD and FMN) also act as carriers of hydrogen atoms during oxidation and reduction reactions but their actions differ in several ways:

- the flavin coenzyme is attached firmly to the enzyme molecule, and for this reason it is a prosthetic group;
- the flavin can accept either one hydrogen atom (to form FADH or FMNH) or two hydrogen atoms (to form $FADH_2$ or $FMNH_2$) – Figure 6.12.

Subsequently $FADH_2$ can be converted back to FAD when the hydrogen atoms are passed to the electron transport chain.

ATP and ADP

Adenosine triphosphate (ATP) and adenosine diphosphate (ADP) are also two very important coenzymes. They are not vitamins, however, as they can be synthesized by animals and plants. Many of the reactions in biochemical pathways (particularly those involving sugars) take place using substrates that are linked to phosphate groups, and it is mainly from the terminal (γ) phosphate group of ATP that these are derived (Figure 6.13). In addition the conversion of ATP into ADP is accompanied by the release of considerable amounts of energy which can be used to drive reactions.

All reactions reach an equilibrium:

$$aA + bB \rightleftharpoons cC + dD \qquad \text{Equation 6.6}$$

The equilibrium constant (K_{eq}) is given by:

$$K_{eq} = \frac{[C]^c[D]^d}{[A]^a[B]^b} \qquad \text{Equation 6.7}$$

Where A and B are reactants, C and D the products. Some reactions however are essentially irreversible and, for all practical purposes, take place in one direction only. In these cases, at equilibrium, the concentrations of the products [C] and [D] are much greater than those of the reactants [A] and [B]. Thus, K_{eq} is very large.

The free energy of a reaction is the energy difference between products and reactants. When this is measured under standard conditions of concentration (1 M), temperature (25° C) and pressure (1 atm) this is referred to as

ΔG^0, the standard free energy change. A solution with a 1 M hydrogen ion concentration has a pH of zero, but in biochemistry it is more usual to discuss energy changes taking place at pH 7, as these approximate more closely to conditions in the cell. The standard free energy change taking place under these conditions is called $\Delta G^{0'}$ and it is related to the equilibrium constant for a reaction by the equation:

$$\Delta G^{0'} = -RT\ln K_{eq}' \qquad \text{Equation 6.8}$$

where R is the gas constant and T is the absolute temperature.

Values of $\Delta G^{0'}$ calculated from this equation for different values of K'_{eq} are given in Table 6.3.

Thus irreversible reactions have large negative values of $\Delta G^{0'}$ (This is a rather simplified view of energetics. An irreversible reaction actually has a large negative value for ΔG, the free energy change accompanying the reaction under the actual conditions in the cell, not $\Delta G^{0'}$, which assumes that all reactants and products are present at 1 M concentrations. However comparison of $\Delta G^{0'}$ values is sufficiently rigorous for the purposes of this text.)

One very useful characteristic of phosphate esters and other phosphate compounds is their free energy of hydrolysis, i.e. the free energy change which accompanies hydrolysis of a mole of the compound. Some typical values are given in Table 6.4. These compounds can be roughly divided into two groups:

- those with a normal free energy of hydrolysis between 0 and –20 kJ mol^{-1}
- high energy compounds, i.e. $\Delta G^{0'}$, more negative than –20 kJ mol^{-1}(ATP and above in Table 6.4).

Also $\Delta G^{0'}$ values are additive.

An especially useful feature of Table 6.4 is that the energy released during a reaction high up the table can be used to drive the reversal of reactions lower down. Thus a phosphate group can effectively be transferred from the high-energy compounds to other molecules, to produce lower-energy phosphate esters. These points can be illustrated by

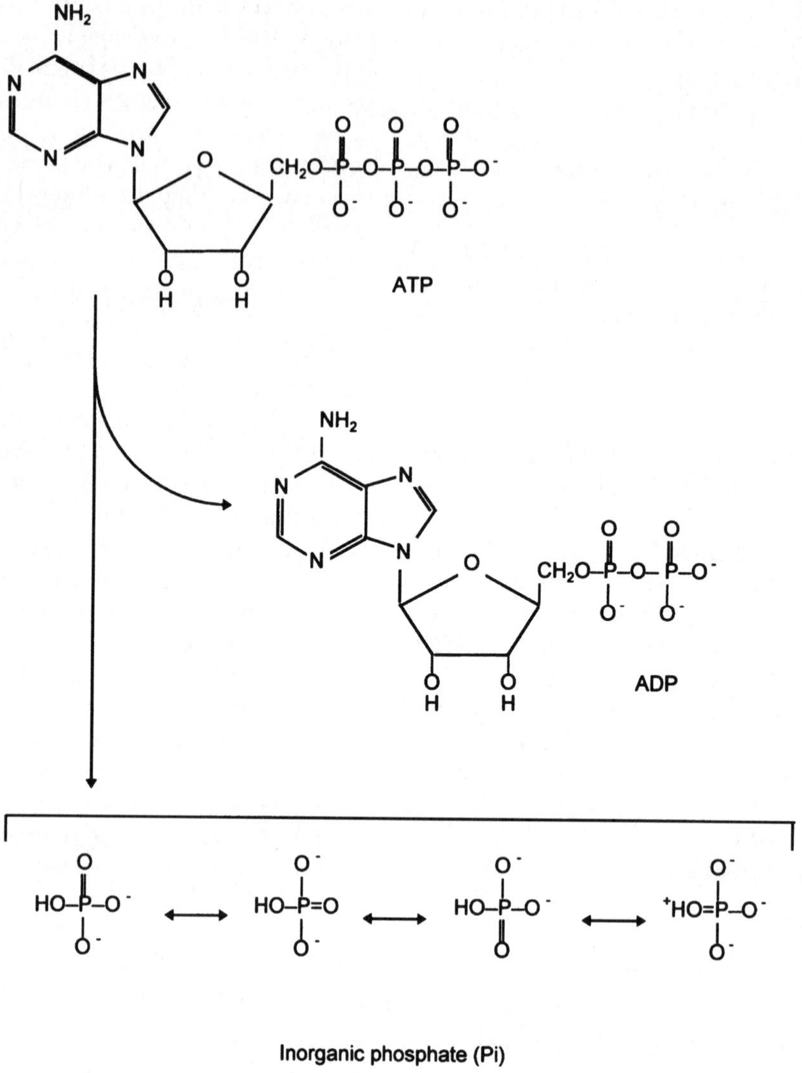

Figure 6.13 The structure of ATP and its breakdown products. Hydrolysis of ATP normally produces ADP and inorganic phosphate (P_i). The reaction releases a considerable amount of energy because of the relative instability of ATP and stability of P_i.

the first step in glycolysis, the conversion of glucose into glucose-6-phosphate, which will be discussed in more detail in Chapter 10.

This reaction might occur by simple condensation of glucose with a molecule of inorganic phosphate (P_i) as follows:

$$glucose + P_i \rightleftharpoons glucose\text{-}6\text{-}phosphate + H_2O$$
$$\Delta G^{0\prime} = +13.8 \text{ kJ mol}^{-1}$$

However this reaction is unfavourable and would not occur spontaneously because of the size of the free energy change. At equilibrium the reaction mixture would contain mainly glucose and P_i ($K_{eq} \sim 0.01$).

An alternative reaction for the production of glucose-6-phosphate from glucose is as follows:

$$glucose + ATP \rightleftharpoons glucose\text{-}6\text{-}phosphate + ADP$$

Table 6.3 Free energy changes and equilibrium constants

K'_{eq}	$\Delta G^{0'}$	
	(kJ mol^{-1})	(kcal mol^{-1})
0.00001	28.5	6.82
0.001	17.1	4.09
0.1	5.7	1.36
1.0	0.0	0.00
10.0	–5.7	–1.36
1000.0	–17.1	–4.09
100 000.0	–28.5	–6.82 ← IRREVERSIBLE

Table 6.4 Free energy of hydrolysis of phosphate esters

Reaction	$\Delta G^{0'}$ (kJ mol^{-1})
Phosphoenolpyruvate + H_2O → pyruvate + P$_i$	–61.9
1,3 Diphosphoglycerate + H_2O → 3 phosphoglycerate + P$_i$	–49.3
Creatine phosphate + H_2O → creatine + P$_i$	–43.1
Acetyl phosphate + H_2O → acetate + P$_i$	–42.2
ATP + H_2O → ADP + P$_i$	–30.5
Glucose-1-phosphate + H_2O → glucose + P$_i$	–20.9
Glucose-6-phosphate + H_2O → glucose + P$_i$	–13.8
Glycerol-1-phosphate + H_2O → glycerol + P$_i$	–9.2

This reaction can be considered in two parts which can be summed to obtain the overall equation:

$$glucose + P_i \rightleftharpoons glucose\text{-}6\text{-}phosphate + H_2O$$
$$\Delta G^{0'} = +13.8 \text{ kJ mol}^{-1}$$

$$ATP + H_2O \rightleftharpoons ADP + P_i$$
$$\Delta G^{0'} = -30.5 \text{ kJ mol}^{-1}$$

$$glucose + ATP \rightleftharpoons glucose\text{-}6\text{-}phosphate + ADP$$
$$\Delta G^{0'} = -16.7 \text{ kJ mol}^{-1} \quad \text{Equation 6.9}$$

For this reaction $K_{eq} \sim 1000$ so the equilibrium position favours production of glucose-6-phosphate, which takes place readily. A consequence of this, however, is that the reverse reaction is extremely unfavourable. Hence in gluconeogenesis (see Chapter 18), glucose-6-phosphate cannot be converted into glucose by transferring its phosphate group to ADP. This transformation is easily achieved by the energetically favourable process of simple hydrolysis as follows:

$$glucose\text{-}6\text{-}phosphate + H_2O \rightleftharpoons glucose + P_i$$
$$\Delta G_0' = -13.8 \text{ kJ mol}^{-1} \quad \text{Equation 6.10}$$

This example illustrates the very general point that reactions in pathways for the biosynthesis and breakdown of biochemical compounds usually differ from one another at key points, as this allows the energetics of both steps to be favourable. It also allows independent control of the rate of both pathways. The continuous synthesis of glucose-6-phosphate from glucose and its subsequent hydrolysis back to glucose by the reactions described above represents a 'futile cycle' in which ATP is wastefully hydrolysed. This can be prevented in the cell by ensuring that only glucose-6-phosphate synthesis or its breakdown takes place rapidly at one time, which is achieved by regulating the activity of enzymes catalysing the reactions.

By far the commonest compound which acts as the donor of phosphate groups, and which therefore drives reactions in this way, is ATP. The vital role of this compound in phosphate group transfer, and in providing the chemical energy needed to make otherwise energetically unfavourable reactions take place, is determined by two main factors.

- The free energy of hydrolysis of ATP (−30.5 kJ mol^{-1}) is high, so that when the conversion of ATP into ADP occurs, sufficient energy is made available to drive other reactions by ensuring that the overall $\Delta G^{0\prime}$ value is negative.
- The free energy of hydrolysis of ATP is not excessive. More energy could be provided to drive reactions by one of the compounds higher up Table 6.4, but towards the top of the list the compounds become more and more difficult to make because they require a great deal of energy for their synthesis.

The particular value of ATP therefore is that it is near the middle of the table and, although a high-energy compound, is relatively easily made by linking its synthesis to other reactions releasing even more energy, such as conversion of 1,3-diphosphoglycerate to 3-phosphoglycerate, hydrolysis of phosphenol pyruvate, or processes occurring during electron transport.

The free energy of hydrolysis is the difference in energy between the starting compounds (ATP) and the products (ADP + P$_i$). In the case of ATP, hydrolysis yields a considerable amount of energy (Figure 6.13) for the following reasons.

- ATP is unstable as a result of the four negative charges which it carries at normal pH, which repel one another. This repulsion is much reduced in ADP because there are fewer charges.
- Phosphate is especially stable because of the existence of resonance structures.

In most reactions in which ATP is involved, the terminal (γ) phosphate is transferred either to water (hydrolysis) or to some other compound. This is called orthophosphate cleavage, but in some cases the two terminal phosphates are removed together in the process of pyrophosphate cleavage.

$$ATP \rightleftharpoons AMP + PP_i \qquad \Delta G^{0\prime} = -41.8 \text{ kJ mol}^{-1}$$

where AMP = adenosine monophosphate and PP$_i$ = pyrophosphate. This can provide more push for the reaction because $\Delta G^{0\prime}$ is higher and because tissues contain pyrophosphatases which catalyse the hydrolysis of pyrophosphate to two molecules of inorganic phosphate. This irreversible process removes the product of the first reaction and hence pulls the equilibrium further to the right.

Reactions in which pyrophosphate cleavage of ATP occurs include:

- activation of fatty acids, which precedes β-oxidation
  RCO$_2$H + ATP + CoASH → AMP + PP$_i$ + R-CO-SCoA
- formation of amino acyl tRNAs in protein synthesis
  amino acid + tRNA + ATP → aminoacyl tRNA + AMP + PP$_i$
- formation of nucleoside diphosphate sugars in di- and polysaccharide synthesis
  ATP + G-1-P → ADP-glucose + PP$_i$

6.6 ALLOSTERIC ENZYMES

The enzymes discussed so far are simple enzymes which obey the Michaelis–Menten kinetic equations. The derivation of these equations assumes that each enzyme molecule behaves independently of others. However some enzymes, known as allosteric enzymes, consist of a number of subunits, each composed of one or more polypeptide chains. The subunits are often identical in structure but may be different. These subunits carry the active site, to which the substrate binds, and one or more regulatory sites which bind signal molecules (effectors). The binding of substrate to an active site or the binding of an effector to a regulatory site induces a conformational change in the subunit which may influence other active sites and either increase or decrease their affinity for the substrate. These effects can be transmitted from one subunit to another so that the active site and regulatory sites may be on different subunits. Where the binding of substrates affects other active sites the kinetics are usually complex, and often a plot of v_0 against [S] will not be a hyperbola but a sigmoid curve (Figure 6.14). This interaction between subunits is called co-operativity. Where binding of an effector affects the active site, complex patterns of either inhibition or activation may be seen as a result of positive or negative co-operativity. Inhibition of the activity of allosteric enzymes by the end-product of metabolic pathways is particularly important in the control of many metabolic pathways (see Chapter 22).

Models developed to describe the mechanism of allosteric enzymes propose that the enzyme subunits exist in two forms with either low or high affinity for the substrate, and that they may be switched between these forms by binding of substrate or effector (Figure 6.15).

6.7 MOLECULAR RECOGNITION

One of the most characteristic and important properties of proteins is their ability to bind or 'recognise' other molecules. The action and specificity of enzymes depends on this property, but many other proteins which are not enzymes, such as receptors and antibodies, also show specific binding.

6.7.1 RECEPTORS

A receptor is a molecule which binds, and thereby detects the presence of, another molecule. Thus for example the binding of a hormone to its receptor may initiate a response which will result in the changes which are characteristic of that hormone. Binding to receptors is analogous to binding of substrates to enzymes and displays high affinity and specificity. The nature of receptors for some hormones is discussed further in Chapter 20.

6.7.2 ANTIBODIES

Antibodies are proteins which animals produce in response to materials (antigens), usu-

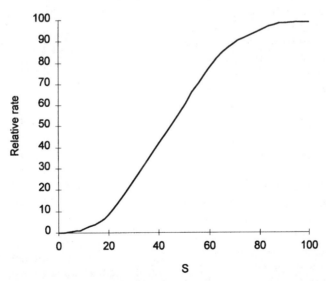

Figure 6.14 Kinetics of an allosteric enzyme. Allosteric enzymes often show sigmoid kinetics because of cooperation between the subunits of which they are composed.

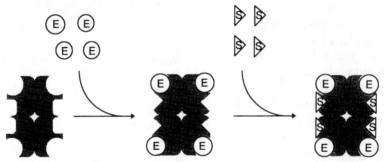

Figure 6.15 Structure of an allosteric enzyme showing its conversion from a low- to a high-activity form by binding an effector (E). Binding of E to a regulatory site causes a conformation change, increasing the affinity of the catalytic site for substrate (S).

ally from outside the body. Antibodies form part of the defence mechanism of the immune system. They may for example be produced in response to bacteria or virus coat proteins. Antibodies form a complex with antigens and neutralize their effects as a precursor to their destruction, thus protecting the animal.

Antibodies are extremely specific and can differentiate not only between proteins from different species but also between the same protein from different individuals of the same species.

Antibodies in general belong to a class of proteins known as immunoglobulins. Within this class there are five main groups known as IgA, IgD, IgE, IgG and IgM. Each of these has a well-defined role in combating the invasion of foreign materials. IgG is the antibody which circulates in the blood, whilst IgA acts in the secretions of the digestive, urinary, respiratory and reproductive tracts. The discussion in this section will be limited to IgG.

Each antibody molecule is composed of a number of individual protein chains which are classified according to their molecular weight as heavy or light. Each IgG molecule is made up of two heavy chains and four light ones (Figure 6.16). The molecule has a characteristic Y shape with two heavy chains forming the stalk of the Y upon which are hinged the two pairs of light chains that each form the arms. The two heavy chains are held together with a pair of disulphide bridges. The light chains in each of the

arms are held together by a single bridge. The part of the molecule which will recognise and react with the antigen is contained in the light

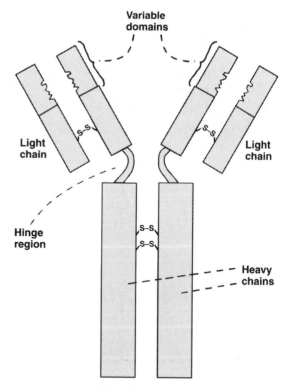

Figure 6.16 The structure of an antibody. Antibodies consist of light and heavy chains, linked together through a 'hinge'. The light chains contain the part of the molecule which recognises and reacts with the antigen.

chains. This area is similar in its binding and specificity to the active site of an enzyme.

The means by which cells are able to synthesize a different antibody in response to almost any challenge encountered is a source of much current debate. Basically the two extreme possibilities are either that all cells contain the genetic information to produce each type of antigen protein; or that each antigen causes a change in the shape of a 'universal' antibody in such a way that it induces complementarity in the antibody.

Recent discoveries concerning the roles of proteins in the formation of the tertiary structure of other proteins, as is found in prions and with the heat-shock proteins, add weight to the possible roles of the antigens in determining the form of their appropriate antibodies.

PURINES, PYRIMIDINES AND NUCLEIC ACIDS

7.1 INTRODUCTION

The properties of all proteins are determined by the sequence of the 20 different amino acids which they contain. Any errors in these sequences usually prevent them from performing their normal biological functions. The order in which the amino acids are assembled is determined by the structure of the DNA (deoxyribonucleic acid) molecules in the cells. DNA therefore acts as the 'blueprint' which determines the nature and function of proteins. DNA also has the vital function of transmitting this information from one generation to the next, as the information in the DNA can be copied and handed on to daughter cells.

DNA is a type of nucleic acid. Although the information required for the synthesis of proteins is carried in DNA, another type of nucleic acid called RNA (ribonucleic acid) is also required for the translation of the information in the DNA into the sequence of amino acids in a protein.

Both DNA and RNA consist of a backbone of alternating sugar and phosphate groups forming very long chains. In the case of RNA the sugar is ribose and in DNA it is deoxyribose. Each of the sugars carries a purine or pyrimidine base, and it is the sequence in which these bases occur that provides the information needed to code for the amino acids in proteins.

7.2 PURINES AND PYRIMIDINES

Purines and pyrimidines are small, nitrogen-containing, organic bases. Pyrimidines consist of a single ring whilst purines contain two fused rings. The purine and pyrimidine bases are rarely found as free compounds; they are normally attached to sugar groups to form nucleosides, or to sugar phosphates (as in the nucleic acids) to form nucleotides (Table 7.1). The structures of the purine and pyrimidine bases which are found in most RNA and DNA are shown in Figure 7.1. In addition a number of unusual bases are found in some specific types of RNA. As well as their role in nucleic acids, some purine and pyrimidine derivatives (e.g. ATP, NADH) are important coenzymes and are involved in energy metabolism.

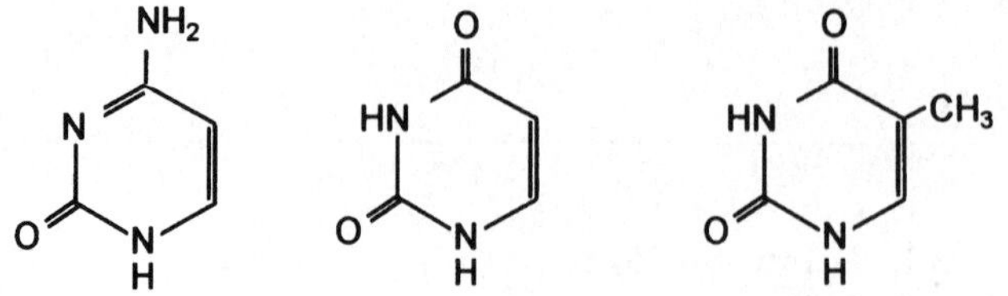

Cytosine   Uracil   Thymine

Pyrimidine bases

Adenine   Guanine

Purine bases

Figure 7.1 Structure of the purine and pyrimidine bases which are commonly found in RNA and DNA. Pyrimidine bases: cytosine, uracil, thymine; purine bases: adenine, guanine.

7.3 DEOXYRIBONUCLEIC ACID (DNA)

7.3.1 CHEMICAL NATURE OF DNA

The basic structure of DNA is shown in Figure 7.2. The sugar in DNA is deoxyribose. Each sugar carries a base attached to the 1′ position. The base may be either a purine or a pyrimidine. The purines are adenine (A) or guanine (G), and the pyrimidines cytosine (C) or thymine (T). DNA is mostly double stranded. The two strands run alongside but in opposite directions and are described as antiparallel.

Table 7.1 Some purine and pyrimidine bases, nucleosides and nucleotides

Base	Nucleoside	Nucleotide
Adenine	Adenosine (A)	AMP (Adenosine monophosphate)
Guanine	Guanosine (G)	GMP (Guanosine monophosphate)
Hypoxanthine	Inosine (I)	IMP (Inosine monophosphate)
Uracil	Uridine (U)	UMP (Uridine monophosphate)
Thymine	Thymidine (T)	TMP (Thymidine monophosphate)
Cytosine	Cytidine (C)	CMP (Cytidine monophosphate)

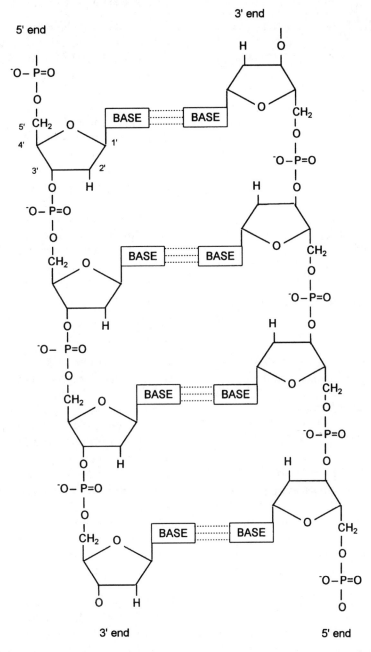

Figure 7.2 Diagrammatic representation of the structure of DNA. Each chain consists of a backbone of alternating deoxyribose sugar and phosphate groups. The sugars carry a purine or pyrimidine base which forms hydrogen bonds with complementary bases in the second chain. The two chains run in opposite directions and are therefore antiparallel.

The two chains are held together by hydrogen bonds formed between adenine and thymine or between cytosine and guanine (Figure 7.3).

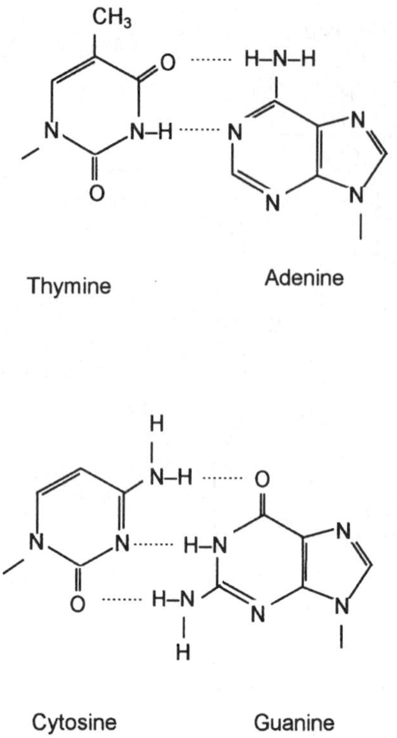

Thymine   Adenine

Cytosine   Guanine

Figure 7.3 Hydrogen bonding between pairs of bases in DNA. The bonding is very specific: two bonds form between adenine and thymine and three between cytosine and guanine. In each pair one base is a purine and one a pyrimidine.

This hydrogen bonding (base pairing) is extremely specific, always involving these particular combinations of purine–pyrimidine pairs. This means that in double-stranded DNA there are always roughly equal quantities of purines and pyrimidines. Three hydrogen bonds can be formed between cytosine and guanine but only two between adenine and thymine. DNA containing a high proportion of C and G is more stable to heat than that containing a lower proportion. The formation of these base pairs can only occur when the chains are arranged in an antiparallel formation. The two chains of the DNA are said to be complementary, as the bases of one are determined by, and are complementary to, the other.

7.3.2 THE DNA DOUBLE HELIX

The structure of DNA forms a double helix in which the two sugar phosphate chains make up the backbone of the molecule and are twisted around one another in a spiral. The bases, which are paired together, point inwards towards one another and lie perpendicular to the axis of the helix. They thus resemble the rungs on a spiral ladder. One turn of the helix occurs roughly every ten nucleotides and occupies about 3.4 nm of the length of the chain. Two grooves, the narrow groove and the wide groove, can be seen along the surface of the helix and these provide possible positions for interaction of the DNA with other molecules such as proteins (Figure 7.4). The DNA helix itself may undergo a type of coiling called supercoiling.

7.3.3 STRUCTURE OF DNA IN PROKARYOTES AND EUKARYOTES

There are differences in the organisation of DNA in prokaryotes and eukaryotes. In bacteria such as *Escherichia coli* the DNA is circular and found in the nucleoid region of the cytoplasm, where it is associated with small amounts of proteins and folded to form about 100 supercoiled loops. In eukaryotes DNA exists in the form of well-defined chromosomes within the nucleus. These chromosomes are made up of chromatin in which DNA forms a complex with large quantities of basic proteins including histones. The composition of chromatin is given in Table 7.2.

Histones are small, basic proteins of molecular weight about 10–20 kDa containing a large proportion of basic amino acids such as arginine and lysine. They have large numbers of positive charges and readily form complexes with negatively charged nucleic acids. Lysine residues in the histones may be acetylated or methylated, and serines may be phosphorylated. Each of these changes reduces the net positive charge and may reduce binding to DNA.

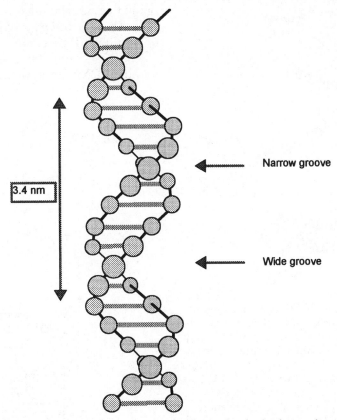

Figure 7.4 The DNA double helix. The two chains are held together by hydrogen bonds between their complementary bases and are twisted into a helix. There are approximately 10 nucleotides to each turn which occupies about 3.4 nm along the length of the chain. A wide and a narrow groove run along the surface of the helix.

The non-histone proteins may play a role in switching on and off genes in some parts of the chromatin. They may do this for example in response to hormones which activate specific genes.

Through the electron microscope, repeating structures called nucleosomes can be identified within the chromosomes. In a nucleosome, DNA is wound around eight histone subunits and these are themselves coiled around a hollow core so that in the chromosome the DNA forms part of a coiled coil (Figure 7.5).

Nucleosomes themselves may occur in relatively open or closely packed arrangements

Table 7.2 The composition of chromatin

Component	Relative amount (DNA=100)
DNA	100
Histones	114
Non-histone proteins	33
RNA	7

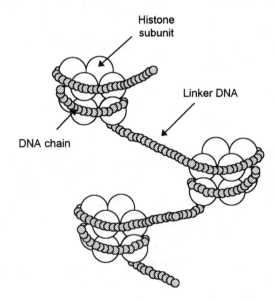

Figure 7.5 Nucleosome structure. In eukaryotes, nucleosomes are observed at at intervals along the DNA. Each nucleosome consists of approximately 140 base pairs of DNA wrapped around eight histone subunits. The region between nucleosomes is called linker DNA and is between 20 and 100 base pairs long, depending on the species. Nucleosomes may themselves adopt a helical arrangement to form fibres or solenoids.

and it is believed that in the latter genes cannot be expressed, whilst in the former they may be activated under some circumstances.

7.3.4 ORGANELLE DNA

In addition to the DNA found in the nucleus, small quantities are also found in chloroplasts and mitochondria. The information carried in this DNA codes for some of the proteins of these organelles. This DNA resembles prokaryotic DNA. Organelle genomes are discussed in more detail in Chapter 21.

7.4 RIBONUCLEIC ACID (RNA)

Chemically RNA is very similar to DNA but there are some important differences.

Like DNA, it consists of a backbone made up of alternating sugar and phosphate groups,

but in this case the sugar is ribose not deoxyribose. There is an OH group at the 2' position of the ribose (Figure 7.6).

The bases also differ slightly from those found in DNA. In particular, thymine is not found in RNA, but the pyrimidine uracil

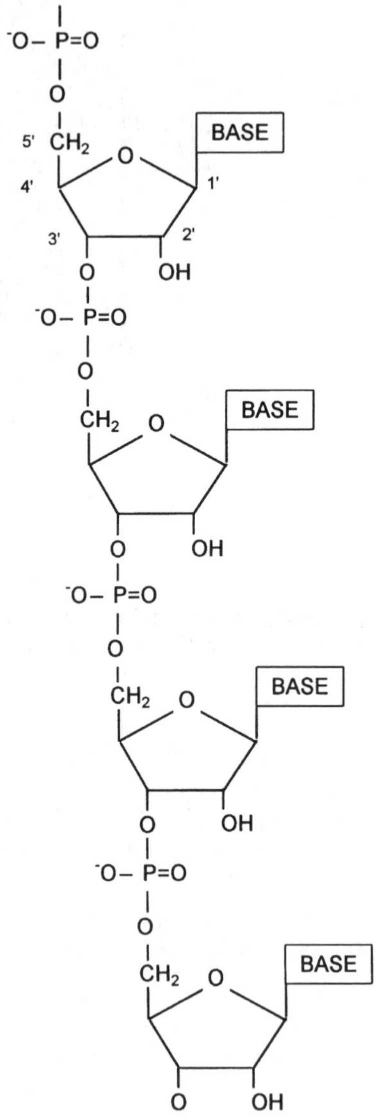

Figure 7.6 Diagrammatic representation of the structure of RNA. The backbone consists of alternating ribose sugars and phosphate groups. Each sugar carries either a purine or pyrimidine base. RNA is normally single-stranded.

occurs instead. The bases found in most RNA molecules are thus adenine, guanine, cytosine and uracil.

Most RNA is single-stranded, but within its structure short regions may be coiled to form base-paired regions.

There are three distinct types of RNA:

- messenger RNA (mRNA)
- transfer RNA (tRNA)
- ribosomal RNA (rRNA).

These differ from one another in their size and function as will be seen below, but chemically they are very similar. The synthesis of all three types is catalysed by enzymes called RNA polymerases. Unlike DNA synthesis, the formation of RNA is not restricted to times of cell division but takes place throughout the life of the cell.

The sequence in which bases are assembled into RNA is determined by the order of bases in DNA. Before RNA synthesis can occur the strands of DNA must separate from one another. RNA is synthesized on the template strand of DNA which is therefore complementary to this strand. The other strand has the same sequence of bases as the RNA produced, except that T replaces U and thus is called the coding strand. The bases in the template strand of DNA and the complementary ones inserted into RNA are noted in Table 7.3.

7.4.1 MESSENGER RNA (mRNA)

mRNA makes up about 2% of the total RNA. It contains the sequence of bases needed to determine the order in which amino acids will be assembled into each type of protein. There

Table 7.3 Relationship between bases in DNA and RNA

Template strand of DNA	Complementary RNA
Adenine	Uracil
Thymine	Adenine
Cytosine	Guanine
Guanine	Cytosine

are several thousand types of mRNA in most cells. In eukaryotes one type of mRNA molecule corresponds to each type of protein, but in prokaryotes some mRNA is polycistronic. In this case a mRNA molecule may carry the code for several different proteins. The molecular weight of mRNA is thus very variable, particularly in prokaryotes.

In prokaryotic cells mRNA is made on the DNA within the nucleoid region of the cytoplasm and is then used immediately to code for protein synthesis. In eukaryotes, however, most mRNA is made in the nucleus, from where it passes out through the nuclear envelope to the cytoplasm. Before transfer to the cytoplasm introns may be removed and 3' or 5' caps may be added (see Chapter 21).

In prokaryotes, mRNA molecules turn over very rapidly. This means that each one is read by a ribosome only a few times before it is destroyed. This allows the cell to respond very rapidly to the changing environment to which it is exposed. On average each mRNA molecule is broken down after only 2–3 minutes. In eukaryotes the stability of different mRNA molecules varies, but on the whole they are much more stable than in prokaryotes. The average half-life of eukaryotic mRNA is about 10 hours but some types persist much longer than this. In reticulocytes, for example, haemoglobin continues to be synthesized even after the nucleus disappears because these cells contain stable mRNA which codes for this protein.

Some mRNA is synthesized in the chloroplasts and the mitochondria as part of the process of protein synthesis occurring in these organelles.

7.4.2 TRANSFER RNA (tRNA)

tRNA makes up about 16% of the total RNA. tRNA molecules are small and uniform in size, having molecular weights in the range of 23–30 kDa. This corresponds to between 75 and 90 nucleotides.

The system for coupling amino acids together to form proteins cannot recognise the different amino acids directly. Instead it uses tRNA as an intermediate carrier. Each molecule of tRNA has a site for attachment of amino acids and another which is complementary to, and recognises, sections of mRNA molecules. tRNA thus brings amino acids to the ribosomes where they are assembled into proteins. There is at least one type of tRNA molecule for each of the 20 amino acids which occur in proteins, and for some amino acids there are several. There are about 50 different types of tRNA in any one cell, at least one type for each amino acid found in proteins.

In addition to the four major bases, tRNA also contains up to 10% of unusual bases such as pseudouridine, ribothymidine or dihydrouracil. Although the exact sequence of bases varies from one type of tRNA to another, each conforms to a common overall structure which can be drawn out as a clover leaf. This has several important parts (Figure 7.7):

- an anticodon, consisting of three bases, which can base pair with complementary bases in mRNA;
- an amino acid attachment site, to which a specific amino acid becomes attached (Chapter 21);
- other arms which may help binding to the ribosomes; the number of nucleotides in all of the arms except the variable one is fixed (Figure 7.7).

7.4.3 RIBOSOMAL RNA (rRNA)

rRNA makes up more than 80% of the total RNA in the cell. It is a major component of the ribosomes, the organelles on which protein synthesis takes place.

Ribosomes consist of two subunits: a large one and a small one. They contain about 65% rRNA and 35% protein. The ribosomes in prokaryotes differ in size from those in eukaryotes. It is convenient to use their sedimentation constant, which is measured in Svedberg units (S), as a measure of their size. This measures how rapidly they sediment in a centrifuge when subjected to a particular gravitational force. Large particles have large S values. They are not additive, in other words joining together a 60S particle with a 40S particle in eukaryotic ribosomes produces an 80S rather than a 100S particle.

70S ribosomes are characteristic of prokaryotes and of the ribosomes found in the mitochondria and chloroplasts of eukaryotes, whilst 80S ribosomes are found in the cytoplasm of eukaryotic organisms. Whatever their size, ribosomes consist of several pieces of RNA and a number of proteins. The composition of 70S and 80S ribosomes is described in Figure 7.8.

The differences between 70S and 80S ribosomes are important because many antibiotics inhibit bacterial protein synthesis on 70S ribosomes without affecting that taking place on 80S ribosomes. They thus selectively kill bacteria and other prokaryotes. More information about the effects of antibiotics on protein synthesis can be found in Chapter 30.

rRNA is made in a specialized region of the nucleus called the nucleolus. The rRNA for one ribosome is made in one long piece. It then complexes with proteins before being split into several smaller pieces which make up the ribosome.

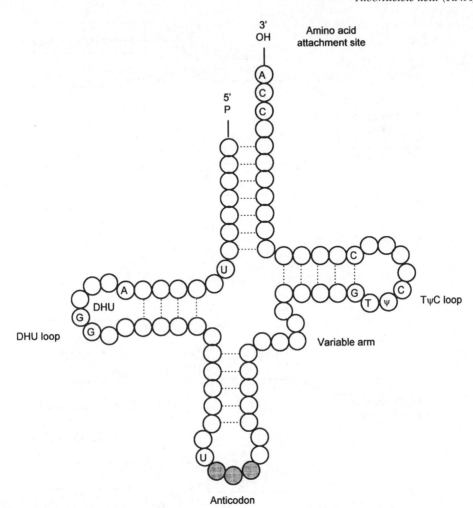

Figure 7.7 The clover-leaf structure of yeast alanine tRNA. Other tRNA molecules also adopt the clover-leaf structure. The regions which are common to all tRNAs are indicated. The amino acid becomes attached to the 3'-OH group of the terminal adenosine residue, and the sequence of bases in the anti-codon is unique to each type of tRNA.

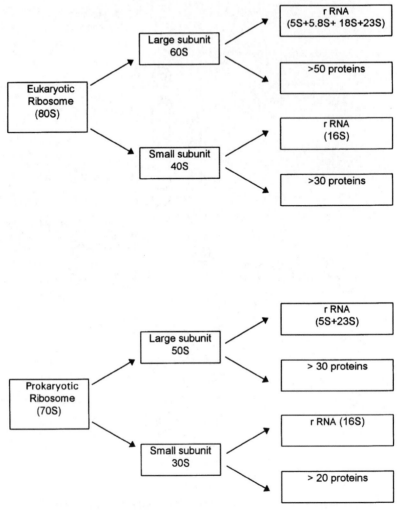

Figure 7.8 Composition of 80S ribosomes which are found in the cytoplasm of eukaryotes and 70S ribosomes found in prokaryotes and eukaryotic organelles. Both types consist of two subunits which contain both RNA and proteins, but 80S ribosomes are generally more complex.

8.1 VITAMINS IN BIOCHEMISTRY

8.1.1 INTRODUCTION

A detailed knowledge of the structure and the biochemical and physiological roles of vitamins has been accumulated over the past 50 years. However, before this time diseases were recognised which were related to diet. We now know that many of these diseases were caused by vitamin deficiencies.

In 1920, before the chemical composition of vitamins was known, it was proposed that they be identified simply by letters of the alphabet. This system worked satisfactorily for a number of years and eventually nine vitamins were discovered: A, B, C, D, E, F, G, H and I. However, it was soon discovered that F was not a true vitamin and that vitamin B was, in fact, a mixture of several vitamins. As a result the lettering system, though still in use, is now disjointed and a number of vitamins are referred to by their chemical names. In addition, the vitamins can be classified into two groups according to their solubility properties. Thus the vitamins of the B complex and vitamin C are known as the water-soluble vitamins, while the remainder are known as the fat-soluble vitamins (Table 8.1).

In general, fat-soluble vitamins contain only carbon, hydrogen and oxygen, whereas the water-soluble vitamins may also contain varying proportions of nitrogen, sulphur or cobalt. Fat-soluble vitamins can occur in plant tissue in the form of provitamins: vitamin precursors which can be converted into vitamins in the animal body. No provitamins are known for the water-soluble vitamins.

The body has little capacity to store water-soluble vitamins and animals require a regular supply of small quantities of them. Amounts consumed in excess of requirements are excreted, usually via the urine. Thus, supranormal intakes of water-soluble vitamins have little effect on long-term body reserves. The body's requirement for vitamins changes with age, physiological status and during illness and infection. Under these

circumstances an increase in intake may be beneficial. There is little evidence of toxicity associated with excessive intake of water-soluble vitamins.

In contrast, fat-soluble vitamins are stored in body tissues. As their name suggests they are found in the fat depots of the body, which may act as a reserve for times of inadequate intake. The liver is also rich in fat-soluble vitamins. Because the body has the ability to build up reserves of these vitamins it is possible for them to accumulate to toxic concentrations.

8.1.2 THIAMIN (VITAMIN B_1)

Biochemical functions

In cells thiamin occurs largely as its active coenzyme form, thiamin pyrophosphate (formerly cocarboxylase) (Figure 8.1). It acts as a coenzyme for two types of enzyme-catalysed reactions involved in carbohydrate metabolism. The first is the decarboxylation of α-keto acids. This type of reaction is important in the

Table 8.1 Vitamins, their coenzyme forms and main functions

Type	Coenzyme or active form	Main Function
Water-soluble		
Thiamine (B_1)	Thiamine pyrophosphate (TPP)	Aldehyde group transfer
Riboflavin (B_2)	Flavin mononucleotide (FMN)	Hydrogen atom (electron) transfer
	Flavin adenine dinucleotide (FAD)	Hydrogen atom (electron) transfer
Nicotinic acid	Nicotinamide adenine dinucleotide (NAD)	Hydrogen atom (electron) transfer
	Nicotinamide adenine dinucleotide phosphate (NADP)	Hydrogen atom (electron) transfer
Pantothenic acid	Coenzyme A (CoA), acyl carrier protein (ACP)	Acyl group transfer
Pyridoxine (B_6)	Pyridoxal phosphate	Amino group transfer
Biotin	Biocytin	Carboxyl group transfer
Folic acid	Tetrahydrofolic acid	One-carbon group transfer
Cyanocobalamin (B_{12})	Coenzyme B_{12}	1,2 shift of hydrogen atoms
Ascorbic acid (C)		Coenzyme in hydroxylation
Fat-soluble		
Vitamin A	11-cis-Retinal	Visual cycle, bone growth, cell differentiation
Vitamin D	1,25-Dihydroxycholecalciferol	Calcium and phosphate metabolism
Vitamin E	α-tocopherol	Antioxidant
Vitamin K		Prothrombin biosynthesis

catabolism of sugars and the TCA cycle (Chapter 11) and examples are:-

- The conversion of pyruvate to acetyl-CoA catalysed by pyruvate dehydrogenase.

$$\text{Pyruvate} + NAD^+ + CoA \rightarrow$$
$$\text{acetyl-CoA} + NADH + H^+ + CO_2$$

- The conversion of α-ketoglutarate into succinyl-CoA catalysed by the enzyme α-ketoglutarate dehydrogenase.

$$\alpha\text{-ketoglutarate} + NAD^+ + CoA \rightarrow$$
$$\text{succinyl-CoA} + NADH + H^+ + CO_2$$

The second type of reaction involving thiamin pyrophosphate occurs in the pentose phosphate pathway and is catalysed by transketolase enzymes (Chapter 15).

$$\text{Xylulose-5-P} + \text{ribose-5-P} \rightarrow$$
$$\text{sedoheptulose-7-P} + \text{glyceraldehyde-3-P}$$

Metabolism of thiamin

The metabolism of thiamin can be inhibited by the presence of synthetic analogues (Figure 8.1). Pyrithiamin blocks the phosphorylation of thiamin preventing the formation of the active coenzyme. Oxythiamin competes with thiamin pyrophosphate for binding sites on enzymes. Amprolium is an anticoccidial agent used in the poultry industry. Its effect is to disrupt thiamin metabolism in coccidia but, if given in high concentration, it can affect the host animal by inhibiting thiamin absorption and blocking phosphorylation.

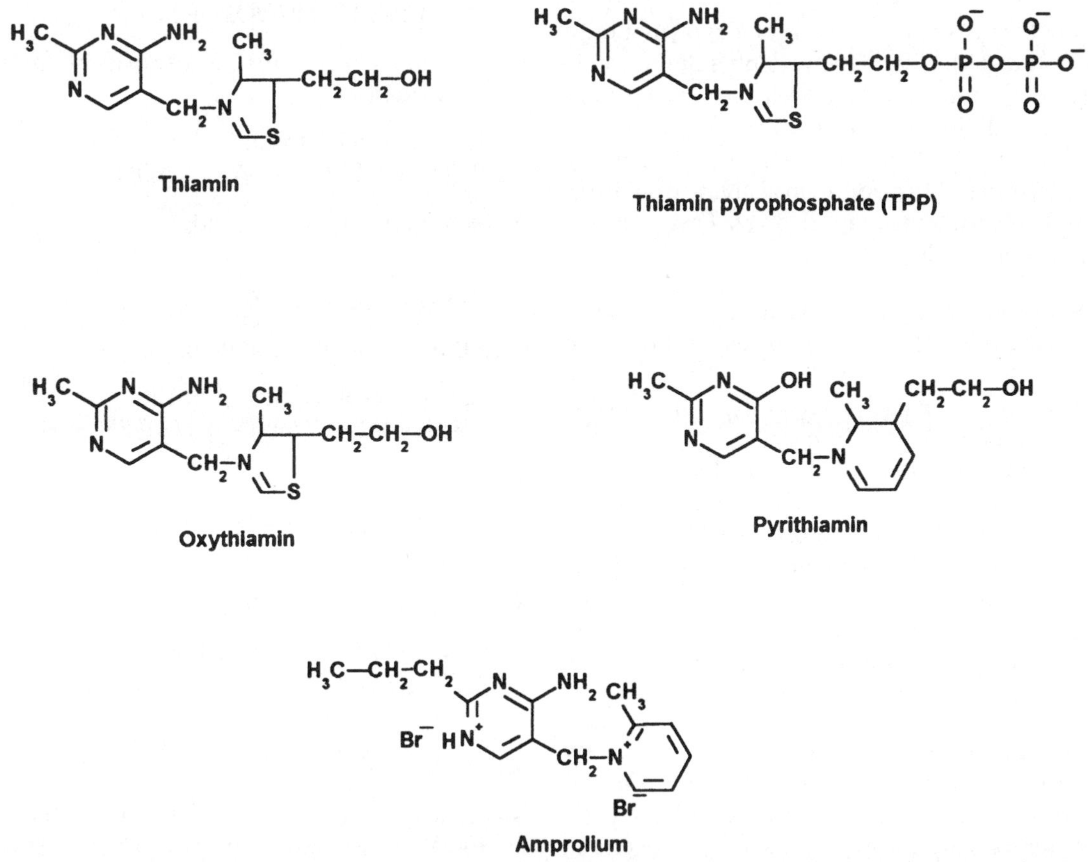

Figure 8.1 The structures of thiamin, the coenzyme thiamin pyrophosphate (TPP), and the thiamin analogues oxythiamin, pyrithiamin and amprolium.

Thiaminases are found in a number of microorganisms and plants, and in certain types of raw fish. They split thiamin into its component pyrimidine and thiazole moieties and thus destroy its activity. Non-enzymic destruction of thiamin may also result from the presence of reactive substances such as caffeic acid, tannins, rutin and quercitin in feedstuffs.

The biochemical bases of the symptoms of thiamin deficiency are not clearly understood. A summary of the deficiency symptoms in different species is given in Table 8.2.

8.1.3 RIBOFLAVIN (VITAMIN B₂)

Biochemical functions

Riboflavin is a component of flavin mononucleotide (FMN) and flavin adenine dinucleotide (FAD) (Figure 8.2), which are coenzymes involved in oxidation/reduction reactions (Chapter 6). Numerous enzymes containing FMN and FAD are found in all living organisms and are usually referred to as flavoproteins. Examples are listed below:

- D- and L- amino acid oxidases which oxidize amino acids to their corresponding keto acids;
- xanthine oxidase, involved in the conversion of purines to uric acid;
- the flavoproteins of the mitochondrial electron transport chain which links substrate oxidation to the synthesis of ATP;
- succinate dehydrogenase, the enzyme responsible for the conversion of succinate to fumarate in the TCA cycle;
- acyl-CoA dehydrogenases involved in the β-oxidation of fatty acids.

Because riboflavin is involved in many important biochemical processes, it is not surprising that its deficiency is reflected in a wide range of symptoms which vary from species to species (see Table 8.2).

8.1.4 NICOTINIC ACID (NIACIN, FORMERLY CALLED VITAMIN B₅)

Biochemical functions

The physiologically active form of nicotinic acid is nicotinamide (Figure 8.3) and as such it is an important component of two coenzymes, NAD^+ and $NADP^+$. These coenzymes are essential for enzyme-catalysed reactions in carbohydrate, protein and lipid metabolism. They function in oxidation–reduction reactions by acting as hydrogen transfer agents (Chapter 6). Examples of reactions involving NAD^+ and $NADP^+$ are listed below:

- interconversion of pyruvate and lactate

$$CH_3COCOOH + NADH + H^+ \rightarrow CH_3CHOHCOOH + NAD^+$$

- oxidation of 3-hydroxyacyl-CoAs in fatty acid oxidation

$$RCH_2CHOHCH_2COSCoA + NAD^+ \rightarrow RCH_2COCH_2COSCoA + NADH + H^+$$

- deamination of amino acids

$$\text{amino acid} + NAD^+ + H_2O \rightarrow \alpha\text{-keto acid} + NADH + H^+ + NH_4+$$

- in the pentose phosphate pathway

$$\text{glucose-6-phosphate} + NADP^+ \rightarrow 6\text{-phosphogluconolactone} + NADPH + H^+$$

- in fatty acid synthesis

$$RCOCH_2COACP + NADPH + H^+ \rightarrow RCHOHCH_2COACP + NADP^+$$

- in amino acid biosynthesis:

$$\text{phenylalanine} + NADPH + H^+ \rightarrow \text{tyrosine} + NADP^+ + H_2O$$

Metabolism of nicotinic acid

Nicotinic acid deficiency in humans results in the development of the disease pellagra. This disease is seen in populations that consume

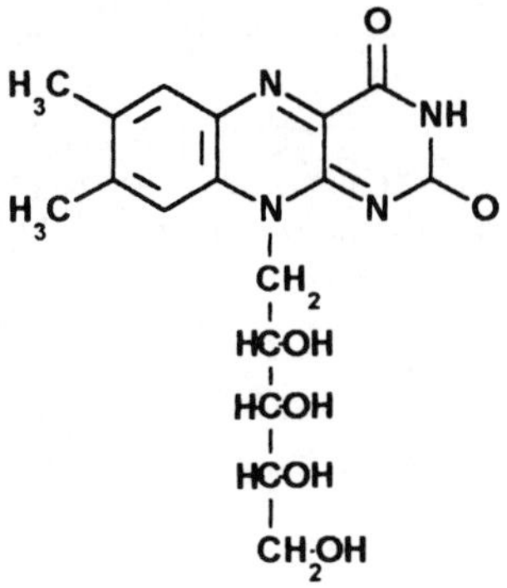

Riboflavin

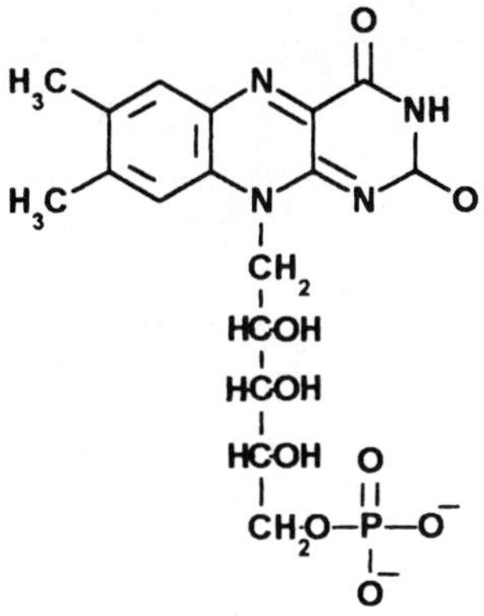

Flavin mononucleotide (FMN)

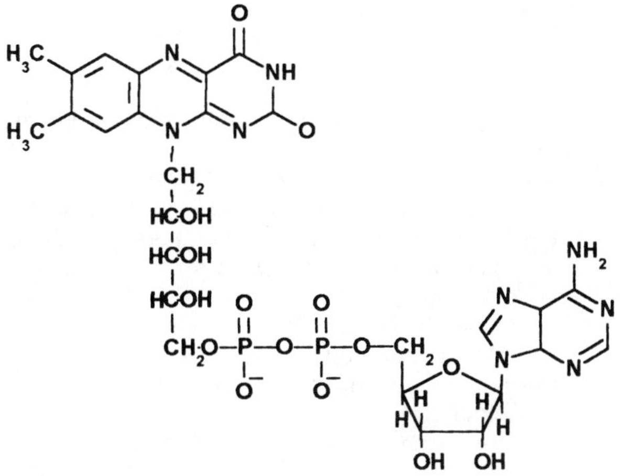

Flavin adenine dinucleotide (FAD)

Figure 8.2 The structures of riboflavin and the coenzymes flavin mononucleotide (FMN) and flavin adenine dinucleotide (FAD).

diets containing large amounts of maize. The deficiency can be overcome by dietary supplementation with nicotinic acid, high-quality protein or the amino acid tryptophan. The relationship between protein intake and nicotinic acid results from the conversion of tryptophan to nicotinic acid via kynurenine and quinolinic acid, probably in the intestinal mucosa. Both riboflavin and vitamin B_6 are required for this conversion, thus both affect the efficiency of tryptophan use. Many animals can synthesize sufficient nicotinic acid to meet their requirements if the dietary protein supply is sufficiently high in tryptophan.

Deficiency symptoms associated with nicotinic acid are summarized in Table 8.2.

Table 8.2 Summary of vitamin deficiencies in humans and animals

Vitamin	Species	Deficiency symptoms
Thiamin	Humans	'Beri-beri': rapid weight loss, muscle weakness, loss of reflexes, tachycardia, elevated blood pyruvate and lactate concentrations, reduced erythrocyte transketolase activity
	Poultry	Polyneuritis: nerve degeneration, paralysis, head retraction (opisthotonus), bradycardia
	Pigs	Loss of weight, weakness, enlargement of the heart, bradycardia
	Ruminants	Cerebrocortical necrosis (CCN), also called polioencephalomalacia (PEM): muscle weakness, loss of reflexes, head retraction, bradycardia
Riboflavin	Humans	Seborrheic dermatitis around nose and mouth, soreness and burning of lips and tongue, photophobia, corneal opacity
	Poultry	'Curled-toe paralysis' associated with degenerative changes in the myelin sheath, in severe cases causing pinching of sciatic nerve; reduced egg production and poor hatchability
	Pigs	Thickened skin, scaly dermatitis, loss of hair, lens opacities and cataracts
	Ruminants	In pre-ruminant animals: loss of appetite and hair, excessive salivation and tear production, lesions in the mouth and reddening of buccal cavity
Nicotinic acid	Humans	'Pellagra': dermal lesions and darkening of skin, nausea, vomiting and diarrhoea
	Cats and dogs	Excessive production of thick saliva, oral lesions, ulceration and bleeding of the digestive tract ('blacktongue' in dogs)
	Pigs	Loss of appetite and body weight, poor growth, rough or starring coat, anaemia and degenerative changes of the digestive tract and nervous system
	Poultry	Poor appetite and growth, red coloration of the crop, upper oesophagus and mouth, dermatitis
	Ruminants	In pre-ruminant animals, deficiency leads to anorexia, diarrhoea and dehydration
Pantothenic acid	Humans	Fatigue, headache, muscle weakness, 'burning feet syndrome'
	Poultry	Retardation of growth and feather production, dermatitis; eye lids become granular and stick together; damage to the liver and spinal cord
	Pigs	Scurvy skin and thin hair, a brownish secretion around the eyes, gastrointestinal disorders and slow growth; a characteristic symptom is the locomotor disorder which leads to stiff, jerky, exaggerated movement of the hind legs called 'goose-stepping'; *post-mortem* examination often reveals signs of nerve degeneration, fat infiltration of the liver and enlargement of the adrenals and heart
	Ruminants	In pre-ruminant animals, anorexia, reduced growth, rough coat, dermatitis
Vitamin B_6	Humans	Seborrheic lesions around eyes, nose and mouth, anaemia, vomiting, hyperirritability, convulsions
	Poultry	Poor growth and impaired plumage development, reduced egg production and hatchability, unusual excitability, trembling and stiff jerky movements eventually leading to convulsions
	Pigs	Anorexia, slow growth, anaemia and convulsions; degeneration of peripheral nerves and deposition of a dark yellow pigment seen *post-mortem*

	Ruminants	In pre-ruminant animals loss of appetite and, in severe cases, onset of convulsions
Biotin	Humans	Dermatitis, somnolence, muscle pains, hyperaesthesia
	Poultry	Classical biotin deficiency causes irregular bone growth resulting in leg abnormalities, particularly enlargement and deformation of the hock joint (perosis); severe lesions of the underside of the feet can occur; dry scaly skin is also found on the upperside of feet and legs, beak growth is abnormal; fatty liver and kidney syndrome (FLKS) in rapidly growing birds subjected to nutritional or environmental stress, associated with reduced activity of hepatic pyruvate carboxylase which limits the ability to synthesize glucose when the dietary supply of glucose is disrupted; increase in tissue palmitoleic acid (C16:1) content and a decrease in a number of the PUFAs probably caused by a reduced availability of malonyl-CoA for fatty acid elongation – birds affected by FLKS rarely show classical signs of biotin deficiency
	Pigs	Dermatitis characterized by shedding of large pieces of skin ('greasy maize flakes'); a characteristic soft hoof condition seen in pigs is 'concrete disease' so-called because the feet of animals kept on abrasive surfaces such as concrete develop severe cracking, often with secondary infections
	Ruminants	No clear link between biotin status and lameness
	Horses	Cracked and weak hooves
	Rats and dogs	Ascending paralysis, cessation of growth and a spectacled eye condition
Folic acid	All species	Anaemias, typically megaloblastic anaemia in which the red blood cells are large and immature; leucopenia, a condition in which the number of white blood cell declines; increased urinary excretion of formiminoglutamic acid (FIGLU)
Vitamin B_{12}	Humans	Pernicious anaemia, weakness, lethargy and progressive paralysis; urinary excretion of FIGLU is elevated following a histidine loading test; increased excretion of methylmalonic acid
	Ruminants and pigs	Poor growth, anaemia characterized by the presence of megaloblasts and, in severe cases, paleness of the mucus membranes; lack of posterior coordination and unsteadiness of gait; urinary excretion of FIGLU is elevated following a histidine loading test and is often accompanied by urinary excretion of methylmalonic acid – marginal vitamin B_{12} deficiency is difficult to detect and is typified by animals that are described as 'unthrifty'; in ruminants, similar symptoms can arise due to cobalt deficiency
Vitamin C	Humans, other primates and guinea pigs	'Scurvy': swollen, bleeding and ulcerated gums, loose teeth, weak bones, fragile capillaries leading to widespread haemorrhaging, increased susceptibility to infection
	Pigs and ruminants	Neonates unable to synthesize this vitamin in the first few weeks of life, and may show increased susceptibility to infection if milk supply is inadequate
	Poultry	Deficiency not normally seen; in heat-stressed birds supplementation can increase growth, egg quality and hatchability
Choline	Humans	Possible liver damage
	Other species	Fatty infiltration of the liver, necrosis or haemorrhagic lesions of the liver, kidney and joints
Carnitine	All species	Elevated levels of plasma non-esterified fatty acids and triacylglycerols suggesting impaired fatty acid oxidation and redirection of fatty acids to alternative pathways of metabolism; in severe cases the increase in

		intracellular concentrations of acyl-CoAs may impair the function of other metabolic pathways leading to liver dysfunction
Vitamin A	Humans	Night blindness, xerophthalmia, follicular hyperkeratosis
	Pigs	Uncoordination, spasm and paralysis; failure of oestrus, foetal and neonatal deformities, foetal resorption
	Poultry	Staggering gait, ruffled plummage, reduced egg production and hatchability, keratinization of mucus epithelium
	Ruminants	Nightblindness, xerophthalmia; abnormal bone remodelling leading to blindness and deafness; low conception rates, foetal resorption and retained placenta; keratinization of digestive, respiratory and reproductive tract epithelium; reduced sperm viability
Vitamin D	Humans	Rickets in growing children, osteomalacia in adults
	Pigs	Rickets in growing animals, osteomalacia in mature animals
	Poultry	Rickets, soft beak and claws, reduced egg production and hatchability, thin egg shells
	Ruminants	Rickets in growing animals, osteomalacia in mature animals; reduced milk production and inhibition of oestrus
Vitamin E	Humans	Haemolytic anaemia in infants, increased erythrocyte haemolysis in adults; few other symptoms
	Pigs	Nutritional microangiopathy characterized by haemorrhagic lesions of the heart, also known as 'mulberry heart disease'; hepatic necrosis
	Poultry	Exudative diathesis (subcutaneous oedema resulting from increased capillary permeability); encephalomalacia ('crazy chick disease'); muscular dystrophy
	Ruminants	Muscular dystrophy affecting both skeletal and cardiac muscle ('white muscle disease' in cattle, 'stiff lamb disease' in sheep
Vitamin K	All species	Reduced blood prothrombin, increased blood clotting time; in farm animals, extensive subcutaenous and internal haemorrhaging and anaemia; in ruminants 'sweet clover disease' due to presence of metabolic antagonist, dicumarol, in mouldy sweet clover

8.1.5 PANTOTHENIC ACID

Biochemical functions

Pantothenic acid (Figure 8.3) is a constituent of coenzyme A and of acyl carrier protein (ACP), both of which contain the same functional group, 4'-phosphopantotheine, formed from pantothenic acid and β-mercaptoethylamine. The terminal SH group of 4'-phosphopan-totheine forms a thioester bond with car-boxylic acids to produce coenzyme A or ACP derivatives. One coenzyme A derivative, acetyl-CoA, occupies a particularly important role in intermediary metabolism (see Chapters 11 and 19).

Pantothenic acid has a variety of metabolic functions, and deficiency results in a wide range of symptoms including growth and reproductive failures, skin and hair lesions, gastrointestinal disorders and lesions of the nervous system (see Table 8.2).

8.1.6 PYRIDOXINE, PYRIDOXAL AND PYRIDOXAMINE (VITAMIN B$_6$)

This vitamin exists in three forms which are interconvertible in body tissues. The parent compound is pyridoxine which is converted to pyridoxal and pyridoxamine (Figure 8.3).

Biochemical functions

Pyridoxal phosphate and pyridoxamine phos-phate act as coenzymes in many enzymes of amino acid metabolism, e.g. transaminases, amino acid deaminases and amino acid decar-boxylases. In transaminase reactions, pyridox-al phosphate acts as the carrier of the amino

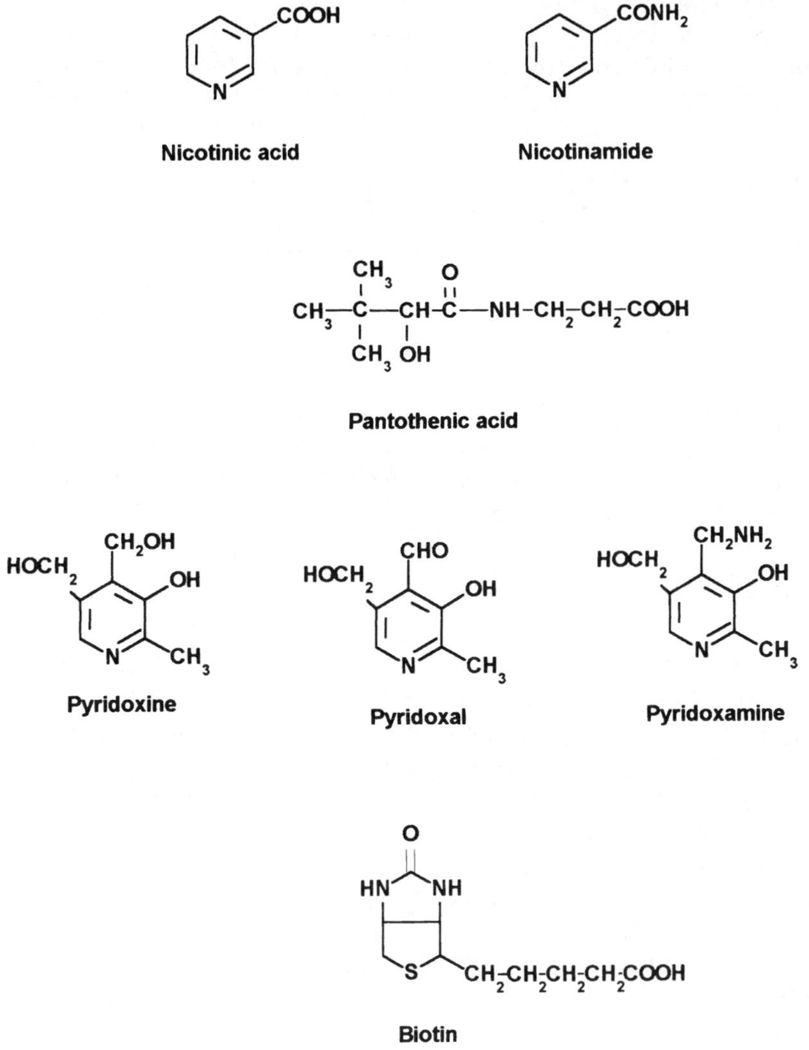

Figure 8.3 The structures of nicotinic acid, nicotinamide, pantothenic acid, pyridoxine, pyridixal, pyridoxamine and biotin.

group as it is transferred from an amino acid to an α-ketoacid. In so doing it is transformed to pyridoxamine phosphate.

- transaminase (aminotransferase) reactions:

  L-alanine + α-ketoglutarate →
  pyruvate + L-glutamate

- non-oxidative deamination reactions:

  serine → pyruvate + NH_3
  threonine → α-ketobutyrate + NH_3

- non-oxidative decarboxylation reactions:

  histidine → histamine + CO_2

Vitamin B_6 has a number of other important biochemical functions. For example:

- in the synthesis of neurotransmitters adrenaline, noradrenaline, dopamine, histamine and γ-aminobutyric acid;
- in the synthesis of protoporphyrin IX found in haemoglobin and myoglobin;

- in glycogenolysis where it is a coenzyme for glycogen phosphorylase;
- in the formation of nicotinic acid from tryptophan.

Because many enzymatic reactions are dependent on vitamin B_6 a wide variety of abnormalities are observed in the deficient animal (see Table 8.2).

8.1.7 BIOTIN

Biochemical functions

Biotin (Figure 8.3) functions as a coenzyme in carboxyl group transfer reactions. In addition, biotin may have functions in the synthesis of proteins, particularly keratins, which cannot be explained entirely on the basis of its role in carboxylation reactions.

Metabolism of biotin

Biotin is found in nature both as the free form and in a form bound to an amino acid, usually lysine (biocytin). In animals, it is readily absorbed from the small intestine in either the free or bound form. Little is known about the metabolism of biotin in humans or animals. Several biotin-binding proteins are produced by microorganisms, e.g. streptavidin and stravidin from *Saccharomyces avidinii*. Egg white contains a similar biotin-binding protein, avidin. Administration of these binding proteins by mouth greatly reduces the availability of dietary biotin, apparently by preventing absorption across the intestinal mucosa.

Deficiency can easily be induced in most animals by feeding raw egg white. Symptoms of biotin deficiency common to all species are retarded growth, dermatitis, loss of hair and disturbances of the nervous system (see Table 8.2).

8.1.8 FOLIC ACID

The structure of folic acid is shown in Figure 8.4. It consists of a pteridine ring linked via carbon number 9 to a *p*-aminobenzoyl residue which is, in turn, linked to poly-γ-glutamate, where the number of glutamate residues varies up to a maximum of eight. The term folic acid or folate is commonly used to mean any one of the various conjugated (glutamylated) forms, although by strict definition folic acid contains only one glutamate residue. The pteridine ring is usually in a reduced form, with hydrogens on carbons 6 and 7 and nitrogens 5 and 8 giving tetrahydrofolic acid.

Biochemical functions

Folic acid has a number of coenzyme forms which occur mainly as polyglutamates in animal tissues. The main functions of tetrahydrofolic acid are in the transfer of one-carbon units in degradative and biosynthetic reactions. The coenzyme forms found in tissues are 5-methyl-tetrahydrofolate; 5,10-methylene-tetrahydrofolate; 5,10-methenyl-tetrahydrofolate; 5-formyl-tetrahydrofolate; 10-formyl-tetrahydrofolate and 5-formimino-tetrahydrofolate.

Some of the processes which require the transfer of one-carbon groups mediated by tetrahydrofolate are:

- glycine, histidine and tryptophan degradation;
- conversion of serine to methionine;
- synthesis of inosine, a precursor of purines, adenine and guanine;
- synthesis of the pyrimidine thymidine (Figure 8.4).

Metabolism of folic acid

Dietary folic acid occurs mainly as polyglutamates. The vitamin is absorbed in the duodenum and the jejunum, apparently by an active process which is stimulated by glucose. During passage through the intestinal mucosa, the polyglutamate derivatives are hydrolysed to the monoglutamate and may undergo reduction and methylation to produce 5-methyl-tetrahydrofolate. These absorbed forms enter

(a)

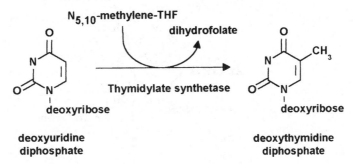

Folic acid

5,6,7,8-tetrahydrofolic acid
(THF)

(b)

N$_{5,10}$-methylene-THF

dihydrofolate

Thymidylate synthetase

CH$_3$

deoxyribose

deoxyribose

deoxyuridine
diphosphate

deoxythymidine
diphosphate

Figure 8.4 The structures of folic acid and tetrahydrofolic acid (THF). Folic acid contains a single glutamate residue. Tetrahydrofolate and its derivatives can occur in the polyglutamate form in which *n* may = 1–8. The conversion of deoxyuridine diphosphate (dUDP) to deoxythymidine diphosphate (dTDP) showing the transfer of a one-carbon unit from $N^{5,10}$-methylene-THF.

the blood stream and are transported to the liver and peripheral tissues. Mammalian tissues can synthesize the pteridine ring and polyglutamate but are unable to link these together through *p*-aminobenzoic acid.

Deficiency symptoms are mainly associated with abnormalities in the formation of red and white cells and disturbances of nucleic acid synthesis (see Table 8.2).

8.1.9 VITAMIN B$_{12}$

Vitamin B$_{12}$ has the most complex structure of all vitamins (Figure 8.5) and was the last to be isolated and identified. The term vitamin B$_{12}$ is a generic term used to refer to a group of compounds characterized by the presence of a tetrapyrrole ring, in which the inner nitrogen atom of each pyrrole ring is coordinated to a single atom of cobalt. The commercially available form of vitamin B_{12} is cyanocobalamin, in which a cyanide group is attached to the cobalt atom. Cyanocobalamin, although naturally occurring, is not an important form of the vitamin. In the three most common forms found in tissues and feedstuffs, the cyanide group is replaced by a 5'-deoxyadenosyl group to produce 5'-deoxyadenosyl cobalamin or coenzyme B_{12}; by a methyl group to produce methylcobalamin; or by a hydroxyl group to produce hydroxocobalamin.

Biochemical functions

In microorganisms, 5'-deoxyadenosylcobalamin is involved in a number of rearrangement reactions. However, in animals only one of this type of vitamin B_{12}-dependent reactions is known to occur. This is the conversion of methylmalonyl-CoA to succinyl-CoA (Figure 8.6). This reaction occurs in all animals and is a useful route by which propionate derived from odd-numbered fatty acid oxidation and amino acid catabolism can be directed to glucose synthesis. It is of particular importance in ruminants in which little or no dietary glucose is absorbed from the digestive tract, and glucose supply is largely dependent on gluconeogenesis from compounds such as propionate and amino acids.

Only one other vitamin B_{12}-dependent reaction occurs in mammalian tissues. In this reaction, catalysed by 5-methyl-tetrahydrofolate homocysteine methyltransferase, methylcobalamin acts as a methyl group carrier in the conversion of homocysteine to methionine (Figure 8.6).

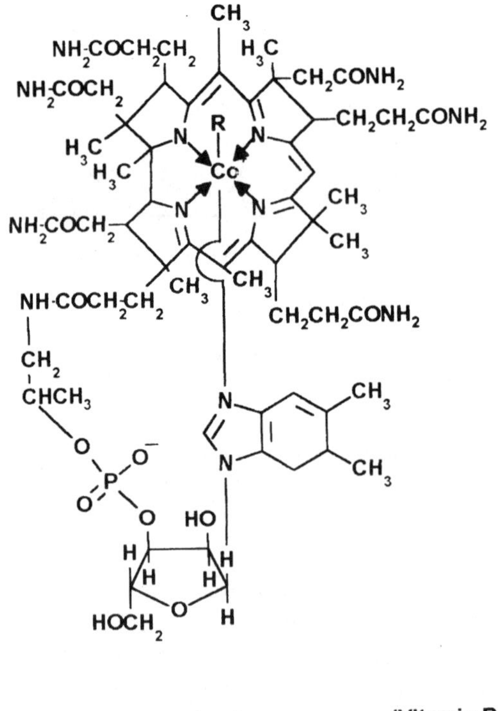

R = CN	Cyanocobalamin	(Vitamin B12)
R = OH	Hydroxocobalamin	(Vitamin B12a)
R = CH3	Methylcobalamin	(methyl B12)
R = 5'-deoxyadenosine	5'-deoxyadenosylcobalamin	(coenzyme B12)

Figure 8.5 Vitamin B_{12} and its derivatives.

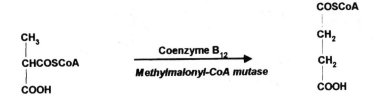

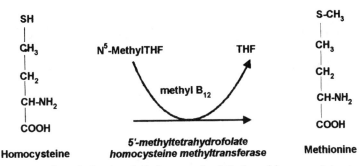

Figure 8.6 The conversion of methylmalonyl-CoA to succinyl-CoA and homocysteine to methionine.

Metabolism of vitamin B_{12}

The vitamin is relatively slowly absorbed from the small intestine and its uptake is an active process dependent on the presence of a glycoprotein called intrinsic factor which is produced by the parietal cells of the stomach (abomasum). Once absorbed into the blood stream it is attached to specific transport proteins called transcobalamins. Transfer of the vitamin from transcobalamins to the intracellular location is receptor-mediated. Within the cell most vitamin B_{12} is found in association with the two vitamin B_{12}-dependent enzymes, methylmalonyl-CoA mutase and 5-methyl-tetrahydrofolate methyltransferase.

Deficiency symptoms associated with vitamin B_{12} are listed in Table 8.2.

8.1.10 VITAMIN C

The active forms of vitamin C are the L isomers of ascorbic and dehydroascorbic acid shown in Figure 8.7.

Biochemical functions

In humans, a deficiency of vitamin C is classically linked with the development of the disease scurvy, in which connective tissue synthesis is impaired. Amino acid analysis of collagen, the principal protein in connective tissue, shows that it contains a significant proportion of the amino acid hydroxyproline. This amino acid is synthesized from proline by an ascorbate-dependent hydroxylation reaction. Ascorbic acid is also involved in the hydroxylation of lysine. The resulting hydroxylysine is further glycosylated and forms cross-links between collagen fibres.

A number of other hydroxylation reactions require ascorbic acid, for example:

- the ascorbate-dependent copper-containing enzyme dopamine β-hydroxylase catalyses the hydroxylation of dopamine in the synthesis of noradrenaline;
- two ascorbate-dependent enzymes, trimethyllysine hydroxylase and γ-butyro-

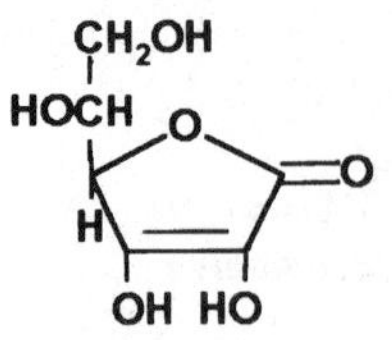

Ascorbic acid

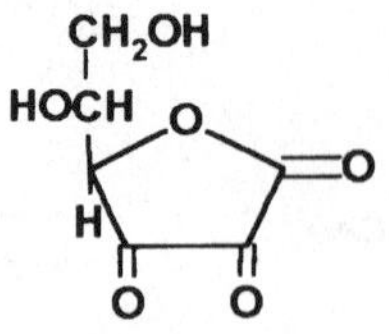

Dehydroascorbic acid

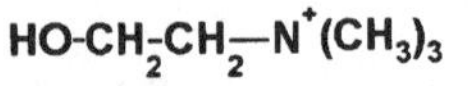

Choline

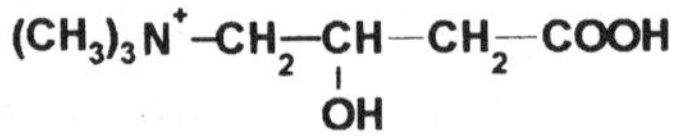

Carnitine

Figure 8.7 The structures of ascorbic acid, dehydroascorbic acid, choline and carnitine.

betaine hydroxylase are required for the synthesis of carnitine;

- ascorbic acid is a coenzyme for peptidyl glycine hydroxylase required for the amidation of a number of neuropeptides.

Other roles of ascorbic acid relate to its ability to act as a reducing agent and antioxidant. Dietary ascorbate has beneficial effects on iron absorption from the small intestine. Dietary iron is absorbed in the Fe(II) state and not as Fe(III). The presence of ascorbate ensures that iron is maintained in its reduced state, and has the added effect that it acts as a chelating agent thereby increasing the efficiency of iron absorption. As an antioxidant ascorbate has two roles. Firstly, it protects against losses of the lipid-phase antioxidant, vitamin E (α-tocopherol), by promoting the regeneration of α-tocopherol from the α-tocopheryl radical produced during peroxidation of unsaturated fatty acids. Secondly, ascorbate is an important aqueous-phase antioxidant, providing protection against free radical species in the cytoplasm.

Metabolism of vitamin C

Vitamin C can be synthesized by all microorganisms and plants, and most animals. The process involves the conversion of hexose sugars (mainly glucose and galactose) to ascorbic acid. Certain insects, fish, birds, flying mammals, guinea pigs and primates lack L-gulonolactone oxidase, the penultimate enzyme in this pathway, and are dependent on an endogenous supply.

The requirement for vitamin C changes with physiological and health status due to increased turnover of the vitamin. For example, increased metabolism occurs during bacterial infection when vitamin C is required to neutralize the free radicals produced by phagocytes to kill microorganisms.

Typical deficiency symptoms associated with vitamin C are listed in Table 8.2

8.1.11 CHOLINE

Choline (Figure 8.7) is commonly classified as a vitamin, although it does not entirely conform to the classical definition of a vitamin as it can be synthesized by most animal species, is required in larger quantities than most vitamins, and has a structural rather than a coenzyme function.

Biochemical functions

Choline has three main biochemical functions.

- It is a part of the phospholipids phosphatidylcholine and sphingomyelin, both of which are structural components of biological membranes. The details of the synthesis of these lipids are described in Chapter 19.
- It is required for the synthesis of the neurotransmitter acetylcholine.
- It is a source of methyl groups for transmethylation reactions, and serves as a source of betaine which performs a similar function.

Metabolism of choline

Choline may be present in the diet both in the free form and as phosphatidylcholine. In the free form about two-thirds is converted to trimethylamine by the intestinal microflora and excreted via the urine. The remaining one-third is absorbed directly into the blood stream via the jejunum and ileum. The digestion of phosphatidylcholine is described in Chapter 28.

Synthesis of phosphatidylcholine and sphingomyelin occurs in most tissues, as described in Chapter 19. Conversion to acetylcholine is a highly active process in nervous tissue and is catalysed by the enzyme choline acetyltransferase (CAT).

choline + acetyl-CoA → acetylcholine + CoA

Betaine synthesis is active in many tissues, particularly in the liver and kidney, and is used in the transmethylation of homocysteine to methionine by a reaction which does not require vitamin B_{12} or folic acid.

8.1.12 CARNITINE

The structure of carnitine (3-hydroxy-4-N-trimethylaminobutyric acid) is shown in Figure 8.7. Like choline, carnitine does not fit the classical definition of a vitamin.

Biochemical functions

Carnitine is required for the oxidation of fatty acids described in detail in Chapter 13. Fatty acids destined for β-oxidation are activated in the cytoplasm to yield fatty acyl-CoAs. As fatty acyl-CoAs cannot cross the inner mitochondrial membrane, they are first converted to fatty acyl-carnitines which are transported into the mitochondrial matrix via a specific membrane translocase, prior to a second exchange reaction which regenerates fatty acyl-CoAs. This is the only known function of carnitine in animal tissues.

Metabolism of carnitine

Carnitine is present in food of animal origin but little or no carnitine is found in plant foods. Little is known about its digestion or bioavailability and it is assumed that it is readily absorbed across the small intestine. The major sites of synthesis in animal tissues are the liver and kidney. The main precursor for synthesis is the amino acid lysine. This pathway is interesting in that it starts from protein-incorporated lysine and not free lysine. The lysine residue undergoes transmethylation to form trimethyllysine which is then released by proteolysis. The subsequent conversion of trimethyllysine to carnitine involves two hydroxylation steps, both of which are ascorbic acid-dependent. Dietary deficiency of lysine, methionine or ascorbic acid may therefore increase the requirement for carnitine supplementation.

8.1.13 VITAMIN A

Vitamin A exists in three forms: retinol, retinal (retinaldehyde), and retinoic acid. The structures shown in Figure 8.8 are the all-*trans* isomers which are quantitatively the most important forms. Vitamin A does not occur in plant materials but is present in its precursor forms, the carotenes. There are a number of different type of carotene and carotene derivatives, some of which have considerable potency as vitamin A precursors; however, quantitatively the most important source of vitamin A is β-carotene (see Figure 8.8).

Biochemical functions

Vitamin A has an important role in vision. It also appears to function as a regulator of the differentiation and metabolism of cells. Its role

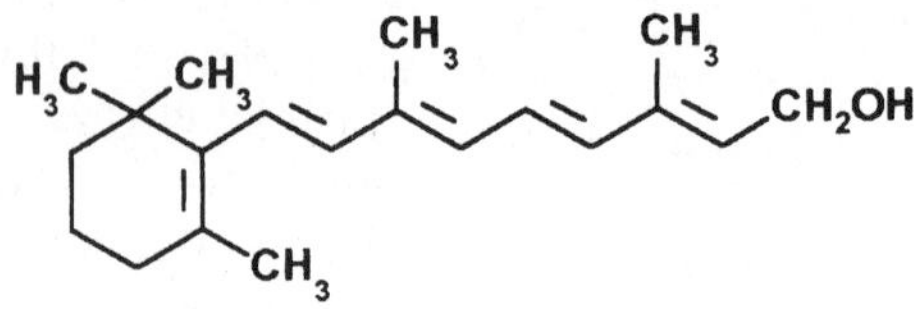

Retinol

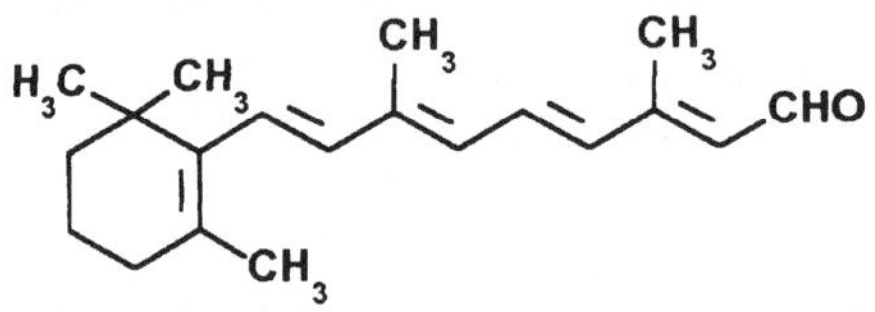

Retinal (Retinaldehyde)

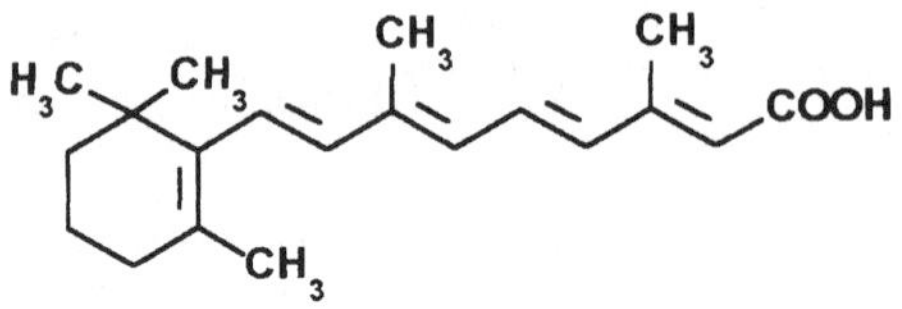

Retinoic acid

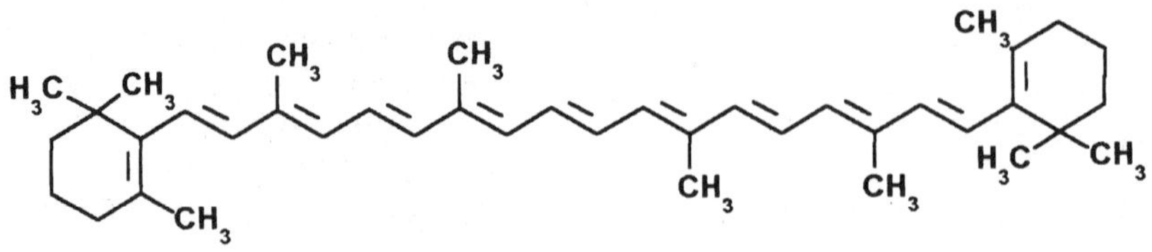

β-Carotene

Figure 8.8 The structures of retinol, retinal (retinaldehyde), retinoic acid and β-carotene.

in visual processes is now well characterized. Its other functions are less well understood.

Vitamin A and vision

The retina contains cells called rods and cones. The former are concerned with vision in dim light, while the latter are involved in vision in bright light and are responsible for colour vision. The characteristic pigments of rods and cones are conjugated carotenoid proteins known, respectively, as rhodopsin and iodopsin. They differ only in the opsin or protein moieties. In rods, opsin (an intrinsic transmembrane protein) makes up about 90% of the proteins in the specialized disk membranes. The carotenoid common to both rhodopsin and iodopsin is 11-*cis*-retinal.

Rhodopsin, which is essential to night vision, is a bright red pigment. On exposure to light it bleaches, resulting in the conversion of 11-*cis*-retinal to all-*trans*-retinal (a yellow compound) and its release from the membrane-bound protein opsin. All-*trans*-retinal is reduced by alcohol dehydrogenase to the colourless all-*trans*-retinol. The latter is isomerized to 11-*cis*-retinol, reduced by alcohol dehydrogenase to 11-cis-retinal and, in the dark, re-incorporated into rhodopsin (Figure 8.9). During these transformations some vitamin A is lost. Thus a deficiency of the vitamin eventually results in the inability to see in dim light and the development of the condition commonly known as night blindness, the earliest symptom of vitamin A deficiency in humans.

Metabolism of vitamin A

Both vitamin A and β-carotene are absorbed from the small intestine with the fat fraction of the diet. Conditions that impair fat absorption decrease the absorption of vitamin A and β-carotene. The majority of the β-carotene absorbed by the intestinal mucosa is cleaved by the enzyme 15,15'-β-carotene dioxygenase to yield two molecules of retinal which are then reduced to retinol by retinaldehyde reductase. Most retinol is then esterified to palmitic acid and incorporated into chylomicrons which are released into the blood stream via the lymphatic system. The ability to transfer absorbed dietary carotenoids from the intestinal mucosa to blood and the tissues varies from species to species. For example, humans, birds and ruminants can absorb both vitamin A and the carotenoids whereas pigs and rats can absorb only vitamin A.

Vitamin A is usually present in foods in an esterified form as retinyl palmitate. These esters are hydrolysed in the lumen of the small intestine. Free retinol is absorbed by the intestinal mucosa where it is usually re-esterified to palmitic acid and incorporated into lymph lipoproteins for transport to the blood stream.

Retinyl esters are stored mainly in the liver. Absorbed carotenoids are stored in liver and adipose tissue. The presence of carotenoids gives rise to a yellow coloration of the fat depots. During mobilization of stored vitamin A from the liver, retinyl esters are hydrolysed enzymatically. The retinol is transferred to a retinol-binding protein (RBP) prior to release into the blood stream for distribution to other tissues. Uptake of RBP-bound retinol is a membrane receptor-mediated process. Once absorbed into the cell the retinol is rapidly transferred to cellular retinol-binding proteins (CRBPs) which may serve to protect the retinol from oxidation and direct its further metabolism.

Deficiency symptoms

Numerous deficiency symptoms have been described in humans and farm animals. In general these fall into three main areas: visual disorders; abnormalities in cell growth and differentiation; and keratinization of epithelial tissues (see Table 8.2).

8.1.14 VITAMIN D

Vitamin D, sometimes called the sunshine vitamin, is primarily known as the vitamin which

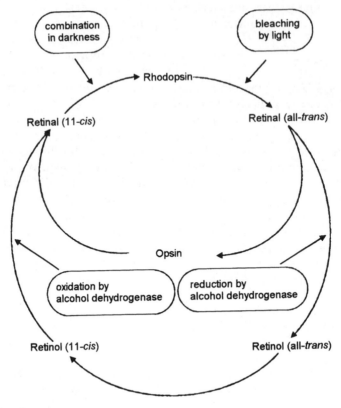

Figure 8.9 The visual cycle.

prevents the development of rickets. There are two forms of vitamin D, ergocalciferol (vitamin D_2) and cholecalciferol (vitamin D_3) and, although it has been shown that about 10 sterol derivatives have some vitamin D activity, these are the two important dietary sources of the vitamin (Figure 8.10).

Biochemical functions

The most important functions of the active metabolite of vitamin D, 1,25-dihydroxycholecalciferol, are in the regulation of calcium and phosphorus metabolism. It stimulates the synthesis of a calcium-binding protein in the intestinal epithelium which promotes the absorption of dietary calcium, acts on the distal renal tubule to increase resorption of calcium, and stimulates the mobilization of calcium

from bone. It is essential for the normal calcification of bone. An adequate supply of vitamin D is particularly important in young animals where bone growth is rapid. It is less important in adult animals where bone mass is relatively stable, but females are particularly susceptible to changes in vitamin D status during pregnancy and lactation.

The mode of action of 1,25-dihydroxy vitamin D is not yet fully understood. It is thought that in the small intestine the vitamin binds to receptor sites in the plasma membrane, and that the vitamin–receptor complex is transported to the nucleus, where it initiates the production of mRNA which codes for the synthesis of the calcium-binding protein calbindin-D. In many respects the action of the vitamin is very similar to that of the steroid hormones, and it has been suggested that 1,25-

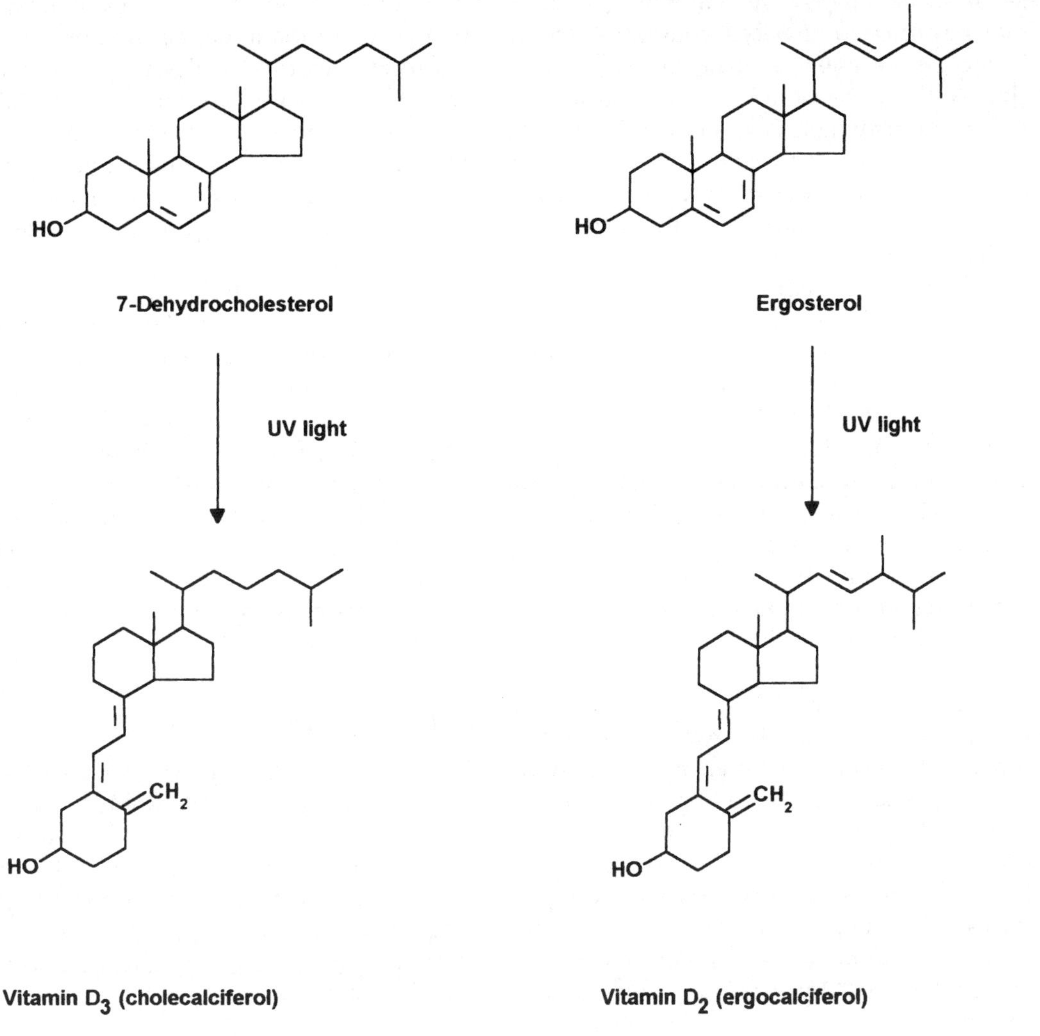

7-Dehydrocholesterol

UV light

Vitamin D₃ (cholecalciferol)

Ergosterol

UV light

Vitamin D₂ (ergocalciferol)

Figure 8.10 The conversion of 7-dehydrocholesterol and ergosterol to vitamin D$_3$ (cholecalciferol) and D$_2$ (ergocalciferol), respectively, under the influence of ultraviolet light.

dihydroxy vitamin D should be reclassified as a hormone and not a vitamin.

Metabolism of vitamin D

The sterols 7-dehydrocholesterol and ergosterol are important precursors of vitamin D. Both the precursors and the vitamin itself (cholecalciferol and ergocalciferol) are absorbed in the small intestine and, like vitamin A, any factor which interferes with or enhances the absorption of fat influences the ability of animals to absorb the sterol precursors or the vitamin. The body has some ability to store the vitamin in liver and adipose tissue but to a much lesser extent than vitamin A.

The effect of sunlight on the body is an important factor in the production of vitamin D. Skin and sebaceous secretions contain 7-dehydrocholesterol and some ergosterol of dietary origin. These sterols are converted to vitamin D$_3$ and D$_2$, respectively, by exposure to ultraviolet light. The vitamins produced may be absorbed directly through the skin or, in the case of many animals, licked from the skin and absorbed through the gastrointestinal tract. Irradiation is less effective in dark-pigmented skin or through a heavy coat of hair. The effectiveness of irradiation depends on the wavelength and intensity of the ultraviolet light incident on the body. Thus, it is most effective in the tropics, in the summer, at noon and at high altitudes. These variations are of great importance in vitamin D nutrition. Animals which are at pasture during the summer never suffer from deficiency even though their diet is practically devoid of the vitamin. This is not the case in winter in temperate climates, when animals may be outside only for short periods, there are generally fewer sunny days and the sunlight is less intense. Thus during winter months in the temperate regions of the world it is necessary to supplement livestock feeds.

Only recently has it become clear that vitamin D is further metabolized. The first step occurs in the liver where the vitamin is hydroxylated by α-25-hydroxylase to produce 25-OH-vitamin D, which is more active than the unhydroxylated form. The second step occurs in the kidney where a second hydroxylation produces 1,25-dihydroxy-vitamin D. This is the most active form of the vitamin. The kidney also contains a 24-hydroxylase which catalyses the formation of 24,25-dihydroxy-vitamin D. The functions of this metabolite are still unclear. It was initially thought to be a mechanism of excretion of excess vitamin D; however, recent research suggests that it has a role in bone mineralization and may suppress the secretion of parathyroid hormone. Elimination of vitamin D from the body requires conversion of 1,25-dihydroxy vitamin D to 1,24,25-trihydroxy vitamin D, followed by formation of more polar derivatives in the liver, transfer to bile, and excretion via the faeces.

The primary symptom of vitamin D deficiency is rickets, which is caused by incomplete calcification of the bones. Bone formation proceeds via mineralization of the organic cartilage matrix. Analysis of rachitic bones reveals a decreased concentration of calcium and phosphorus leading to a reduction in physical strength. In young, rapidly growing animals the bony framework of the body is unable to support the growing muscle mass. This is particularly pronounced in the long bones which tend to remain soft and bend under the weight of the body. The principal cause of rickets is not a defect in the process of calcification but rather a lack of calcium, the major substrate for the mineralization of the bone matrix. Bone is a dynamic tissue and although bone growth is greatest in the early stages of life, even in adult animals bone is in a constant state of turnover. Lack of vitamin D in adult life leads to a loss of calcium from the bone matrix resulting in the condition known as osteomalacia.

8.1.15 VITAMIN E

Vitamin E is the generic term used to describe a group of related compounds, the tocopherols and tocotrienols. Each exists in a number of isomeric forms designated α, β, γ and δ isomers. The α isomers are shown in Figure 8.11. The biopotency of the isomers is expressed in relation to α-tocopherol which is the most abundant and active form.

Biochemical functions

The vitamin was first known as the antisterility vitamin following studies carried out in rats; however, the relationship between vitamin E and fertility does not hold for all species. Vitamin E deficiency has never been linked with fertility problems in humans or ruminants.

Perhaps the most important function of α-tocopherol in plant and animal tissues is as an antioxidant. It is found in membranes in close association with PUFAs, which are particularly susceptible to oxidation. Peroxides arise due to

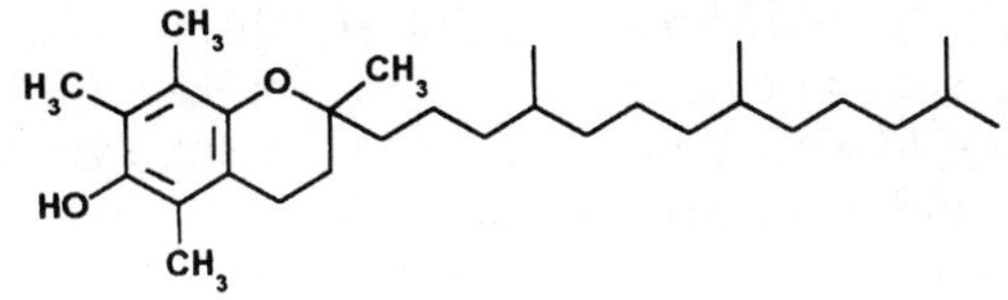

α-Tocopherol

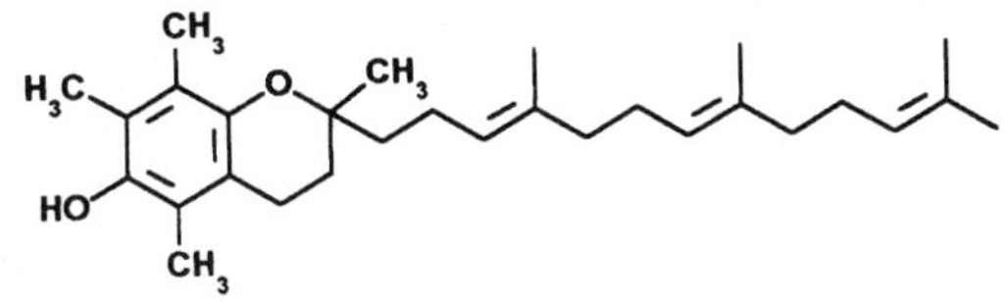

α-Tocotrienol

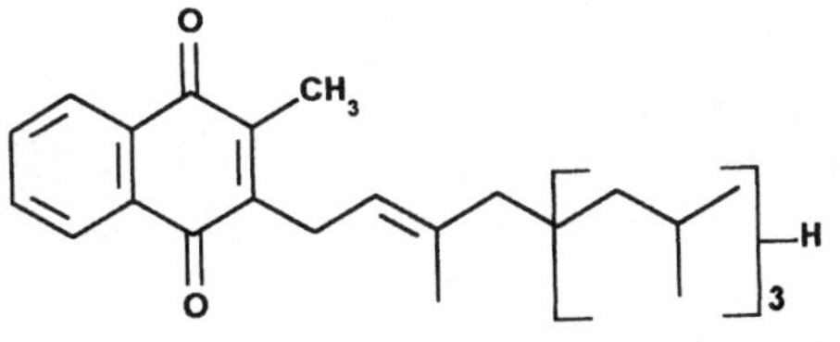

Vitamin K_1 (Phylloquinone)

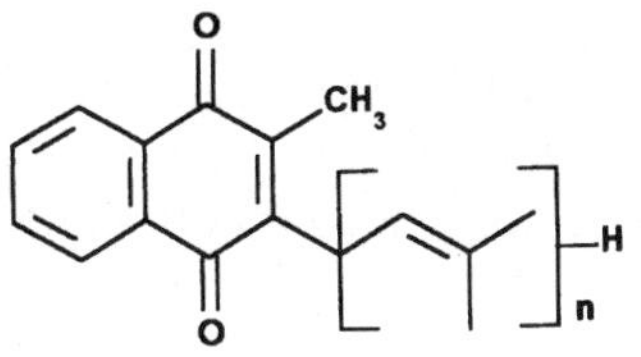

Vitamin K_2 (Menaquinone)

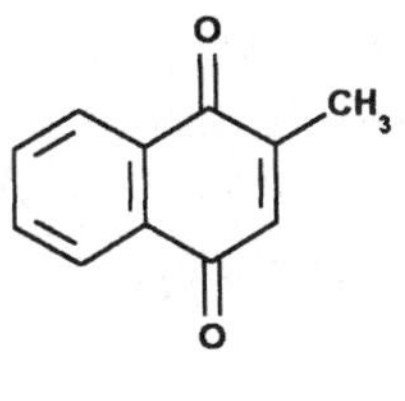

Vitamin K_3 (Menadione)

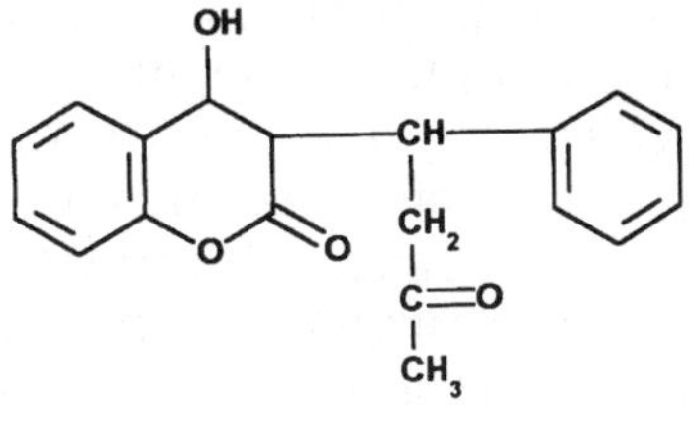

Warfarin

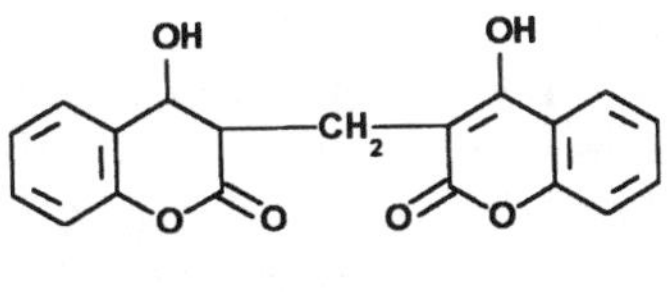

Dicoumarol

Figure 8.11 The structure of α-tocopherol; α-tocotrienol; the various forms of vitamin K: vitamin K_1 (phylloquinone), vitamin K_2 (menaquinone), vitamin K_3 (menadione); warfarin; and dicoumarol.

the reaction of PUFAs with reactive oxygen species such as superoxide ($O_2^{\cdot-}$) and hydroxyl ($OH^{\cdot}$) radicals. They also occur as intermediates in the conversion of PUFAs to eicosanoids. Lipid peroxides are unstable, and spontaneously decompose to form free radicals which initiate an autocatalytic cycle of membrane-lipid oxidation. If uncontrolled, this process can have damaging effects on membrane structure and function *in vivo* and can lead to development of rancidity in lipid-rich foods such as meat and oils. α-Tocopherol prevents uncontrolled lipid peroxidation by acting as a chain-breaking antioxidant which reacts with peroxide-propagating radicals and is converted to the relatively stable α-tocopheroxyl radical.

The antioxidant properties of α-tocopherol are important in agricultural food products. The shelf life of meat can be extended by increasing its α-tocopherol content, which slows the process of tissue fatty acid oxidation and delays the development of 'off flavours' and taints. In addition, dietary supplementation with supranutritional levels of α-tocopherol inhibits the oxidation of meat pigments, principally myoglobin, thereby maintaining the fresh meat colour without the need to add artificial colorants.

There is increasing interest in the role of α-tocopherol in the immune response. Lymphocytes and mononuclear cells have been shown to contain the highest concentration of α-tocopherol of all the cells in the body. Experiments have shown that supplementation of animal diets with α-tocopherol results in an increase in lymphocyte transformation in response to mitogen stimulation, increased antibody production and greater resistance to pathogenic organisms. Some of these effects may be mediated by the influence of α-tocopherol on eicosanoid metabolism and by changes in membrane fluidity.

Metabolism of vitamin E

Tocopherols are absorbed via the small intestine and transported to the blood stream via the lymph system. Any esters of vitamin E,

such as α-tocopherol acetate, are hydrolysed prior to absorption, probably by mucosal esterase. Unlike retinol, there does not appear to be a specific transport protein for α-tocopherol which is found mainly in the lipoprotein fraction. α-Tocopherol is the major form of vitamin E deposited in tissues: major reserves are found in liver, adipose tissue and muscle. There is very little further metabolism of the vitamin.

Vitamin E deficiency results in a wide range of symptoms in farm animals but is poorly characterized in humans except in the neonate, particularly the pre-term infant. The main effects of deficiency are listed in Table 8.2. Because its antioxidant function is part of an integrated antioxidant defence mechanism including glutathione peroxidase (a selenium-containing enzyme), many of the symptoms of vitamin E deficiency can be induced by selenium deficiency and alleviated by selenium supplementation.

8.1.16 VITAMIN K

Vitamin K occurs naturally in two main forms, vitamin K_1 (phylloquinone), found in plants, and vitamin K_2 (menaquinone) present in bacteria and animals. A third compound, menadione, has some vitamin K activity and is sometimes referred to as vitamin K_3. Vitamins K_1 and K_2 contain phytyl and polyisoprenoid side chains, respectively (Figure 8.11).

Biochemical functions

Vitamin K is important in regulating the synthesis of calcium-binding proteins containing γ-carboxyglutamic acid. Some of these are important in blood clotting and have been identified as factors II (prothrombin), VII, IX and X. The synthesis of prothrombin involves the post-translational modification of a precursor protein. In vitamin K deficiency this precursor undergoes incomplete glutamate carboxylation, resulting in the production of preprothrombin. However, in animals with adequate vitamin K, the precursor is converted to prothrombin which contains ten γ-carboxy-glutamic acid residues in the 40 amino acids at the amino-terminal end. The precise role of vitamin K in the enzyme mechanism is still not clear; however, in all species a deficiency of this vitamin leads to a fall in prothrombin concentration in blood and the incidence of widespread haemorrhaging (Table 8.2). Vitamin K may have a similar role in skeletal tissue where the protein osteocalcin also contains γ-carboxyglutamic acid. This protein appears to have a role in the mineralization of bone and is present in higher concentrations in rapidly growing animals.

A number of analogues and antagonists to vitamin K have been developed for medical and pest-control purposes. The best known of these are a family of coumarin derivatives which have been used successfully in anticoagulation therapy. One of the most widely used vitamin K antagonists is the rodenticide warfarin. There is concern that rat populations are now developing resistance to the effects of warfarin, which has led to the development of a number of other potent coumarin-based rodenticides.

Metabolism of vitamin K

The main site of absorption is the small intestine, and the vitamin is transported to the blood stream via the lymph. There is little evidence of any specific carrier protein for transport in the blood, and most vitamin K in blood is associated with the lipoprotein fraction. Bacteria can synthesize menaquinone so that ruminants, and those non-ruminants with well developed caeca, can synthesize much of their requirements. There is little storage of the vitamin in body tissues, but it is found in highest concentrations in the liver.

8.2 MINERALS IN BIOCHEMISTRY

8.2.1 INTRODUCTION

Many mineral elements have important functions in living organisms. In this chapter their functions in animals are described; their role in plants is considered in Chapter 26. Animals require some elements in quite large quantities where they may have a structural role as well as important biochemical functions. These are referred to as the major mineral elements. It is the convention to consider calcium, chlorine, magnesium, phosphorus, potassium, sodium and sulphur as the major mineral elements. These are usually present in tissues in quantities varying from 15 g kg^{-1} tissue for calcium to 0.4 g kg^{-1} for magnesium.

The remaining mineral elements are usually referred to as trace elements. As analytical techniques become more sensitive the list of trace elements present in living tissues is growing. Some of these trace elements have clearly identified biochemical functions: thus it is clear that cobalt, copper, iodine, iron, manganese, molybdenum, selenium and zinc are involved in specific biochemical processes. Other trace elements such as nickel, chromium and vanadium have metabolic roles that are only now being discovered. The concentration of trace elements varies considerably, from iron, which may be present in tissues at between 20 and 80 mg kg^{-1} tissue, to cobalt present at concentrations as low as 0.02–0.1 mg kg^{-1}. Some elements may be toxic at high concentrations. In some cases, such as copper, molybdenum and selenium, there is a fine line between normal and toxic concentrations, and the toxic concentration may show considerable animal species variation. For example, levels of copper normally included in pig diets would be toxic if fed to sheep.

8.2.2 CALCIUM

Calcium is the most abundant mineral found in vertebrates. Bone constitutes the major body reserve which can be mobilized when calcium intake is low.

In addition to its structural role, calcium has numerous metabolic functions. It acts as an important intracellular messenger involved in the regulation of a number of enzymes. For example, in contracting muscle the calcium activation of glycogen phosphorylase, pyruvate dehydrogenase, isocitrate dehydrogenase and α-ketoglutarate dehydrogenase serves to increase glycogen breakdown and glucose oxidation to provide the ATP required for muscle contraction. Many of the functions of calcium are mediated via the multipurpose calcium-binding protein calmodulin, and may be initiated hormonally via activation of plasma membrane G proteins (see Chapter 22).

In nervous tissue, calcium is involved in transmission of nerve impulses. Depolarization of the presynaptic neural membrane results in an influx of calcium through pores known as voltage-gated calcium channels. The increase in intracellular calcium concentration promotes the fusion of acetylcholine vesicles with the presynaptic membrane, resulting in the release of acetylcholine into the synaptic cleft.

Other processes in which calcium is involved are blood clotting, control of gap junction closure between interconnecting cells, and cell-to-cell adhesion.

Abnormalities in calcium metabolism can occur due to poor dietary intake or to physiological and metabolic disturbances. In young humans and animals, rickets results from incomplete mineralization of bones leaving them soft and unable to support the growing body. In adults, abnormal calcium resorption and deposition in mature bones can lead to conditions such as osteoporosis and osteomalacia. In lactating females, particularly high-yielding dairy cows, the sudden loss of calcium in milk at the onset of lactation can lower plasma calcium levels dramatically and lead to the condition known as milk fever (parturient hypocalcaemia) which can be fatal if untreated.

8.2.3 PHOSPHORUS

Phosphorus is widespread in living organisms and has functions in more biochemical processes than any other mineral. It is vital to energy transduction and oxidation–reduction reactions in all organisms due to its role in the phosphorylated nucleotides, ATP and GTP. It is a component of nucleic acids, phosphoproteins and phospholipids. The phosphorus in vertebrate bone constitutes more than 80% of the total body phosphorus, and acts as a useful store which can be used to buffer phosphorus requirements during periods of dietary insufficiency.

The modification of protein function and regulation of enzyme activity are achieved by a number of strategies, one of which is covalent modification by phosphorylation and dephosphorylation. The addition or removal of one or more phosphate residues induces conformational changes in proteins which result in changes in the shape and/or size of binding sites or active sites. Examples of processes where these mechanisms operate are the transport of sodium and potassium across the plasma membrane and control of enzyme activity in glycogen metabolism, both of which are discussed in Chapter 22.

Phosphorus is particularly important in the metabolism of sugars and polysaccharides. Simple sugars such as glucose are relatively inert unless they are converted to phosphate esters. In glycolysis the first step in the metabolism of glucose is its conversion to glucose-6-phosphate. All of the intermediate compounds between glucose and pyruvate, the end product of glycolysis, are phosphate esters.

8.2.4 MAGNESIUM

Magnesium is a minor component of bones (0.5–0.7 mg kg^{-1} bone) and this accounts for approximately 60–70% of the magnesium in the body. The remainder is found in soft tissues.

Almost all of the ATP and ADP found in cells is complexed to magnesium, as shown in Figure 8.12. During enzyme-catalysed phosphorylation reactions, magnesium forms a bridge between the pyrophosphate of ATP and the enzyme molecule and lowers the activation energy of the reaction.

Other enzymes which require magnesium as a coenzyme include dehydrogenases and

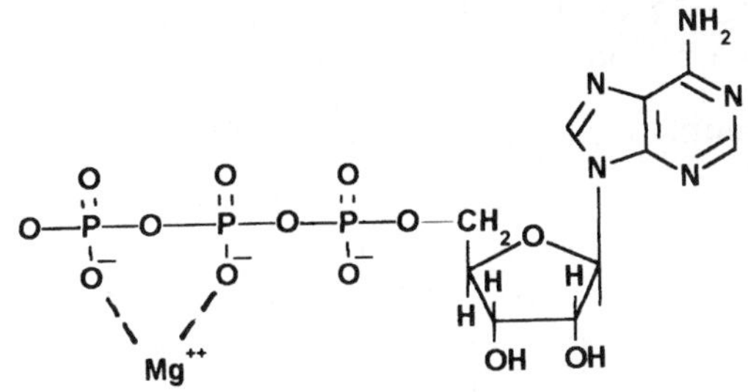

Figure 8.12 The structure of the magnesium–ATP complex.

enolase. The requirement for magnesium is not absolute and it may be replaced by manganese, another divalent cation.

Magnesium acts as a counter-ion to nucleic acids, which exist as polyanions under physiological conditions. Thus, magnesium may influence the transcription of DNA to mRNA and the translocation of mRNA to protein by stabilizing ribosomal structures and activating the transfer of amino acids from amino acyl tRNAs to the growing polypeptide.

8.2.5 SODIUM, CHLORIDE AND POTASSIUM

These three elements provide the major electrolytes of physiological fluids. Sodium constitutes over 90% of the total cations and chloride in excess of 65% of the anions in blood. Potassium is the major intracellular cation. The osmotic pressure of extracellular and intracellular fluid can be markedly changed by excessive intake and loss of one or more of these ions.

The concentration of Na^+, K^+ and Cl^- in a typical mammalian cell is shown in Table 8.3.

Table 8.3 Concentration of sodium, potassium and chloride in intracellular and extracellular fluid in mammalian cells

Ion	Intracellular concentration (mM)	Extracellular concentration (mM)
Na^+	5–15	145–150
K^+	140–145	5–10
Cl^-	5–15	110–115

The importance of maintaining the correct electrolyte balance in intracellular and extracellular fluids is illustrated by the fact that a major portion of the energy expenditure in cells is directed towards ion pumping (Chapters 22 and 29).

The movement of sodium across cell membranes is required for the transmission of nerve impulses and the uptake of nutrients from the digestive tract and into cells. For example, the depolarization and repolarization of neural membranes involves the rapid influx of sodium through voltage-gated sodium channels. In the digestive tract sodium, which is present in high concentrations in digestive fluids, is co-transported with monosaccharides and amino acids across the plasma membranes of cells lining the lumen of the small intestine. In contrast chloride, which is required in high concentrations in gastric secretions, is secreted by the gastric mucosa (Chapter 28). There are few examples of chloride acting as a coenzyme and activator of enzymes, however intestinal amylase is one enzyme which is activated by chloride.

Potassium is required for the transmission of nerve impulses to muscle fibres and in the contraction of muscle fibres. It is also an activator of a number of enzymes, for example, hexokinase, pyruvate kinase and fructokinase.

8.2.6 SULPHUR AND IRON

Sulphur occurs as a component of a wide range of molecules. The sulphur amino acids methionine, cystine and cysteine are components of most proteins. The SH groups of cysteine stabilize the tertiary and quaternary structure of proteins by undergoing oxidation to form disulphide bridges with adjacent SH groups. SH groups also participate in the binding and transformation of substrates and products on the active sites of enzymes. The supply of sulphur amino acids for protein synthesis can be a limiting factor in the growth of both plants and animals.

Sulphur occurs in a number of other biomolecules. It is a component of glycosaminoglycans, chondroitin sulphate, dermatan sulphate, heparan sulphate and keratan sulphate, which form the extracellular matrix surrounding cells in tissues. It is also found in the coenzymes lipoic acid and coenzyme A, in acyl carrier protein and glutathione, and in the vitamins biotin and thiamin. Animal tissue cannot reduce sulphate to the SH form found in most proteins and therefore animals are dependent on reduced sulphur compounds in the diet. Rumen microorganisms can, however, use inorganic sulphur to synthesize amino acids.

Iron is found in a number of proteins in plants and animals. In many cases it is coordinated in the haem prosthetic group, also called Fe protoporphyrin IX. This gives the red colour to the oxygen-binding proteins myoglobin and haemoglobin, found in muscle and blood, respectively. It is a component of the respiratory chain cytochromes. A number of non-haem proteins also contain iron.

In animals transferrin is the major iron-transport protein found in the blood, and it also mediates the uptake of iron by cells. Once in the cell, iron is stored bound to the protein ferritin (phytoferritin in plants) which contains a dense core of ferric hydroxide.

A number of iron–sulphur proteins are components of the mitochondrial electron transport chain. These proteins contain iron–sulphur centres with either two or four iron atoms bound to an equal number of sulphur atoms, some of which are components of the amino acid cysteine. These iron–sulphur centres function as electron carriers. There are at least six different iron–sulphur centres in the respiratory chain, five of which are known to occur in the NADH dehydrogenase complex and at least one in the *b–c* complex (Chapter 12). In chloroplasts the final transfer of electrons from photosystem I to NADP reductase is mediated by another iron–sulphur protein, ferredoxin. Iron–sulphur centres are also found in a number of enzymes such as aconitase and xanthine oxidase.

8.2.7 OTHER ELEMENTS WITH KNOWN BIOCHEMICAL FUNCTIONS

Copper

A number of enzymes require copper. These include cytochrome oxidase, tyrosinase, ascorbate oxidase and polyphenol oxidase. In addition, copper is thought to be involved in the desaturation of fatty acids, and is a component of cytoplasmic superoxide dismutase which catalyses the rapid removal of the toxic superoxide radical anion $O_2^{\cdot-}$. The active centre of this enzyme contains two copper and two zinc atoms.

Copper deficiency blocks the formation of cross-links essential for structural proteins. For example, elastin contains many cross-links between lysine residues in adjacent polypeptide chains which render the protein insoluble. The key copper-containing enzyme required for normal cross-linking is lysyl oxidase. Copper deficiency in pigs results in failure to cross-link, and accumulation of a soluble precursor of elastin has been identified.

In animal nutrition copper, molybdenum and sulphur are often considered together because of the interaction between these elements in the digestive tract, which can reduce their availability. This is due to the formation of highly insoluble copper thiomolybdates.

Cobalt

All the known functions of cobalt are related to its role as a component of vitamin B_{12}. These are discussed above (section 8.1.9). Non-ruminant animals cannot utilize cobalt and require a supply of preformed vitamin B_{12}. Because rumen microorganisms can incorporate inorganic cobalt into the vitamin, these animals are able to take advantage of dietary cobalt.

Iodine

In animals iodine is required for the synthesis of the thyroid hormones, triiodothyronine (T_3) and tetraiodothyronine or thyroxine (T_4). For this reason over 90% of the iodine in the body is concentrated in the thyroid gland. Most dietary iodine is absorbed in the form of iodide (I⁻). It is transferred from the circulation to the thyroid gland by an active transport process associated with the plasma membrane Na^+, K^+-ATPase. In the thyroid gland, iodine is converted to a highly reactive free radical which rapidly reacts with the phenyl groups of the tyrosine moieties of thyroglobulin. The iodinated protein is stored within the gland and is broken down in lysosomes to release T_3 and T_4 when the cells are activated by thyroid-stimulating hormone. A characteristic symptom of iodine deficiency is goitre, an enlargement of the thyroid gland. As the gland is unable to synthesize T_3 and T_4, a hypothyroid state ensues which has profound effects due to the role of thyroid hormones in the regulation of many metabolic processes.

Iodine is found in plant tissues, and marine plants are a particularly rich source. However, iodine does not appear to be an essential element in plant nutrition.

Manganese

Manganese is a divalent ion which can substitute for many of the functions of magnesium. In animals, it is an activator of a number of enzymes such as hydrolases, kinases, decarboxylases and transferases. This activation is not always manganese-specific, and other divalent ions may have a similar effect. However, in the case of glycosyltransferases, manganese is essential for activation. These enzymes are required for the synthesis of polysaccharides, glycoproteins and glycolipids. A deficiency of manganese can lead to bone and joint abnormalities, reflecting the role of glycosyltransferases in the synthesis of bone matrix and cartilage. Mitochondria contain manganese-dependent superoxide dismutase, and increased lipid peroxidation and membrane dysfunction may be related to a decrease in the activity of this enzyme in manganese deficiency. Steroid and terpene biosynthesis may also be affected by manganese deficiency as it may be required for the enzyme farnesyl pyrophosphate synthase.

Molybdenum

A number of molybdenum-dependent enzymes have been identified in animals and plants. In animals it is a component of xanthine oxidase, aldehyde oxidase and sulphite oxidase. In desert irrigation water, the concentration of molybdenum is often high and may lead to a serious risk of secondary copper deficiency.

Selenium

Glutathione peroxidase is the only enzyme known to contain selenium in animals. This enzyme is part of the intracellular antioxidant mechanism and works in concert with superoxide dismutase, catalase, vitamin E and vitamin C. Many of the selenium deficiency symptoms observed in animals are related to tissue damage associated with increased levels of lipid peroxides. A number of other selenoproteins have been identified in animal tissues, some of which may be selenium transport proteins; however, no specific functions have been assigned to these proteins.

Unlike animals, higher plants do not contain glutathione peroxidase and no other selenium-containing enzymes have been identified.

Zinc

Zinc is a component of a large number of enzymes in animal tissues, for example carboxypeptidase, thermolysin, alkaline phosphatase, alcohol dehydrogenase, carbonic anhydrase, lactate dehydrogenase and cytoplasmic superoxide dismutase.

In both plants and animals, zinc also plays an important role in the regulation of DNA and RNA metabolism. Regulatory proteins bind to these nucleic acids through a number of different recognition sites. Some of these proteins contain a so-called zinc finger, a loop in the protein structure which is stabilized by the formation of a tetrahedral complex with zinc and the amino acids histidine and cysteine.

THE COMPOSITION OF AGRICULTURAL PRODUCTS

9.1 INTRODUCTION

In general the composition of animal tissues tends to be restricted to a much narrower range, both of substances and of amounts, than is found in plants. Animals are to a very great extent composed of water, protein, lipid and minerals (mainly calcium phosphates). Within individual animal tissues there are differences in the ratios of the principal components but these are not major. If we exclude the hard parts of bone then the gross compositions of the structural, reproductive and digestive tissues from animals are remarkably similar. These similarities persist throughout the whole of the animal's life cycle. On the other hand, plant tissues vary enormously in composition both between the different plant tissues and between the same tissue at different stages in its development.

The really striking difference between plants and animals lies in their use of carbohydrates. Plants use these as one of their main means for storing energy, whereas the long-term energy reserves of animals are almost completely in the form of lipid. The other important roles of carbohydrates in plants, as the structural elements of the cell and as the building materials for the vascular system, are taken over by proteins and lipids in animals.

The amount of carbohydrate in an animal body is rarely greater than about 0.2% of body weight; of this, over 90% will be present as glycogen. In some plant tissues the carbohydrates account for well over 90% of the dry matter.

9.1.1 ENERGY STORAGE IN ANIMALS AND PLANTS

The difference in choice of energy reservoir in plants and animals is quite understandable in terms of the ways in which plants and animals exist. As a store of energy the carbohydrates have the advantage of being synthesized and used very easily. Lipids bring with them all sorts of problems because of the fact that they are not easily metabolized in the aqueous environment of any biological system. The disadvantage of carbohydrates is that they are not as energy-dense as the lipids. One gram of carbohydrate has a gross energy (the heat of combustion under standard conditions) of about 18 MJ kg^{-1} whereas lipids produce approximately 39 MJ kg⁻¹. In other words, to store a given amount of energy requires about twice the weight of carbohydrate as lipid. To have a heavy fuel storage system is no great disadvantage to the part of a plant which remains rooted in one spot. Animals, on the other hand, are mobile and carrying extra weight would be a distinct handicap. Putting this into human terms, an average male of 70 kg body weight probably has 12 kg of fat. To store this energy as carbohydrate would add a further 13 kg to his mass without giving any compensating advantage.

Seeds may be regarded as the exception to the rule in plant tissues. They are small, mobile entities and some, such as rape seed (canola) and linseed, take advantage of a high energy/weight ratio by storing a large proportion of their energy reserves as oil.

9.2 THE COMPOSITION OF ANIMALS

9.2.1 BODY COMPOSITION

In looking at the composition of the animal body, one component that must be ignored is the contents of the gut – particularly in herbivores where this may account for a quarter of the animal's overall body weight. The gross composition of the carcass (excluding the digestive tract) of young and mature beef animals is shown in Figure 9.1. These proportions may be taken as fairly typical for farm livestock. The only components that are shown are

lipid, protein, minerals and water. Certainly, the animal will contain a small amount of carbohydrate, but this is too small even to show on the diagrams. Looking at the two diagrams, the striking difference between the young (75 kg) and the mature (500 kg) animal is in the amounts of fat and water. The mineral fraction remains quite constant and the protein tends to decrease with age, but the fat content increases more than three-fold. To compensate for the increased fat there is a reduction in the water content (see Chapter 30).

The ratio of the major mineral elements in an animal carcass tends to remain very constant throughout life (see Figure 9.2). The six minerals illustrated form the largest part of those present; the trace elements would not even be visible at the scale of this diagram.

9.2.2 MILK

Milk is the one agricultural product of animal origin that contains substantial quantities of carbohydrate, in the form of the disaccharide lactose. Here, the advantage of the high energy/weight ratio is outweighed by the needs of the young for an instant and easily metabolized energy source derived from carbohydrate. The milks from different species vary considerably in composition (Table 9.1); for instance human milk has a low protein content, and the milk from arctic sea-dwelling mammals such as whales has extremely high levels of fat. The commonest milk in agricultural production

Table 9.1 Composition of milks from various species

Animal	Lactose (g kg⁻¹)	Protein (g kg⁻¹)	Fat (g kg⁻¹)	Solids, not fat (g kg⁻¹)	Calcium (g kg⁻¹)
Holstein cow	46	33	35	85	1.2
Jersey cow	49	36	46	90	1.3
Tropical cow (*Bos indicus*)	46	32	50	85	1.3
Sheep	44	60	75	115	2.0
Goat	42	34	40	87	1.3
Buffalo	49	38	75	90	1.9
Dog	41	83	97	131	3.0
Pig	52	60	83	119	2.7
Horse	63	22	12	85	0.8
Human	68	13	40	90	0.3

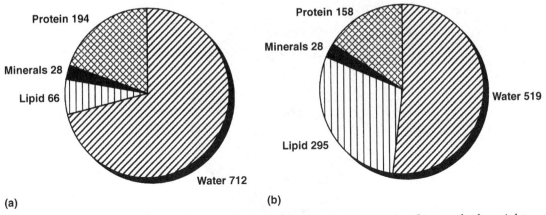

Figure 9.1 Overall composition of the animal body. (a) Calf, body weight 75 kg; (b) steer, body weight 500 kg (figures in g kg^{-1} of carcass).

comes from dairy cattle and the typical composition is illustrated in Figure 9.3. Milks from different breeds of cattle vary slightly in composition, particularly in the fat content, and there are also changes in response to variations in the diet and with season.

9.3 PLANT MATERIALS

The water content of plant materials is even more variable than that in animals. Figure 9.4 shows the amounts of water in a variety of plant products. It can be seen clearly that they form two groups with very high or very low water content. In general, the water content has to be reduced below 150 g kg^{-1} if spoilage is not to occur during storage. Even after successful storage, the water content of materials will fluctuate in response to changes in the moisture content of the environment. For this reason the composition of most agricultural commodities is expressed on a dry matter basis (g kg^{-1} DM or mg kg^{-1} DM, as appropriate). In agricultural terms we can divide plant materials arbitrarily

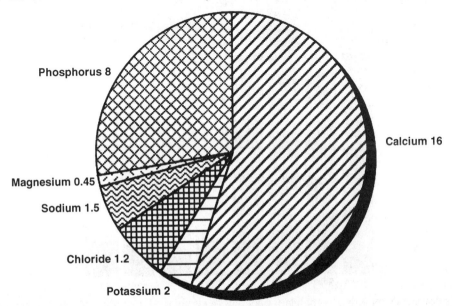

Figure 9.2 Mineral composition of a beef steer (g kg^{-1} of carcass).

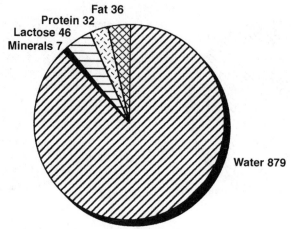

Figure 9.3 The composition of cow's milk (g kg⁻¹ of milk).

9.4 PRINCIPAL NUTRIENTS IN PLANTS AND ANIMALS

upon the basis of their chemical composition and in most cases this will be related to the use that is made of them (Table 9.2).

9.4 PRINCIPAL NUTRIENTS IN PLANTS AND ANIMALS

9.4.1 PROTEINS

One of the main purposes of agriculture is the provision of food. Amongst the nutritional needs of humans is a supply of the essential amino acids in approximately the right proportions. Not surprisingly, the proteins supplied by animals – meat, fish, eggs and milk – usually have ratios of amino acids that are closer to the requirements of humans than are the proteins from plant sources. Figure 9.5 shows the pattern of essential amino acids in the proteins of fish and those from groundnuts. These are compared with the spectrum that is needed in an animal's diet. It is quite clear that some

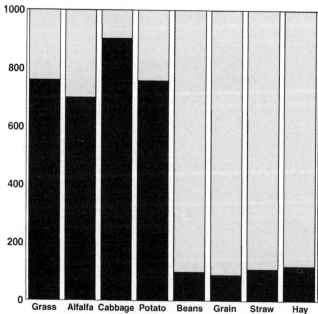

Figure 9.4 Typical water content of some plant materials (g kg⁻¹ of fresh material; solid bar, water; shaded bar, dry matter). Note that the water content may vary greatly depending upon storage conditions.

Table 9.2 Division of plant materials in terms of composition and nutritional use

	Water content	Protein content	Lipid content	Sugar content	Starch content	Fibre content	*Principal food use*
High protein/low-to-medium fibre							
Oil seeds	L	H	H	L	L	M	Edible oil extraction
Oil-seed cakes	L	H	L	L	L	M	Any livestock
Pulses	L	H	M/H	L	M/H	L	Human/non-ruminant animals
High starch/low-to-medium fibre							
Cereal grains	L	L	M/L	L	H	M/L	Human or any livestock
Root vegetables	H	L	M/L	L	H	M/L	Human or any livestock
High fibre/low-to-medium protein							
Grasses	H	L	L	M/L	M/H	H	Ruminant/other herbivore
Forage legumes	H	M/H	L	M/L	M/H	H	Ruminant/other herbivore
High fibre/very low protein							
Cereal residues e.g. straw	L	L	L	L	L	H	Ruminant/other herbivore
High moisture/high sugars							
Green vegetables	H	L	L	H	L	M	Human or any livestock
Fruits	H	L	L	H	M/L	L	Human or any livestock
Brassica roots e.g. turnip	H	L	L	H	L	M	Human or any livestock
High fibre/high sugars							
Sugarcane	H	L	L	H	L	H	Human or any livestock

H = High, M = Medium, L = Low

amino acids such as arginine are over-represented in the plant protein but that, equally, some such as leucine and methionine are in short supply. In order to ensure health, a protein source such as groundnut should be supplemented with another that is adequate in leucine and methionine.

9.4.2 LIPIDS

There are characteristic differences between the lipids of plants and those of terrestrial mammals. The most important of these is in the degree of unsaturation of the fatty acids. In animal milk fat, most of the fatty acids are either saturated or monounsaturated (C_{16}, $C_{18:0}$ and $C_{18:1}$) whereas the fatty acids from a plant such as soyabean are almost all polyunsaturated. In general, lipids from terrestrial animals (especially the ruminants) are solid at room temperatures, whereas plant storage lipids are oils (see Tables 4.3 and 4.4).

9.4.3 CARBOHYDRATES

Plants contain a huge diversity of carbohydrates which reflects the variety of purposes to which they are put. A comparison of two different carbohydrate-rich sources shows that in a grain most of the carbohydrate is either the starch of the grain kernel or the cellulose of the outer protective layers. A green plant such as cabbage has a quite different spectrum of materials, demonstrating that most of the carbohydrates are typically those of the cell walls.

It is perhaps no surprise that there is a difference between grain and cabbage leaf in their carbohydrates, but there are also variations between plants which appear superfi-

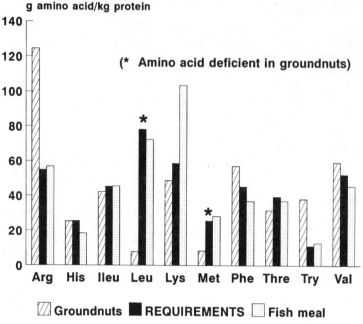

Figure 9.5 Essential amino acids required for growth in non-ruminants compared with the composition of dietary protein sources.

cially similar. One difference which has consequences in agriculture is the difference between temperate and tropical grasses in their carbohydrates (Figure 9.6). The grasses contain similar amounts of cellulose, but the high levels of easily fermented sugars and fructosans in temperate grasses are replaced by hemicellulose in tropical species. This means that it is very difficult to make silage (see Chapter 16) with tropical grasses and they are not as well digested in the ruminant gut.

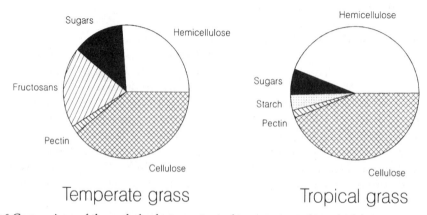

Figure 9.6 Comparison of the carbohydrate content of temperate and tropical grasses.

PART TWO
METABOLISM

GLYCOLYSIS

10.1 INTRODUCTION

Glycolysis, literally the splitting of sugars, occurs in the cytosol of nearly every different cell type of almost all organisms, from bacteria to higher animals. The enzymes are soluble rather than being bound to membranes. It has two main roles:

- oxidizing sugars and, in the process, producing energy for use in other parts of the cell;
- breaking sugars into smaller molecules with two, three or four carbon atoms that can be used as starting materials for the synthesis of other important chemical compounds.

The pathway consists of two sections. Firstly, a molecule of glucose (or of any hexose) is converted to fructose-1,6-bisphosphate (fructose that bears a phosphate group at each end). This six-carbon compound is broken into two three-carbon molecules. The second part of the pathway converts both of these three-carbon residues to pyruvate (pyruvic acid) – still with three carbons.

Pyruvate can form the starting material for a wide range of other pathways. Perhaps strangely for a compound that is so important, cells never accumulate pyruvate so it is not found in large quantities.

10.2 STAGE 1 – PREPARING GLUCOSE FOR SPLITTING INTO TWO THREE-CARBON UNITS

The first few steps of the pathway involve the production of a sugar molecule which has phosphate groups attached to it, as illustrated in Figure 10.1.

10.2.1 GLUCOSE PHOSPHORYLATION

Glucose is taken into cells by active transport and is then phosphorylated at the 6 position. The phosphate group comes from ATP and the reaction is catalysed by one of two enzymes, hexokinase or glucokinase. These belong to a large group of enzymes known as the kinases, which phosphorylate compounds using ATP.

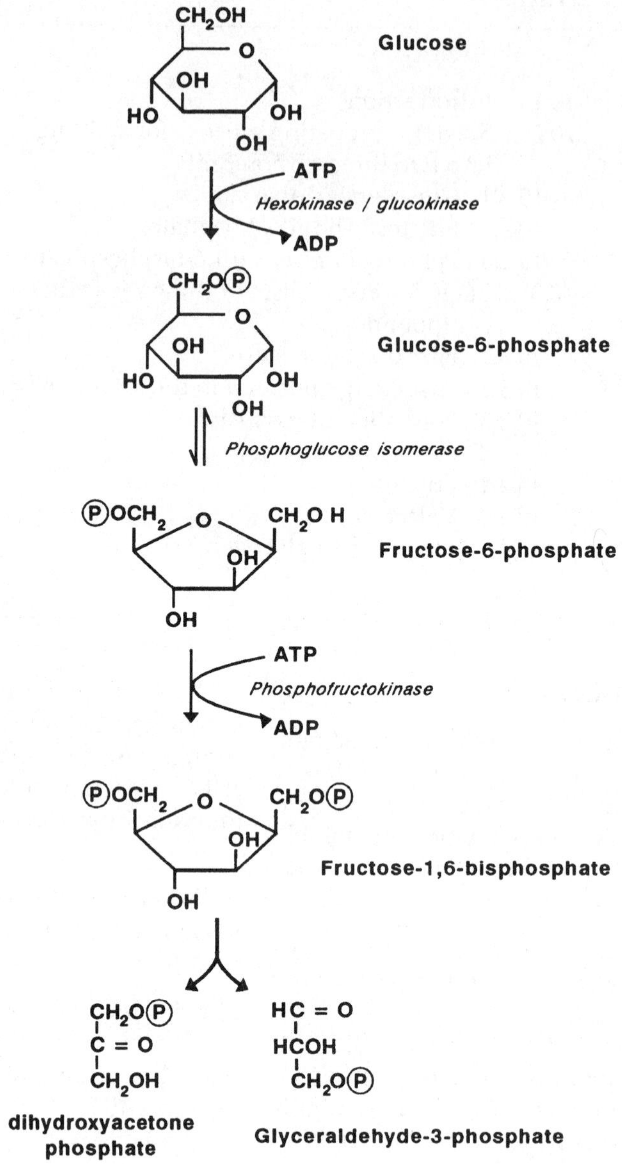

Figure 10.1 The first stages of glycolysis.

Like other kinases, these enzymes have a requirement for Mg^{2+} ions; other ions, particularly Mn^{2+}, will act instead but they are probably not the normal intracellular cofactors.

Both enzymes catalyse the conversion of glucose to glucose-6-phosphate.

$$glucose + ATP \rightarrow$$
$$glucose\text{-}6\text{-}phosphate + ADP$$

Hexokinase is found in muscle and glucokinase in liver. The K_M of hexokinase for glucose is about 0.1 mM and that for glucokinase 10 mM. The concentration of glucose in the blood varies between about 4 and 8 mM as a result of ingestion of carbohydrates. Both of these values are well above the K_M for hexokinase, so that this enzyme is saturated all of the time – it is working at maximum rate and the velocity of formation of glucose-6-phosphate in muscle is independent of ingestion of carbohydrates. However the K_M of glucokinase is similar to the glucose concentration in blood, and changes in blood glucose levels therefore dramatically change the rate of glucose-6-phosphate formation and hence glycogen production in the liver. Formation of glycogen therefore increases substantially after a meal when there is plenty of glucose in the blood.

Whichever enzyme is involved, the overall standard free energy change of –17 kJ mole^{-1} is quite large; the reaction will thus proceed easily and spontaneously in the direction of the formation of glucose-6-phosphate, but it is practically irreversible under biological conditions.

10.2.2 FRUCTOSE AND ITS PHOSPHATES

Glucose-6-phosphate then has to be transformed into its structural isomer, fructose 6-phosphate, which is accomplished by the enzyme glucose phosphate isomerase. In order to change the aldehyde group of glucose into the ketone of fructose the enzyme must convert the sugars into the straight-chain form.

The second phosphate group is then introduced at the opposite end (carbon 1) of the fructose molecule. The product is fructose-1,6-bisphosphate. The enzyme is phosphofructokinase (PFK). Like the previous kinase reaction, the free energy change is large and negative (about –14 kJ mole^{-1}). Again, the reaction will proceed easily in the direction of the formation of fructose-1,6-bisphosphate and is effectively irreversible. The irreversibility of this step forms the basis of the main mechanism regulating the flow through the pathway (see Chapter 22).

10.2.3 SPLITTING OF FRUCTOSE-1,6-BISPHOSPHATE

Fructose-1,6-bisphosphate is broken into two molecules, each with three carbon atoms and a phosphate group. This reaction is catalysed by the enzyme aldolase, which is the last reaction of the sequence shown in Figure 10.1. The free energy change of the reaction is small and thus it can be reversed.

10.3 STAGE 2 – METABOLISM OF THE THREE-CARBON COMPOUNDS

These are phosphorylated versions of the two three-carbon sugars (Chapter 3). They are structural isomers of one another. The two different products of the breakdown of fructose-1,6-bisphosphate can be interconverted by the enzyme triosephosphate isomerase (see Figure 10.2). Like many reactions catalysed by isomerases, the reaction is quite freely reversible. Therefore, each molecule of glucose effectively gives two molecules of glyceraldehyde-3-phosphate.

10.3.1 FIRST OXIDATION STEP

If we are following the products formed from a molecule of glucose, from now on we must remember to double the throughput of the pathways that follow.

Up to this point the pathway has merely split the hexose sugar but no energy has been extracted. The cell has had to use two molecules of ATP for every molecule of glucose entering the pathway. The next steps represent the first points at which some energy (in the form of ATP) is returned to the cell. Glyceraldehyde-3-phosphate undergoes two changes simultaneously under the influence of the enzyme glyceraldehyde-3-phosphate dehydrogenase. As its name implies, this enzyme takes hydrogen away from its substrate; it also adds another phosphate group to form 1,3-bisphosphoglycerate. This time the phosphorylation proceeds without using ATP, and inorganic phosphate is the donor. The

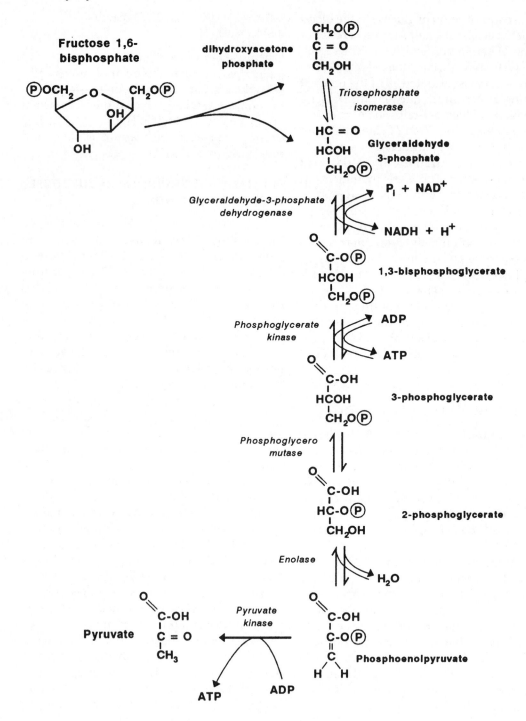

Figure 10.2 The final stages of glycolysis.

hydrogen is removed by the coenzyme NAD⁺. For this process to continue, there must be a constant renewal of the supply of NAD⁺; this can be achieved in a number of ways depending upon the presence or absence of oxygen.

10.3.2 FIRST ENERGY RELEASED IN THE FORM OF ATP

The next step yields some profit in terms of energy: one of the phosphate groups is removed and is used to produce ATP from ADP. Phosphoglycerate kinase transfers the phosphate group from the 1 carbon of the substrate, 1,3-bisphosphoglycerate, to ADP. It thus has a similar (but reverse) action to the hexokinase that added phosphate groups to glucose. For this reason it is also a kinase enzyme, but one which is working in the opposite direction. The free energy change for the reaction is quite small and thus the reaction is reversible – its very name, phosphoglycerate kinase, implies that it works in the opposite direction to the one we are considering at the moment. The product of this step, 3-phosphoglycerate, is one point at which other pathways can interact with glycolysis; this intermediate is the point of departure for the synthesis and breakdown of glycerol.

Each molecule of glucose used in this pathway required two molecules of ATP to get as far as fructose 1,6-bisphosphate. But each molecule of glucose gave two molecules of 1,3 bisphosphoglycerate which each gave one

molecule of ATP. The 'balance sheet' (Table 10.1) for phosphate groups is now equal.

10.3.3 FORMATION OF PYRUVATE

The final part of the glycolytic pathway is the formation of pyruvate from 3-phosphoglycerate. The sequence of reactions starts with the rearrangement of 3-phosphoglycerate to give 2-phosphoglycerate. The reaction is catalysed by the enzyme phosphoglyceromutase. The free energy change in the reaction is quite small, so it is freely reversible.

The next step, which is also freely reversible, involves the removal of the hydroxyl group on the carbon 3 and a hydrogen at carbon 2 to form a double bond, producing water and phosphoenolpyruvate This is an important intermediate in its own right, particularly in reactions that synthesize glucose.

Pyruvate is produced from phosphoenolpyruvate by the transfer of the phosphate group to ADP. The reaction, which is catalysed by pyruvate kinase, has a large negative free energy change, rendering it irreversible.

The overall balance sheet (Table 10.2) now shows that more energy has been extracted in the form of ATP than had to be put in at the start.

10.4 THE ENTRY OF OTHER SUGARS

Although glucose is important, it is not the only sugar that is encountered in animals and

Table 10.1 Energy balance for the production of 3-phosphoglycerate from glucose

Entering pathway	*Leaving pathway*
One molecule of glucose	Two molecules of 3-phosphoglycerate
Two molecules of ATP	Two molecules of ATP
Two molecules of NAD⁺	Two molecules of NADH

Table 10.2 Energy balance for the production of pyruvate from glucose

Entering pathway	Leaving pathway
One molecule of glucose	Two molecules of pyruvate
Two molecules of ATP	Four molecules of ATP
Two molecules of NAD$^+$	Two molecules of NADH

plants. As examples, many plants both produce and use large quantities of fructose. The suckling mammal obtains half of its sugar from the galactose of milk lactose.

10.4.1 FRUCTOSE

It is not difficult to see how fructose enters glycolysis, it merely needs to be phosphorylated to fructose-6-phosphate to enter the pathway. Many plants and animals have a specific enzyme, fructokinase, which catalyses the reaction:

$$\text{fructose} + \text{ATP} \rightarrow$$
$$\text{fructose-6-phosphate} + \text{ADP}$$

In animals, hexokinase is able to phosphorylate fructose as well as glucose.

10.4.2 GALACTOSE

The entry of galactose is not as simple as that of fructose: it has to be converted into glucose-6-phosphate before it can be used. During many sugar transformations the sugar molecule has to be attached to a nucleotide, this pattern is seen in the synthesis of disaccharides such as lactose and in the formation of starch (Chapter 18). The nucleotide used in these transformations is uridine diphosphate (UDP) although the coupling is not a simple one.

Galactose is first phosphorylated directly at the 1 position:

$$\text{galactose} + \text{ATP} \leftrightarrows$$
$$\text{galactose-1-phosphate} + \text{ADP}$$

In the next step there is an exchange in activating groups between glucose and galactose (changing UDP for phosphate) leading to the production of UDP-galactose and glucose-1-phosphate (Figure 10.3):

$$\text{UDP-glucose} + \text{galactose-1-phosphate} \leftrightarrows$$
$$\text{UDP-galactose} + \text{glucose-1-phosphate}$$

The glucose-1-phosphate can enter the glycolytic pathway after its conversion to glucose-6-phosphate by the enzyme phosphoglucomutase.

To complete the cycle of reactions, UDP-glucose can be regenerated from UDP-galactose by the enzyme UDP-galactose-4-epimerase.

$$\text{UDP-galactose} \leftrightarrows \text{UDP-glucose}$$

The overall effect is that one molecule of ATP is used to produce one molecule of glucose-1-phosphate from a molecule of galactose.

10.4.3 THE ENTRY OF GLYCOGEN

The pathway for the breakdown of glycogen is designed to provide a rapidly available source of energy. Under the influence of the enzyme glycogen phosphorylase, glucose units are removed from the non-reducing ends of the chains. The sugars removed are therefore those with a free hydroxyl group at the 4 position. The process is phosphorolysis rather than hydrolysis, involving the attack of a phosphate group on the glycosidic link to yield a shortened version of the glycogen and a molecule of glucose-1-phosphate. This can then be transformed into glucose-6-phosphate ready for entry into the glycolytic pathway (phosphoglucomutase). In the case of liver glycogen, the

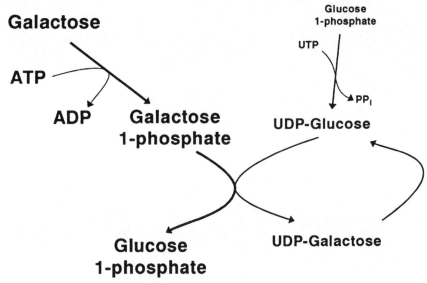

Figure 10.3 Conversion of galactose to glucose-1-phosphate prior to its metabolism by the glycolytic pathway.

glucose-6-phosphate can be dephosphorylated to yield free glucose, to be used in supplying the peripheral tissues of the body.

$$(\text{glycogen})_n + P_i \rightarrow$$
$$(\text{glycogen})_{n-1} + \text{glucose-1-phosphate}$$

The branched structure of glycogen brings with it a great advantage in promoting rapid utilization. If glycogen were a simple, linear, unbranched molecule then only one glucose group per molecule would be available to glycogen phosphorylase at any one time. The branched structure means that a large number of non-reducing glucose units are always available.

THE TRICARBOXYLIC ACID CYCLE 11

11.1 INTRODUCTION

Like all organic compounds, glucose will burn in air to produce carbon dioxide and heat. This oxidation process has been harnessed by living organisms not to produce heat, but to trap chemical energy in the form of ATP which can be used for a variety of cellular functions.

We have already seen that the first stage in the oxidation of glucose, glycolysis, can produce a small quantity of useful energy. For each molecule of glucose converted to pyruvate, two molecules of ATP are produced directly, together with two molecules of NADH which can be used to produce ATP in the electron transport chain (see Chapter 12).

Under aerobic conditions much more energy can be extracted from glucose if the pyruvate produced by glycolysis is completely oxidized to carbon dioxide and water. This is achieved by the tricarboxylic acid (TCA) cycle, also known as the citric acid or Krebs cycle. This cycle can be considered as the hub of

intermediary metabolism. A glance at a map of metabolic pathways will quickly reveal that the TCA cycle has many inputs and serves to oxidize not only pyruvate, but also the products of amino acid and fatty acid catabolism. Furthermore, it provides intermediates for many biosynthetic pathways such as amino acid synthesis, fatty acid synthesis and the production of glucose via gluconeogenesis.

Although it is generally agreed that in plants and animals the reactions of the cycle are the same, regulation of the pathway differs.

11.2 THE REACTIONS OF THE TCA CYCLE

11.2.1 PRODUCTION OF ACETYL-COA

In theory a cycle has no beginning or end, but it is necessary to consider the reactions of the TCA cycle in sequence, and it is accepted convention to describe the cycle from the point at which it is fed by the glycolytic pathway. The production of pyruvate by glycolysis occurs in the cytoplasm. The metabolism of pyruvate via the TCA cycle occurs in the mitochondrial matrix, thus pyruvate must be transported into the mitochondrion before being further oxidized. The mitochondrion is bounded by a double membrane. The outer membrane is relatively permeable; however, the inner membrane is not and pyruvate must cross this membrane via a specific carrier system, the pyruvate translocase.

Pyruvate does not feed directly into the TCA cycle but must first be converted to acetyl coenzyme A, usually referred to as acetyl-CoA (Figure 11.1).

This conversion is catalysed by pyruvate dehydrogenase (PDH). This enzyme complex catalyses the irreversible oxidative decarboxylation of pyruvate to yield acetyl-CoA, CO_2 and NADH. It consists of three different enzymes: pyruvate dehydrogenase, containing thiamin pyrophosphate as a coenzyme; dihydrolipoyl transacetylase; and dihydrolipoyl dehydrogenase, a flavoprotein containing FAD. The PDH complex controls the flux of carbon into the TCA cycle from glucose.

11.2.2 REACTIONS LEADING TO THE PRODUCTION OF CO_2

The first reaction *per se* of the TCA cycle is the condensation of the two-carbon compound, acetyl-CoA, with the four-carbon compound, oxaloacetate, to form a new six-carbon compound, citrate. This reaction is catalysed by the enzyme citrate synthase. The hydrolysis of the thioester bond in acetyl-CoA during the condensation is highly exergonic, making

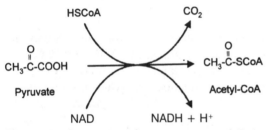

Figure 11.1 Conversion of pyruvate to acetyl-CoA by the pyruvate dehydrogenase complex.

this reaction effectively irreversible (Figure 11.2).

Citrate undergoes a rearrangement to form isocitrate via *cis*-aconitate. The reaction is catalysed by the enzyme aconitase (Figure 11.3). *cis*-Aconitate remains bound to the enzyme and is, therefore, found in only very low concentrations in the cell.

The conversion of isocitrate to α-ketoglutarate is the first of two oxidative decarboxylation reactions of the TCA cycle. Isocitrate loses a carbon atom as CO_2 and is oxidized by the transfer of hydrogen to NAD^+ or $NADP^+$ to form NADH or NADPH. This reaction is catalysed by isocitrate dehydrogenase (Figure 11.4). Two forms of this enzyme occur in most microorganisms, plants and animals: (a) NAD^+-dependent isocitrate dehydrogenase, and (b) $NADP^+$-dependent isocitrate dehydrogenase. The NAD^+-dependent enzyme occurs only in mitochondria, whereas the $NADP^+$-dependent enzyme is found in both the mitochondrial matrix and the cytoplasm. It is generally accepted that the NAD^+-dependent enzyme is the major catalyst of isocitrate oxidation in the TCA cycle whereas the $NADP^+$-dependent enzyme is important in the synthesis of NADPH in the cytoplasm (see Chapter 19 on fatty acid synthesis).

A second carbon atom is lost as CO_2 in the next reaction of the cycle in which α-ketoglutarate undergoes oxidative decarboxylation to produce succinyl-CoA (Figure 11.5). This reaction is catalysed by α-ketoglutarate dehydrogenase, which is similar in both structure and mechanism to the pyruvate dehydrogenase complex. It consists of three enzymes, α-ketoglutarate dehydrogenase, dihydrolipoyl succinyltransferase and dihydrolipoyl dehydrogenase.

11.2.3 REACTIONS LEADING BACK TO THE FORMATION OF OXALOACETATE

Reference to Figure 11.6 shows that at this point in the cycle two molecules of CO_2 have been produced. The cycle started with the

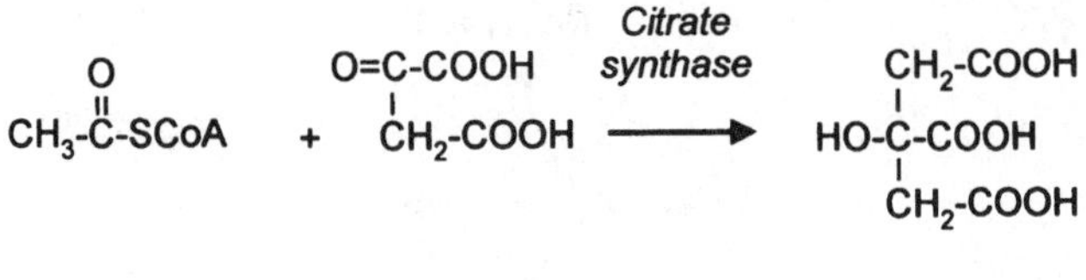

$$CH_3-\overset{\overset{\text{O}}{\|}}{C}-SCoA \;+\; \overset{O=C-COOH}{\underset{CH_2-COOH}{|}} \;\xrightarrow{\;\;\overset{\textit{Citrate}}{\textit{synthase}}\;\;}\; \overset{CH_2-COOH}{\underset{\underset{CH_2-COOH}{|}}{\overset{|}{HO-C-COOH}}}$$

Acetyl-CoA   Oxaloacetate   Citrate

Figure 11.2 The condensation of acetyl-CoA with oxaloacetate to yield citrate.

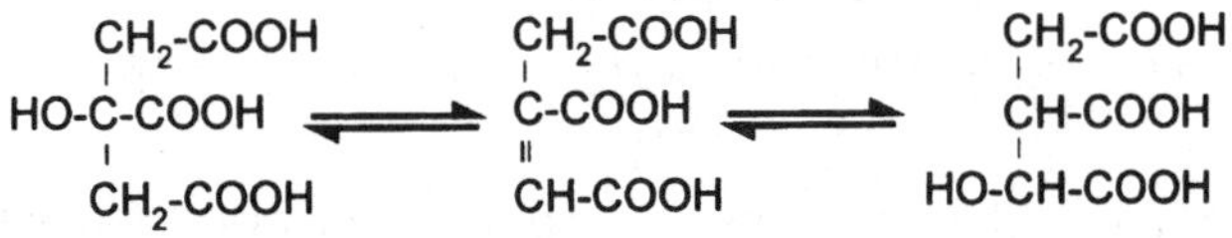

$$\overset{CH_2-COOH}{\underset{\underset{CH_2-COOH}{|}}{\overset{|}{HO-C-COOH}}} \;\rightleftharpoons\; \overset{CH_2-COOH}{\underset{\underset{CH-COOH}{\|}}{\overset{|}{C-COOH}}} \;\rightleftharpoons\; \overset{CH_2-COOH}{\underset{\underset{HO-CH-COOH}{|}}{\overset{|}{CH-COOH}}}$$

Citrate   *cis*-Aconitate   Isocitrate

Figure 11.3 The conversion of citrate to isocitrate via *cis*-aconitate.

input of the two-carbon compound acetyl-CoA. Thus, there has been a net oxidation of acetyl-CoA to carbon dioxide. The remaining reactions of the cycle not only bring about the regeneration of oxaloacetate but also produce the high-energy compounds GTP (ATP in plants and bacteria), $FADH_2$ and NADH which can subsequently be used to synthesize ATP.

GTP is synthesized by the hydrolysis of the energy-rich thioester bond in succinyl-CoA to yield succinate. The reaction is catalysed by the enzyme succinyl-CoA synthetase, sometimes also referred to as succinate thiokinase (Figure 11.7).

Succinate is subsequently oxidized to fumarate by the action of succinate dehydrogenase, a flavoprotein containing FAD which

is reduced to $FADH_2$ during the course of the reaction (Figure 11.8). The enzyme is located on the surface of the inner mitochondrial membrane where it interfaces with the mitochondrial electron transport chain, allowing electrons from $FADH_2$ to pass along the chain to oxygen, with the consequent synthesis of ATP (Chapter 12).

The next reaction of the cycle is the addition of a molecule of water across the *trans* double bond of fumarate to produce L-malate. The enzyme responsible for this conversion is fumarase (Figure 11.9). This type of stereospecific addition of water occurs elsewhere in metabolic pathways. For example, the enolase reaction of fatty acid oxidation in which the addition of water across the *trans* double bond of the enoyl-CoA intermediate produces an L-3-hydroxyacyl-CoA intermediate. Fumarase is specific for the *trans* double bond of fumarate and will not act on the *cis* isomer of fumarate, known as maleate.

The final reaction of the cycle is the oxidation of malate to form oxaloacetate and NADH. This reaction is catalysed by NAD+-dependent malate dehydrogenase (Figure 11.10). The reaction is freely reversible, but

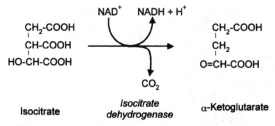

Figure 11.4 The conversion of isocitrate to α-keto-glutarate by isocitrate dehydrogenase.

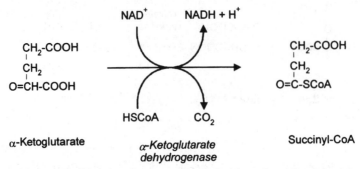

Figure 11.5 The conversion of α-ketoglutarate to succinyl-CoA by α-ketoglutarate dehydrogenase.

because oxaloacetate is rapidly utilized for the synthesis of citrate, the reaction is 'pulled' in the direction of oxaloacetate synthesis. NAD⁺-dependent malate dehydrogenase is also found in the cytoplasm where it catalyses the formation of malate from oxaloacetate. Also found in the cytoplasm is an NADP⁺-dependent malate dehydrogenase which catalyses

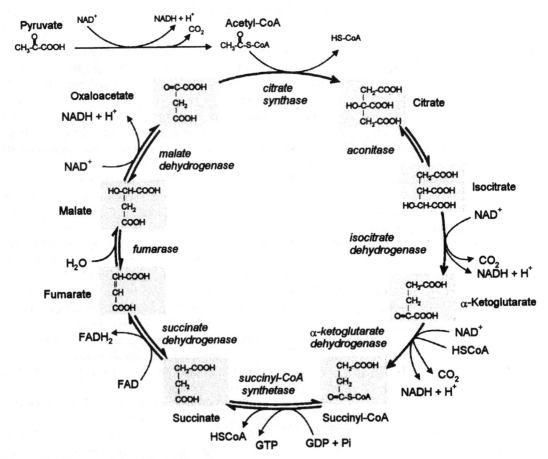

Figure 11.6 The reactions of the TCA cycle.

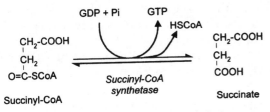

Figure 11.7 The conversion of succinyl-CoA to succinate by succinate synthetase.

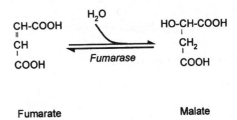

Figure 11.9 The conversion of fumarate to malate.

the oxidative decarboxylation of malate to pyruvate. To distinguish the two forms of the enzyme, the NADP⁺-dependent form is often referred to as the malic enzyme.

11.2.4 OVERALL REACTIONS OF THE TRICARBOXYLIC ACID CYCLE

In the absence of inputs to the cycle other than acetyl-CoA, or of the removal of intermediates for biosynthetic purposes, the net effect of the TCA cycle can be summarized as shown below:

$$CH_3COSCoA + GDP + Pi + 3NAD^+ FAD + 2H_2O \rightarrow 2CO_2 + GTP + 3NADH + FADH_2 + 2H^+ + CoASH$$

Thus, acetyl-CoA is oxidized to two molecules of CO_2. Oxaloacetate does not appear in the equation as, although it condenses with acetyl-CoA to produce citrate, it is regenerated in the last reaction of the cycle. The production of GTP by the cycle leads to the production of ATP through the action of the enzyme nucleoside diphosphate kinase and thus adds to the energy stores of the cell (Figure 11.11).

The largest contribution of the cycle to energy production comes from the further

processing of the NADH and $FADH_2$ which have been synthesized. These must be re-oxidized to NAD^+ and FAD if the cycle is to continue to function and it is in this oxidation, by a process called oxidative phosphorylation, that useful quantities of ATP are produced. The complete oxidation of one mole of glucose via glycolysis and the TCA cycle leads to the production of 38 moles of ATP (Table 11.1).

11.3 LINKS WITH OTHER METABOLIC PATHWAYS

Under most circumstances in both plants and animals the TCA cycle is fed by pyruvate derived from glucose, but it has many links with other metabolic pathways, and many catabolic processes produce end products which can feed directly into the cycle. In addition, intermediates of the cycle can be 'siphoned off' for use as the starting point of numerous biosynthetic pathways.

Quantitatively, the most important input into the TCA cycle is pyruvate produced by the glycolytic pathway. However, fatty acid oxidation can also make an important contribution to the supply of acetyl-CoA required

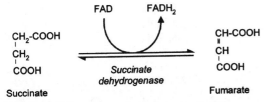

Figure 11.8 The conversion of succinate to fumarate.

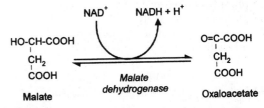

Figure 11.10 The conversion of malate to oxaloacetate.

$$GTP + ADP \underset{Mg^{2+}}{\rightleftharpoons} GDP + ATP$$

Figure 11.11 The conversion of GTP to ATP.

for TCA cycle function. This is particularly important in animals that are in negative energy balance. In this situation glucose oxidation is greatly reduced in response to a fall in blood glucose concentrations. When this occurs, fatty acids are mobilized from adipose tissue depots and undergo β-oxidation in mitochondria to form acetyl-CoA which enters the TCA cycle (see Chapter 13). This has important benefits because it allows those tissues which can readily oxidize fatty acids to meet their ATP requirements, while sparing glucose for those tissues that are essentially dependent on glucose as an energy source (e.g. the brain and nervous tissue) or those tissues that have a specific demand for glucose for biosynthetic processes (e.g. lactose synthesis in mammary tissue). In plants, there is little evidence that storage lipids are oxidized via the TCA cycle in significant quantities. Fatty acid oxidation in plants occurs mainly in the glyoxysomes not in the mitochondria, and the acetyl-CoA produced enters the glyoxylate cycle rather than the TCA cycle (see section 11.6). This mecha-

nism allows the conversion of storage lipid to sucrose which is more readily transported around the plant.

Amino acids can also supply fuel for the TCA cycle and feed into the cycle at various points. Thus, for example, glutamate and aspartate undergo transamination to produce α-ketoglutarate and oxaloacetate, respectively; valine and isoleucine are metabolized to succinyl-CoA; and leucine and lysine produce acetyl-CoA. The relationship between amino acids and the TCA cycle is discussed in Chapter 14. In plants, it is unlikely that much amino acid is metabolized via the TCA cycle. Most evidence suggests that amino acids released by protein breakdown are transported to other tissue sites and stored or used directly for protein synthesis.

The main outflow from the TCA cycle in animals is directed towards the synthesis of fatty acids, amino acids and glucose. In non-ruminant animals, the main source of the acetyl-CoA required for fatty acid synthesis is glucose. This undergoes partial oxidation via the glycolytic pathway to produce pyruvate, which is then converted to mitochondrial acetyl-CoA. Translocation of acetyl-CoA to the cytoplasm occurs via citrate which is split into its two component parts, acetyl-CoA and oxaloacetate,

Table 11.1 ATP production from the oxidation of glucose via glycolysis and the TCA cycle

Substrate	Cofactor produced	ATP consumed	ATP produced
glucose ⟶ glucose-6-phosphate		1	
fructose-6-phosphate ⟶ fructose-1,6-bisphosphate		1	
2 x glyceraldehyde-3-phosphate ⟶ 1,3-bisphosphoglycerate	2 NADH		6
2 x 1,3-bisphosphoglycerate ⟶ 3-phosphoglycerate			2
2 x phosphoenolpyruvate ⟶ pyruvate			2
2 x pyruvate ⟶ acetyl-CoA	2 NADH		6
2 x isocitrate ⟶ α-ketoglutarate	2 NADH		6
2 x α-ketoglutarate ⟶ succinyl-CoA	2 NADH		6
2 x succinyl-CoA ⟶ succinate	2 GTP		2
2 x succinate ⟶ fumarate	2 FADH$_2$		4
2 x malate ⟶ oxaloacetate	2 NADH		6
		2	40
Net synthesis			38

by the cytoplasmic enzyme ATP-citrate lyase. The acetyl-CoA can then be utilized for fatty acid synthesis (see Figure 19.1). In plants, the importance of TCA cycle intermediates in fatty acid and isoprenoid synthesis is less clear. It has been suggested that mitochondrial acetyl-CoA is hydrolysed by acetyl-CoA hydrolase to free acetate, which diffuses from the mitochondria to chloroplasts where it can be reactivated and used for fatty acid synthesis. However, evidence for the existence of free acetate in plant cells is not convincing, and the ATP-citrate lyase route may be the more important source of chloroplast acetyl-CoA.

TCA cycle intermediates can be used for the synthesis of many amino acids, for example the direct transamination of pyruvate, oxaloacetate and α-ketoglutarate yields alanine, aspartate and glutamate, respectively. Glutamate derived from the TCA cycle can then be converted to arginine, proline and glutamine. Oxaloacetate can also be converted via phosphoenolpyruvate to serine, glycine, cysteine, phenylalanine, tyrosine and tryptophan. The links between the TCA cycle and the synthesis of amino acids are discussed in Chapter 20.

TCA cycle intermediates can be directed towards the synthesis of carbohydrate, mainly glucose. This is achieved by conversion of oxaloacetate to phosphoenolpyruvate which feeds into the gluconeogenic pathway (see Chapter 18).

11.4 REPLENISHMENT OF TCA CYCLE INTERMEDIATES

When the TCA cycle serves as a source of precursors for biosynthetic reactions the concentration of cycle intermediates is depleted. Unless the supply of these intermediates is replenished, the flux of carbon through the TCA cycle slows. However, under most conditions the concentration of cycle intermediates remains relatively constant. This balance is achieved by anaplerotic reactions which feed intermediates back into the TCA cycle at spe-

cific points. In animals the most important of these reactions is the carboxylation of pyruvate to produce oxaloacetate. This reaction is catalysed by the allosteric enzyme pyruvate carboxylase. The main activator of this enzyme is acetyl-CoA, and when the intramitochondrial concentration of acetyl-CoA is low the enzyme is virtually inactive. However, if the flux through the TCA cycle slows and the concentration of acetyl-CoA increases, pyruvate carboxylase is activated, making more oxaloacetate available for condensation with acetyl-CoA and thereby increasing the flux through the cycle. A number of other reactions also replenish the supplies of TCA cycle intermediates. These include conversion of phosphoenolpyruvate to oxaloacetate by phosphoenolpyruvate carboxykinase in animals and by phosphoenolpyruvate carboxylase in plants. An anaplerotic reaction common to both plants and animals is the conversion of pyruvate to malate by the malic enzyme. These reactions are shown in Figure 11.12.

11.5 CONVERSION OF PROPIONATE TO GLUCOSE VIA THE TCA CYCLE

Propionate is a three-carbon compound which can be converted to glucose. This conversion is particularly important in ruminant animals where, because of the fermentation of dietary carbohydrate in the rumen, there is little or no supply of glucose to the animal directly from the digestive tract. The conversion of propionate to glucose is discussed in more detail in Chapter 18.

11.6 REGULATION OF THE TCA CYCLE

Because of its central position in the pathways of intermediary metabolism, the TCA cycle is subject to stringent regulation at a number of points. The main entry point into the cycle is via acetyl-CoA which can be supplied from glucose via pyruvate or by fatty acid oxidation. The conversion of pyruvate to either acetyl-CoA or oxaloacetate is controlled by the

$$\text{Pyruvate} + HCO_3^+ + ATP \xrightleftharpoons{\textit{pyruvate carboxylase}} \text{oxaloacetate} + ADP + Pi$$

$$\text{Phosphoenolpyruvate} + CO_2 + GDP \xrightleftharpoons[]{\substack{\textit{phosphoenolpyruvate} \\ \textit{carboxykinase}}} \text{oxaloacetate} + GTP$$

$$\text{Phosphoenolpyruvate} + HCO_3^- \xrightleftharpoons{\textit{pyruvate carboxylase}} \text{oxaloacetate} + Pi$$

$$\text{Pyruvate} + HCO_3^- + NADPH + H^+ \xrightleftharpoons{\textit{malic enzyme}} \text{malate} + NADP^+$$

Figure 11.12 Reactions which replenish the supply of TCA cycle intermediates.

allosteric regulation of the PDH complex or pyruvate carboxylase, respectively. The PDH complex is activated by its substrate, pyruvate. It is inhibited by its product, acetyl-CoA, and by long-chain fatty acyl-CoAs. This pattern of inhibition provides a mechanism which limits the use of glucose for ATP synthesis when fatty acids are available. Pyruvate carboxylase, on the other hand, is activated by acetyl-CoA. The opposing effects of acetyl-CoA on these two enzymes have a number of benefits. For example, when the TCA cycle functions in an aerobic tissue, primarily to supply ATP, there is little requirement for additional input of oxaloacetate into the cycle, and the acetyl-CoA concentration in the mitochondria remains relatively low due to its constant oxidation. If, however, cycle intermediates are removed to supply biosynthetic pathways, the supply of oxaloacetate will be reduced so that acetyl-CoA may not be utilized as rapidly as it is produced. This can result in an increase in the mitochondrial concentration of acetyl-CoA. The higher acetyl-CoA concentration reduces the activity of the PDH complex and increases the activity of pyruvate carboxylase, thereby restoring oxaloacetate availability.

In addition to allosteric modification, the PDH complex may also be regulated by phosphorylation/dephosphorylation. Its activity is

reduced by phosphorylation which occurs in response to increased concentrations of NADH and acetyl-CoA.

The three enzymes in the TCA cycle which catalyse irreversible reactions, citrate synthase, isocitrate dehydrogenase and α-ketoglutarate dehydrogenase, also regulate the activity of the cycle. Their activity is modified by changes in the energy status of the cell which are reflected in the ratios of the concentrations of ATP/ADP and NADH/NAD$^+$. In general, high concentrations of end product, ATP and/or NADH, have inhibitory effects.

11.7 THE GLYOXYLATE CYCLE

Located in special subcellular organelles, the glyoxysomes, this pathway occurs in most bacteria, protozoa, fungi, algae and higher plants. The cycle provides a means by which sugars and other important cellular metabolites can be synthesized from two-carbon compounds such as acetate and ethanol.

The conversions of phosphoenolpyruvate to pyruvate, and of pyruvate to acetyl-CoA, have large negative free energy changes and are therefore effectively irreversible. Thus, acetyl-CoA derived, for example, from fatty acid oxidation cannot be converted directly to pyruvate and on to glucose. Instead, glucose synthesis is achieved from a variety of gluco-genic intermediary metabolites which feed into the gluconeogenic pathway at a number of different points. The most important of these glucogenic metabolites is oxaloacetate, which is converted to phosphoenolpyruvate by phosphoenolpyruvate carboxykinase.

As has been seen, the main fate of mito-chondrial acetyl-CoA is to undergo condensa-tion with oxaloacetate to form citrate. The net synthesis of glucose from acetyl-CoA is impos-sible because two carbon atoms are lost as CO_2 in the reactions catalysed by isocitrate dehy-drogenase and α-ketoglutarate dehydroge-nase. In order for glucose synthesis to occur from acetyl-CoA, the two reactions which pro-duce CO_2 must be by-passed. This is what occurs in the glyoxylate cycle.

The reactions of the glyoxylate cycle are shown in Figure 11.13. The first two reactions are identical to those of the TCA cycle, but isocitrate does not then undergo oxidative decarboxylation to α-ketoglutarate. Instead, it is split by the enzyme isocitrate lyase into suc-cinate and the two-carbon compound, glyoxy-late. Glyoxylate then condenses with a mole-cule of acetyl-CoA to form malate which can be converted to oxaloacetate. Since no CO_2 is produced in these reactions, the overall effect of the glyoxylate cycle is the utilization of two molecules of acetyl-CoA for the net synthesis of oxaloacetate. This oxaloacetate may then be converted to phosphoenolpyruvate and on to glucose in the cytoplasm.

The succinate produced in glyoxysomes can also be used to produce oxaloacetate. However, the glyoxysomes do not contain the TCA cycle enzymes required to bring about this conver-sion, and the succinate must be transferred to the mitochondria before it can be metabolized to oxaloacetate. In order to maintain the supply of oxaloacetate for the glyoxylate cycle, mito-chondrial oxaloacetate must be transferred back to the glyoxysomes. A barrier to this transfer is the relative impermeability of the inner mito-chondrial membrane to oxaloacetate. In order to overcome this problem oxaloacetate is transaminated to aspartate, which is then trans-ported to the glyoxysomes where it is convert-ed back to oxaloacetate (Figure 11.14). This cycle is important during the germination of many types of seeds where significant quanti-ties of lipid are stored. It enables these concen-trated energy reserves to be converted to sugars which can be used to provide the building blocks for the growing plant.

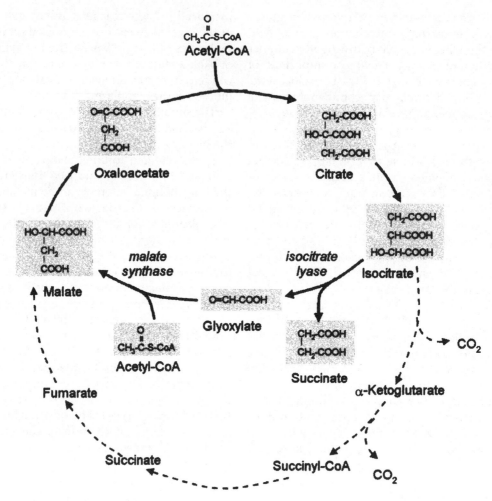

Figure 11.13 The glyoxylate cycle.

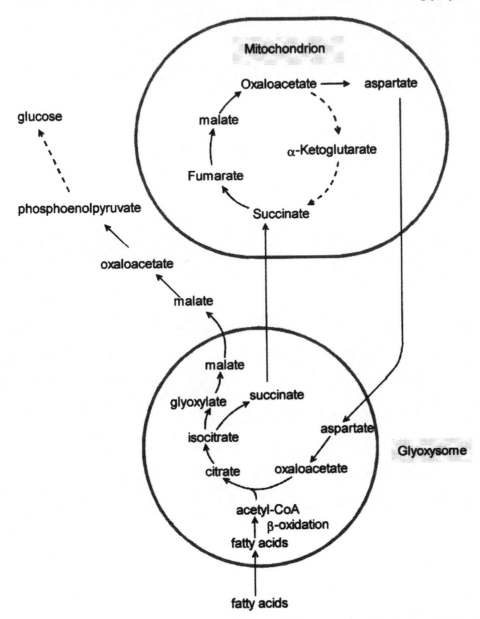

Figure 11.14 The relationship between the metabolism of glyoxysomes and mitochondria. Isocitrate is split into glyoxylate and succinate in the glyoxysomes. Glyoxylate is converted to malate, which is transported to the cytoplasm and used for glucose synthesis. Because glyoxysomes do not contain the enzymes needed to regenerate oxaloacetate, succinate is transported to the mitochondria and converted to oxaloacete which is transported back to the glyoxysomes as aspartate. In the glyoxysomes, aspartate is converted back to oxaloacetate.

ELECTRON TRANSPORT AND OXIDATIVE PHOSPHORYLATION

12.1 INTRODUCTION

In previous chapters we have seen that the catabolic processes such as glycolysis, the TCA cycle and fatty acid oxidation bring about oxidation of fuel molecules in order to release the 'trapped' chemical energy they contain. The oxidation reactions usually involve the transfer of electrons and/or hydrogen to acceptor molecules. A recurrent theme of oxidation pathways is the reduction of NAD^+ to NADH $+ H^+$ and FAD to $FADH_2$. The cell contains a finite amount of NAD^+ and FAD. As they are essential for oxidative metabolism, it follows that once they are converted to their reduced forms, intermediary metabolism will grind to a halt unless there is a mechanism for their re-oxidation. One such mechanism is the conversion of pyruvate to lactate under anaerobic conditions (Figure 12.1) (see also Chapter 16).

However, under aerobic conditions the re-oxidation of NADH and $FADH_2$ is brought about by a complex series of reactions located in the inner mitochondrial membrane and known as the electron transport chain. The overall effect of this process is the transfer of electrons from NADH and $FADH_2$ to oxygen, which is reduced to water:

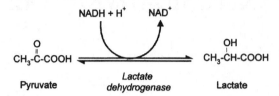

Figure 12.1 The conversion of pyruvate to lactate by lactate dehydrogenase.

$$NADH + H^+ + \tfrac{1}{2}O_2 \rightarrow H_2O + NAD^+$$
$$FADH_2 + \tfrac{1}{2}O_2 \rightarrow H_2O + FAD$$

Electron transfer to oxygen is not direct but takes place in an ordered sequence of discrete steps during which there is considerable release of free energy. This energy is not wasted but is utilized by the mitochondrion to bring about the phosphorylation of ADP to ATP. The overall process of synthesis of ATP via the electron transport chain is referred to as oxidative phosphorylation, and is summarized in Figure 12.2.

The nature of the link between electron transport and ATP synthesis puzzled biochemists for many years, and it was not until the pioneering work of Peter Mitchell in the 1960s on the movement of protons across the inner mitochondrial membrane that a credible mechanism linking electron transport and ATP synthesis began to develop. In the intervening years Mitchell's original ideas have been extensively tested, and it is now accepted that the electrochemical energy of a proton gradient created across the inner mitochondrial membrane by the electron transport chain is used to drive ATP synthesis. This mechanism is referred to as the chemiosmotic model of ATP synthesis. In fact, it is now known that chemiosmotic mechanisms are not only involved in ATP synthesis, but also in a number of energy-dependent processes such as active transport across membranes.

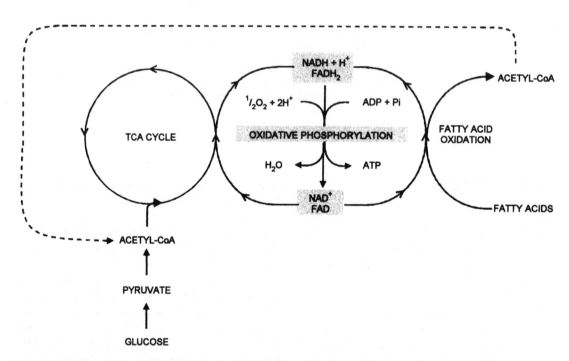

Figure 12.2 The regeneration of NAD^+ and FAD by the mitochondrial electron transport chain. Mitochondria contain a limited supply of NAD^+ and FAD. The reduced coenzymes produced by fatty acid oxidation and the TCA cycle feed into the electron transport chain where they are oxidized. In the process, ATP is synthesized. The oxidized coenzymes are re-used for oxidative metabolism.

12.2 THE MITOCHONDRION

Mitochondria are surrounded by two membranes. The outer membrane has little biochemical activity but acts as a partial barrier to the movement of molecules. It contains a protein, porin, which forms channels through the membrane allowing free movement of ions and molecules up to a molecular weight of approximately 10 000. The inner mitochondrial membrane, on the other hand, is a highly selective permeability barrier and the movement of ions and polar molecules across this membrane is tightly regulated by specific translocases. The ratio of protein to lipid in this membrane (about 3:1) is higher than in any other cellular membrane and is a reflection of the biochemical activity of the membrane. When observed under the electron microscope it can be seen that the inner mitochondrial membrane is highly convoluted. The folds of the membrane are referred to as cristae. It is probable that this extensive folding evolved to increase the membrane surface area to allow for rapid exchange between the central compartment of the mitochondria, the matrix, and the intermembrane space which has a similar composition to the cytoplasm. The processes of electron transport and oxidative phosphorylation are located in this membrane. The mitochondrial matrix is the site of the TCA cycle, fatty acid oxidation and amino acid metabolism.

12.3 COMPONENTS OF THE ELECTRON TRANSPORT CHAIN

Extensive analysis of the inner mitochondrial membrane has revealed much about the components of the electron transport chain. In broad terms, it is possible to group its components into four main categories. Each will be considered before describing the transfer of protons from one component to another, and how the various components are arranged into a functional electron transport chain.

Mitochondria are a rich source of cytochromes and it is these proteins which give them their brown colour. A number of different cytochromes have been identified in the electron transport chain. Cytochromes are globular proteins all of which contain iron porphyrin groups similar to those in haem. The different cytochromes, a, a_3, b_{560}, b_{562}, c and c_1 are characterized by the different subtypes of porphyrins they contain. The b cytochromes contain iron-protoporphyrin IX, which is identical to that found in haem and myoglobin. In cytochrome types a and c the substituent groups attached to the porphyrin ring structure are slightly different, the structures being referred to as haem A and haem C, respectively.

The important feature of the cytochromes, which enables them to act as electron carriers, is the presence of the iron atom which can undergo oxidation and reduction between the Fe (II) and Fe (III) states (Figure 12.3).

The a-type cytochromes also contain copper, which can participate in electron transfer reactions through transition between the Cu (I) and Cu (II) oxidation states. These copper atoms play an important role in the final stage of transfer of electrons to oxygen.

12.3.1 FLAVOPROTEINS

The dehydrogenase enzymes which initiate the movement of electrons into the electron transport chain are flavoproteins. NADH dehydrogenase, which catalyses the oxidation of NADH, is a membrane-bound enzyme which contains flavin mononucleotide (FMN) as a prosthetic group. This enzyme accepts electrons from NADH produced in the mitochondrial matrix. Succinate dehydrogenase is the only enzyme of the TCA cycle which is directly linked to the electron transport chain. It has FAD as its prosthetic group and is located on the inner surface of the inner membrane.

$$Fe\ (III) + e^- \rightleftharpoons Fe\ (II)$$

Figure 12.3 The interconversion of iron between oxidation states.

There are a number of other flavin-containing enzymes which feed electrons into the electron transport chain. Acyl-CoA dehydrogenase, which catalyses the first reaction in the β-oxidation of fatty acids, contains FAD and is located in the mitochondrial matrix. Glycerol-3-phosphate dehydrogenase is located on the outer surface of the inner mitochondrial membrane. This enzyme also contains FAD and catalyses the oxidation of glycerol-3-phosphate to dihydroxyacetone phosphate.

12.3.2 THE IRON–SULPHUR PROTEINS

These proteins differ from the cytochromes in that, although they contain iron, they do not contain haem. Instead, the iron atoms are located between sulphur atoms of cysteine residues of the protein and inorganic sulphide to form a complex. The iron atoms function in the same way as in cytochromes, that is, they act as electron carriers by undergoing cyclical oxidation and reduction. The iron and sulphur occur as clusters or centres within the protein. In at least one case, that of succinate dehydrogenase, these iron–sulphur centres occur in proteins which also contain other electron carriers, e.g. FAD, and may assist in the transfer of electrons from FAD.

12.3.3 UBIQUINONE

Ubiquinone, also referred to as coenzyme Q, has a benzoquinone structure and is a lipid-soluble electron carrier. A number of different ubiquinones have been identified which differ in the length of their isoprenoid side chains. The structure of ubiquinone is shown in Figure 4.10.

As will become clear below, ubiquinone acts as an electron sink which links the transfer of electrons from NADH and from $FADH_2$ in succinate dehydrogenase, acyl-CoA dehydrogenase and glycerol-3-phosphate dehydrogenase to the rest of the electron transport chain.

12.4 THE ELECTRON TRANSPORT CHAIN COMPLEXES

Studies of the inner mitochondrial membrane have shown that electron transport is a highly organised series of reactions which occur in a fixed sequence. The various component parts of the chain have been examined in detail, and it has become clear that the process can be split into four discrete stages which contain adjacent groups of electron carriers and are referred to as complexes I, II, III and IV.

12.4.1 COMPLEX I – THE NADH–DEHYDROGENASE COMPLEX

Also known as NADH–ubiquinone oxidoreductase, this complex is the largest of the four. It contains approximately 26 polypeptides and at least seven iron–sulphur centres. It catalyses the transfer of electrons from NADH to the FMN prosthetic group of NADH dehydrogenase, their subsequent passage through the iron–sulphur centres, and finally transfer to the mobile carrier ubiquinone. The orientation of complex I in the inner mitochondrial membrane allows it to accept the hydrogens from NADH which are transferred to FMN, reducing it to $FMNH_2$. At this point the hydrogens and their associated electrons part company. The protons are released into the intermembrane space while the electrons are passed via the iron–sulphur proteins to ubiquinone, where a further two protons are taken from the matrix to yield UQH_2, reduced ubiquinone (Figure 12.4).

In this way, complex I makes the first contribution to the development of the proton gradient across the inner mitochondrial membrane.

12.4.2 COMPLEX II – THE SUCCINATE DEHYDROGENASE COMPLEX

This complex, which is also known as succinate–ubiquinone oxidoreductase, is linked directly to the TCA cycle via the enzyme succi-

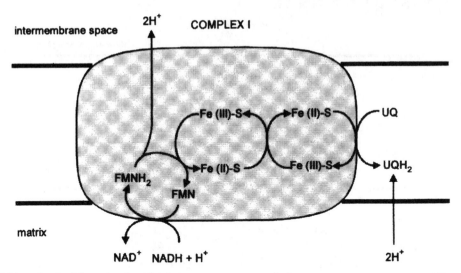

Figure 12.4 Complex I of the mitochondrial electron transport chain.

nate dehydrogenase. The complex contains two iron–sulphur proteins which accept electrons from the $FADH_2$ in succinate dehydrogenase and pass them on to ubiquinone. The transfer of electrons to ubiquinone is accompanied by the uptake of two protons from the matrix. However, there is no evidence that this complex contributes to the transmembrane proton gradient, and while it is generally accepted that complexes I, III and IV span the inner membrane, it is likely that complex II does not (Figure 12.5).

12.4.3 COMPLEX III – THE CYTOCHROME B, c_1 COMPLEX

This complex is also known as ubiquinone cytochrome c oxidoreductase. Its net effect is to transfer electrons from reduced ubiquinone to cytochrome c. The complex contains three cytochromes, b_{560}, b_{562} and c_1, an iron–sulphur protein, and at least six other proteins which together span the membrane. The transfer of electrons from ubiquinone to cytochrome c involves a rather circuitous series of reactions known as the Q cycle, involving two molecules of ubiquinone, the transfer of electrons through the cytochrome

complex in two one-electron steps, and the transfer of two protons into the intermembrane space. This complex makes an important contribution to the transmembrane potential. Finally, electrons pass from cytochrome c_1 to cytochrome c, which is not part of complex III, but is a peripheral protein located on the outer surface of the inner membrane and acts as the link between complexes III and IV (Figure 12.6).

12.4.4 COMPLEX IV – CYTOCHROME OXIDASE

This complex contains cytochromes a and a_3. It differs from the other cytochrome-containing complex, complex III, in that its cytochromes contain not only Fe atoms but also Cu atoms, which are essential in the final reduction of oxygen to water. The stoichiometry of the reaction suggests that complex IV catalyses the transfer of four electrons. This is important because it allows the complete reduction of oxygen to water without the generation of reactive intermediate products such as hydrogen peroxide, H_2O_2, which can cause considerable damage within the cell.

$$O_2 + 4H^+ + 4e^- \rightarrow 2H_2O$$

COMPLEX II

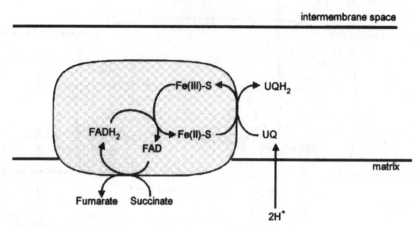

Figure 12.5 Complex II of the mitochondrial electron transport chain.

This complex contributes to the proton gradient in two ways: firstly by the consumption of matrix protons during the reduction of oxygen, and secondly by the direct transfer of protons across the membrane (Figure 12.7).

The overall arrangement of electron transport complexes is shown in Figure 12.8.

12.5 COUPLING OF ELECTRON TRANSPORT TO ATP SYNTHESIS

The movement of electrons along the electron transport chain results in a large release of free energy. The energy is not used directly to synthesize ATP, but is conserved as the transmembrane proton gradient which is

COMPLEX III

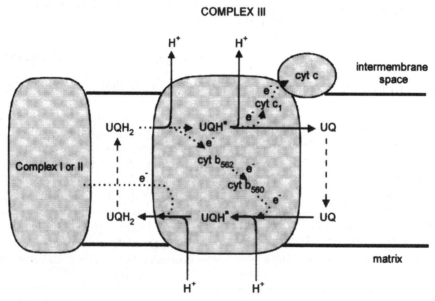

Figure 12.6 Complex III of the mitochondrial electron transport chain.

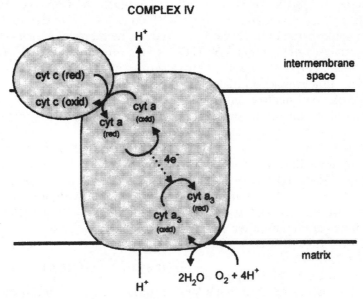

Figure 12.7 Complex IV of the mitochondrial electron transport chain.

coupled to the synthesis of ATP. The movement of protons across the inner mitochondrial membrane has two major effects. Firstly, it creates a chemical gradient of protons across the membrane which can be measured in terms of a difference in pH between the mitochondrial matrix and the intermembrane space. Thus, the matrix is alkaline with respect to the intermembrane space. Secondly, because protons are charged ions, the intermembrane space is positively charged compared with the matrix. These two factors combine to create an electrochemical gradient across the inner mitochondrial membrane which is referred to as the proton motive force (PMF).

The PMF is maintained because the inner mitochondrial membrane is impermeable to protons and does not allow them to re-enter the matrix down their concentration gradient.

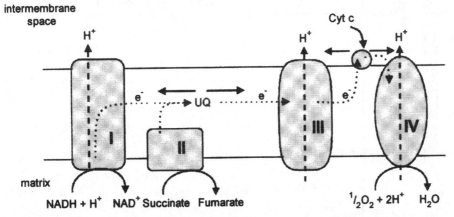

Figure 12.8 The organisation of the mitochondrial electron transport chain in the inner mitochondrial membrane, showing how protons are 'pumped' into the intermembrane space.

Instead the energy of the PMF is channelled through an enzyme complex in the inner mitochondrial membrane known as the F_0F_1-ATPase, which catalyses the synthesis of ATP from ADP and inorganic phosphate. Examination of the inner mitochondrial membrane under the electron microscope reveals an array of projections on its inner surface. Each projection has the appearance of a spherical unit linked to the membrane via a stalk. Initially it was thought that these structures might be artefacts produced by the fixation process required to visualize the mitochondrial membrane, but since it has been possible to isolate them, studies to examine their properties have confirmed that they correspond to the F_0F_1-ATPase (Figure 12.9).

The structure of the enzyme complex is not fully understood but it is recognised that it has two main components. The F_0 unit (so named because it confers oligomycin sensitivity on the enzyme) acts as an anchor point within the membrane for the F_1 unit and forms a channel through which protons can return to the mitochondrial matrix.

The F_1 unit contains the catalytic site for ATP synthesis. When isolated from the membrane, the F_1 unit cannot synthesize ATP but catalyses the hydrolysis of ATP to ADP and P_i. When isolated F_0 and F_1 units are reconstructed in membrane vesicles they combine to form a functional enzyme complex capable of synthesizing ATP in the presence of a proton gradient.

The precise mechanism by which the F_0F_1-ATPase catalyses the synthesis of ATP is not known. It has been suggested that once synthesized, ATP remains tightly bound to the enzyme active site, blocking any further synthesis of ATP. The proton gradient appears to provide the driving force to release ATP from the enzyme, allowing further synthesis to take place.

12.6 THE YIELD OF ATP

Estimates of the amount of ATP synthesized when electrons pass from NADH to oxygen vary slightly. It is generally agreed that around 10 protons are transported across the inner mitochondrial membrane for each molecule of NADH oxidized. It is estimated that three protons must pass through the F_0F_1-ATPase during the synthesis of each molecule of ATP. In addition, one proton is involved in the membrane transport of ADP and P_i into the mitochondrial matrix, and of ATP into the cytoplasm. Thus the number of molecules of ATP synthesized per molecule of NADH oxidized can be calculated at between 2.5 and 3.3 (depending on whether the number of protons used is taken as three or four). Similarly, during the transfer of electrons from $FADH_2$ to oxygen, six protons are transported into the intermembrane space, giving an estimated yield of ATP per molecule of $FADH_2$ oxidized of 1.5–2.0. These figures agree well with the laboratory measurements. For most purposes it is normally assumed that the ATP yield for NADH and $FADH_2$ is three and two, respectively. This fits well with the view that there are three coupling sites in the electron transport chain between complexes I and IV, and only two coupling sites between complexes II and IV.

intermembrane space

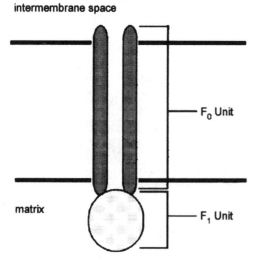

F_0 Unit

matrix

F_1 Unit

Figure 12.9 The structure of F_1F_0-ATPase.

12.7 NADH PRODUCED IN THE CYTOPLASM ENTERS THE ELECTRON TRANSPORT CHAIN VIA SHUTTLE REACTIONS

Not all of the NADH oxidized by the electron transport chain is produced in the mitochondrial matrix; some is produced in the cytoplasm. However, the inner mitochondrial membrane is impermeable to NADH. This problem is overcome by use of shuttle reactions.

In the malate–aspartate shuttle, oxaloacetate is reduced to malate in the cytoplasm by malate dehydrogenase. In this reaction, cytoplasmic NADH is oxidized to NAD^+. The malate is transported into the mitochondrial matrix via a membrane carrier which exchanges cytoplasmic malate for mitochondrial α-ketoglutarate. Once in the mitochondria, NADH is regenerated by conversion of malate back to oxaloacetate by mitochondrial malate dehydrogenase. Thus the net effect of this part of the shuttle is the transfer of NADH from the cytoplasm to the mitochondrial matrix. The shuttle cycle is completed by a transamination reaction in which α-ketoglutarate is converted to glutamate and oxaloacetate is converted to aspartate. This reaction is catalysed by aspartate aminotransferase. The aspartate produced is transported to the cytoplasm in exchange for glutamate, and is used to regenerate cytoplasmic oxaloacetate by a reversal of the transamination reaction catalysed by cytoplasmic aspartate aminotransferase (Figure 12.10).

Cytoplasmic NADH can also feed into the electron transport chain via the glycerol-3-phosphate shuttle, in which cytoplasmic glycerol-3-phosphate dehydrogenase catalyses the reduction of dihydroxyacetone phosphate to glycerol-3-phosphate. Glycerol-3-phosphate can pass freely from the cytoplasm to the mitochondrial intermembrane space, where it undergoes oxidation to dihydroxyacetone phosphate by a mitochondrial glycerol-3-phosphate dehydrogenase located on the outer surface of the inner mitochondrial membrane. The enzyme contains FAD as its prosthetic group and links directly to complex III of the electron transport chain (Figure 12.11).

NADH entering the mitochondria via the malate–aspartate shuttle feeds into the electron transport chain via complex I and can therefore yield three molecules of ATP, whereas protons from NADH passing via the glycerol-3-phosphate shuttle enter the electron transport chain via complex III, resulting in the production of only two molecules of ATP.

12.8 REGULATION OF OXIDATIVE PHOSPHORYLATION BY ADP/ATP SUPPLY

Measurement of the P/O ratio of mitochondria demonstrates the tight coupling between electron transport and ATP synthesis in freshly prepared mitochondria. This ratio, measured using an oxygen electrode, indicates the quantity of ATP synthesized per atom of oxygen reduced to water. When mitochondria are given an oxidizable substrate such as malate or succinate and an excess of inorganic phosphate, they will consume oxygen at a rate dependent on the supply of ADP.

When ADP supply is very low the rate of oxygen consumption is also very low. However, if ADP is added to the mitochondrial suspension there is a burst of oxygen consumption which lasts until the ADP has been consumed, i.e. converted to ATP. This regulation by ADP supply is called respiratory control, and it depends on the maintenance of the integrity of the inner mitochondrial membrane. As mitochondrial preparations age, respiratory control is gradually lost as the inner membrane becomes 'leaky' and the proton gradient is dissipated. Similarly, if mitochondrial membranes are damaged by chemical or mechanical treatment, respiratory control is lost.

Mitochondria that exhibit tight respiratory control are said to be 'coupled', that is the process of electron transport is governed by the rate at which ATP is synthesized, which in turn is proportional to the [ATP]/[ADP] ratio of the cell. Many compounds have been identified which in some way break the link

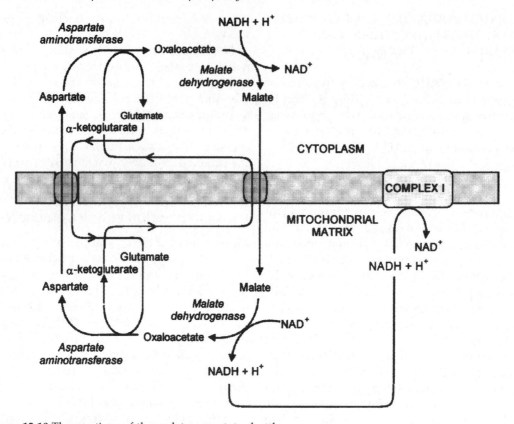

Figure 12.10 The reactions of the malate–aspartate shuttle.

between electron transport and ATP synthesis. These compounds are referred to as 'uncouplers'. They are distinct from compounds which specifically inhibit electron transport or ATP synthesis, in that their common mode of action is to destroy the transmembrane proton

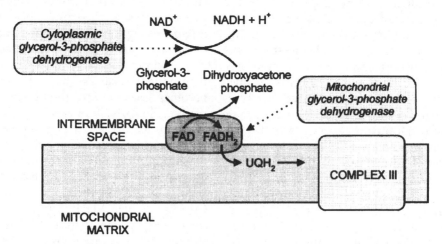

Figure 12.11 The reactions of the glycerol-3-phosphate shuttle and its relationship to complex III.

gradient (Figure 12.12). A classical example of a synthetic uncoupler is 2,4-dinitrophenol, a lipid-soluble weak acid which can penetrate the inner mitochondrial membrane, and acts as a proton shuttle. The ionophore valinomycin can also act as an uncoupler, not by allowing protons to re-enter the matrix, but by providing a mechanism for the influx of K^+ ions into the matrix. This influx of positively charged ions reduces the electrical potential across the membrane and reduces the effectiveness of the PMF.

Thermogenin, a naturally occurring uncoupler, has received much attention. It is found in the mitochondria of brown adipose tissue which is important in the process of non-shivering thermogenesis, particularly in neonatal animals. This type of adipose tissue is brown due to the presence of unusually high numbers of mitochondria. The function of this tissue is to oxidize fatty acids, not for ATP synthesis, but for heat production. This is achieved by the presence of thermogenin in the inner mitochondrial membrane, which provides hydrophilic channels in the membrane through which protons can pass. The energy released by electron transport in these mitochondria is dissipated as heat via the extensive network of capillaries which perfuse the tissue.

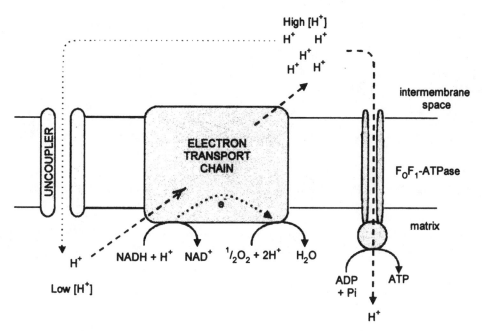

Figure 12.12 The overall process of oxidative phosphorylation showing how ATP synthesis occurs when protons in the intermembrane space may flow back to the matrix via F_0F_1-ATPase, and how uncouplers destroy the proton gradient.

FATTY ACID OXIDATION AND LIPID BREAKDOWN

13.1 INTRODUCTION

Fatty acids can undergo a number of oxidative modifications. Quantitatively the most important pathway of fatty acid oxidation is referred to as the β-oxidation pathway. This process occurs in all organisms and is the main route by which fatty acids are utilized for energy production. In β-oxidation, fatty acids are degraded to smaller compounds, usually acetyl-CoA, which can then be oxidized via the TCA cycle. The conversion of long-chain fatty acids to acetyl-CoA and its subsequent oxidation in the TCA cycle leads to the production of large quantities of the reduced coenzymes, NADH and $FADH_2$, which are used for the production of ATP. Additionally, in plants, the β-oxidation of fatty acids can be used to provide acetyl-CoA for carbohydrate synthesis via the glyoxylate pathway (Chapter 11). Two other oxidation processes, known as α-oxidation and ω-oxidation, are involved in the specific modification of fatty acids, the former required to oxidize fatty acids at carbon 2, the latter oxidizing fatty acids at the methyl terminal or ω-carbon.

Another type of fatty acid oxidation occurs during the synthesis of the eicosanoids. This group of fatty acid derivatives, considered to be 'local hormones' in animals, are produced from the fatty acids all-*cis* $\Delta^{8,11,14}$ C20:3, all-*cis* $\Delta^{5,8,11,14}$ C20:4 and all-*cis* $\Delta^{5,8,11,14,17}$ C20:5 by the action of either a lipoxygenase or a cyclo-oxygenase.

Unsaturated fatty acids also undergo non-enzymic peroxidation. This process is usually free radial-mediated, and the susceptibility of fatty acids to peroxidative breakdown is directly proportional to the number of double bonds present. Peroxidation can be both detrimental and beneficial. The 'off' smells and flavours associated with fatty foods are due to the build-up of short-chain aldehydes and ketones with unpleasant odours. However, some of the pleasant aromas and tastes associated with fresh fruit and vegetables are also attributed to volatile derivatives of fatty acids.

13.2 β-OXIDATION

Most tissues can oxidize fatty acids by β-oxidation. There are two main sites of β-oxidation in the cell. In animals, the mitochondrial matrix is the site where fatty acids are completely oxidized to acetyl-CoA. Peroxisomal oxidation of fatty acids is important in the liver and kidney. β-Oxidation in peroxisomes appears to be a mechanism for shortening the chain length of fatty acids, which may then be further oxidized in the mitochondrial matrix. In plant leaf tissue, peroxisomes rather than mitochondria are the main sites of fatty acid oxidation, whereas in seeds, glyoxysomes are the major sites of fatty acid oxidation where the product of oxidation, acetyl-CoA, can feed directly into the glyoxylate pathway.

13.2.1 MITOCHONDRIAL β-OXIDATION IN ANIMAL TISSUES

Fatty acids metabolized by the β-oxidation pathway may come from both exogenous sources (i.e. dietary) or endogenous sources (i.e. from adipose tissue and liver). During transport around the body in the blood they may be present as non-esterified fatty acids (NEFAs) bound to the carrier protein albumin, or contained in phospholipids and triacyglycerols incorporated in lipoproteins, mainly chylomicrons and very low-density lipoproteins (VLDL).

Fatty acids in lipoprotein-bound lipids are released into the blood by the action of lipoprotein lipase found on the luminal surface of the capillary endothelial cells. NEFAs pass from the blood into adjacent cells both by diffusion and apparently by a membrane-mediated process.

In the cytoplasm, fatty acids must be activated to their CoA ester derivatives before they can be further metabolized. The reaction requires ATP and is catalysed by enzymes called acyl-CoA synthetases (Figure 13.1). There are a number of different enzymes all of which catalyse the same general reaction but which have differing chain-length specificities. The two most important in fatty acid oxidation are:

- medium-chain acyl-CoA synthetase (C4–C12);
- long-chain acyl-CoA synthetase (C10+).

Long-chain acyl-CoA synthetase is a membrane-bound enzyme found on the peroxisome membrane, the endoplasmic reticulum and the outer mitochondrial membrane. This distribution allows fatty acids to be metabolized by various pathways at different subcellular sites, i.e. chain shortening in peroxisomes, phospholipid and triacylglycerol synthesis on the endoplasmic reticulum, and β-oxidation in the mitochondria.

The inner mitochondrial membrane presents a barrier to fatty acyl-CoAs destined for oxidation. Neither CoA itself nor esters of CoA can pass through this membrane. This is largely due to the large size and charged nature of the CoA molecule (see Figure 19.6).

To facilitate the movement of fatty acids across the inner mitochondrial membrane to the mitochondrial matrix, acyl-CoAs are con-

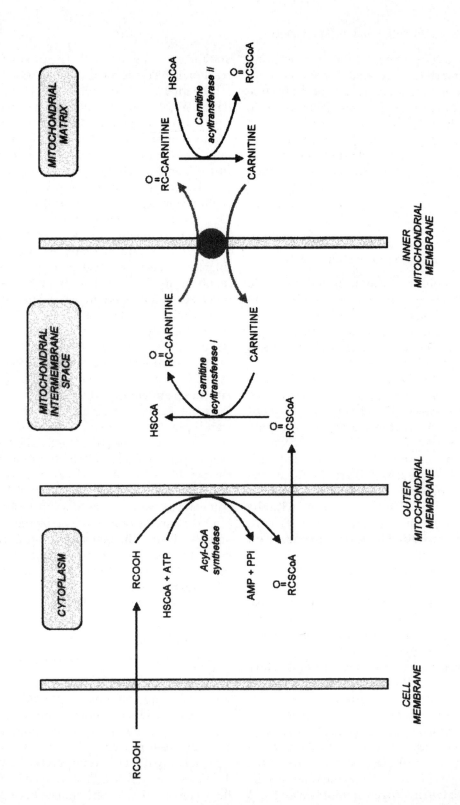

Figure 13.1 Activation and transport of fatty acids to the mitchondrial matrix.

verted to acyl-carnitines by the enzyme carnitine acyltransferase I located in the mitochondrial intermembrane space. Fatty acyl-carnitine derivatives cross the inner membrane in exchange for carnitine by means of a transport protein, acyl-carnitine translocase. In the mitochondrial matrix fatty acyl-carnitines are converted back to CoA derivatives by carnitine acyltransferase II. Once in the mitochondrial matrix as CoA esters, fatty acids can undergo β-oxidation (Figure 13.1).

13.2.2 THE REACTIONS OF β-OXIDATION

The β-oxidation pathway uses a sequence of four reactions to remove a two-carbon unit from the carboxyl end of the fatty acid. This reaction sequence is repeated until the fatty acid is completely oxidized. This process is summarized in Figure 13.2.

The first step in the β-oxidation of a fatty acyl-CoA is the removal of a hydrogen atom from each of carbons 2 and 3 to yield a Δ^2 *trans*-enoyl-CoA. The hydrogen atoms are accepted by FAD to give $FADH_2$ and the reaction is catalysed by a group of enzymes called acyl-CoA dehydrogenases (Figure 13.2). In the complete oxidation of palmitic acid, this reaction takes place seven times. To cope with the decrease in the chain length of the fatty acyl-CoA as it is oxidized at least three acyl-CoA dehydrogenases are required.

The next step in the oxidation process is the addition of a molecule of water across the double bond (Figure 13.2). The enzyme responsible for this reaction is enoyl-CoA hydratase. This enzyme, which has a broad chain-length specificity, catalyses the stereospecific addition of water across the double bond. Thus, the hydration of a *trans* double bond yields the L-3-hydroxyacyl-CoA (Figure 13.2).

The third enzyme in the sequence, L-3-hydroxyacyl-CoA dehydrogenase, is relatively non-specific with respect to chain length but is absolutely specific for the L-stereoisomer of the 3-hydroxyacyl-CoA. The enzyme catalyses the removal of two hydrogen atoms from carbon 1

and their transfer to NAD^+. The product of the reaction is 3-ketoacyl-CoA (Figure 13.2).

The final step in the reaction cycle of β-oxidation is the thiolytic cleavage of 3-ketoacyl-CoA between carbons 2 and 3. The result is the production of a molecule of acetyl-CoA and a fatty acyl-CoA two carbon atoms shorter than the fatty acid which started the cycle. The reaction is catalysed by the enzyme acetyl-CoA acyltransferase, also known as β-ketothiolase. This enzyme is active against substrates varying in chain length from 4 to 18 carbons. In addition, a second enzyme, acetoacetyl-CoA thiolase, is also found in mitochondria. It is involved in the metabolism of the ketone body, acetoacetyl-CoA, in tissues such as heart muscle which use ketone bodies as an energy source, and in liver, a major site of ketone body synthesis (Figure 13.2).

The two major fuels used by cells to synthesize ATP are glucose and fatty acids. Long-chain fatty acids are highly reduced compounds and therefore their complete oxidation to CO_2 and water can lead to the production of large amounts of ATP. The complete β-oxidation of 1 mole of palmitic acid results in the production of 8 moles of acetyl-CoA which can be further oxidized via the TCA cycle. The ATP yield from palmitic acid oxidation is shown in Table 13.1.

One molecule of ATP is hydrolysed to AMP and PP_i during the activation of a fatty acid prior to transport into the mitochondrial matrix. This is equivalent to the utilization of 2 ATP, thus there is a net synthesis of 129 moles of ATP from the β-oxidation of 1 mole of palmitic acid.

13.2.3 β-OXIDATION OF ODD-NUMBERED ACIDS

The β-oxidation of odd-numbered straight-chain fatty acids is almost identical to those containing an even number of carbon atoms. The process proceeds by the sequential removal of two carbon units, acetyl-CoA, until a five-carbon intermediate remains. This is then converted into one molecule of acetyl-

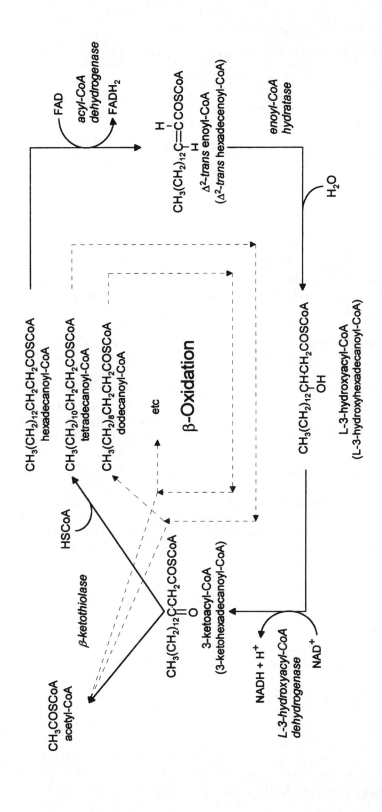

Figure 13.2 The overall scheme of β-oxidation.

Table 13.1 ATP yield from the oxidation of palmitic acid

Pathway	Net conversion	Cofactors produced	ATP yield
β-oxidation	palmitic acid to 8 acetyl-CoA	7 NADH	≡ 21 ATP
		7 FADH$_2$	≡ 14 ATP
TCA cycle	8 acetyl-CoA to CO$_2$ and water	24 NADH	≡ 72 ATP
		8 FADH$_2$	≡ 16 ATP
		8 GTP	≡ 8 ATP
		Total	= 131 ATP

CoA and one molecule of propionyl-CoA. Since propionyl-CoA is a three-carbon compound it can be utilized for glucose synthesis; however, the quantity of odd-numbered fatty acids found in animal tissues is usually small (1–2%) and therefore the contribution they make to glucose production is negligible compared with other glucogenic substrates. This is one of the few occasions in animals when a product of fatty acid oxidation can be converted to glucose.

13.2.4 β-OXIDATION OF UNSATURATED ACIDS

A slightly modified version of the β-oxidation scheme is required to oxidize unsaturated fatty acids. Figure 13.2 shows that the unsaturated intermediate produced by acyl-CoA dehydrogenase has the Δ^2-*trans* configuration, whereas most naturally occurring unsaturated fatty acids contain double bonds with the *cis* configuration. As unsaturated fatty acids are degraded by removal of two carbon units from the carboxyl end, the position of the double bonds moves relative to the carboxyl carbon. The carboxyl carbon is always carbon 1, so as the fatty acid becomes shorter the position of any double bond approaches carbon 1. This has two important consequences in the β-oxidation of unsaturated fatty acids.

For a fatty acid like oleic acid with one *cis* double bond between carbons 9 and 10, one additional reaction is required to complete β-oxidation. The first stage in the process involves the removal of six carbon atoms (three molecules of acetyl-CoA) by the nor-mal pathway of β-oxidation. This results in the production of a 12-carbon enoyl-CoA intermediate (Δ^3-*cis* dodecenoyl-CoA) in which the double bond is in the wrong position, Δ^3 instead of Δ^2, and in the wrong configuration, *cis* instead of *trans*. The position and configuration of the double bond are modified by the action of the enzyme enoyl-CoA isomerase. This enzyme converts the Δ^3 *cis* double bond into a Δ^2 trans double bond which allows β-oxidation to continue to completion (Figure 13.3).

The β-oxidation of linoleic acid ($\Delta^{9,12}$C18:2) requires yet another enzyme. The oxidation process proceeds as for oleic acid (described above), however, following the action of enoyl-CoA isomerase only one further cycle of β-oxidation takes place before a 10-carbon intermediate with a *cis* double bond between carbons 4 and 5 (Δ^4 decenoyl-CoA) is formed. This undergoes dehydrogenation in the first step of the β-oxidation cycle, catalysed by acyl-CoA dehydrogenase, to produce Δ^2 trans, Δ^4 *cis*-decadienoyl-CoA. This conjugated double-bond system is then transformed by the enzyme 2,4-dienoyl-CoA reductase to yield Δ^3 *trans*-decenoyl-CoA. Finally, the Δ^3 trans double bond is converted to a Δ^2 trans double bond by enoyl-CoA isomerase, which allows β-oxidation to continue to completion (Figure 13.4).

By the use of these two additional enzymes, enoyl-CoA isomerase and 2,4-dienoyl-CoA reductase, almost all naturally occurring unsaturated fatty acids can be oxidized via the β-oxidation pathway.

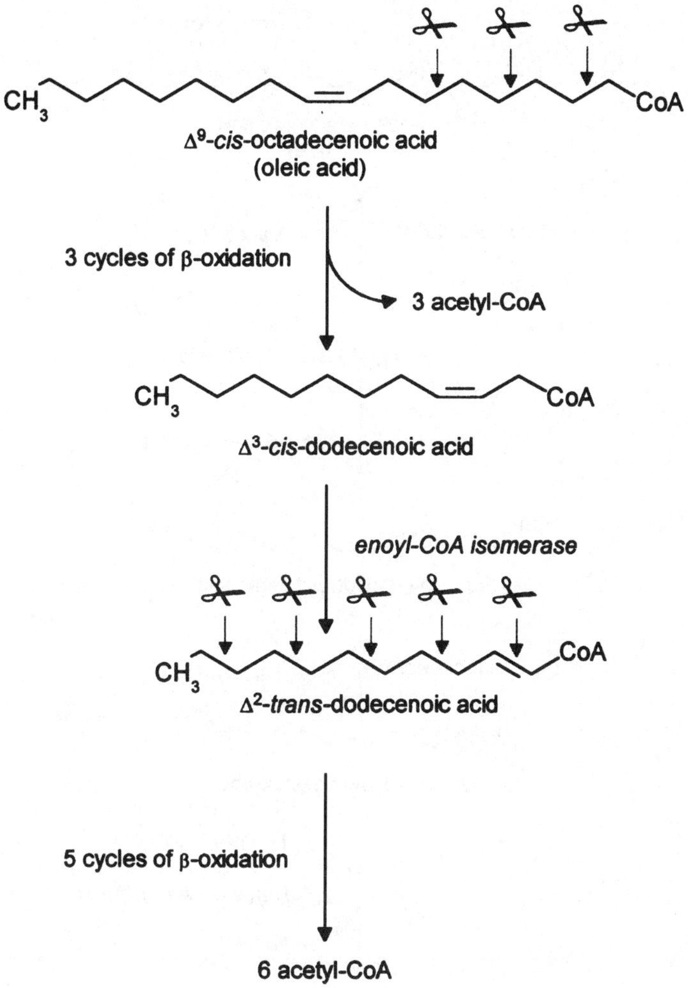

Figure 13.3 The β-oxidation of oleoyl-CoA.

13.2.5 β-OXIDATION IN PEROXISOMES AND GLYOXISOMES

The β-oxidation of fatty acids in peroxisomes or glyoxisomes differs from that in mitochondria. The first reaction is catalysed by acyl-CoA oxidase in peroxisomes and glyoxisomes. This reaction produces hydrogen peroxide, which is then broken down by the action of catalase. The remaining reactions of the β-oxidation cycle appear to be the same as those in the mitochondria, although there may be structural and functional differences between the enzymes in the different subcellular organelles. The membranes of peroxisomes and glyoxisomes do not present a permeability barrier to the CoA derivatives of fatty acids as does the inner mitochondrial membrane, thus there is no requirement for fatty acids to be converted to their carnitine derivatives before they are oxidized. Unlike mitochondria, peroxisomes and glyoxisomes do not contain an electron transport chain capable of utilizing the NADH produced by L-3-hydroxyacyl-CoA dehydrogenase. In order to ensure a supply of NAD^+ for continued fatty acid oxidation,

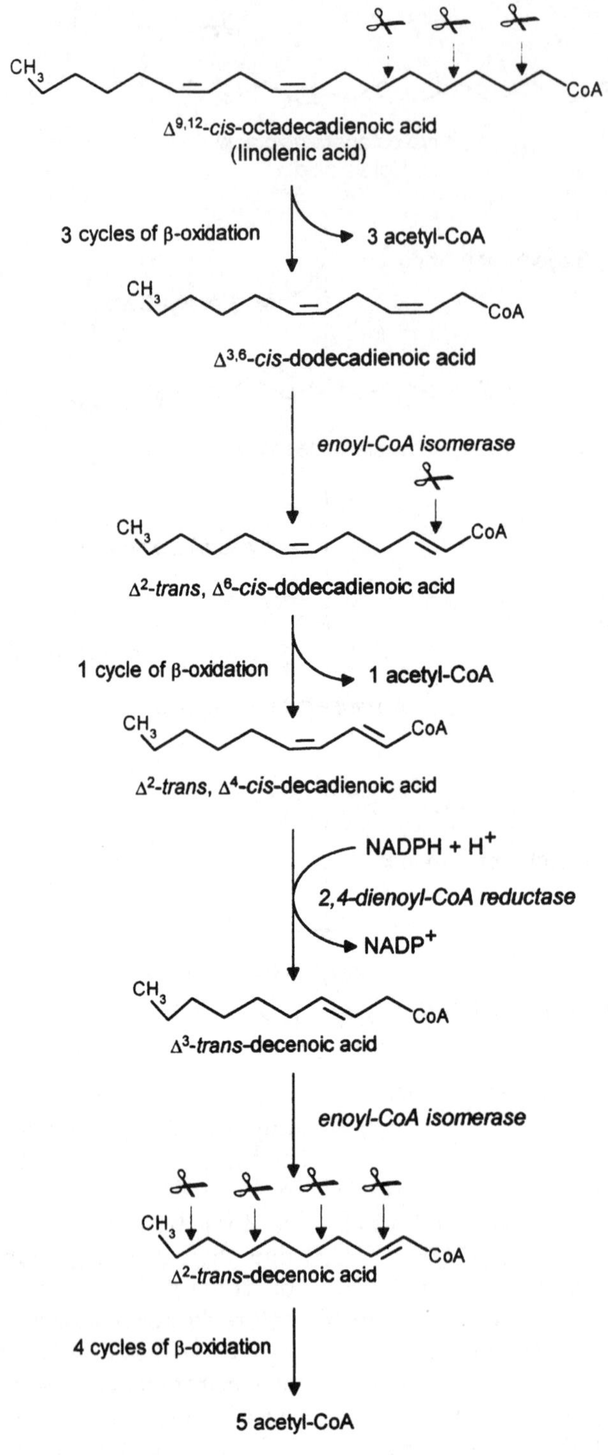

Figure 13.4 The β-oxidation of linoleoyl-CoA.

NADH is transported to the cytoplasm in exchange for NAD$^+$.

In animals, peroxisomal oxidation makes a significant contribution to overall fatty acid oxidation: in liver this may be as high as 50% of the total fatty acid oxidation. However, only partial oxidation of fatty acids occurs in peroxisomes. These organelles appear to be particularly important in conversion of long-chain fatty acids to medium-chain-length products which are then converted to acyl-carnitine derivatives and transported to the mitochondrial matrix, where the oxidation process is completed. The acetyl-CoA produced in the peroxisome is also transported to the mitochondrial matrix where it can be oxidized in the TCA cycle or converted to ketone bodies. The movement of acetate between these subcellular organelles is via a carnitine-mediated mechanism.

In plants, glyoxisomes in the seeds and peroxisomes in the leaf are the primary sites of β-oxidation. Indeed, it is now thought that mitochondrial β-oxidation makes only a very small contribution to the oxidation of fatty acids in plants, and it is clear that both glyoxisomes and peroxisomes are capable of complete oxidation of fatty acids. In some germinating seeds the quantity of fatty acids mobilized from stored fat and oxidized via the glyoxisomes is large, particularly in oil-seed varieties. These specialized subcellular organelles contain both the enzymes required to oxidize fatty acids to acetyl-CoA, and the enzymes of the glyoxylate pathway, isocitrate lyase and malate synthase, which enable plants to convert the acetyl-CoA to oxaloacetate and on to glucose.

13.2.6 THE FORMATION OF KETONE BODIES

There are three principal ketone bodies, β-hydroxybutyrate, acetoacetate and acetone (Figure 13.5). These compounds are normally present in blood in trace amounts, except in ruminants where β-hydroxybutyrate is a product of butyrate metabolism as it passes across the rumen wall. When their plasma concentrations increase significantly above the norm,

animals are said to be suffering from ketosis. The synthesis of ketone bodies, referred to as ketogenesis, occurs mainly in the liver.

Ketogenesis occurs when the supply of oxaloacetate is insufficient to allow all of the acetyl-CoA produced in the mitochondria to enter the TCA cycle. It is a mechanism which allows acetyl-CoA to be converted to useful compounds which can be exported from the liver and used by other tissues as an energy source. However, in certain circumstances which can arise in pregnant sheep and lactating cows, the concentration of ketone bodies in the blood becomes too high and can lead to the development of clinical conditions known as pregnancy toxaemia and bovine ketosis.

The pathway of ketone body synthesis occurs mainly in the mitochondrial matrix. The initial steps of ketone body production are identical to those of the synthesis of cholesterol described in Chapter 19. Two molecules of acetyl-CoA condense to form acetoacetyl-CoA by a reversal of the β-ketothiolase reaction of β-oxidation. The acetoacetyl-CoA then reacts with another molecule of acetyl-CoA to form hydroxymethylglutaryl-CoA (HMG-CoA). This HMG-CoA is hydrolysed to yield one molecule of acetoacetate and one molecule of acetyl-CoA. Some acetoacetate is reduced to β-hydroxybutyrate by β-hydroxybutyrate dehydrogenase. The extent of this reduction is determined largely by the ratio of NAD$^+$ to NADH in the liver. Both acetoacetate and β-hydroxybutyrate can leave the mitochondria and enter the

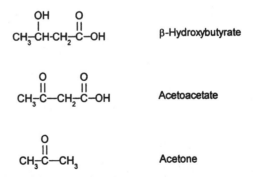

Figure 13.5 The structure of the ketone bodies: (a) β-hydroxybutyrate; (b) acetoacetate; (c) acetone.

bloodstream. Acetoacetate may undergo spontaneous decomposition to produce acetone, which can often be smelled on the breath of ketotic animals (Figure 13.6).

13.3 α-OXIDATION

As its name suggests, this process is the oxidation of fatty acids at the α-carbon (carbon 2). It results in either the removal of a single carbon atom from the carboxyl end of a fatty acid or the production of α-hydroxy fatty acids (Figure 13.7).

The metabolic significance of α-oxidation in animal tissues is still not understood. There are three areas in which it may play an role. Firstly, it may be the mechanism whereby odd-numbered fatty acids are synthesized, i.e. by the removal of a carbon atom from the carboxyl end of an even-numbered fatty acid. Secondly, it may be involved in the synthesis

and/or breakdown of the α-hydroxy fatty acids found in certain types of tissue lipids, particularly in brain tissue where the sphingolipid fraction contains a large amount of this type of fatty acid. Thirdly, it may act in concert with β-oxidation, to facilitate the oxidation of fatty acids with structural features which prevent oxidation by β-oxidation alone. An example of this combined action of α- and β-oxidation of fatty acids is the degradation of phytanic acid (3,7,11,15-tetramethyl palmitic acid) derived from phytol. The presence of the methyl substituent on carbon 3 inhibits β-oxidation; however, if the fatty acid is shortened at the carboxyl end by one carbon, β-oxidation can proceed. Due to the position of the methyl groups, the products of phytanic acid oxidation are alternate molecules of propionyl-CoA and acetyl-CoA. It is not clear whether fatty acids undergo α-oxidation in the free form or as CoA derivatives, or what is the

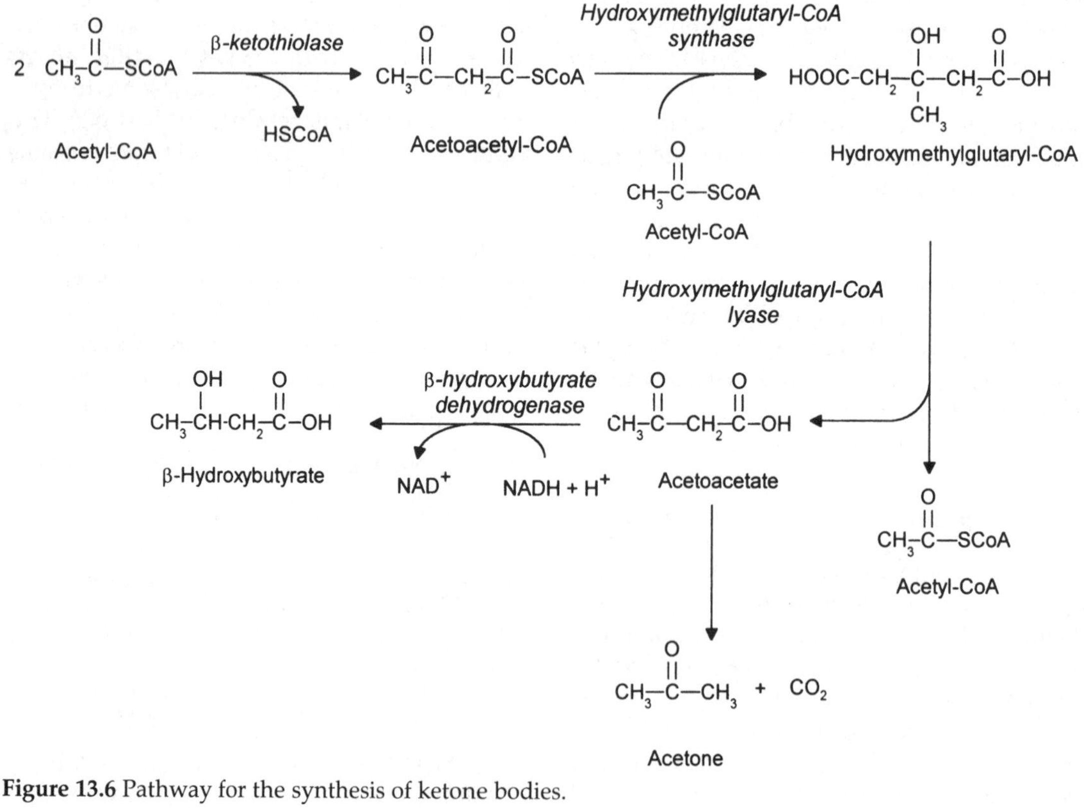

Figure 13.6 Pathway for the synthesis of ketone bodies.

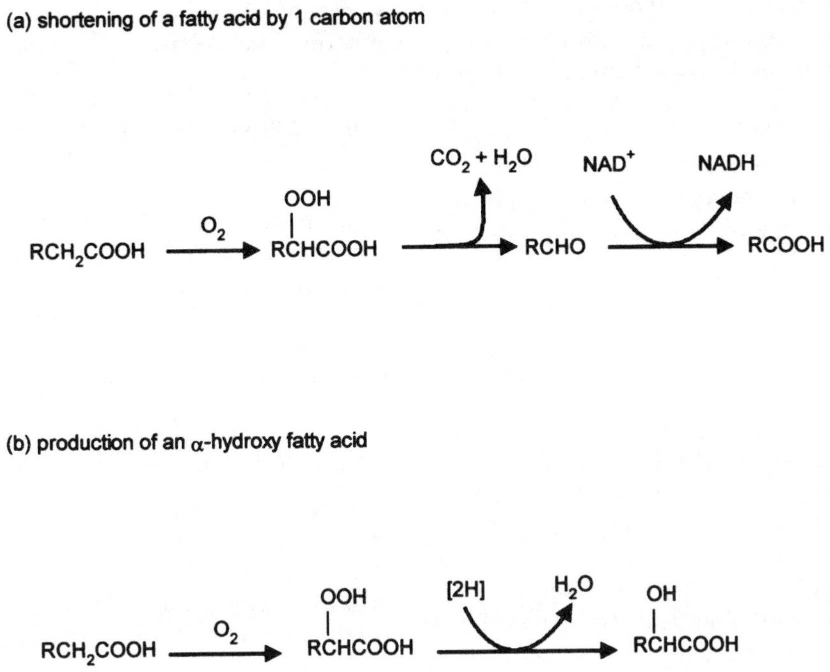

(a) shortening of a fatty acid by 1 carbon atom

$$RCH_2COOH \xrightarrow{O_2} \overset{\overset{\displaystyle OOH}{\displaystyle |}}{RCHCOOH} \xrightarrow{\quad CO_2 + H_2O \quad} RCHO \xrightarrow{\quad NAD^+ \quad NADH \quad} RCOOH$$

(b) production of an α-hydroxy fatty acid

$$RCH_2COOH \xrightarrow{O_2} \overset{\overset{\displaystyle OOH}{\displaystyle |}}{RCHCOOH} \xrightarrow{\quad [2H] \quad H_2O \quad} \overset{\overset{\displaystyle OH}{\displaystyle |}}{RCHCOOH}$$

Figure 13.7 α-oxidation of fatty acids.

exact subcellular location of the process. Nevertheless, it seems likely that the process is associated with the microsomal fraction.

The importance of α-oxidation in plants is also unclear but it has been demonstrated that, with the exception of the germinating seed where β-oxidation is very active, α-oxidation may be the most important pathway of fatty acid oxidation. In addition to its role in degradation of fatty acids, α-oxidation is almost certainly involved in the production of the long-chain fatty alcohols and hydrocarbons found in the cutin and suberin components of the cuticle.

13.4 ω-OXIDATION

In this process, fatty acids undergo oxidation at the ω- or methyl carbon to form dicarboxylic acids and ω-hydroxy fatty acids. In animals, the enzyme responsible appears to be a mixed-function oxidase associated with the endoplasmic reticulum and probably involving a specialized cytochrome (P_{450}). In plants, the involvement of the cytochrome is in doubt. ω-Oxidation may be an essential step in the oxidation of fatty acids where the carboxyl end is unavailable for β-oxidation, as the production of a carboxyl group from the methyl carbon of a fatty acid may allow the β-oxidation process to start from the opposite end of the molecule. The production of ω-hydroxy fatty acids may have a role in the formation of cutin and suberin.

13.5 PEROXIDATION OF FATTY ACIDS

Food that is not stored under the correct conditions will rapidly develop unpleasant organoleptic properties characterized by 'off' flavours and smells. This is an extremely complex process involving the interaction of many physical, chemical and biological factors, the combined effects of which are perceived primarily as changes in the taste, smell and texture of food.

The deterioration in food quality is often linked to enzymic and non-enzymic modifications of lipids and, in particular, to the fatty acids they contain. Unsaturated fatty acids, particularly PUFAs, are susceptible to non-enzymic peroxidation. This process contributes to the development of unpleasant odours and flavours associated with fatty foods (rancidity).

13.5.1 CHEMISTRY OF LIPID PEROXIDATION

Peroxidation is directly proportional to substrate concentration and the partial pressure of oxygen. The rate of peroxidation also increases with the extent of existing fatty acid oxidation, indicating that the process is autocatalytic; hence it is often referred to as autoxidation. Once started, a chain reaction is set up which perpetuates and accelerates the peroxidation process. The susceptibility of unsaturated fatty acids to peroxidation is proportional to the number of double bonds they contain; thus monounsaturated fatty acids undergo peroxidation at a slower rate than PUFAs. The relationship between the degree of unsaturation of a fatty acid and its susceptibility to peroxidation can be seen most clearly with pure fatty acids. For fatty acids with 1, 2, 3, 4, 5 and 6 double bonds, the relative rates of oxidation are approximately 1, 40, 80, 160, 240 and 320, respectively. In fats or oils with a mixed fatty acid composition, indicators of the degree of unsaturation, such as the iodine value, can be useful predictors of its likely susceptibility to peroxidation.

The production of a fatty acid hydroperoxide proceeds via the formation of a fatty acyl free radical, which arises due to the reaction of an unsaturated fatty acid with other free radical species. Fatty acyl free radicals can themselves interact with other unsaturated fatty acids to promote further free radical formation.

Three phases have been identified in the peroxidation of fatty acids:

- initiation phase
- propagation phase
- termination phase.

In the initiation phase molecular oxygen, or a free radical such as the highly reactive hydroxyl radical OH$^{\bullet}$, reacts with an unsaturated fatty acid to form a fatty acyl free radical by abstraction of hydrogen from a methylene group adjacent to the double bond in the fatty acid. The presence of pro-oxidant factors such as transition metals (e.g. iron and copper), ultraviolet or ionizing radiation promote this initial stage of peroxidation:

$$RH + O_2 \rightarrow R^{\bullet} + HO_2^{\bullet}$$
$$RH + OH^{\bullet} \rightarrow R^{\bullet} + H_2O$$

Propagation occurs when the fatty acyl radical reacts with molecular oxygen to form a fatty acyl peroxide radical, which in turn initiates new radical formation and results in the production of a fatty acyl hydroperoxide.

$$R^{\bullet} + O_2 \rightarrow ROO^{\bullet}$$
$$ROO^{\bullet} + RH \rightarrow ROOH + R^{\bullet}$$

Fatty acyl hydroperoxides decompose either to produce fatty acyl peroxide radicals which further initiate peroxidation, or may be converted to alkoxy radicals. These transformations are greatly enhanced by the presence of transition metals.

$$ROOH + M^{(n+1)+} \rightarrow ROO^{\bullet} + M^{n+} + H^+$$
$$ROOH + M^{n+} \rightarrow RO^{\bullet} + OH^- + M^{(n+1)+}$$

Alkoxy radicals are unstable compounds which undergo chain fragmentation, termed β-scission, to produce alkyl radicals and low molecular-weight volatile aldehydes such as octanal, nonanal, 2-decenal and 2-undecenal (Figure 13.8). Some of these compounds are responsible for the unpleasant smells and tastes characteristic of rancid fatty foods.

Termination occurs when radicals react together or with chain-breaking compounds to produce stable molecules which do not initiate or propagate further peroxidation. Most natural and synthetic antioxidants are chain-breaking compounds and are themselves oxi-

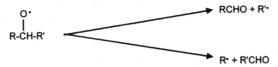

Figure 13.8 β-Scission of an alkoxy radical to produce an alkyl radical and an aldehyde.

dized and destroyed in the process. The length of the induction phase of oxidation can be extended by the addition of antioxidants, propagation being delayed until the antioxidant activity is reduced.

13.5.2 PREVENTION OF FATTY ACID PEROXIDATION

In biological systems, natural antioxidants are present which minimize the extent to which fatty acid peroxidation takes place. The most important of these are the group of compounds known collectively as vitamin E. In animals, the most commonly occurring of these compounds is α-tocopherol. In plants, and in particular in seed oils, the other isomers of tocopherol and tocotrienol are present in significant amounts; however the vitamin E activity of these isomers is usually expressed as α-tocopherol equivalents (Table 13.2).

The nature of the complex interactions of α-tocopherol with unsaturated fatty acids and with other intracellular antioxidants such as ascorbic acid are slowly being unravelled. It is clear that α-tocopherol plays an important role in the protection of membrane unsaturated fatty acids.

The lipophilic phytyl side chain of α-tocopherol is located in the hydrophobic lipid bilayer region of membranes (Figure 13.9). The role of α-tocopherol appears to be two-fold: firstly, to neutralize free radical species in the region of the membrane and thereby prevent initiation of fatty acid peroxidation; and secondly, when fatty acid peroxide radicals have been formed, to break the chain of autoxidation by converting the peroxide radical to a hydroperoxide. It is estimated that about 90% of the fatty acid peroxide radicals formed in membranes are neutralized by α-tocopherol before they react with other PUFAs. Both of these functions require that the α-tocopherol has a higher affinity for free radicals than do unsaturated fatty acids.

A number of synthetic antioxidants are used by the animal feed and human food industries to minimize oxidative deterioration of lipids during processing and storage. The structures of the most commonly used synthetic antioxidants are shown in Figure 13.10.

13.5.3 DETECTION AND MEASUREMENT OF LIPID PEROXIDATION

Numerous methods have been developed to detect and quantify lipid peroxidation. Some,

Table 13.2 Tocopherol and tocotrienol content of common fats and oils

Fat/oil	Tocopherols (mg)				Tocotrienols (mg)		Vitamin E activity as α-tocopherol (mg)
	α	β	γ	δ	α	β	
Corn	112	50	602	19	–	–	198
Cottonseed	390	–	387	–	–	–	428
Palm	256	–	316	70	146	3	335
Soyabean	75	15	800	266	2	–	171
Wheat germ	1330	710	260	271	26	18	1736
Cod liver	220	–	–	–	–	–	220
Tallow	27	–	–	–	–	–	27

Adapted from Gunstone, F.D., Harwood, J.L. and Padley, F.B. (1994) *The Lipid Handbook*, 2nd edn, Chapman & Hall, London.

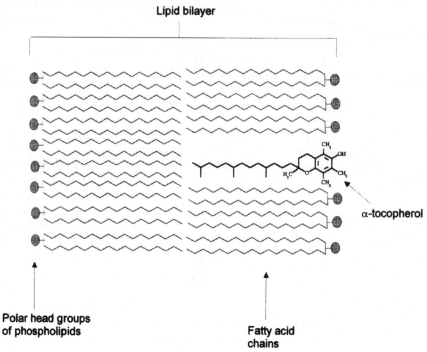

Figure 13.9 Orientation of α-tocopherol in the membrane lipid bilayer.

such as measurement of changes in UV absorbance due to migration of double bonds and formation of conjugated dienes during peroxidation, or estimation of the peroxide value (PV), measure peroxide formation. Others, such as the thiobarbituric acid test or determination of the anisidine value, rely on measurement of the products of hydroperoxide breakdown.

13.5.4 EFFECTS OF PEROXIDATION IN LIVING ORGANISMS

There is growing interest in the effects of lipid peroxidation in living organisms. It has been known for some time that peroxidized, highly unsaturated oils are toxic in a number of animal species, and that the toxicity is proportional to peroxide content. For example, rats fed diets containing fat with a peroxide value (PV) of 100 showed no adverse effects, whereas the growth of those fed diets containing fat with PVs between 400 and 800 was severely retarded. Diets containing fats with a PV in excess of 1200 induced rapid death. In poultry, a number of reported cases of toxicity have involved the feeding to birds of oils which have been heated at high temperatures for prolonged periods. This leads not only to the formation of peroxides, but also to the production of polymerized fatty acid derivatives which are toxic. The use of recycled cooking oil has been linked with increased mortality, and with the incidence of burnt hocks as a result of the production of sticky excreta caused by poor digestion of oxidized and polymerized oil.

Membrane phospholipids and glycolipids are rich in PUFAs which are susceptible to oxidative damage. Aerobic metabolism within cells generates active oxygen species, such as hydrogen peroxide, hydroxyl radicals and superoxide radicals, which can react with membrane lipids producing peroxides and peroxide-breakdown products. These oxidative changes can cause alterations in mem-

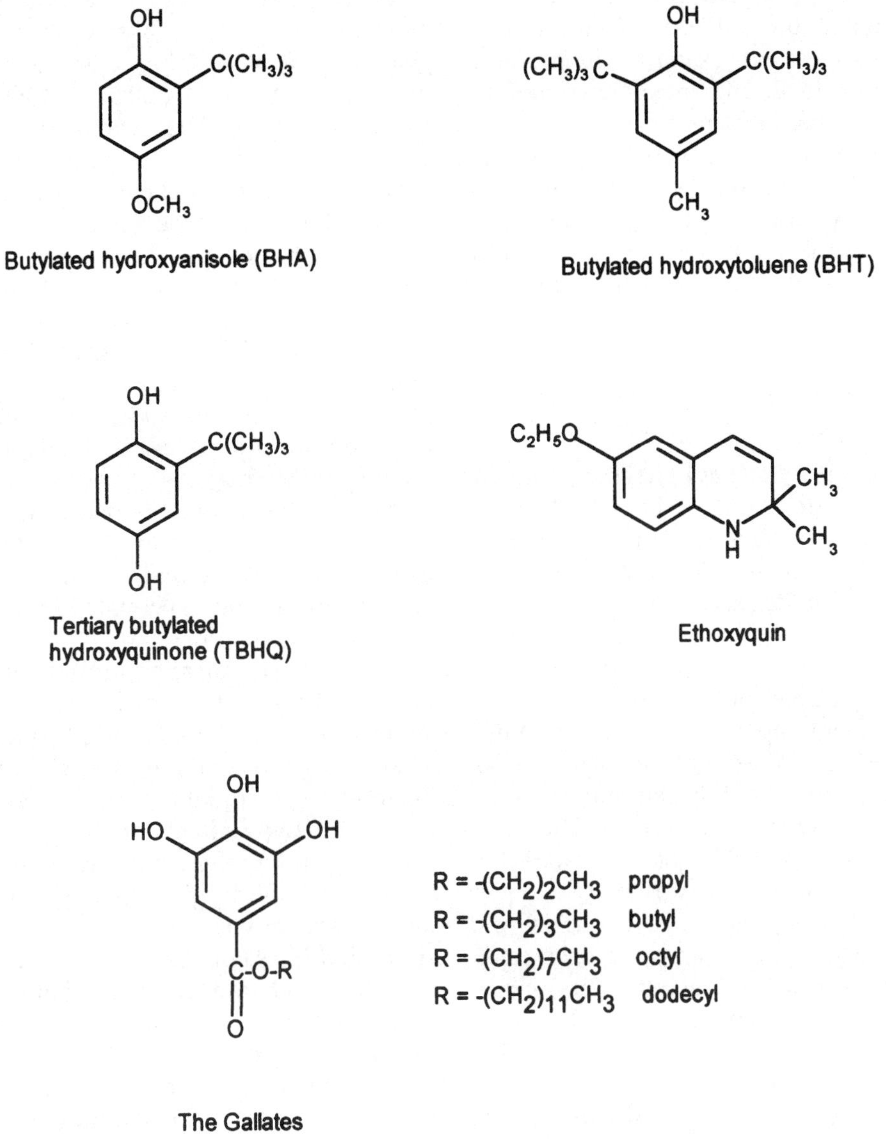

Butylated hydroxyanisole (BHA)

Butylated hydroxytoluene (BHT)

Tertiary butylated
hydroxyquinone (TBHQ)

Ethoxyquin

$R = -(CH_2)_2CH_3$ propyl
$R = -(CH_2)_3CH_3$ butyl
$R = -(CH_2)_7CH_3$ octyl
$R = -(CH_2)_{11}CH_3$ dodecyl

The Gallates

Figure 13.10 Commonly used synthetic antioxidants: butylated hydroxyanisole (BHA); butylated hydroxytoluene (BHT); tertiary butylated hydroxyquinone (TBHQ); ethoxyquin; the gallates.

brane structure and integrity which may affect cellular function. For example, the reaction of fatty acid peroxides with membrane proteins may result in changes in receptor and transport protein-binding characteristics. The release and conversion of PUFAs to prostaglandins and other eicosanoids may also be adversely affected. In the immune system, the function of and communication between the different cell types involved in the coordinated immune response is mediated via the plasma membrane. Infection and disease often stimulate the production of free radicals which may lead to an increase in membrane lipid peroxidation and impairment of the immune response. A number of defence mechanisms are present within the cell to combat the adverse effects of these reactive species, for example antioxidants such as vitamin E, and metalloenzyme systems such as glutathione peroxidase, superoxide dismutase and catalase. The immune response in diseased animals may be enhanced by nutritional supplementation with α-tocopherol and essential metalloenzyme minerals such as selenium, zinc and manganese.

13.5.5 LIPOXYGENASE AND CYCLO-OXYGENASE

A number of lipoxygenase enzymes, containing non-haem iron, occur in plant and animal tissues. These enzymes catalyse the oxidation of fatty acids by molecular oxygen. The immediate product of the reaction is a hydroperoxide which may be further metabolized to a variety of hydroxy derivatives and low-molecular-weight volatile compounds.

In some plants, lipoxygenases are present in large amounts, for example in yellow bean and soyabean seeds (family Leguminosae) and potato tubers (family Solanaceae). The soyabean enzyme is probably the best characterized. Lipoxygenase activity is of two types: type I is active against free fatty acids, and type II which is most active against esterbound fatty acids. These enzymes are important in the development of tastes and flavours in plant tissues. These occur due to the further metabolism of the initial hydroperoxide product of lipoxygenase to low-molecular-weight aldehydes and organic acids such as hexanal, 2-nonenal and 9-oxo-nonanoic acid, produced from linoleic acid. It has been suggested that short-chain products and other derivatives of hydroperoxides may act as growth regulators, for example, in fruit ripening and seed germination. Post-harvest, uncontrolled lipoxygenase activity can lead to the development of 'off' flavours, and heat treatment is often necessary to inactivate the enzymes.

In mammals, lipoxygenase and cyclo-oxygenase catalyse the initial steps in the conversion of all-*cis* $\Delta^{8,11,14}$ C20:3, all-*cis* $\Delta^{5,8,11,14}$ C20:4 and all-*cis* $\Delta^{5,8,11,14,17}$ C20:5 to leukotrienes, prostaglandins, prostacyclin and thromboxanes, compounds collectively known as eicosanoids. Their structures are shown in Figure 13.11. Leukotrienes are produced by the action of lipoxygenase, whereas cyclo-oxygenase modification of PUFAs leads to the synthesis of prostaglandins, prostacyclin and thromboxanes. The latter enzyme is a component of the multifunctional protein, prostaglandin endoperoxide synthetase (PES). PES inhibitors include several non-steroidal anti-inflammatory drugs, e.g. aspirin and indomethacin, which compete with the fatty acid substrates for the enzyme active site. In the case of aspirin (acetylsalicylic acid) the active site is irreversibly acetylated resulting in enzyme inactivation.

The biochemical and physiological functions of these fatty acid derivatives are the subject of considerable ongoing research. They appear to function as locally produced hormones or signalling molecules, which may have some effects at their site of production, but more generally are released into the extracellular fluid and have effects in the local tissue environment. They have very short half-lives, in the order of seconds to minutes, and can have profound effects upon cellular activity at concentrations ranging from 10^{-6} to 10^{-15} M. Most body tissues and fluids contain minute quantities of eicosanoids, although human and sheep seminal fluid are particularly rich sources of prostaglandins, the most common being PGE_2 and $PGF_{2\alpha}$.

The main physiological effects elicited by eicosanoids are summarized in Table 13.3. Because of the potential medical and veterinary uses of eicosanoids, there is considerable interest in the pharmaceutical industry in the development of longer-lived analogues. Probably the best known use of prostaglandin

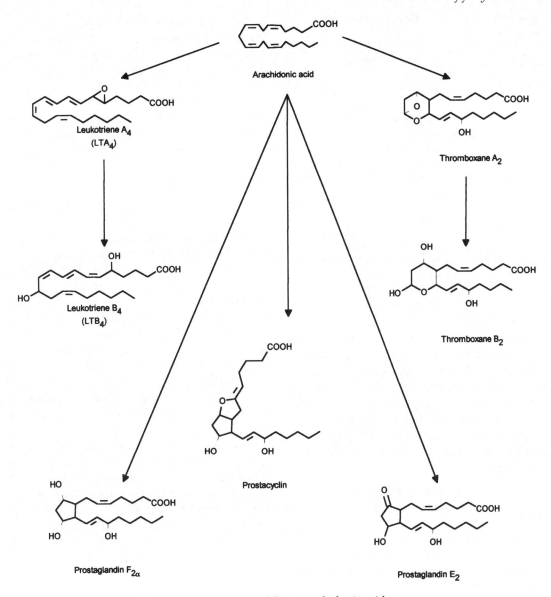

Figure 13.11 Leukotrienes and eicosanoids derived from arachidonic acid.

analogues in agriculture is in the synchroniza-
tion of oestrus in farm animals. For example,
the synthetic analogues of $PGF_{2\alpha}$ such as clo-
prostenol, given in two successive injections at
intervals of 10–12 days, result in the onset of
oestrus about 3 days after the second injection.

13.6 BREAKDOWN OF LIPIDS

Both storage and structural lipids are in a constant state of turnover – they are continually
being synthesized and broken down. This is a
complex, regulated process which, within

Table 13.3 Summary of the main effects of eicosanoids

Eicosanoid	Effects
PGEs	Contraction of intestinal longitudinal muscle; relaxation of sphincters; dilation of the cervix
$PGF_{2\alpha}$	Degeneration of the corpus luteum; contraction of uterine smooth muscle during parturition
Prostacyclin (PGI_2)	Vasodilator; inhibition of platelet aggregation
Thromboxane A_2	Vasoconstrictor; stimulation of platelet aggregation
Leukotrienes	Chemotaxis, inflammatory responses, bronchoconstriction, vasoconstriction, vasodilation

highly organized organisms such as plants and animals, allows the type and quantity of individual lipids to be modified in response to changes in nutritional and physiological circumstances. The biosynthesis of complex lipids is discussed in Chapter 19. In the following section the breakdown of triacylglycerols, phospholipids and glycolipids are described.

13.6.1 TRIACYLGLYCEROL BREAKDOWN

Plants

In some plants, large quantities of triacylglycerol are stored in seeds or fruits. In dry seeds prior to germination there is little enzymic activity, however during imbibition (the uptake of water by seeds during germination) there is an increase in the activity of a number of enzymes including triacylglycerol lipases. Their substrates are contained in oil droplets (oleosomes) within the seed and the enzymes act at the surface of the droplet, probably with the help of binding proteins to facilitate the attachment process. The lipases catalyse the release of fatty acids esterified to the 1 and 3 position of triacylglycerols, to yield a monoacylglycerol. Complete release of all three fatty acids is achieved by the migration of fatty acid from the 2 position to the 1 position and its subsequent release. The fatty acids released are taken up by the glyoxysomes where they are oxidized. The mechanism of transfer of fatty acids between the sites of storage and of oxidation is poorly understood, but in some seeds this may be achieved by direct contact between the oil droplet and the glyoxysome.

Animals

In adipose tissue, fatty acids are released from triacylglycerols by lipolysis. The release of fatty acids and their transport for use elsewhere in the body is often referred to as fatty acid mobilization. This requires the complete hydrolysis of triacylglycerols to fatty acids and glycerol. The fatty acids are transferred from the adipocyte to the bloodstream where they become bound to albumin and are then referred to as non-esterified fatty acids (NEFAs). As fatty acids released from adipose tissue triacylglycerols may be re-esterified without being released into the circulation, the rate at which NEFAs enter the blood depends not only on the rate of lipolysis, but also on the rate of esterification of fatty acids into triacylglycerols. Both lipolysis and esterification are subject to acute and long-term regulation in response to changes in nutritional and physiological status. In very simple terms, fatty acids are deposited as triacylglycerols in adipose tissue when the rate of esterification exceeds that of lipolysis, whereas fatty acids are released into the blood when the rate of lipolysis exceeds that of esterification.

The hydrolysis of a triacylglycerol occurs in three stages. The rate-limiting step is the initial cleavage of triacylglycerol to diacylglycerol and a fatty acid. This step is catalysed by the enzyme triacylglycerol lipase, sometimes also called hormone-sensitive lipase, which is the main site of regulation of lipolysis (see further detail in Chapter 30). Diacylglycerols produced in this reaction are further metabolized by other lipases to give fatty acids and glycerol (Figure 13.12).

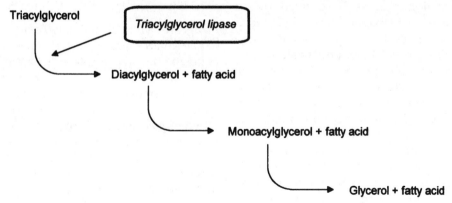

Figure 13.12 The sequential breakdown of triacylglycerols.

13.6.2 PHOSPHOLIPID BREAKDOWN

Enzymes which are involved in the breakdown and remodelling of phospholipids are called phospholipases. In plant and animal tissues, a number of phospholipases have been identified which differ in the position at which they act on the phospholipid molecule. These enzymes and their sites of action are shown in Figure 13.13.

Phospholipases A_1 and A_2 are widely distributed and are responsible for the removal of fatty acids from intact phospholipids. The resulting compound which contains only one fatty acid, in either the 1 or 2 position, is called a lysophospholipid. These enzymes are involved in the modification of the fatty acid composition of existing membrane phospholipids. In animals, phospholipase A_2 is particularly important in the release of arachidonic acid required for eicosanoid synthesis. Both enzymes are found in the secretions of the digestive tract and are required for the digestion of dietary phospholipids.

Phospholipase B appears to be present only in microorganisms, and because it acts at both the 1 and 2 positions, it is able to act on both intact phospholipids and lysophospholipids.

Phospholipase C has an important role in the control of enzyme activity. Many enzymes are regulated by a calcium-dependent phosphorylation/dephosphorylation mechanism.

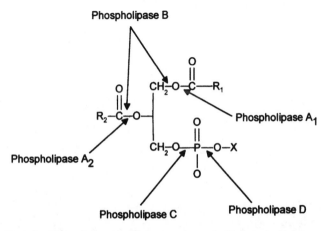

Figure 13.13 The sites of action of phospholipases on a typical phospholipid.

The regulation process is initiated by an increase in the activity phospholipase C which catalyses the hydrolysis of plasma membrane phosphatidylinositol-4,5-bisphosphate to produce inositol trisphosphate and diacylglycerol. The former causes release of sequestered calcium. In addition to this regulatory function, phospholipase C is also involved in the general remodelling of tissue phospholipids.

Although phospholipase D is found in animal tissues, the main source of this enzyme is plant tissue. It is a very active enzyme involved in phosphatidate exchange reactions; however, it is not clear what its role is in phospholipid metabolism.

13.6.3 BREAKDOWN OF GLYCOLIPIDS

Plants are rich in glycolipids. Their catabolism is particularly active during senescence and following damage to tissues. Complete breakdown occurs in two stages: firstly, the fatty acids are removed by acyl hydrolase, and secondly, the sugar residues are removed by the action of galactosidases. In grazing ruminants a large proportion of the ingested lipid is in the form of glycolipids. These lipids are rapidly broken down by the action of bacterial acyl hydrolases and glycosidases.

BREAKDOWN OF PROTEINS AND THE OXIDATION OF AMINO ACIDS

14

14.1 INTRODUCTION

In living cells, proteins are generally unstable compounds, and are continually broken down and resynthesized. Proteins are easily denatured and even the most robust of them lose their activity over time. The needs of many cells vary from minute to minute and cells of all organisms adapt to this by changing, to varying extents, the proteins which they make and hence the functions which they are able to perform.

There are other circumstances in which proteins must be broken down. Animals obtain amino acids in their diets in the form of proteins. However they cannot directly use dietary proteins of either plant or animal origin to make their own proteins. Instead almost all dietary proteins have to be broken down in the gut before absorption as amino acids. These amino acids then undergo processing to ensure that they are available in the proportions required by the animal, which are often not the same as those present in the diet. As animals are frequently supplied with more amino acids than are required for protein synthesis and they have little capacity to store them, the excess must be degraded. Animals do this by oxidizing amino acids, obtaining energy from the carbon skeletons, and excreting the nitrogen in the form of compounds such as urea. Animals may also degrade proteins to produce amino acids from which they can make glucose if they cannot obtain sufficient carbohydrates from their diets or from stored glycogen (see Chapter 18).

In contrast to animals, plants show indeterminate growth, that is they do not stop growing when they reach a particular size. Often their growth is limited by the amount of nitrogen available from the soil or the atmosphere and therefore they only rarely have to deal with an excess of nitrogen. Thus amino acid oxidation occurs to a smaller extent than in animals. Plants do, however, need to break down amino acids produced as a result of protein turnover or from the breakdown of storage proteins, for example in seeds. Under these circumstances proteins are broken down to amino acids which may be degraded and subsequently give rise to other amino acids according to the needs of the cells.

14.2 BREAKDOWN OF PROTEINS

The first stage in the breakdown of proteins is their hydrolysis to amino acids. There is a range of proteolytic enzymes which catalyse these reactions. Some of these are relatively non-specific in their action, but others will only hydrolyse bonds next to specific, individual amino acids within the protein sequence. Hydrolysis of proteins takes place in the digestive system and in cells during protein turnover.

14.2.1 DIGESTION OF PROTEINS

Extensive degradation of proteins occurs during digestion. This process is discussed in Chapter 28.

14.2.2 PROTEIN TURNOVER

In all organisms, proteins are constantly synthesized from amino acids and then broken down again. A proportion of the amino acids released are oxidized and completely degraded. This may seem wasteful but it maintains a supply of active proteins and allows the organism to adapt to changing conditions which it encounters.

Not all proteins turn over at the same rate. In general, structural or storage proteins turn over more slowly than those involved in metabolism (Table 14.1). Proteins containing regions rich in the amino acids proline, glutamate, serine and threonine appear to have short half lives, and the presence of certain amino acids at the amino terminus also appears to accelerate breakdown. In addition, abnormal or damaged proteins also break down rapidly, probably because they cannot adopt the stable conformation of the normal protein. Once these proteins have been recognised they become conjugated to the small

protein ubiquitin (76 amino acids), by formation of covalent bonds between lysine residues in the proteins and the C-terminal COOH group of the ubiquitin. The conjugates are then rapidly degraded by the action of specific proteases. On the other hand, formation of complexes with molecules such as heat-shock proteins may stabilize proteins against attack by proteolytic enzymes.

The turnover of proteins in a 70 kg man result in release of about 400 g of amino acids per day, and up to 25% of these may be oxidized or used to synthesize sugars.

14.3 BREAKDOWN OF AMINO ACIDS

In both ruminants and non-ruminants, the amino acids released during digestion are absorbed in the small intestine and pass to the liver. The liver is the main site of amino acid metabolism but kidney also has some capacity to degrade amino acids. Some amino acids may be metabolized in the liver or may pass to other tissues where they are used to synthesize new proteins. Any excess amino acids are oxidized. Amino acids produced as a result of protein turnover also pass to the liver for processing.

Although each of the 20 amino acids found in proteins is oxidized by a separate pathway, the first step in their oxidation is usually the removal of the amino group. This may occur by either transamination or oxidative deamination.

14.3.1 TRANSAMINATION REACTIONS

Transamination involves removal of the α-amino group of the amino acid and its transfer

Table 14.1 Turnover of different classes of proteins

Protein	Half life (days)
Proteins of blood serum, liver and kidney	2–10
Haemoglobin	30
Muscle proteins	180
Collagen	1000

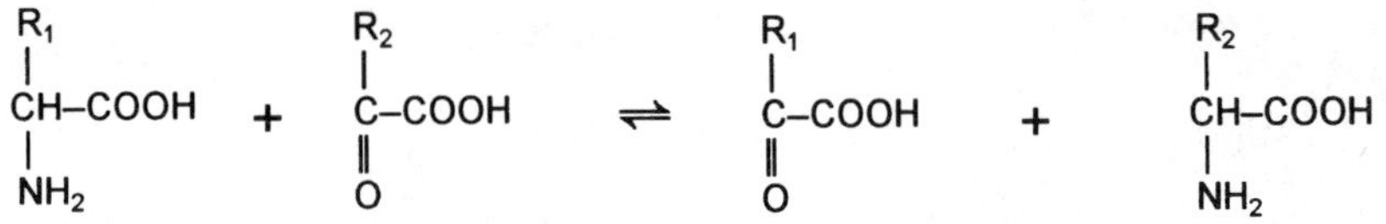

$$
\begin{array}{ccccccc}
\overset{\displaystyle R_1}{\underset{\displaystyle NH_2}{\overset{|}{\underset{|}{CH-COOH}}}} & + & \overset{\displaystyle R_2}{\underset{\displaystyle O}{\overset{|}{\underset{\parallel}{C-COOH}}}} & \rightleftharpoons & \overset{\displaystyle R_1}{\underset{\displaystyle O}{\overset{|}{\underset{\parallel}{C-COOH}}}} & + & \overset{\displaystyle R_2}{\underset{\displaystyle NH_2}{\overset{|}{\underset{|}{CH-COOH}}}}
\end{array}
$$

α-amino acid 1 + α-ketoacid 2 ⇌ α-keto acid 1 + α-amino acid 2

α-amino acid + α-ketoglutarate ⇌ α-keto acid + glutamate

Figure 14.1 Transamination reactions take place between an amino acid and an α-keto acid. The amino group is transferred to the ketoacid producing a new amino acid and a new ketoacid. The reactions are readily reversed and are used in both breakdown and synthesis of amino acids. The amino group acceptor is often α-ketoglutarate which is converted to glutamate.

to an α-keto acid. The amino acid itself is converted into the corresponding α-keto acid. The reaction is shown in Figure 14.1. Reactions of this type are catalysed by transaminases, also known as aminotransferases.

Although a number of keto acids can take part in this reaction, α-ketoglutarate is often used and glutamate is produced (Figure 14.1). This reaction is catalysed by glutamate transaminase.

Transamination reactions are completely reversible and, as will be seen in Chapter 20, can also be used to synthesize amino acids. Many amino acids can take part in this reaction, their amino groups are incorporated into glutamate. Thus the nitrogen from the amino acids is concentrated as the amino group of glutamate.

All transaminases function in a similar way and have pyridoxal phosphate as a tightly bound coenzyme. When the amino group is removed from the amino acid it is initially transferred to the pyridoxal group bound to the enzyme (converting it to pyridoxamine phosphate) but it is subsequently transferred to the keto acid, regenerating pyridoxal phosphate.

14.3.2 DEAMINATION

Because transamination replaces one amino acid by another, it does not actually result in

net amino acid breakdown. Thus the process of oxidative deamination is essential for release of ammonia from amino acids. In this way the α-amino group of the amino acid is converted directly to ammonia. The oxidative deamination of glutamate, catalysed by glutamate dehydrogenase, occurs very commonly (Figure 14.2). This results in the production of ammonia which is then converted rapidly into urea.

14.3.3 OXIDATION OF CARBON SKELETONS OF AMINO ACIDS

Removal of the amino groups from amino acids by transamination or deamination results in the production of the corresponding α-keto acids. These can be further oxidized by a series of pathways, resulting in the production of metabolites which include pyruvate, acetyl-CoA and the TCA-cycle intermediates. They may be completely oxidized to carbon dioxide, or in some cases, converted into sugars such as glucose by gluconeogenesis.

In plants, the keto acids formed by breakdown of amino acids are used mainly in synthesis of new amino acids and are only degraded by oxidation to a very minor degree.

The pathways by which keto acids are broken down are complex and will not be considered in detail in this text. However some

$$
\begin{array}{ccc}
\begin{array}{c}
\text{COOH} \\
| \\
\text{CH}_2 \\
| \\
\text{CH}_2 \\
| \\
\text{CH-NH}_2 \\
| \\
\text{COOH}
\end{array}
& + \text{NAD}^+ + \text{H}_2\text{O} \longrightarrow &
\begin{array}{c}
\text{COOH} \\
| \\
\text{CH}_2 \\
| \\
\text{CH}_2 \\
| \\
\text{C=O} \\
| \\
\text{COOH}
\end{array}
& + \text{NH}_3 \quad + \text{NADH} \quad + \text{H}^+
\end{array}
$$

Glutamate α-ketoglutarate

Figure 14.2 Oxidative deamination of glutamate catalysed by glutamate dehydrogenase. Oxidation of glutamate is accompanied by reduction of NAD$^+$ to NADH and is irreversible.

important principles can be understood by considering the amino acids to belong to families according to the end product of their breakdown. The end products of breakdown of the carbon skeletons of amino acids are shown in Figure 14.3.

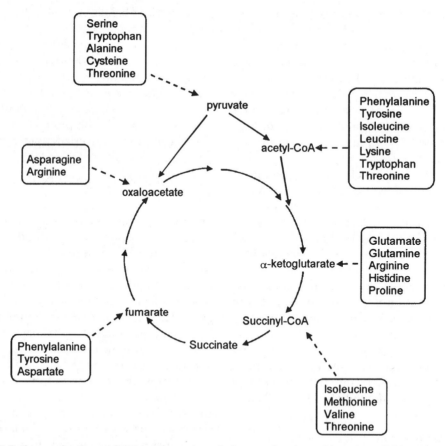

Figure 14.3 Pathways for the oxidation of the carbon skeletons of amino acids to form TCA-cycle intermediates or related compounds. These may be further oxidized or used in biosynthetic reactions. Amino acids which are converted into three- or four-carbon compounds can give rise to sugars and are glucogenic.

Pyruvate or intermediates of the TCA cycle containing three or more carbon atoms can be converted into oxaloacetate and then into glucose. Hence amino acids which are broken down into these compounds are called glucogenic amino acids. On the other hand, compounds such as acetate (in the form of acetyl CoA or acetoacetyl CoA, which itself breaks down to acetyl CoA) do not result in oxaloacetate production and therefore cannot be used as precursors of glucose. Instead, the acetyl CoA is converted into a range of other compounds, including the ketone bodies. Amino acids forming acetyl CoA are therefore called the ketogenic amino acids. A third group of amino acids break down into several fragments, some of which are ketogenic and some glucogenic. These acids must be considered to be both glucogenic and ketogenic.

14.3.4 THE FATE OF AMMONIA

Ammonia is very toxic and, in most organisms, any formed by oxidative deamination is rapidly converted into less-toxic organic compounds. Where water is plentiful, ammonia may be excreted directly, as in protozoa, nematodes and aquatic amphibians. In most terrestrial animals ammonia is converted into urea by the urea cycle, before being excreted. In birds and many terrestrial reptiles, nitrogen is excreted predominantly as uric acid (Figure 14.4) which is synthesized from ammonia via the pathway which forms purines (see Chapter 21).

14.3.5 THE UREA CYCLE

The urea cycle is the pathway by which urea is synthesized from ammonia and takes place mainly in the liver. Urea has two amino groups, one comes directly from ammonia and the other from the amino group of aspartate. The pathway for the synthesis of urea is a cyclic one, and was largely discovered in 1932 by H.A. Krebs and his colleagues, the same people who elucidated the TCA cycle. The pathway itself is shown in Figure 14.5.

Ammonia first reacts with carbon dioxide and two molecules of ATP to produce a compound called carbamoyl phosphate. Only one of the ATPs is needed to provide the phosphate group, the other provides the energy to drive the reaction. The next four intermediate compounds are all amino acids but only one of them, arginine, is ever found as part of proteins. Carbamoyl phosphate reacts with the amino acid, ornithine, to produce another amino acid, citrulline. This then reacts with a molecule of aspartate to produce argininosuccinate. This step requires a large amount of energy so that ATP has to be broken down to AMP and pyrophosphate (PP$_i$), using the equivalent of two high-energy phosphate bonds. The amino group of the aspartate is derived from glutamate by transamination.

The arginine produced at this stage is then hydrolysed, so that the end of the molecule is liberated as urea. This regenerates ornithine, with which the cycle began. Thus both of the

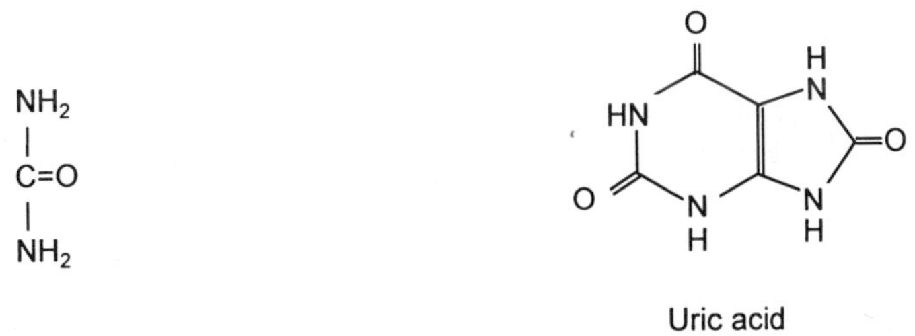

Urea

Uric acid

Figure 14.4 The structures of urea and uric acid, end products of nitrogen metabolism in most terrestrial animals.

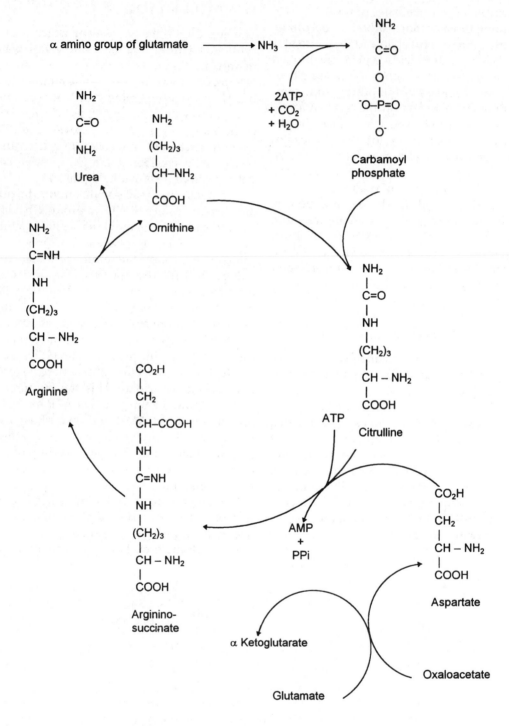

Figure 14.5 The urea cycle takes place mainly in the liver and converts ammonia into urea. Four high-energy phosphate bonds are broken for each turn of the cycle.

amino groups of the urea are ultimately derived from the amino group of glutamate.

The excretion of ammonia by way of urea requires a considerable amount of energy. Each turn of the cycle requires the expenditure of four high-energy phosphate bonds (two ATPs change to ADP and one changes to AMP). Urea is excreted through the kidneys in the urine. This means that there is also a high requirement for water in the process. As the main end product of their amino acid metabolism, birds and reptiles excrete uric acid in the form of a solid suspension which reduces the need for water in their excretory processes.

In addition to urea, mammals also excrete small amounts of uric acid, not as an end product of amino acid catabolism, but formed by the breakdown of purines.

14.3.6 THE FATE OF UREA IN RUMINANTS

In ruminant animals some of the urea that is produced in the liver is not captured by the kidneys but instead may be returned to the rumen in saliva or directly through the rumen wall. Microbes in the rumen can use the urea to make their own cell proteins. The microbial cells then pass further down the digestive tract to the abomasum (true stomach) and small intestine where the proteins that they contain are digested. In this way ruminants reduce their dependence on external supplies of protein.

14.4 PRECURSOR FUNCTIONS OF AMINO ACIDS

As well as being constituents of proteins, many amino acids are also precursors of other important compounds. Some of these functions are indicated in Table 14.2.

Table 14.2 Examples of precursor functions of amino acids

Amino acid	Product
Aspartate	Pyrimidines
Glycine	Purines; glutathione (tripeptide: γ-glutamyl cysteinyl glycine); porphyrins e.g. haem, chlorophyll
Histidine	Histamine – a biologically active amine released from some animal tissues in response to inflammation or as an allergic reaction causing dilation of blood vessels
Glutamic acid	γ-aminobutyric acid (GABA) – neurotransmitter in brain
Tyrosine	Adrenalin; morphine; thyroid hormone
Tryptophan	Nicotinic acid (component of NAD^+ etc.); serotonin (neurotransmitter); indole acetic acid (plant hormone)
Valine	Pantothenic acid (component of coenzyme A)

A common reaction is the decarboxylation of amino acids to form amines, e.g. histamine and γ-aminobutyric acid. These reactions are catalysed by pyridoxal-requiring decarboxylase enzymes. In the case of many alkaloids, further oxidation of amines to aldehydes and condensation of these with amines gives rise to the heterocyclic rings which are found in these compounds.

15.1 INTRODUCTION

The pentose phosphate pathway (also known as the phosphogluconate pathway or the hexose monophosphate shunt) provides a route by which glucose can be oxidized to carbon dioxide. As in the case of glycolysis, the substrate for the pathway is glucose-6-phosphate. An outline of the pathway is shown in Figure 15.1. The pathway can be considered in two parts: oxidative reactions and rearrangement reactions.

15.2 OXIDATIVE REACTIONS

In these reactions, glucose-6-phosphate is oxidized by successive removal of two pairs of hydrogen atoms. The first step, catalysed by the enzyme glucose-6-phosphate dehydrogenase [1], forms 6-phosphogluconolactone (Figure 15.2). This then undergoes hydrolysis, catalysed by lactonase [2], producing 6-phosphogluconate as a result of ring opening. Oxidative decarboxylation of 6-phosphogluconate, catalysed by 6-phosphogluconate dehydrogenase [3], yields the five-carbon sugar ribulose-5-phosphate. In both oxidation steps, hydrogen atoms removed from the substrates are transferred to NADP$^+$, reducing it to NADPH.

15.3 REARRANGEMENT REACTIONS

The remainder of the pathway consists of a series of reactions in which sugars undergo rearrangement and transfer reactions, regenerating some glucose-6-phosphate from the ribulose-5-phosphate produced in the oxidation steps (Figure 15.3).

Ribulose-5-phosphate is converted into other pentose sugars: ribose-5-phosphate by the action of ribulose phosphate isomerase [5], and xylulose-5-phosphate by the action of ribulose phosphate-3-epimerase [4] (Figure 15.1). These sugars are then acted on by transketolase [6] (Figures 15.1 and 15.3) which transfers a –CO–CH$_2$OH group from the ketose sugar (xylulose-5-phosphate) to the aldose (ribose-5-phosphate). This reaction results in the production of a seven-carbon sugar (sedoheptulose-7-phosphate) and a three-carbon one (glyceraldehyde-3-phosphate). These two sugars then react together in a step catalysed by transaldolase [7] (Figures 15.1 and 15.3) in which a three-carbon fragment (–CHOH–CO–CH$_2$OH) is transferred from sedoheptulose-7-phosphate to the glyceraldehyde-3-phosphate, resulting in six-carbon (fructose-6-phosphate) and four-carbon (erythrose-4-phosphate) sugars. Transket-olase is not specific in the acceptor it uses for its transfer reaction, and also catalyses transfer of a two-carbon fragment from xylulose-5-phosphate to erythrose-4-phosphate, forming six- and three-carbon products (fructose-6-phosphate and glyceraldehyde-3-phosphate, respectively). Condensation of three-carbon fragments, catalysed by triose phosphate isomerase [8], aldolase [9] and phosphoglucose isomerase [10], results in eventual production of glucose-6-phosphate.

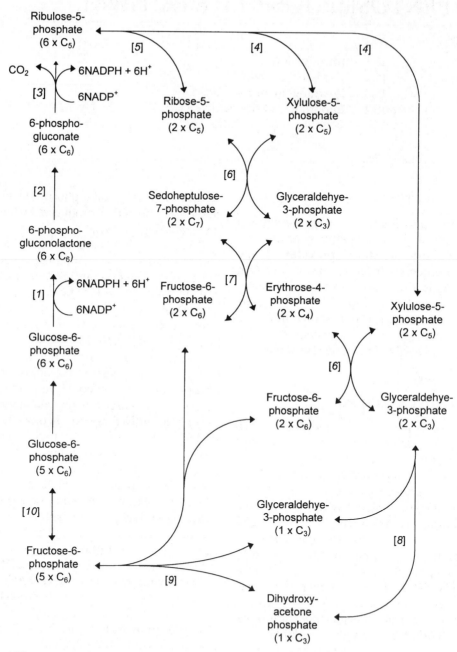

Figure 15.1 The pentose phosphate pathway. Enzymes catalysing individual steps are as follows: [1] glucose-6-phosphate dehydrogenase, [2] lactonase, [3] 6-phosphogluconate dehydrogenase, [4] ribulose-phosphate-3-epimerase, [5] ribulose-phosphate isomerase, [6] transketolase, [7] transaldolase, [8] triosephosphate isomerase, [9] aldolase, [10] phosphoglucose isomerase.

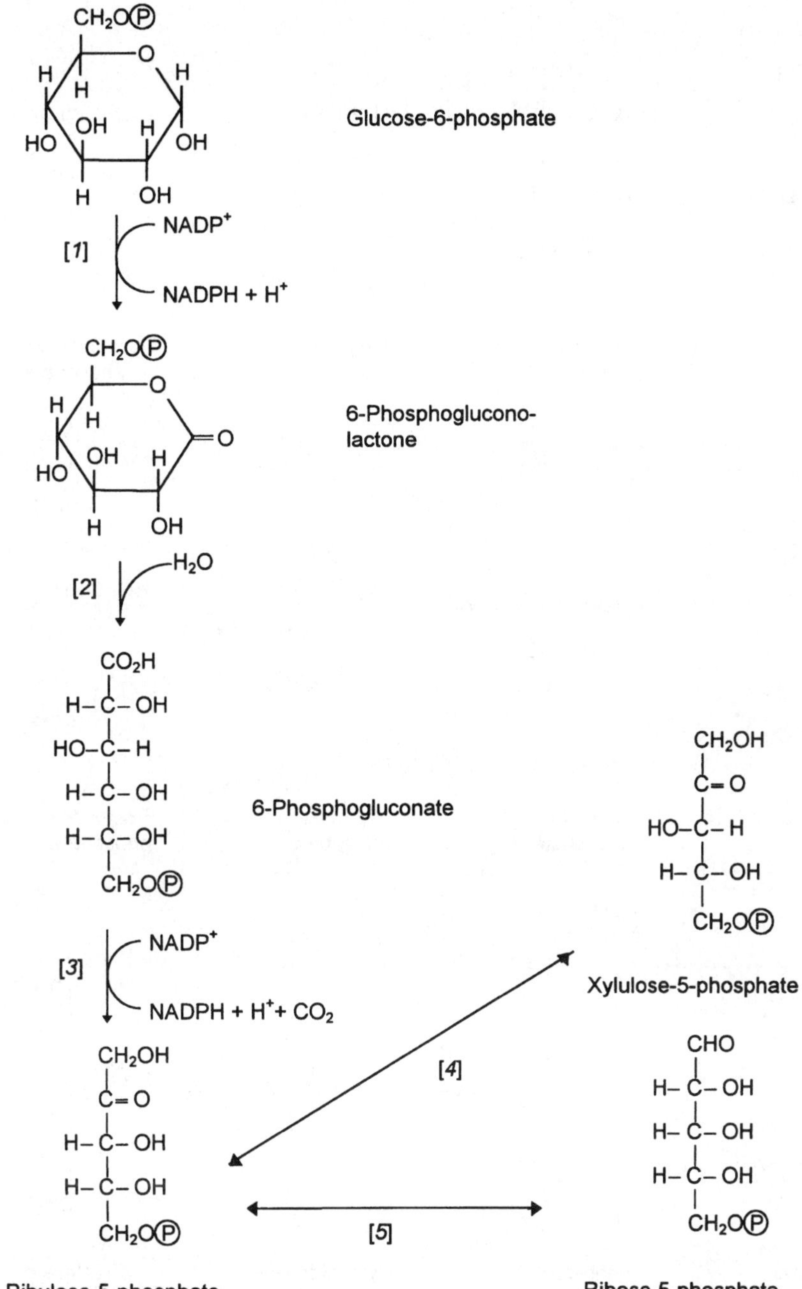

Figure 15.2 Oxidative reactions of the pentose phosphate pathway. Both oxidation steps produce NADPH. The names of the enzymes are given in the legend to Figure 15.1.

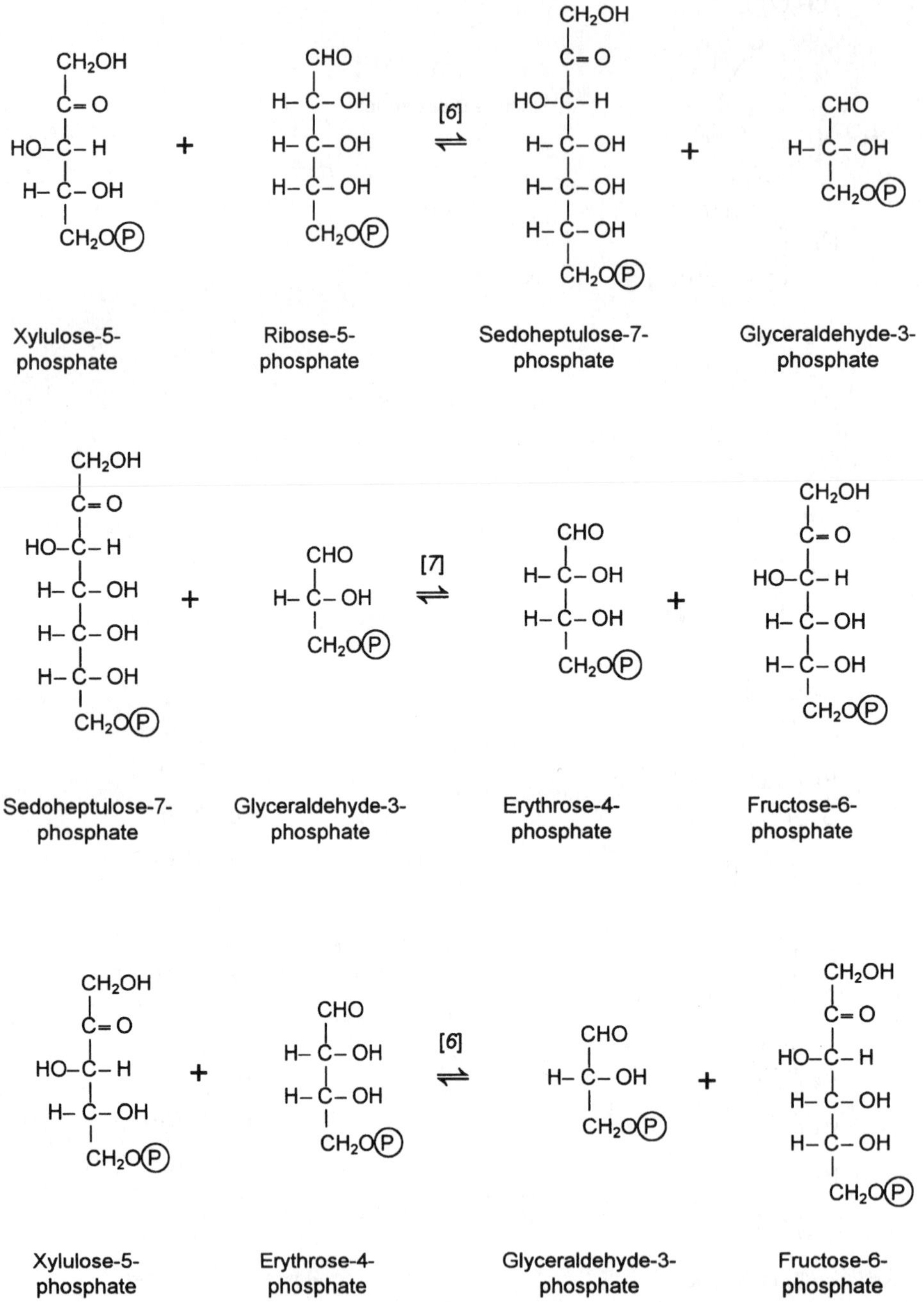

CH₂OH structures...

Xylulose-5-phosphate	Ribose-5-phosphate	Sedoheptulose-7-phosphate	Glyceraldehyde-3-phosphate

$$\text{Xylulose-5-phosphate} + \text{Ribose-5-phosphate} \underset{[6]}{\rightleftharpoons} \text{Sedoheptulose-7-phosphate} + \text{Glyceraldehyde-3-phosphate}$$

$$\text{Sedoheptulose-7-phosphate} + \text{Glyceraldehyde-3-phosphate} \underset{[7]}{\rightleftharpoons} \text{Erythrose-4-phosphate} + \text{Fructose-6-phosphate}$$

$$\text{Xylulose-5-phosphate} + \text{Erythrose-4-phosphate} \underset{[6]}{\rightleftharpoons} \text{Glyceraldehyde-3-phosphate} + \text{Fructose-6-phosphate}$$

Figure 15.3 Rearrangement reactions of the pentose phosphate pathway. These reactions are reversible and allow the interconversion of sugars containing 3-, 4-, 5-, 6- and 7-carbon atoms. The names of the enzymes are given in the legend to Figure 15.1.

The overall equation for this pathway is as follows:

$$\text{Glucose-6-phosphate} + 12\ \text{NADP}^+ + 7\text{H}_2\text{O} \rightarrow$$
$$6\text{CO}_2 + 12\text{NADPH} + 12\text{H}^+ + \text{P}_i$$

15.4 IMPORTANCE OF THE PATHWAY

Although the pentose phosphate pathway and glycolysis/TCA cycle both bring about complete oxidation of glucose to carbon dioxide, they serve different purposes. The pentose phosphate pathway produces NADPH, whilst the glycolysis/TCA cycle produces NADH. As has been noted elsewhere (Chapter 6), NADPH is used primarily as a reducing agent in biosynthetic reactions, whilst NADH serves as a source of hydrogen atoms for the electron transport chain. Thus, whilst glycolysis is mainly concerned with energy metabolism and is active in all tissues, the pentose phosphate pathway is found mainly in tissues in which biosynthetic processes are important. NADPH is required in large amounts for the synthesis of fatty acids and steroids, and is also required for production of deoxyribonucleotides. The pathway is therefore active in liver, lactating mammary gland and adipose tissue. In plants, although NADPH is produced in the chloroplasts during the light reactions, it is not readily available to the cytoplasm and is, in any event, only produced in this way in photosynthetic tissues. Thus the pentose phosphate pathway is also required by plants as a source of NADPH and of ribose and erythrose-4-phosphate.

A further function of NADPH may be to keep glutathione in its reduced form. In animals, deficiencies of the enzyme glucose-6-phosphate dehydrogenase lead to lowered levels of reduced glutathione, so making red blood cells susceptible to damage from hydrogen peroxide.

The pentose phosphate pathway has, as intermediates, sugars containing three-, four-, five-, six- and seven-carbon atoms. It can therefore serve as a source of ribose, required for the production of nucleotides and nucleic acids, and of erythrose-4-phosphate, a precursor of the aromatic rings which form part of some amino acids and phenolics, such as lignin, tannins and anthocyanins (Chapter 5). The latter process does not occur in animals which are unable to synthesize aromatic compounds.

The pentose phosphate pathway has several intermediates in common with glycolysis. The rearrangement steps of the pathway are readily reversible and can be used to make ribose and erythrose from glucose-6-phosphate. The pentose phosphate pathway and glycolysis, operating together, provide a very versatile system which can work in several different ways, depending on the relative requirements of the cell for NADPH and ribose-5-phosphate.

- Much more NADPH than ribose-5-phosphate required: the complete pathway, using both oxidative and rearrangement reactions, leads to oxidation of glucose-6-phosphate to carbon dioxide.
- Much more ribose-5-phosphate than NADPH required: glucose-6-phosphate is converted to fructose-6-phosphate and glyceraldehyde-3-phosphate by glycolysis, and then ribose-5-phosphate is produced by reversal of the rearrangement steps. The oxidative steps of the pathway are not used. This situation arises, for example, in rapidly dividing cells.
- Balanced requirement for NADPH and ribose-5-phosphate: the oxidative steps are used to produce two NADPH molecules for each ribose-5-phosphate. Rearrangement steps are not required.

The pathway can provide a mechanism by which pentose sugars are utilized. These may be present in animal diets as components of hemicellulose, xylans, pectins and nucleic acids. In this way, pentose sugars can be used for biosynthesis or energy production.

15.5 REGULATION OF THE PATHWAY

In view of the variety of ways in which the pathway can operate, it is not surprising that it is subject to tight regulation. The enzyme glucose-6-phosphate dehydrogenase is the main point of control. This enzyme is inhibited by NADPH. If NADPH is used to synthesize fatty acids, the activity of the enzyme, and hence production of NADPH, increase. On the other hand, accumulation of fatty acyl CoAs decreases the activity of the enzyme and NADPH synthesis.

FERMENTATION PATHWAYS

16.1 INTRODUCTION

It has been assumed in previous chapters that there is an abundant supply of oxygen and that most biological oxidations can be thought of as controlled versions of combustion. There are a number of situations in which oxygen supply may be very limited or even non-existent, but many types of cell survive under these conditions. In biological systems, oxidation normally takes place by the addition of oxygen, the removal of hydrogen or the removal of electrons. Hydrogen atoms that are removed are subsequently oxidized to water under aerobic conditions. Even under anaerobic conditions, oxidation is made possible by the movement of electrons.

16.2 ANAEROBIC ENVIRONMENTS IN AGRICULTURE

Anaerobic conditions can develop even in supposedly aerobic organisms such as animals, but principally they are to be found where the environment precludes the penetration of oxygen, as for example in:

- muscle contraction when demand for oxygen is greater than the supply from the circulation
- animal digestive systems
- soils
- waste treatment
- fermentation in milk products
- changes in meat *post mortem*
- changes in crops after harvest
- ensiling of forages
- production of ethanol.

16.3 LACTATE PRODUCTION

The biochemical pathway for lactate production makes use of glycolysis, in which glucose is broken down to pyruvate. This pyruvate is then utilized for lactate synthesis. For each mole of glucose converted to pyruvate, glycolysis produces four moles of ATP and two of NADH, and consumes two moles each of

NAD⁺ and ATP (Chapter 10). In terms of useful biochemical energy, the yield is two moles of ATP from each mole of glucose. This would be a useful way of producing a limited amount of energy from glucose in the absence of oxygen, but cells would soon run out of NAD⁺ and the process would grind to a halt. This problem is overcome by changing the pyruvate into lactate (lactic acid), which changes the two molecules of NADH back into NAD⁺ (Figure 16.1). The reaction, which is easily reversed, is catalysed by the enzyme lactate dehydrogenase. In this way the overall pathway from glucose to lactate can continue to catabolize glucose and gain a small amount of energy in the form of the two molecules of ATP. This is only one of two schemes for lactate production from glucose; it differs from the other in producing only lactate as its product, and it is therefore known as a homolactate fermentation. The yield of ATP from this pathway is low, but the substrate has not been extensively degraded and the product, lactate, is available for other purposes.

16.3.1 MUSCLE METABOLISM

Muscles require large amounts of energy to function, but the work performed varies. Some need to be able to respond very quickly to changing situations, whereas others may be required to maintain a constant work output but at a relatively low intensity. Within muscles a number of different types of fibre have been identified which lie in a spectrum from fast-twitch to slow-twitch: these are often denoted as white and red fibres, respectively.

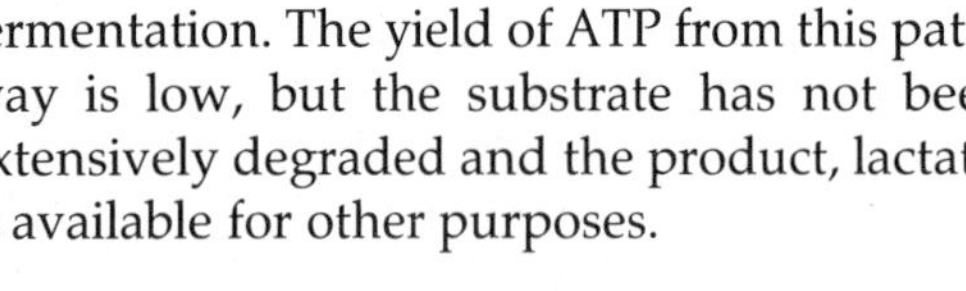

Figure 16.1 Reduction of pyruvate to lactate. Together with the glycolytic pathway, this reaction leads to the production of lactate from glucose; lactate is the only product and thus the process is known the homolactate pathway.

Slow-twitch cells obtain most of their energy directly from the aerobic oxidation processes of the mitochondria with which they are well supplied. The fast-twitch cells can also obtain energy from the anaerobic oxidation of glucose (released from intracellular glycogen stores) to lactate.

16.3.2 REGENERATION OF GLUCOSE

The lactate produced in muscle is passed via the blood to the liver where it is used as a precursor of glucose synthesis (see Chapter 18). The glucose synthesized in the liver can then, in turn, be sent to back to the muscle. This cycle of reactions has been named the Cori cycle after its discoverers C.F. Cori and G.T. Cori (see Figure 16.2).

If lactate starts to accumulate due to sustained physical activity, then it is a common experience in man that the muscles ache – the actual pain may be due to the decrease in intramuscular pH caused by the lactate produced.

16.4 ANIMAL DIGESTIVE SYSTEMS

Anaerobic conditions are found in parts of the gut in all animals. It has only recently been appreciated that the anaerobic fermentation in the human large intestine may have an importance of its own. However, fermentation is essential in herbivores, in the rumen and reticulum of ruminants and tylopods (camels etc.), or in the caecum of the non-ruminant herbivores (rabbits, horses etc.). The major end products of these fermentations are the volatile fatty acids which supply most of the animals' dietary energy. In non-herbivores this role is played by glucose. There are three major volatile fatty acids: acetate, propionate and butyrate, together with a number of minor ones. They are produced by microorganisms – bacteria, fungi and protozoa which live symbiotically in the digestive tract. The microorganisms benefit by gaining their energy from the partial breakdown of the carbohydrates eaten by the animals. The end products

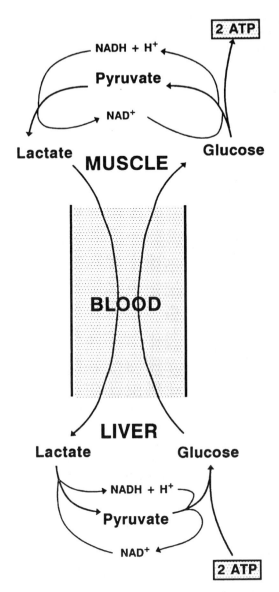

include many (e.g. cellulose and hemicellulose) that cannot be broken down by simple-stomached animals such as man. The microbes are also able to utilize simple sources of nitrogen as precursors for the synthesis of amino acids and proteins. In ruminants, these proteins eventually become available to the animal as they pass further through the digestive tract. Additional benefits are brought to the animal in the form of a supply of vitamins.

The common feature of the pathways for volatile fatty acid synthesis from glucose is that they all involve pyruvate.

16.4.1 ACETATE FORMATION

The simplest route is the pathway for acetate production, in which the pyruvate reacts with coenzyme A forming acetyl-CoA and losing formate (Figure 16.3). The acetyl-CoA loses its coenzyme A with the simultaneous production of ATP, part of the benefit accruing to the microbe in the form of energy. The formate is split into carbon dioxide and gaseous hydrogen. Hydrogen is removed by conversion to methane.

16.4.2 PROPIONATE FORMATION

Propionate is the only one of the three major volatile fatty acids that the ruminant can use as a source of glucose. There are two different pathways for propionate synthesis from pyruvate, each favoured by different types of microbes (Figure 16.4). One of these pathways proceeds by an extension of the route for the synthesis of lactate. Lactate is first activated by the formation of its coenzyme A ester. This loses both CO_2 and water to form propernoyl-CoA, the unsaturated analogue of propionyl-CoA. In turn this is hydrogenated to yield propionyl-CoA, which loses coenzyme A to yield the product. The second pathway for the generation of propionate is virtually a reversal of the route that is used in animal cells, and particularly the liver, to synthesize carbohydrates from propionate (see Chapter 18).

Figure 16.2 The Cori cycle. Glucose passes from the liver to muscle where it is oxidized to lactate. The lactate is then returned to the liver where it is used for glucose synthesis. Under conditions of high energy demand, such as exercise, lactate may temporarily accumulate in muscle.

of microbial metabolism are available to the host animal, which gains because the carbohydrates that are fermented by microorganisms

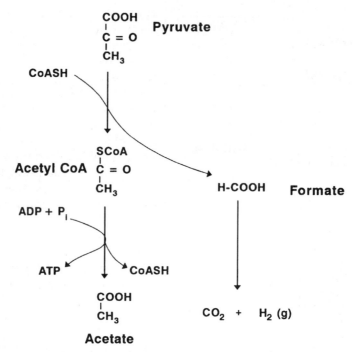

Figure 16.3 The fermentation of glucose to produce acetate. Formate produced in the decarboxylation of pyruvate is broken down to CO_2 and gaseous hydrogen.

16.4.3 BUTYRATE SYNTHESIS

The first three steps of butyrate production are reactions that are commonly encountered under other circumstances (Figure 16.5). The first one is the pyruvate dehydrogenase step, which links glycolysis with the tricarboxylic acid cycle. As a start to the production of butyrate, two molecules of pyruvate must be oxidized to yield two molecules of acetyl-CoA. Carbon dioxide is lost and NAD^+ is reduced to NADH. Acetoacetyl-CoA is formed from two molecules of acetyl-CoA, with one molecule of coenzyme A being liberated. Acetoacetyl-CoA is reduced to yield 3-hydroxybutyryl-CoA, utilizing NADH and liberating NAD^+.

The hydroxybutyryl-CoA then loses water to form an unsaturated compound, crotonyl-CoA. Similar, but not identical, steps are to be found both in fatty acid oxidation and synthesis.

NADH produced in earlier parts of the pathway is used to reduce crotonyl-CoA to yield the coenzyme A ester of butyric acid. The ester is hydrolysed to produce butyrate and free coenzyme A. This reaction is coupled to the phosphorylation of ADP to ATP, which provides more energy for the metabolism of the microorganism.

16.5 SOILS

Soils provide a mixture of aerobic and anaerobic environments; which of these predominates at one time depends on the porosity of the soil, and on the extent to which air movement is impeded by close packing of the soil matrix or by the filling of the pores by water. Much of the decomposition of organic material in soils takes place through aerobic oxidation; even O_2 concentrations of as little as 1% may be sufficient to ensure aerobic degradation. It is for this reason that in many soils the organic matter does not reach very high levels

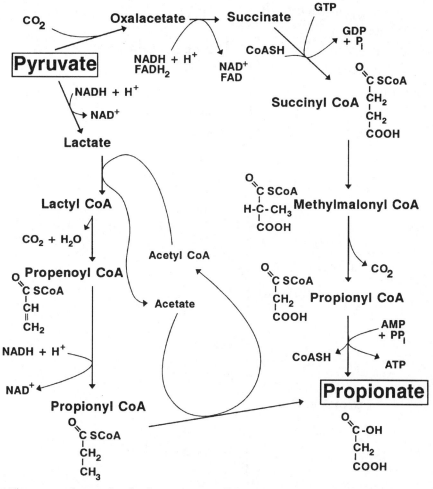

Figure 16.4 The two pathways for the fermentation of glucose to propionate. One pathway passes through lactate and involves a complex 'shuttle' of Coenzyme A via acetyl-CoA. The other pathway, via succinate, is the reverse of the mechanism used by animals to synthesize glucose from propionate.

despite large inputs of decaying vegetation each year. There are systems where decay is severely impeded and anaerobic conditions prevail. The enormous deposits in the peat bogs in cool temperate regions are a testimony to the effect of anaerobic fermentation. In the absence of oxygen, anaerobic organisms can use a whole range of other compounds as receptors for electrons. A good example is the use of sulphur compounds which are reduced to the sulphide. Under acid conditions this associates with hydrogen ions to form hydro-gen sulphide, a common product in anaerobic biological systems.

16.6 WASTE TREATMENT

One of the features of the intensification of agriculture in the 20th century has been an increasing awareness of the problems associated with agricultural wastes – particularly where these are produced close to centres of human population. Basically there are two types of waste treatment, aerobic and anaerobic. In the former

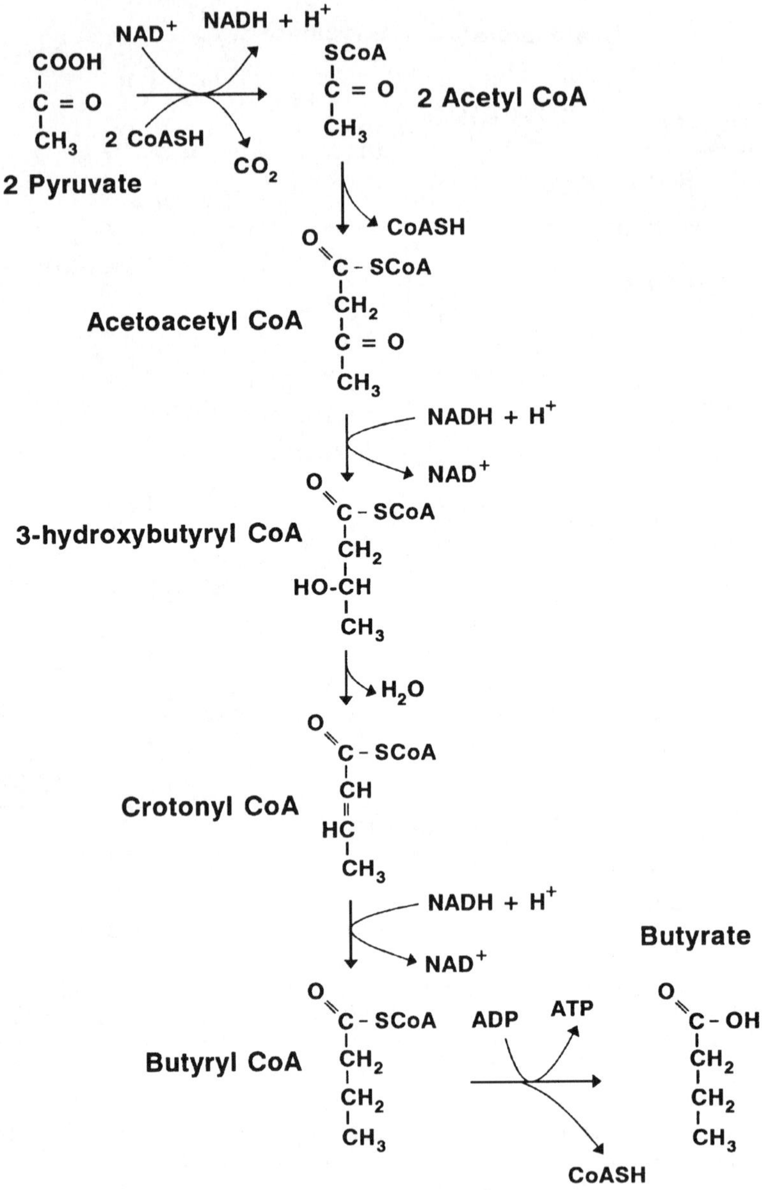

Figure 16.5 The fermentation of glucose to yield butyrate. Although the first step produces NADH, there are two susbsequent steps that absorb it. This allows the pathway to 'soak up' the NADH that is produced in glycolysis.

case, organic wastes are almost completely oxidized, with the eventual production of CO_2 and water. Anaerobic oxidation yields a considerable amount of biomass which may be recycled into the agricultural system. A product that is of growing interest, particularly in the developing world, is the production of methane (biogas) by anaerobic fermentation.

16.7 METHANE PRODUCTION

The reactions of methane synthesis are not confined to biogas generation: they occur to a great extent in the digestive tracts of animals, particularly ruminants. Four major precursors are shown in Figure 16.6. In methane generation from waste the first of these, using acetate, tends to account for the larger part of gas production.

In the rumen, the reactions that use hydrogen, formate or methanol are the most important. Methane represents a fairly inert material that overcomes the potentially toxic effects of high concentrations of any of these precursors. On average, about 8% of the energy in diets eaten by ruminants is released in the form of methane. This has led to concern that animals may make a major contribution to the 'greenhouse' gases thought to be responsible for global warming.

Many anaerobic pathways in the rumen produce hydrogen as an end product which may be used for the hydrogenation of fatty acids in the rumen.

16.8 DAIRY PRODUCTS

Refrigeration is a comparatively recent technique for preserving dairy products, but the use of anaerobic fermentation to yield acidophilic milk, yoghurt and cheese goes back to the dawn of civilization. Most of the initial fermentative changes in milk involve the homolactate pathway described for muscle.

16.9 MEAT

Once an animal is dead the supply of oxygen to the muscles ceases abruptly. Anaerobic catabolism then takes place in the tissues and

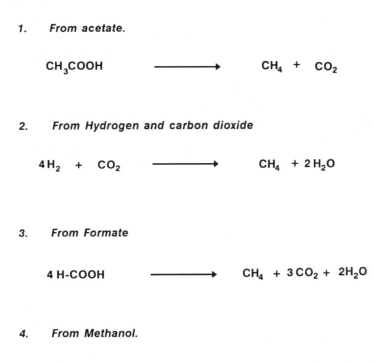

1. From acetate.

$$CH_3COOH \longrightarrow CH_4 + CO_2$$

2. From Hydrogen and carbon dioxide

$$4H_2 + CO_2 \longrightarrow CH_4 + 2H_2O$$

3. From Formate

$$4 H\text{-}COOH \longrightarrow CH_4 + 3CO_2 + 2H_2O$$

4. From Methanol.

$$4 CH_3OH \longrightarrow 3CH_4 + CO_2 + 2H_2O$$

Figure 16.6 Four different pathways for the synthesis of methane. The pathway used varies from microorganism to microorganism.

to a great extent determines the quality of the meat product. Glycogen, present in the muscle at death, is oxidized anaerobically, leading to a build-up of lactate in the muscle which reduces the pH within the tissue from its normal (living) value of about 7.3 to around 5.4. If animals are stressed in some way prior to slaughter, the level of glycogen in their muscles is reduced. For instance, in a chicken that is struggling at the time of death, glycogen values may only be half of those for a stunned animal. In turn, lactate production is reduced and the pH values obtained are higher, perhaps as high as 6.5. The character of the meat is different: it is drier, darker and more firmly textured than normal. In the beef industry this dark, firm, dry (DFD) meat can cause significant financial losses (see Chapter 32).

16.10 FERMENTATION IN HERBAGES

In most parts of the world the production of fodder occurs on a seasonal basis. This may be because in winter the temperatures are too low for crop growth (as in temperate regions) or because water is in short supply during a dry season (many tropical areas). Fodder crops from the growing season are preserved for use during the shortage. One method is drying, but this presents technical problems in many environments. A large proportion of farmers now effectively pickle their crop by allowing it to ferment. The acids which are produced lower the pH to such an extent that a well preserved and palatable product is formed. Almost all of the methods for silage-making rely on anaerobic fermentation of carbohydrates (sugars or starch), these may be present in the crop as in temperate grasses or in maize, or they may have to be added, possibly as molasses, in order to ferment the tropical grasses. Whatever the source of carbohydrate the processes are quite similar. The most important condition that must be achieved is an absence of oxygen to ensure anaerobic fermentation (see Chapter 24).

There are many pathways to be found in fermenting herbage but the most important ones are those that produce lactate. In general, good silages are those in which the commonest acid is lactic acid and the levels of the volatile fatty acids are very low. One pathway for lactate production (the homolactate pathway) has been considered already. Another (the heterolactate pathway) takes place in silage (Figure 16.7). This involves the prior breakdown of hexose sugars to pentoses (ribulose-5-phosphate and xylulose-5-phosphate); these steps are also found in the first part of the pentose phosphate pathway. Xylulose-5-phosphate is cleaved into two compounds, one with three carbons and the other with two. The three-carbon glyceraldehyde-3-phosphate is metabolized via pyruvate (as in glycolysis) to lactate. The two-carbon acetyl phosphate is merely dephosphorylated (producing ATP) to give acetate.

The main objective in making silage is to reduce the pH to a low value, preferably in the region of 4.5, as quickly as possible. In the presence of oxygen, the fermentation pathways move towards the production of large quantities of butyric acid, which leads to an unpleasant-smelling product with a pH of 5.5 or higher. Under these conditions the nutritive value of the product is very low and animals are extremely reluctant to eat it. All effective silage-making procedures depend on excluding air by compressing the herbage and sealing the silage carefully with an impermeable material such as plastic sheet.

16.11 ETHANOL PRODUCTION

Some crops are grown specifically as the raw material for alcohol fermentation. They all produce large quantities of sugars (e.g. sugarcane) or starch (grains and some roots).

The biochemical pathway is very simple: in essence it is an alternative use for the pyruvate produced during glycolysis. There are only two extra steps (Figure 16.8). In the first step, pyruvate loses CO_2 (decarboxylation) to yield acetaldehyde. The reaction occurs irreversibly

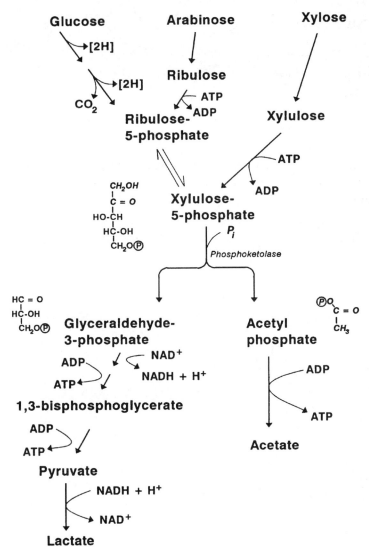

Figure 16.7 The heterolactate pathway which operates during ensiling forages. This differs from the homolactate pathway of Figure 16.1 in that a mixture of products, lactate and acetate, is formed.

under the influence of pyruvate decarboxylase. This enzyme shares with the pyruvate dehydrogenase complex (Chapter 11) a requirement for thiamin pyrophosphate (TPP) as a coenzyme. The pyruvate is attached to the molecule of TPP, whilst the decarboxylation is taking place.

The second step is reversible and involves the reduction of acetaldehyde to yield ethanol.

The reducing power comes from NADH produced in the earlier stages of glycolysis. The NAD$^+$ produced in this pathway can then be recycled to glycolysis, allowing it to continue.

Mammalian systems may sometimes have to metabolize ethanol which has either been directly administered or produced by fermentation in the gut of herbivores. The first stage reverses the alcohol dehydrogenase reaction

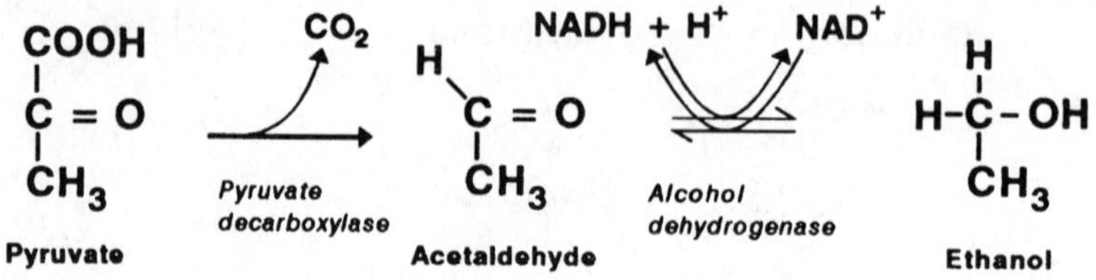

Figure 16.8 Reactions leading to the synthesis of ethanol from pyruvate.

to produce acetaldehyde. An aldehyde dehydrogenase then uses NAD^+ as a hydrogen acceptor to oxidize the acetaldehyde to acetate, which can be further metabolized through the TCA cycle or as a precursor for fatty acid synthesis.

17.1 INTRODUCTION

Photosynthesis is the process in which the energy of light is used to bring about synthesis of complex organic molecules, such as sugars, from carbon dioxide and water. It is of enormous importance as it is the principal means by which energy can enter biological systems. Photosynthesis taking place in the past produced the fossil fuels which form most of our current energy reserves.

The scale on which photosynthesis occurs is staggering. The amount of carbon fixed per year has been estimated to be of the order of 100 billion tons (1×10^{11} tons) – equivalent to about 200 billion tons of organic matter as CH_2O. Even temperate grassland, with its modest growth rates, may produce 600 g of organic matter per m^2 per year.

It is generally considered that about two-thirds of carbon dioxide fixation takes place on land and about one-third in the oceans, but the contribution made by the latter has probably been underestimated. The quantity of carbon dioxide directly available to plants from the atmosphere makes up only about 0.001% of the total carbon on earth. Over 99% exists in the form of rocks and sediments. About 13% of the carbon dioxide in the atmosphere is used in photosynthesis and released by respiration each year, and an approximately equal quantity exchanges with carbon dioxide dissolved in the oceans.

Photosynthesis occurs in a wide range of organisms, from green plants and algae to some bacteria. The processes themselves are always associated with biological membranes in which pigments are embedded. In eukaryotes such as green plants and green algae, these membranes are found in organelles called chloroplasts, but in prokaryotes such as photosynthetic bacteria and blue-green algae, the pigments are present in specialized membranes within the cytoplasm.

The process of photosynthesis can be divided into two parts.

- Light reactions, in which light is absorbed and its energy used to make ATP and NADPH from ADP and NADP+, respectively. As described in more detail later in this chapter, the light reactions are brought about by the flow of electrons through two photosystems: photosystem I (PS I) and photosystem II (PS II). This electron flow generates a proton gradient across the membranes and this in turn drives the synthesis of ATP by an enzyme called ATP synthase.
- Dark reactions or light-independent reactions, in which ATP and NADPH are used to reduce carbon dioxide to produce sugars. These reactions are brought about by the operation of a metabolic pathway usually known as the Calvin cycle.

17.2 CHLOROPLASTS

Chloroplasts are disc-shaped organelles about 5 μm long. They are surrounded by a double-membrane system (Figure 17.1). The outer membrane is relatively permeable, but the inner one is selectively permeable and regulates the influx and efflux of many metabolites. Between the membranes is an intermembrane space. Inside the chloroplast is the stroma, a protein-rich, gel-like material which contains the enzymes responsible for carbon dioxide fixation and sugar synthesis. Running through the stroma a series of membranes called thylakoids can be seen, and in some regions rounded tongues of thylakoid membranes are stacked on top of one another to form grana. Parts of the grana thylakoids, which are in close contact with one another but separated from the stroma, are known as the appressed

regions of the grana. It is in these areas that PS II is mainly located. The parts of the grana at the top and bottom of the stacks are called the non-appressed regions. PS I and ATP synthase are restricted to these regions and to the stroma thylakoids, both of which are in close contact with the stroma. The thylakoid membranes enclose a continuous cavity, the thylakoid space, which they therefore separate from the stroma.

17.3 THE LIGHT REACTIONS

Pigment molecules are responsible for absorbing light and form part of discrete structural components of the membranes called photosystems, which are embedded in the thylakoid membranes.

17.3.1 PHOTOSYNTHETIC PIGMENTS

Chlorophylls

The chlorophylls are the main light-absorbing pigments. They contain a magnesium atom chelated by the pyrrole nitrogen atoms of a porphyrin ring. Porphyrins are formed from succinyl-CoA (an intermediate in the TCA cycle) and glycine (an amino acid) via an intermediate called δ-aminolevulinic acid (ALA) which gives rise to each of the four pyrrole rings. One of the COOH groups of chlorophyll is esterified with MeOH and the other with the hydrophobic hydrocarbon phytol. Chlorophyll *b* differs from chlorophyll *a* only in having a –CHO group in ring II where chlorophyll *a* has a CH_3 (Figure 17.2)

Because of the long series of conjugated double bonds (i.e. alternating double and single bonds) in the rings, chlorophylls absorb visible light very strongly indeed, their molar extinction coefficients being amongst the highest of any organic compounds. Chlorophyll *a* and chlorophyll *b* absorb blue and red light particularly strongly. However their absorption spectra do differ slightly from one another so that they complement each

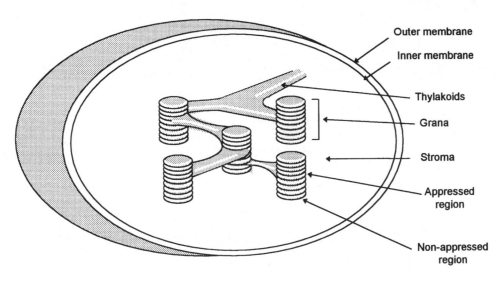

Figure 17.1 Structure of a chloroplast. Thylakoid membranes run through the stroma. In some regions tongues of thylakoid membranes are stacked on top of one another to form grana. The thylakoid membranes enclose a continuous cavity called the thylakoid space which connects the interiors of the grana to one another.

other and allow a wider range of wavelengths of light to be absorbed (Figure 17.3). In higher plants, the average chlorophyll *a*/chlorophyll *b* ratio is 3; it is higher than this for PS I and lower for PS II. Chlorophyll *b* is replaced by chlorophyll *c* in brown algae and by chlorophyll *d* in red algae. Photosynthetic bacteria do not contain chlorophyll *a* but have bacteriochlorophylls *a* or *b*.

Carotenoids

The carotenoids are polyisoprenoid compounds with long series of conjugated double bonds (Figure 17.2b).

There are two classes of carotenoids:

- carotenes such as β-carotene which contain no oxygen;
- xanthophylls such as lutein which contain oxygen.

Carotenoids absorb light of wavelengths between about 400 and 500 nm and therefore have absorption spectra which are complementary to the chlorophylls and help to trap more light. The carotenoids are also important antioxidants, preventing photo-oxidation of chlorophyll at high light intensity

In solution, chlorophyll *a* absorbs maximally in the red region of the spectrum at 676 nm and chlorophyll *b* at 642 nm. However, within the thylakoid membranes chlorophyll molecules are associated with membrane proteins, and the wavelength of the absorption maxima of individual molecules may shift because of this. *In vivo*, therefore, individual chlorophyll *a* molecules may have absorption maxima between about 660 and 700 nm as a result of their different environments.

17.3.2 LIGHT ABSORPTION

When a molecule absorbs a photon of light, an electron becomes excited to a higher energy level. The molecule in its excited state may lose this extra energy in one of several ways:

- emitting a photon of lower energy (longer wavelength) than that absorbed, by the process of fluorescence;

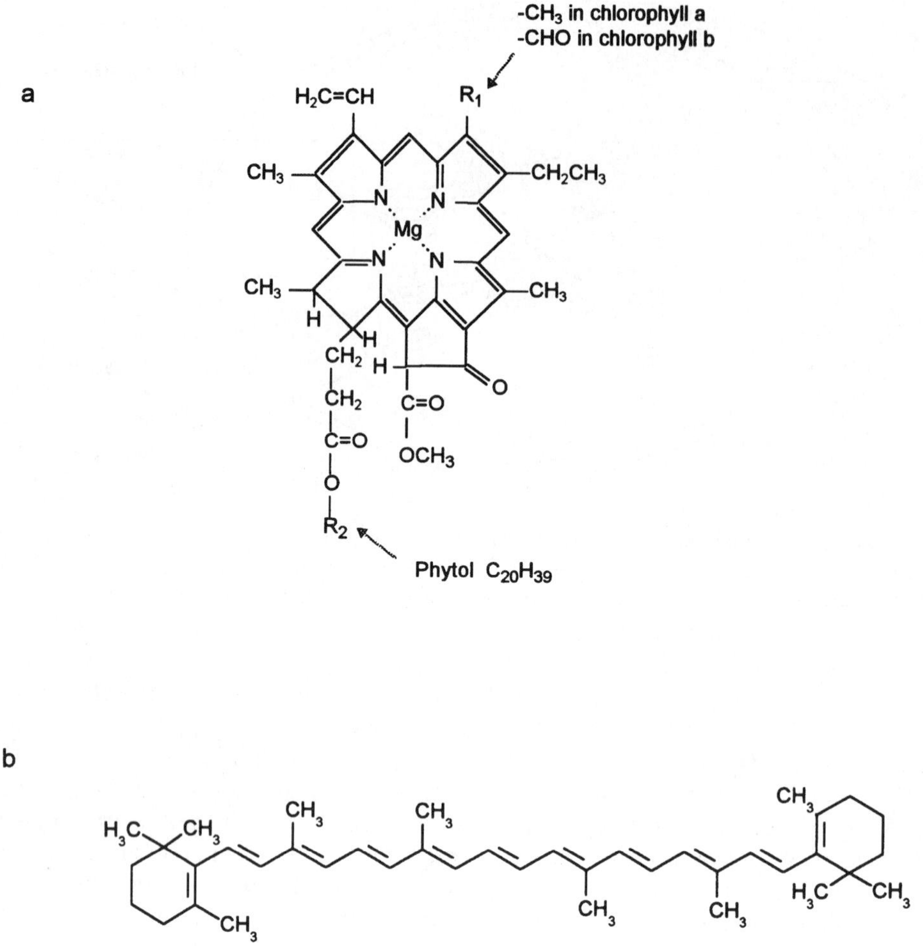

Figure 17.2 Structure of photosynthetic pigments: (a) chlorophyll (chlorophylls *a* and *b* differ only in the nature of R_1);(b) β-carotene. Both types of pigment contain long systems of conjugated double bonds which enable them to absorb visible light.

- colliding with a solvent molecule;
- transferring energy to another molecule by resonance energy transfer (RET);
- losing an electron in a chemical reaction.

Excited molecules in solution usually collide with, and pass on their energy to, solvent (often water) molecules before any other process can occur. Thus the absorbed light energy is converted into heat and the solution becomes warmer. When pigment molecules are embedded in membranes, as in chloroplasts, they are protected from collision with solvent and may be close and at a specific orientation with respect to other molecules. These factors favour energy transfer by the process of RET.

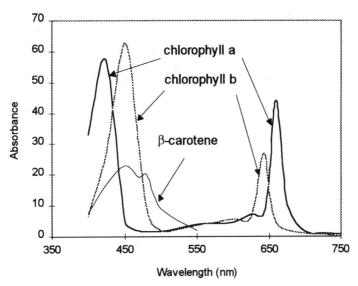

Figure 17.3 Absorption spectra of photosynthetic pigments. Both chlorophyll *a* and *b* absorb blue light (wavelength 400 to 470 nm) and red light (600 to 670 nm) but the spectra differ slightly. Carotenoids absorb mainly between 400 and 500 nm.

17.3.3 RESONANCE ENERGY TRANSFER

A molecule (the donor) may transfer its excitation energy to another nearby molecule (the acceptor) provided that the acceptor absorbs light of longer wavelength (lower energy) than the donor. The migration of energy between pigment molecules in the chloroplast membranes takes place by this process.

The majority of chlorophyll molecules are bound to proteins and form part of light-harvesting or antenna complexes. These complexes contain chlorophyll, together with accessory pigments such as carotenoids. The pigment molecules have different environments and therefore different absorption maxima as a result of their interaction with proteins of the complexes. Light absorbed by a molecule in the complex can migrate to nearby molecules which absorb at longer wavelength, and the process can continue until the molecule which absorbs at the longest wavelength becomes excited. In this way all the energy absorbed anywhere in the complex will eventually reach this special molecule which therefore

acts as an energy sink or reaction centre for the complex. This molecule cannot lose energy by RET and instead loses an energetic electron which is transferred to another molecule termed the **primary electron acceptor**.

17.3.4 THE ELECTRON TRANSPORT SYSTEM

Numerous studies have shown that photosynthesis involves the co-operation of PS I and PS II. Electrons are transported between these via an electron transport chain, and the energy of the electrons is increased as a result of light energy absorbed by each photosystem.

The manner in which the two photosystems interact with one another is represented by the **Z-scheme**, which is shown in Figure 17.4. In this diagram the vertical axis represents the redox potential, values towards the top are increasingly negative and those towards the bottom increasing positive. The redox potential indicates an electron's energy and how easily it can be transferred to another compound. As the addition of an electron is a reduction reaction, compounds high on the scheme are strong reducing agents and can transfer electrons to those lower down, reducing them in the process.

Water molecules are split by PS II. This releases electrons which are excited by the light absorbed by PS II and passed via a system of electron carriers to PS I. The electron transport process is coupled to the synthesis of ATP from ADP and P_i. Electrons reaching PS I receive further energy, absorbed by this photosystem, and pass into another part of the electron transport chain where they eventually reduce $NADP^+$ to NADPH.

In a physical sense the electron transport chain consists of several structural units, which include the two photosystems and the cytochrome b–f complex, which form part of the chloroplast membranes.

17.3.5 PHOTOSYSTEM I

The units of PS I are found only in the stroma thylakoids and non-appressed regions of the grana that face the stroma (Figure 17.1). PS I is a trans-membrane complex consisting of at least 13 polypeptides. It consists of a core com-

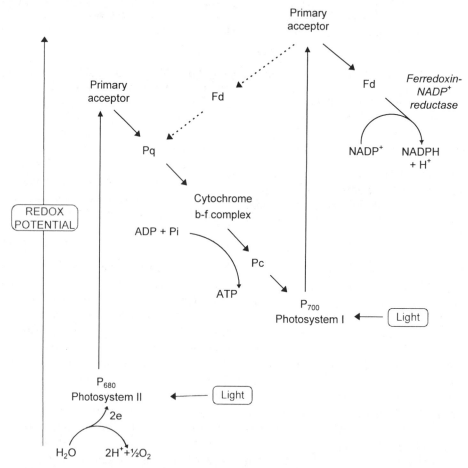

Figure 17.4 Photosynthetic Z-scheme. Electrons derived from the splitting of water pass through photosystem II, through the cytochrome b–f complex and through photosystem I before being used to reduce $NADP^+$. Absorption of light by each photosystem increases the energy of the electrons. ATP is produced during the passage of electrons between the two photosystems. Electrons from photosystem I can be recycled via ferredoxin and the cytochrome b–f complex in cyclic photophosphorylation.

plex and a light-harvesting complex (LHC I), which absorbs and channels energy towards the reaction centre (Figure 17.5). The reaction centre of PS I is a chlorophyll *a* molecule (or possibly a pair of chlorophyll *a* molecules) in a special environment and is called P_{700}.

The core complex

The main components of the core complex are two nearly identical proteins: psaA (83 kDa) and psaB (82 kDa) proteins. The P_{700} reaction centre, a quinone, an Fe–S complex and some chlorophyll *a* and β-carotene molecules also form part of the core complex. The primary acceptor from P_{700} is called A_0 and is the most powerful known biological reducing agent. It is thought to be a special chlorophyll *a* molecule. From here the electron is transferred to a quinone, then to an Fe–S centre and via other carriers to ferredoxin on the stroma side of the membrane. Ferredoxin is a red protein of low molecular weight (approximately 11.6 kDa). It is a non-haem iron protein in which the Fe is associated with sulphur. Spinach ferredoxin has two Fe atoms bound to two S atoms. One of the Fe atoms undergoes $Fe^{III} \rightarrow Fe^{II}$ changes – in the reduced state it is a very powerful reducing agent, able to reduce $NADP^+$ under the influence of ferredoxin–$NADP^+$ reductase, a flavoprotein. During this last reaction protons are consumed on the stroma side of the membrane.

LHC I

This light-harvesting antenna complex surrounds the core and contains about 80 molecules of chlorophyll *a* and 20 of chlorophyll *b*. These absorb light and, through RET, direct the energy towards the pigment molecules of the core complex and hence to the reaction centre.

17.3.6 PHOTOSYSTEM II

The units of PS II are concentrated in the appressed regions of the grana thylakoids, with only small amounts being found in the non-appressed regions of the grana and the stroma thylakoids. PS II consists of a complex of about 10 polypeptides which span the membranes in which they are embedded. It is made up of light-harvesting complexes (LHC II), a reaction centre and a water-splitting unit (Figure 17.5). The reaction centre is a chlorophyll *a* molecule (or possibly a pair of chlorophyll *a* molecules) in a special environment and is called P_{680}.

LHC II

This consists of a polypeptide (approximately 26 kDa), about six or seven chlorophyll *a* molecules, about the same number of chlorophyll *b* molecules and a few carotenoids.

Reaction centre and water splitting

The core of PS II consists of about six polypeptides. These include:

- D1 and D2 (each approximately 32 kDa) to which quinones and manganese ions bind and which bind the P_{680} reaction centre itself;
- 33-, 17- and 23-kDa proteins which are involved in splitting of water;
- 43- and 47-kDa proteins containing chlorophyll, which also serve as antenna molecules.

Each P_{680} molecule receives the energy absorbed by about 250 chlorophyll molecules, consisting of roughly equal numbers of chlorophyll *a* and chlorophyll *b*. The primary acceptor of PS II is thought to be pheophytin, a chlorophyll molecule lacking a magnesium atom at its centre.

PS II is also responsible for the water-splitting reaction which can be summarized as follows:

$$2H_2O \rightarrow O_2 + 4H^+ + 4e^-$$

This process is very poorly understood but both Mn^{II} and Cl^- are required. Synthetic electron donors such as semicarbazide and hydroxylamine can substitute for water in this reaction.

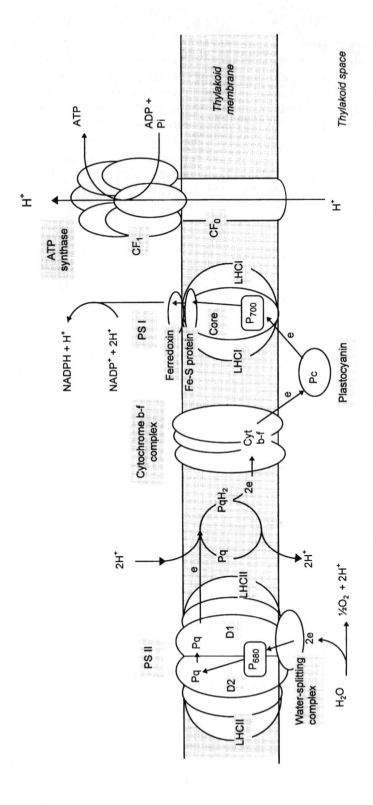

Figure 17.5 Diagrammatic representation of the complexes involved in photosynthesis. Passage of electrons along the electron transport chain moves protons from the stroma into the thylakoid space and thus creates a proton gradient. This is used to drive the synthesis of ATP. Plastoquinone, plastocyanin and ferredoxin are mobile and can transport electrons between the complexes.

The part of the unit responsible for the reaction contains four manganese atoms which undergo a cycle of oxidation changes as electrons are extracted from water and passed on to the P_{680} molecule, which has lost its electron. The 33-, 17- and 23-kDa proteins and a tyrosine residue on the D1 protein appear to be involved in this process.

The primary acceptor for PS II, to which an electron is transferred from P_{680}, is pheophytin. From here the electron is transferred to a molecule of plastoquinone (Pq) which is permanently bound to D2 and which becomes partially reduced. The electron from this molecule is passed to another plastoquinone molecule bound to D1. This molecule remains bound until it has accepted a second electron to fully reduce it, after which it is released. By this means the two electrons needed for the reduction of plastoquinone can be supplied, one at a time, by pheophytin.

17.3.7 CYTOCHROME b–f COMPLEX

This consists of a complex of four subunits: cytochrome f (33 kDa), cytochrome b_{563} (23 kDa, containing two haems, also called cytochrome b_6), Fe–S protein (20 kDa) and another 17-kDa chain. The complex spans the membrane and pumps protons across it as electrons flow from plastoquinone to plastocyanin. Cytochrome b allows two electrons to be accepted from plastoquinone and handed on one at a time to plastocyanin.

17.3.8 PLASTOCYANIN

This is a blue 11-kDa copper-containing protein. The copper atom co-ordinates with the S atoms of a methionine and cysteine residue and with two histidine residues.

17.4 INTEGRATION OF THE ELECTRON TRANSPORT SYSTEM

Operation of the electron transport chain requires PS I and PS II to work together. Removal of an excited electron from PS I to reduce $NADP^+$ leaves PS I short of an electron (i.e. oxidized) and it is dependent on PS II to replace it. Until this has happened no further electrons can be lost. Similarly, the electron lost from PS II must be replaced by electrons released during the splitting of water. Thus, the chain behaves as though electrons flow through it from water to $NADP^+$.

The locations of the PS I, PS II and ATP synthase complexes are relatively fixed: they are located in different regions of the thylakoid membranes. However, some of the electron carriers are mobile and are able to transfer electrons between the fixed complexes. Thus reduced plastoquinone, plastocyanin and ferredoxin can all move. The cytochrome b–f complex may also be mobile and move between stacked and unstacked regions of the thylakoids.

17.5 ATP PRODUCTION

The components of the electron transport chains are embedded in the thylakoid membranes in such a way that reactions in which H^+ are used up take place on the stroma side of the membrane, and those in which H^+ are released take place on the thylakoid space side. This results in development of a gradient of $[H^+]$ across the membrane, as shown in Figure 17.5. The pH of the thylakoid space may fall to around pH 4.

In addition to the electron transporting complexes described above, an additional complex, ATP synthase, is also embedded in the thylakoid membranes. This resembles the corresponding enzyme present in mitochondria and consists of CF_0 and CF_1 components. CF_0 spans the membrane and CF_1 protrudes on the stromal side, and is the part which actually synthesizes ATP from ADP and P_i. The proton gradient, generated during electron transport, is used to drive the synthesis of ATP from ADP and Pi. These processes are entirely analogous to those which result in ATP production during electron transport in mitochondria (Chapter 12).

Although it is known that the flow of electrons between PS II and PS I generates the proton gradient needed to produce ATP, it is not easy to determine the number of ATP molecules produced for each pair of electrons transported. This is because two processes can give rise to ATP:

- non-cyclic photophosphorylation;
- cyclic photophosphorylation.

The non-cyclic process is that already described, in which electrons are transported from water by PS II to PS I and used to reduce $NADP^+$. If insufficient $NADP^+$ is available to accept electrons from PS I then they may be diverted into a cyclic process during which they are passed back to the cytochrome b–f complex and hence back to PS I. This results in production of ATP but no NADPH is formed. This process is represented by the dotted line in Figure 17.4.

The dark reactions of photosynthesis require three molecules of ATP and two molecules of NADPH for each molecule of carbon dioxide fixed. Cyclic photophosphorylation appears to operate to a variable extent, and this may allow the proportion of ATP and NADPH produced to be adjusted to meet the demands of the dark reactions.

17.6 THE DARK REACTIONS (CALVIN CYCLE)

Operation of the light reactions results in the release of NADPH and ATP into the chloroplast stroma. These compounds are used to convert carbon dioxide into sugars, in a process which does not itself depend on light. The reactions involved make up the Calvin cycle, named after its discoverer Melvin Calvin. This cycle is shown in Figure 17.6.

Reactions of the Calvin cycle are catalysed by enzymes in solution in the stroma or on the stromal side of the thylakoid membranes. The crucial first step, in which carbon dioxide reacts with ribulose-1,5-bisphosphate (RUBP), converts inorganic carbon into an organic form and is hence referred to as carbon fixation. The reaction is catalysed by the enzyme RUBP carboxylase/oxygenase, which is usually known as Rubisco. Rubisco catalyses the reaction of ribulose-1,5-bisphosphate with carbon dioxide to form, initially, an unstable, six-carbon compound, which rapidly breaks down to form two molecules of 3-phosphoglyceric acid (Figure 17.7). Rubisco is located on the stromal surface of the thylakoid membranes. It is the commonest protein on earth, making up about 50% of the soluble protein in leaves. A hectare of lawn contains between 5 and 7 kg of pure enzyme. The reaction which it catalyses requires the presence of magnesium. The enzyme is not totally specific for carbon dioxide and will also catalyse a reaction with O_2 instead of CO_2. This reduces photosynthetic efficiency, as discussed later in this chapter.

The next two steps use ATP and NADPH and depend on the light reactions for production of these compounds. These reactions produce glyceraldehyde-3-phosphate. Some of this is used to regenerate ribulose-1,5-bisphosphate by a rather complex series of reactions so that the cycle can continue. These reactions are catalysed by the enzymes transaldolase and transketolase, and the steps are virtually identical to parts of the pentose phosphate pathway. Six molecules of ribulose-1,5-bisphosphate (6×5 carbon atoms) and six molecules of carbon dioxide (6×1 carbon atoms) give rise to 12 molecules of glyceraldehyde-3-phosphate (12×3 carbon atoms). Ten molecules of glyceraldehyde-3-phosphate are used to re-form six molecules of ribulose-1,5-bisphosphate, and two molecules are left over. These represent the net output of the cycle.

The sugars produced by the Calvin cycle are either converted into starch within the chloroplasts, or exported as triose phosphate (glyceraldehyde-3-phosphate) into the cytoplasm. Starch accumulates during periods of illumination but may be broken down and the carbon exported into the cytoplasm during darkness. The starch therefore acts as a temporary store of carbon during times of rapid carbon fixation.

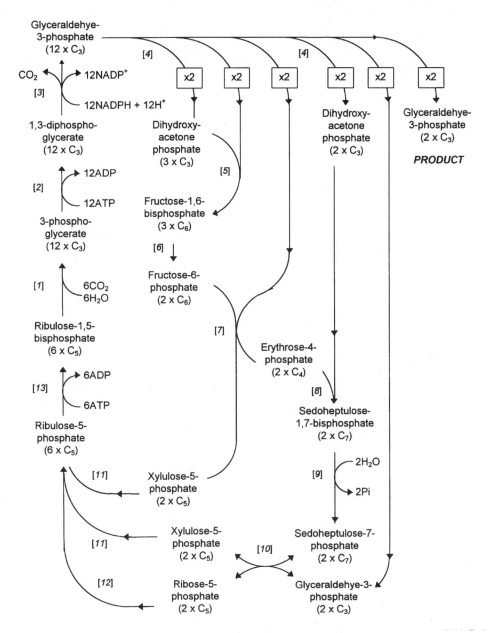

Figure 17.6 The Calvin cycle. Reaction of ribulose-1,5-bisphosphate with carbon dioxide yields 3-phosphoglycerate which is phosphorylated and reduced to form glyceraldehyde-3-phosphate. Most of this is used to regenerate ribulose-1,5-bisphosphate by reactions which are very similar to those of the pentose phosphate pathway, but the remainder is converted to sugars. Enzymes catalysing the reactions are as follows: [1] ribulose 1,5-bisphosphate carboxylase/oxygenase (Rubisco), [2] phosphoglycerate kinase, [3] glyceraldehyde-3-phosphate dehydrogenase, [4] triose phosphate isomerase, [5] fructose bisphosphate aldolase, [6] fructose bisphosphatase, [7] transketolase, [8] aldolase, [9] sedoheptulose bisphosphatase, [10] transketolase, [11] ribulose phosphate 3-epimerase, [12] ribulose phosphate isomerase, [13] phosphoribulokinase.

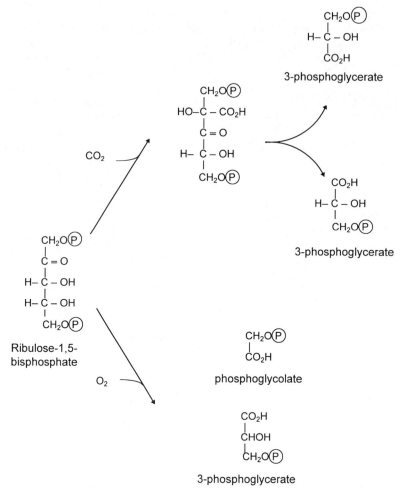

Figure 17.7 The reactions catalysed by Rubisco. This enzyme can catalyse reaction of ribulose-1,5-bisphosphate with either carbon dioxide or oxygen. The former reaction produces two molecules of 3-phosphoglycerate as part of the Calvin cycle. The latter reaction produces one molecule of 3-phosphoglycerate and one of phosphoglycolate which is used in photorespiration.

The inner membrane of the chloroplast is practically impermeable to polar molecules such as free sugars, P_i and triose phosphates, and specific carriers are needed to transport some of them across the membrane. A carrier which facilitates the counter-transport of triose phosphates and P_i appears to be particularly important in regulating the movement of the products of photosynthesis out of the chloroplast into the cytoplasm. P_i is required for the synthesis of organic molecules by the Calvin cycle and it must enter the chloroplasts as triose phosphates move into the cytoplasm. When P_i is not available from the cytoplasm, starch accumulation in the chloroplasts is favoured. This process liberates P_i which can be re-used in the Calvin cycle.

Some of the glyceraldehyde-3-phosphate may be converted to dihydroxyacetone phosphate. These two molecules then condense to form fructose-1,6,-bisphosphate, from which fructose-6-phosphate, glucose, starch and other sugars may arise by reactions described in Chapter 18.

Operation of the pathway may be considered to bring about conversion of six molecules of carbon dioxide to one molecule of fructose-6-phosphate. The overall equation for operation of the Calvin cycle can be obtained by adding together the steps shown in the cycle (Figure 17.6). The equation which results is:

$$6CO_2 + 12NADPH + 12H^+ + 18ATP +$$
$$11H_2O \rightarrow \text{fructose-6-phosphate} + 12NADP^+$$
$$+ 18ADP + 17P_i$$

17.7 CONTROL OF PHOTOSYNTHESIS

Many factors influence the rate of photosynthesis. Light is the most important of these, and some of its effects are listed below:

- a direct effect on the light reactions, leading to increased availability of ATP, NADPH, etc.;
- a long-term effect on the development of chloroplasts and the induction of photosynthetic enzymes;
- effects on the activity of Calvin cycle and other enzymes.

At least five Calvin cycle enzymes (Rubisco, glyceraldehyde-3-phosphate dehydrogenase, fructose bisphosphatase, sedoheptulose bisphosphatase and phosphoribulokinase) are activated by light. Several factors may contribute to this activation; for example, changes in pH resulting from proton movement from the stroma into the lumen of the chloroplasts may activate enzymes such as Rubisco. In addition, compounds such as NADPH and ATP formed in the light reactions are positive allosteric effectors of Rubisco and of glyceraldehyde-3-phosphate dehydrogenase. A further very important effect is brought about through the action of light effect mediators. These are proteins containing cysteine residues that can exist either in the –SH form or in the oxidized –S–S– form. The best known of these is called thioredoxin. In the light, reduced ferredoxin converts thioredoxin into its reduced (–SH) form. This activates a number of Calvin cycle enzymes, including glyceraldehyde-3-phosphate dehydrogenase, fructose bisphosphatase, sedoheptulose bisphosphatase and phosphoribulokinase. It may also inhibit enzymes involved in sugar breakdown.

The rate of conversion of triose phosphates into sucrose in the cytoplasm is also regulated by fructose 2,6-bisphosphate, which has a similar role in control of glycolysis and gluconeogenesis in animals. This regulation appears to be complex, but in general the compound accelerates conversion of fructose-6-phosphate into fructose-1,6-bisphosphate and inhibits the reverse reaction catalysed by fructose-1,6-bisphosphatase. By this mechanism the relative rates of sugar synthesis and breakdown are changed.

17.8 PHOTORESPIRATION

Plant cells contain mitochondria and therefore respire to produce ATP. In the process they consume oxygen and produce carbon dioxide. This occurs in both green and non-green tissues and is not dependent on light (except indirectly, in that photosynthesis provides sugars which are substrates for respiration). However, in many plants another type of respiration, photorespiration, also takes place. Photorespiration is dependent on light and is often much faster than mitochondrial respiration. Photorespiration involves the production and metabolism of glycolate by a process which is sometimes called the glycolate cycle. Photorespiration is closely linked to the Calvin cycle, as the enzyme Rubisco is common to both pathways.

Rubisco can catalyse the reaction of either carbon dioxide or oxygen with ribulose 1,5-bisphosphate. Reaction with carbon dioxide (carboxylase activity) produces two molecules of 3-phosphoglycerate, but reaction with oxygen (oxygenase activity) produces 3-phosphoglycerate and phosphoglycolate (Figure 17.7). The former can be used for the synthesis of sugars but the latter undergoes oxidation in which carbon dioxide is produced. This oxidation involves the chloroplasts, peroxisomes and mitochondria, as shown in Figure 17.8.

During photorespiration oxygen is consumed and carbon dioxide is produced. Some of the ribulose 1,5-bisphosphate which could have been used in the Calvin cycle to produce sugars is converted back to carbon dioxide. Photorespiration therefore acts in the opposite sense to photosynthesis, and the operation of this pathway reduces the net efficiency of conversion of carbon dioxide into sugars. Under many circumstances photorespiration occurs at 20–25% of the rate of photosynthesis, and it is a major factor which limits the photosynthetic efficiency of many plants.

Carbon dioxide and oxygen compete for reaction at the active site of Rubisco. High levels of carbon dioxide favour the Calvin cycle, whilst high levels of oxygen favour photorespiration. In the atmosphere the ratio of $CO_2:O_2$ is very low (0.02:20%) so it may be inevitable that some ribulose bisphosphate oxygenase activity will occur. Photorespiration may, under these circumstances, be an attempt by the cell to convert some phosphoglycolate back into sugars.

In photorespiration, one molecule of NH_4^+ is produced from glycine for each molecule of CO_2 released. This NH_4^+ is rapidly reassimilated into organic compounds by the reactions described in Chapter 20, and this constitutes the photorespiratory nitrogen cycle. Much of the ammonia converted into organic compounds originates in this way.

17.8.1 FACTORS AFFECTING RATES OF PHOTORESPIRATION

Reducing the oxygen content of the air reduces the rate of photorespiration because Rubisco has a low affinity for oxygen and is not saturated even at concentrations of oxygen found in the air. In mitochondrial respiration, however, cytochrome oxidase, which catalyses the reaction between cytochromes and molecular oxygen, has a very high affinity for oxygen and is completely saturated even at very low oxygen concentrations. Thus at lowered oxygen concentrations, rates of photorespiration can be reduced without affecting mitochondrial respiration. Young plants grown in 2–5% O_2 instead of the normal 20% therefore grow about twice as fast as normal. Carbon dioxide enrichment of the atmosphere is commonly used in growing protected crops. It not only increases rates of photosynthesis, but decreases rates of photorespiration by competing with oxygen for Rubisco.

The relative importance of photorespiration increases with increasing light intensity, temperature and pH. As temperature rises the solubility of carbon dioxide in water decreases faster than that of oxygen. Increasing pH decreases the availability of carbon dioxide as it is converted to HCO_3^- at high pH. In bright light the pH of the stroma is increased and this increases the rate of photorespiration, as described above. Photorespiration is thus a particular problem for plants growing in bright light and at high temperatures, and many plants show specialized adaptations to growth under these circumstances.

The fact that plants grow faster when photorespiration is reduced, and that some of the most productive plants do not carry out the process at all, suggests that it may be possible to increase the efficiency of crop growth by decreasing rates of photorespiration. This might be achieved through using chemical inhibitors of glycolate oxidase such as α-hydroxypyridine methanesulphonate or 2-hydroxy-3-butynoic acid. It might also be achieved by changing the structure, and therefore the specificity, of the active site of Rubisco by conventional breeding or by genetic engineering. By specific modification of individual bases in the genes which code for Rubisco it is possible to change the amino acids at the active site of the enzyme (see Chapter 21), and this may alter its specificity so that it reacts less readily with oxygen. However this is difficult to achieve because of the complexity of Rubisco,

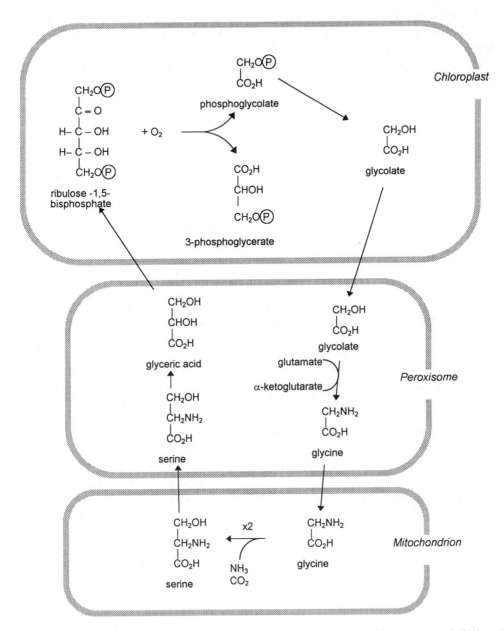

Figure 17.8 The pathway of photorespiration. Phosphoglycolate is produced by reaction of ribulose-1,5-bisphosphate with oxygen and is then oxidized, with production of carbon dioxide, in a series of reactions which involve the chloroplast, peroxisome and mitochondrion. One effect of this process is to degrade ribulose-1,5-bisphosphate which would otherwise be used in the Calvin cycle, and thus to reduce net photosynthetic sugar production.

which has a molecular weight of about 560 kDa. It consists of eight large subunits, of about 56 kDa each, and eight small subunits, each about 14 kDa. The sequence of amino acids in the large subunits is encoded in the chloroplastic DNA and the small subunits in nuclear DNA. The latter proteins are therefore synthesized in the cytoplasm and must be imported into the chloroplasts where they form a complex with large subunits to form active enzyme. Not only must the genes be modified in the correct way, but they must also direct their products to the correct location in the cell.

17.9 PHOTOSYNTHESIS IN C_4 PLANTS

The major crop plants of temperate regions of the world fix carbon dioxide using the Calvin cycle. Photorespiration occurs in these plants so that a significant proportion of their potential yield is lost.

In other plants, typically tropical grasses which grow under high light intensities and at high temperatures, a modified type of photosynthesis takes place and photorespiration is virtually undetectable, so yields of organic matter are much higher.

These tropical plants use the Hatch and Slack or C_4 pathway for photosynthesis. Here the initial products of carbon dioxide fixation are four-carbon compounds, hence the name C_4 pathway. In contrast, plants using the Calvin cycle are called C_3 plants as three-carbon compounds (3-phosphoglycerate and glyceraldehyde-3-phosphate) are the initial products of carbon dioxide fixation.

The C_4 pathway involves the cooperation of two types of cell – mesophyll cells and bundle sheath cells. The mesophyll cells occur towards the outside of the leaf and are freely accessible to the air. In these cells carbon dioxide is initially fixed into C_4 compounds, which are then transported into the bundle sheath before being broken down to re-release carbon dioxide which enters the Calvin cycle (Figure 17.9).

The chloroplasts of the mesophyll cells contain the enzyme phosphoenolpyruvate carboxylase which catalyses the reaction between phosphoenolpyruvate and carbon dioxide. The oxaloacetate produced is then reduced by NADPH (produced in the light reactions) to

malate by the action of malate dehydrogenase. The malate moves to the bundle-sheath cells, where it is acted on by malic enzyme to produce pyruvate and CO_2. $NADP^+$ is reduced to NADPH. The carbon dioxide released reacts with ribulose 1,5-bisphosphate in the Calvin cycle using the enzyme Rubisco. The pyruvate passes back to the mesophyll cells and is converted to phosphoenolpyruvate at the expense of two high-energy phosphate bonds in a reaction catalysed by pyruvate phosphate dikinase.

This system acts as a 'pump' for carbon dioxide. In contrast to Rubisco, phosphoenolpyruvate carboxylase has a very high affinity for carbon dioxide and none for oxygen. The uptake of carbon dioxide is thus rapid and specific. When the carbon dioxide is delivered to the Rubisco, oxygen is excluded, eliminating formation of phosphoglycolate and photorespiration.

In the Calvin cycle each molecule of carbon dioxide fixed requires two NADPH and three ATP molecules. In the C_4 mechanism a further two high-energy phosphate bonds are broken in forming phosphoenolpyruvate (ATP is converted to AMP). In plants growing at high light intensities ATP is readily available from the light reactions, so the extra ATP requirement of the C_4 pathway may be less important than losses resulting from photorespiration. In many tropical crops, therefore, C_4 photosynthesis is favoured. In temperate climates light is more likely to be a limiting factor so it may be more efficient for plants to economise on use of ATP, and under these conditions the C_3 pathway may be preferred.

Another advantage to the plant results from use of the C_4 route. As phosphoenolpyruvate carboxylase has a very high affinity for carbon dioxide it is able to obtain sufficient carbon dioxide even when the stomata are very nearly closed. Under arid conditions this allows the plants to conserve water more effectively than if they were using the C_3 pathway.

The efficiency of the C_4 process is seen when the rates of photosynthesis of different plants are compared (Table 17.1).

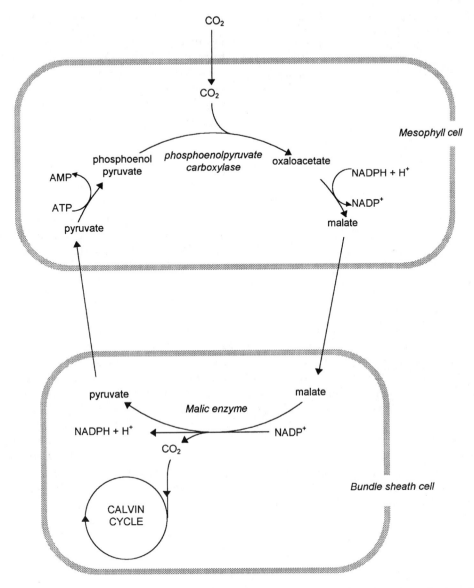

Figure 17.9 Photosynthesis in C$_4$ plants. This involves initial reaction of carbon dioxide with phosphoenol pyruvate in the mesophyll cells in a reaction catalysed by phosphoenolpyruvate carboxylase. The four-carbon compound, malate, moves to the bundle sheath cells where it is degraded to release carbon dioxide which enters the Calvin cycle. The process requires more ATP than C$_3$ photosynthesis but, because of the high specificity of phosphoenolpyruvate carboxylase, prevents photorespiration from taking place.

17.10 CRASSULACEAN ACID METABOLISM

This type of photosynthesis is common amongst succulent plants and some bromeliads, lilies and orchids. These plants have been known for a long time to increase their acid content at night and to decrease it in the day. Normally the stomata are open during the night but closed in the day. At night, carbon dioxide is fixed in the chloroplast-containing

Table 17.1 Typical rates of carbon dioxide fixation in different types of plant

Type	Rate of CO_2 fixation ($mg\ CO_2\ dm^{-2}\ h^{-1}$)
Slow-growing desert species, orchids etc	1–10
Tropical and subtropical evergreens, temperate evergreens	5–15
Most temperate herbs and deciduous trees	15–30
Rapidly growing temperate agronomic crops (wheat, soyabean, etc.)	20–40
C_4 tropical grasses (maize, sugarcane etc.)	50–90

cells of the photosynthetic leaf or stem tissue, and converted to oxaloacetate through the action of phosphoenolpyruvate carboxylase. Much of this is converted to L-malic acid which is stored in the large vacuoles that are characteristic of these plants. During the day malic acid is decarboxylated to form pyruvate or phosphoenolpyruvate under the action of one of several enzymes, which differ from species to species. The carbon dioxide produced enters the Calvin cycle, where it is used to produce starch and sucrose in the normal way (Figure 17.10).

Pineapple is the main agricultural crop using this type of photosynthesis. This mechanism allows the stomata to remain closed during the day, which conserves water. This achieves the same purpose as C_4 photosynthesis, but in this case the Calvin cycle and initial carbon dioxide fixation are separated in time rather than space.

17.11 HERBICIDES AND PHOTOSYNTHESIS

Because photosynthesis is of vital importance to plants but does not occur in animals it is a very obvious target for the development of potential herbicides, and a number of important herbicides function by inhibiting the process.

The dipyridinium herbicides such as diquat and paraquat act on PS I. They divert electrons away from $NADP^+$ in the region of ferredoxin, and the herbicide molecules themselves become reduced in the process. As the reduced herbicide is reoxidized, superoxide radicals are generated and these are very toxic, causing extensive damage by reacting with a wide range of molecules present in the cell.

The D1 protein of PS II is thought to be the site of action of substituted urea herbicides such as monuron (now superseded) and diuron, which thus prevent plastoquinone reduction. This is of considerable interest to genetic engineers, as alteration of the structure of this protein may make crop plants resistant to these herbicides. Weeds containing the normal protein would remain susceptible to the herbicides. Such modified D1 proteins may be in found herbicide-resistant plants or may be produced by deliberate modification of the gene coding for the protein. Introduction of such modified genes into crops would allow the use of potent, wide-spectrum herbicides to control weeds in these crops.

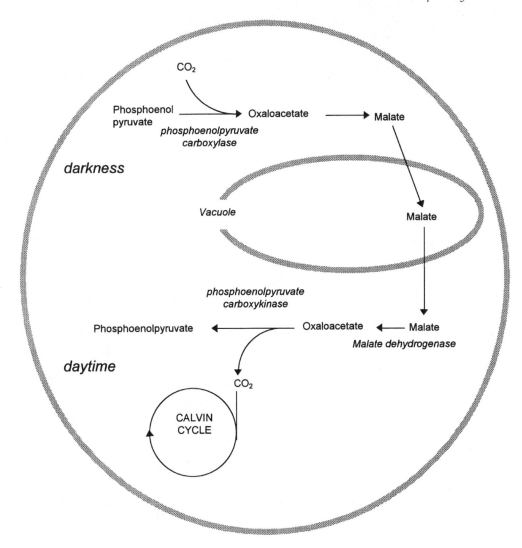

Figure 17.10 Crassulacean acid metabolism. In some plants, carbon dioxide is converted into four-carbon acids such as malate during darkness. It is released from these compounds in the light and is converted into sugars by the Calvin cycle.

GLUCONEOGENESIS AND CARBOHYDRATE SYNTHESIS

18.1 INTRODUCTION

Gluconeogenesis is the process which produces glucose from simpler molecules. In plants it is one of the principal routes leading to the eventual synthesis of complex carbohydrates. Animals divide into two groups as far as the importance of gluconeogenesis is concerned. Simple-stomached animals can usually rely on a supply of glucose in their food from the breakdown of α-linked carbohydrates such as starch. Apart from a small amount of glycogen synthesis, their glucose metabolism is principally one in which it is oxidized. On the other hand, ruminant animals receive very limited supplies of glucose from the gut and rely on synthesis from the volatile fatty acids, mainly propionate. The pathway is very active because ruminants have a large requirement for glucose as an energy source in the tissues and, in the case of the lactating animal, for making lactose.

18.1.1 STARTING MATERIALS

Many metabolites can be used as starting materials for gluconeogenesis, but some simple compounds cannot. Three-carbon compounds such as pyruvate, phosphoenolpyruvate or triose phosphates are all capable of being transformed into glucose and are termed glucogenic materials. The reaction in which acetyl-CoA is produced from pyruvate (pyruvate dehydrogenase) is irreversible in organisms lacking the glyoxylate cycle. Thus acetyl-CoA and related compounds (fatty acids, ketone bodies, ketogenic amino acids) cannot be used for glucose synthesis.

18.1.2 OUTLINE OF THE PATHWAY

In essence the pathway is very similar to a reversal of glycolysis. It starts with two molecules of a three-carbon compound. After a series of transformations, these become one molecule each of dihydroxyacetone phosphate and glyceraldehyde-3-phosphate which are linked to give fructose-1,6-bisphosphate (Figure 18.1). The final part of the pathway is a rearrangement to yield glucose-6-phosphate (Figure 18.2).

18.1.3 DIFFERENCES BETWEEN GLUCONEOGENESIS AND GLYCOLYSIS

There are three steps in glycolysis that are essentially irreversible because they are accompanied by large, negative, free energy changes:

- hexokinase/glucokinase
- phosphofructokinase
- pyruvate kinase.

The pathway for glucose synthesis has to avoid these steps by a series of 'by-pass reactions'. These steps have a further use in providing points at which the flow through the pathway can be controlled.

18.2 GLUCONEOGENESIS VIA PYRUVATE

Under normal circumstances pyruvate itself is not found in high concentrations in cells, but other precursors such as malate, lactate and glucogenic amino acids must pass through this metabolite. Due to the irreversibility of the enzyme pyruvate kinase, the direct production of phosphoenolpyruvate from pyruvate is not possible, and a two-stage pathway is used to by-pass the obstruction (Figure 18.3). This short sequence is complicated by the fact that it takes place in different parts of the cell. Pyruvate, produced in the cytosol, has first to be taken up by the mitochondrion where it is carboxylated using bicarbonate ions to form oxaloacetate under the influence of the enzyme pyruvate carboxylase. This reaction requires the hydrolysis of a molecule of ATP in order to overcome part of the energy barrier to the pathway.

In the next step, phosphoenolpyruvate carboxykinase catalyses the decarboxylation of oxaloacetate to produce phosphoenolpyruvate. This step also requires energy to drive it, in this case in the form of GTP. This means that the conversion of pyruvate to phosphoenolpyruvate uses two high-energy phosphate groups, whereas the reverse reaction produces only one. If both of these sequences of reactions were active at the same time, the overall effect would be that one mole of ATP and one of GTP would be used, and only one mole of ATP would be produced. This leads to the potential for a futile cycle, as illustrated in Figure 18.3b. (The term 'futile cycle' is well-accepted in biochemistry but it may be a misnomer – under some circumstances these reaction sequences can have important functions in the overall energy metabolism of warm-blooded animals; see Chapter 29.)

18.2.1 THE CONTROL OF PYRUVATE PRODUCTION AND USE

The most important controls of the steps illustrated in Figure 18.3 are exerted through the activities of the allosteric enzymes pyruvate kinase and pyruvate carboxylase. The activity of pyruvate kinase is greatly inhibited by the presence of high levels of ATP or NADH. On the other hand, pyruvate carboxylase is almost inactive in the absence of acetyl-CoA.

18.3 THE PRODUCTION OF FRUCTOSE-1,6-BISPHOSPHATE

The conversion of phosphoenolpyruvate to fructose-1,6-bisphosphate is an exact reversal of the equivalent steps in glycolysis (see Figure 18.1). The free energy changes for all of these reactions are quite small.

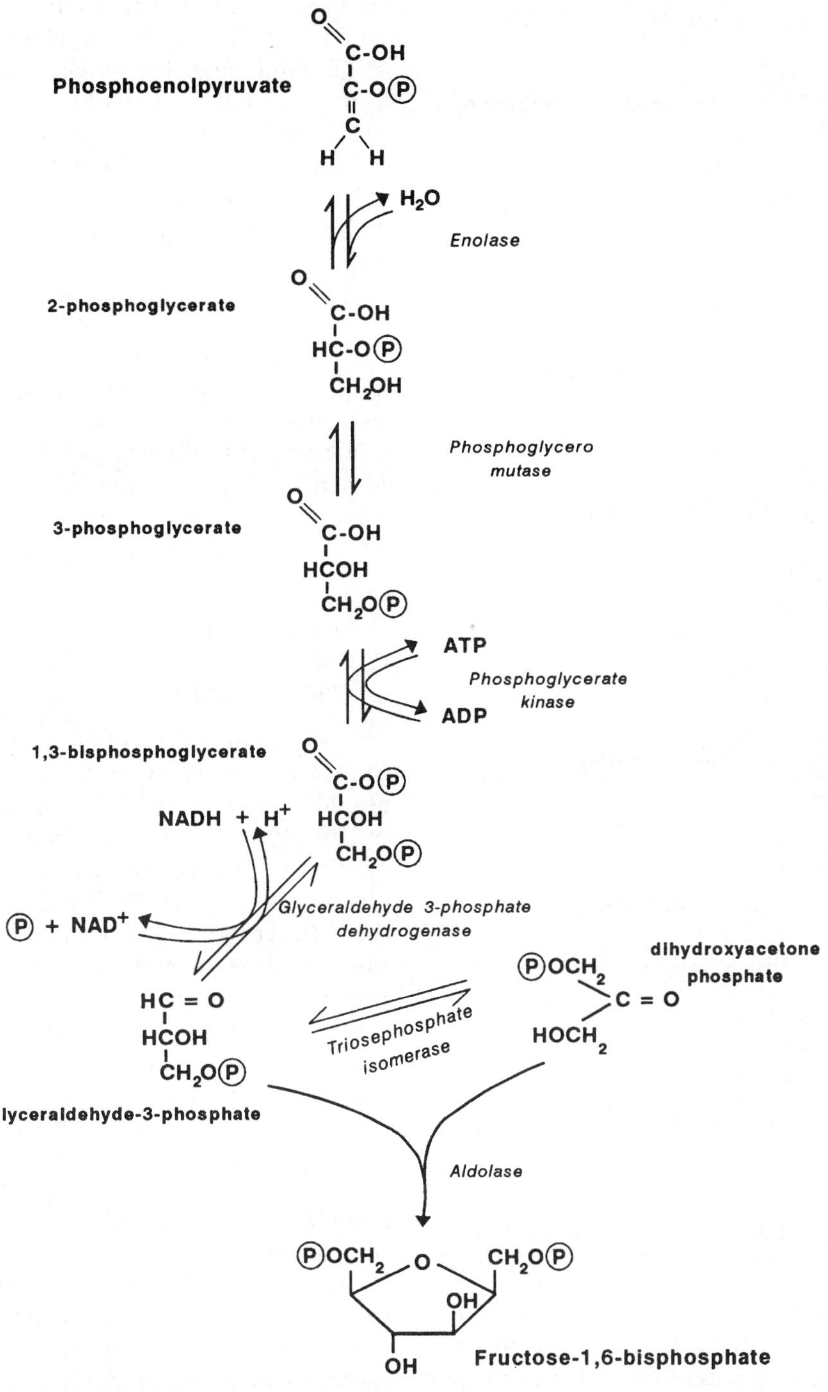

Figure 18.1 Pathway for the synthesis of fructose-1,6-bisphosphate from phosphoenolpyruvate. The steps are essentially a reversal of the equivalent ones in glycolysis (3-GPA = 3 phosphoglycerate).

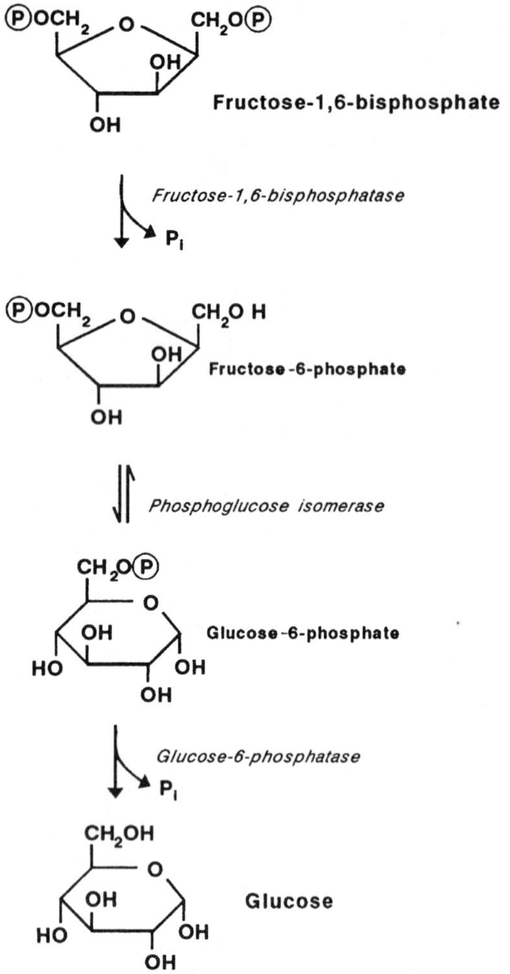

Figure 18.2 Pathway for the conversion of fructose-1,6-bisphosphate into glucose-6-phosphate. The steps catalysed by phosphatases overcome the unfavourable energy changes which would be needed to reverse the kinase-catalysed reactions of glycolysis.

18.4 THE HYDROLYSIS OF FRUCTOSE-1,6-BISPHOSPHATE

In glycolysis, phosphofructokinase uses ATP to add the phosphate group to fructose-6-phosphate. In glucogenesis, the reverse reaction involves the simple hydrolysis of a phosphate group by the enzyme fructose-1,6-bisphosphatase. This reaction does not phosphorylate

ADP. Figure 18.4 illustrates the futile cycle that could result if both of these reactions occurred at the same time. The overall reaction, shown as the sum of the two reactions, would proceed very readily because the free energy change for the breakdown of ATP to ADP and phosphate is large and negative.

This pair of reactions provides the most important control mechanism for the flow of material between glucose and pyruvate. The reactions catalysed by phosphofructokinase and by fructose-1,6-bisphosphatase are never allowed to take place at the same time. If phosphofructokinase is operative then glycolysis will take place; if fructose-1,6-bisphosphatase is active then the metabolic flow will be towards glucose synthesis. Both of the enzymes are allosteric.

18.4.1 THE CONTROL OF PHOSPHOFRUCTOKINASE AND FRUCTOSE-1,6-BISPHOSPHATASE

The step catalysed by phosphofructokinase is the first one in which the substrate is absolutely committed to being broken down by glycolysis. Prior to this step the phosphorylation of glucose and the rearrangement of glucose-6-phosphate to fructose-6-phosphate are both reactions that are used for other purposes, such as the interchange of sugars or the pentose phosphate pathway. Inhibiting the activity of phosphofructokinase will therefore selectively 'switch off' the process of glycolysis. In looking at possible control mechanisms, it is clear that the main purpose of glycolysis is the eventual production of energy in the form of ATP. If ATP is abundant then there is little point in the pathway functioning. However, if the converse is true and most ATP has been hydrolysed to ADP, or even AMP, then the cell may be regarded as being deficient in energy. Accumulation of citrate indicates that there is a low demand for the oxidation of glucose by glycolysis. The levels of ATP, ADP, AMP and citrate define the circumstances when glycolysis ought to be inactive or active.

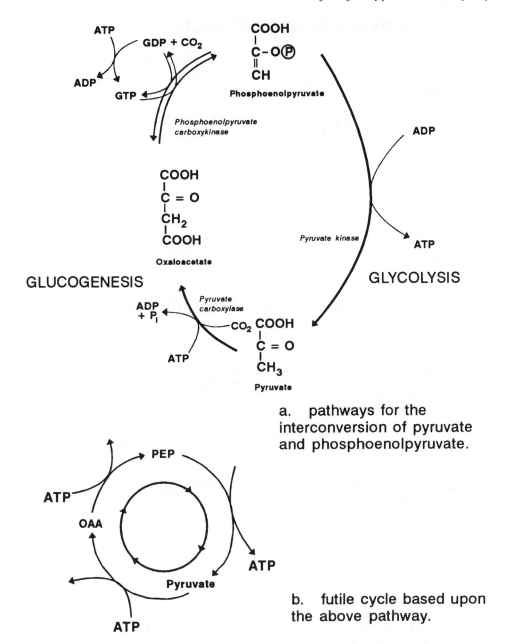

a. pathways for the interconversion of pyruvate and phosphoenolpyruvate.

b. futile cycle based upon the above pathway.

Figure 18.3 Pathways for the interconversion of pyruvate and phosphoenolpyruvate. If all of the enzymes were active at the same time, these reactions would form a futile cycle by hydrolysing two molecules of ATP but forming only one.

Accordingly, phosphofructokinase activity is insignificant when ATP or citrate concentrations are high and AMP and/or ADP are low. Phosphofructokinase is active when ATP and citrate levels are low, and AMP and/or ADP are high.

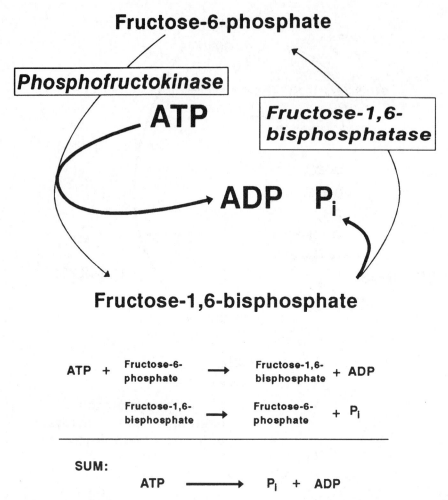

Figure 18.4 Futile cycle resulting from the simultaneous activity of phosphofructokinase and fructose-1,6-bisphosphatase. The net effect is the hydrolysis of ATP to ADP and the release of large amounts of energy as heat.

The purpose of gluconeogenesis is to synthesize sugars and through them the more complex carbohydrates which act as energy reservoirs. Cells can store energy only if there is energy to spare. The process of gluconeogenesis should act only when ATP levels are high and AMP and/or ADP levels are low. It appears that AMP is the principal inhibitor of fructose-1,6-bisphosphatase, which is stimulated by citrate. Hormonal control of these enzymes occurs through their interaction with fructose-2,6-bisphosphate (see Chapter 22).

18.4.2 THE FATE OF FRUCTOSE-6-PHOSPHATE

In some circumstances the gluconeogenesis pathway stops with fructose which is found in a number of cells; for instance mammalian semen has high levels of this sugar. Fructose is liberated from its phosphate under the influence of a phosphatase.

Alternatively, if fructose is to be further processed the enzyme phosphoglucose isomerase, which takes part in both glycolysis and gluconeogenesis, changes fructose-6-phosphate into glucose-6-phosphate.

18.5 THE UTILIZATION OF GLUCOSE-6-PHOSPHATE

In liver and kidney, glucose is produced in its free form for export to other tissues. The enzyme responsible, glucose-6-phosphatase, simply hydrolyses the phosphate group to yield the free sugar and an orthophosphate group. The standard free energy change is large and negative. In glycolysis the equivalent step, using hexokinase, involves the transfer of orthophosphate from ATP – the free energy change is also large and negative. Once again this raises the possibility of a futile cycle that achieves only the hydrolysis of ATP to ADP and orthophosphate.

18.6 GLUCONEOGENESIS FROM PROPIONATE

One pathway which is of outstanding importance in animal agriculture is the one which provides most of the glucose for ruminant animals. In non-ruminants the major energy supply is in the form of glucose; in ruminants this role is undertaken by a group of volatile fatty acids (VFAs). Despite the importance of the VFAs in the nutrition of ruminants, the animals still have an absolute need for glucose for a number of important functions such as supplying energy to the brain and nervous tissues. The importance of glucose rises steeply during lactation (see Chapter 31). Only one of the major VFAs, propionate, is capable of serving as a precursor for the synthesis of glucose.

The first stage of the pathway, the conversion of propionate into propionyl-CoA, is common to the metabolism of all free fatty acids (Chapter 13). Thus, the only unusual steps are those that convert propionate into succinyl-CoA (Figure 18.5). Firstly, carboxylation under the influence of propionyl-CoA carboxylase leads to the formation of methylmalonyl-CoA. The reaction mechanism is very similar to one of the first stages in fatty acid synthesis (see Chapter 19) where CO_2 is added to acetyl-CoA to produce malonyl-CoA (acetyl-CoA carboxylase). The reaction, which is irreversible, requires energy in the form of ATP, and uses biotin as its coenzyme.

The methylmalonyl-CoA formed is rearranged to form succinyl-CoA by the enzyme methylmalonyl-CoA mutase. This enzyme is unusual because it is one of very few that use vitamin B_{12} as a cofactor. An effect of cobalt deficiency in ruminants is to interfere with the operation of this pathway, which is of critical importance to the ruminant. Succinyl-CoA is used to provide phosphoenolpyruvate, via oxaloacetate, which is then used to synthesize glucose-6-phosphate.

18.7 THE SYNTHESIS OF COMPLEX CARBOHYDRATES

18.7.1 DISACCHARIDES

Many disaccharides are found in nature. Some, such as maltose, exist more commonly as breakdown products of polysaccharides. But two compounds are outstanding in their economic importance in agriculture: in animals, lactose is formed in the lactating mammary gland from galactose and glucose; in plants, sucrose, composed of fructose and glucose, is of great commercial importance.

Lactose synthesis

In the dairy cow, lactose synthesis is a large-scale operation. A high-yielding cow (producing, say, 50 kg milk per day) will secrete 2 kg lactose each day and in a whole lactation of 10 months will produce nearly half a tonne. This gives some indication of the massive strain that lactation imposes on the cow's biochemical resources.

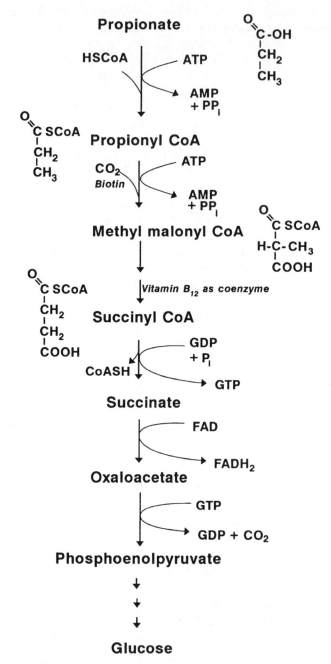

Figure 18.5 Glucogenesis from propionate. Essentially, these reactions are the reverse of those used by microorganisms in the formation of propionate. A shortage of vitamin B_{12} or of the cobalt required to synthesize it has serious consequences in sheep which rely greatly on this pathway for their supply of glucose.

The mammary gland takes up glucose from blood for lactose synthesis. In outline, for every molecule of lactose produced the pathway needs two molecules of glucose, half of which is converted into galactose before being used. The first part of the pathway therefore consists of the production of galactose in an activated form.

As with almost all intracellular reactions of glucose, the first step is phosphorylation at the 6 position under the influence of hexokinase.

$$\text{glucose} + \text{ATP} \rightarrow$$
$$\text{glucose-6-phosphate} + \text{ADP}$$

The next step is to move the phosphate group from the 6 position to the 1 position under the influence of phosphoglucomutase. The reaction is freely reversible and is also involved in the metabolism of many other sugars. Glucose-1-phosphate also has a role in the oxidation of galactose via glucose, and is a key intermediate in the synthesis of starch, glycogen and cellulose.

$$\text{glucose-6-phosphate} \rightleftharpoons \text{glucose-1-phosphate}$$

Glucose-1-phosphate is then activated by conversion to UDP-glucose which uses the enzyme UDP-glucose pyrophosphorylase.

$$\text{glucose-1-phosphate} + \text{UTP} \rightleftharpoons$$
$$\text{UDP-glucose} + \text{PP}_i$$

The enzyme UDP-glucose-4-epimerase catalyses the transformation of glucose to galactose by changing the orientation of the groups on carbon 4.

$$\text{UDP-glucose} \rightleftharpoons \text{UDP-galactose}$$

The next step in the pathway involves an enzyme complex with unusual properties. In a whole range of species, the enzyme galactosyl transferase is bound to the Golgi bodies. This normally catalyses the transfer of a galactose group from UDP-galactose to a sugar derivative known as *N*-acetyl-D-glucosamine (see Figure 3.5f) but not to glucose. This reaction is a normal component of the process for the synthesis of glycoproteins. During pregnancy, when the mammary gland is developing, it starts to produce proteins that are characteristic of milk and are not found under other circumstances. Amongst these proteins is α-lact-

albumin. This protein links itself to galactosyl transferase and modifies its action so that its specificity changes. Instead of *N*-acetyl-D-glucosamine as its principal substrate, it now uses glucose. Once the enzyme is bound to α-lactalbumin, the complex is known as lactose synthase.

$$\text{UDP-galactose} + \text{glucose} \rightleftharpoons \text{lactose} + \text{UDP}$$

This reaction is unusual for another reason – because the second molecule of glucose is incorporated into lactose without first being phosphorylated. In almost all other circumstances, both in higher plants and in animals, glucose is phosphorylated as soon as it enters cells. The overall sequence of reaction in lactose synthesis is shown in Figure 18.6.

The synthesis of sucrose

Sucrose is usually found as an end product of photosynthesis, indeed its major commercial source is sugarcane which is the most efficient photosynthetic plant known. Once again the starting material for the pathway is UDP-glucose, formed under the influence of UDP-glucose pyrophosphorylase. The glucose unit is then transferred to the 1 carbon of fructose-6-phosphate, yielding sucrose-6-phosphate:

$$\text{UDP-glucose} + \text{fructose-6-phosphate} \rightarrow$$
$$\text{sucrose-6-phosphate} + \text{UDP}$$

The reaction is catalysed by sucrose phosphate synthase, a large (360–400 kDa) allosteric enzyme which is inhibited by sucrose. The sucrose-6-phosphate is then hydrolysed to liberate the sucrose:

$$\text{sucrose-6-phosphate} \rightarrow \text{sucrose} + \text{P}_i$$

The relationship between these steps is shown in Figure 18.7.

Despite the role of sucrose as the ultimate end point of photosynthesis, the enzymes UDP-glucose pyrophosphorylase and sucrose phosphate synthase are found not in the chloroplast but in the cytoplasm. The transfer between the pathways of photosynthesis and

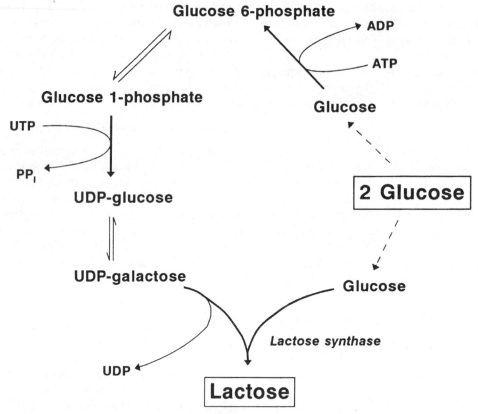

Figure 18.6 Overall pathway for the synthesis of lactose from two molecules of glucose. One of these enters the pathway in the form of glucose-6-phosphate but, unusually, the pathway also uses one molecule of glucose directly.

sugar formation takes place through the compound 3-phosphoglycerate. This is produced in the chloroplast and actively transferred out by a membrane-bound system that simultaneously carries phosphate in.

18.7.2 SYNTHESIS OF POLYSACCHARIDES

Glycogen synthesis

The starting material for glycogen synthesis is UDP-glucose, from which glucose is added sequentially to the growing saccharide chain. The enzyme catalysing the reaction is glycogen synthetase, which elongates the chain by adding glucose groups at carbon 4 of the last glucose in the chain (α-1,4 bonding).

Approximately every tenth glucose unit there are branching points at which there is also an α-1,6 linkage to a second chain of glucose residues. A separate, branching enzyme is responsible for the formation of these branches.

Starch synthesis

Synthesis of starch takes place in two different locations within plants: the first is in the chloroplasts of green tissues, and the second is in specialized intracellular organelles, the amyloplasts, which are found in reserve tissues such as seeds and roots. Amyloplasts resemble chloroplasts in the double-layer form of their surrounding membranes, but they do

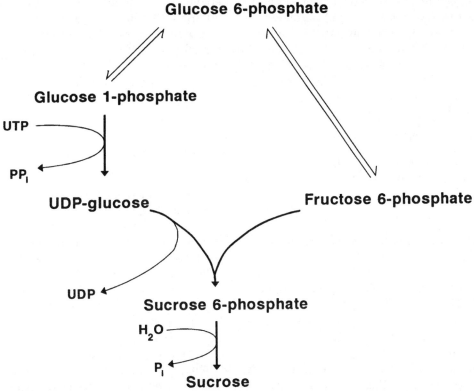

Figure 18.7 Pathway for the synthesis of sucrose. Sucrose is actually produced in the form of its phosphate which is then dephosphorylated to yield free sugar.

not contain any of the specialized enzymes of photosynthesis. Starch synthesis takes place by a mechanism that is broadly similar to that of glycogen. In essence, activated glucose is formed by a pyrophosphorylase enzyme and this is used to elongate the growing chain of sugar units. The difference between this reaction sequence and that for other polysaccharides is that it uses ATP instead of UDP.

$$\text{ATP} + \alpha\text{-glucose-1-phosphate} \rightleftharpoons$$
$$\text{ADP-glucose} + \text{PP}_i$$

The enzyme is ADP-glucose pyrophosphorylase.

Starch synthase then catalyses the transfer of the sugar from ADP-glucose to the growing polysaccharide. On its own, this enzyme is responsible for the formation of amylose; the synthesis of amylopectin also requires the

action of a branching enzyme. Starch synthase has both a catalytic and a controlling function in starch synthesis – it is activated by 3-phosphoglycerate and inhibited by high concentrations of orthophosphate. This is easily explained in chloroplasts because high levels of 3-phosphoglycerate are to be expected in the light when photosynthesis is at its maximum. At the same time, the levels of P_i will be low due to the fact that most phosphate will be incorporated into ATP. This ensures high rates of starch synthesis in chloroplasts when 3-phosphoglycerate is formed in photosynthesis.

It is easy to understand the role of 3-phosphoglycerate (a direct product of photosynthesis) in chloroplasts, but its importance in amyloplasts is not as obvious. The main precursor for starch synthesis in amyloplasts is sucrose, previously synthesized in the chloroplasts. Two mechanisms have been suggested for sucrose utilization. The first is the hydrolysis of sucrose to glucose and fructose by invertase. Both glucose and fructose can then be used as precursors for the formation of glucose-1-phosphate. The second is a reversal of the reaction catalysed by sucrose synthase, yielding UDP-glucose.

$$\text{sucrose} + \text{UDP} \rightleftharpoons \text{UDP-glucose} + \text{fructose}$$

The UDP-glucose is then hydrolysed to liberate glucose-1-phosphate.

$$\text{UDP-glucose} + H_2O \rightleftharpoons$$
$$\text{UMP} + \text{glucose-1-phosphate}$$

UDP-glucose, ADP-glucose and glucose-1-phosphate cannot enter the amyloplast without modification. Sugar for starch synthesis thus enters the amyloplast as triosephosphate. The transport of 3-phosphoglycerate is active and involves phosphate groups being simultaneously pumped out of the organelle (Figure 18.8).

The rationale can now be seen for the activation of starch synthase by high concentrations of 3-phosphoglycerate and low levels of P_i.

Within the amyloplast, the enzymes of glucogenesis take over and resynthesize glucose-6-phosphate. Glucose-6-phosphate is converted to glucose-1-phosphate, at which point it is available as a substrate for the pyrophosphorylase reaction.

In the synthesis of amylopectin there is an enzyme which is able to transfer an oligosaccharide unit onto the 6-carbon of a limited number of glucoses in the polysaccharide chain. These α-1,6 linkages form the branching points which lead to the characteristic structure of amylopectin.

Cellulose synthesis

The main purpose of cellulose is a structural one, and for that reason its synthesis takes place in such a way that the cellulose is produced ready for use in microfibrils. The enzymes that elongate the chains function by adding UDP-glucose units in a similar way to glycogen synthesis. The pathway first involves the formation of UDP-glucose by a pyrophosphorylase reaction but, unlike the situation in glycogen synthesis, the sugar is in the β orientation.

$$\text{UTP} + \text{β-glucose-1-phosphate} \rightleftharpoons \text{UDP-}$$
$$\text{glucose} + PP_i$$

This activated sugar is then used as the unit for elongating the polysaccharide chain. One of the main problems in studying cellulose synthesis has been the fact that it is not possible to isolate the separate parts of the pathway and still retain high levels of the right sort of activity. Cellulose is a β-1,4 chain, but in isolated systems the enzymes of cellulose synthesis produce a different polysaccharide called callose, in which the glucose units are arranged β-1,3. Callose is naturally produced when plants and trees are physically damaged. The enzyme responsible for elongation is cellulose synthase which is located within the plasma membrane of the cells. Cellulose synthase molecules are found in groups arranged in a rosette form. Electron microscopy following fracture of the cell wall has revealed growing microfibrils of cellulose attached to the enzyme complexes. One theory to explain the fact that intact cells produce cellulose, whereas damaged ones form callose, involves a modifier protein. Cellulose-producing cells contain a relatively small (18 kDa) protein which appears to be needed for activity by the membrane-bound cellulose synthase (see Figure 18.9).

If the cells are intact they will have low intracellular levels of calcium ions and a negative electrical potential across the membrane. Under these conditions the enzyme complex will synthesize cellulose. However once cells are damaged, higher levels of calcium will be encountered near the enzyme, the electrical potential across the membrane will be lost and the solu-

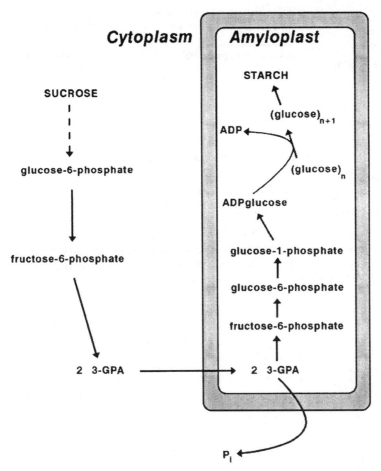

Figure 18.8 Pathway for the transport of sugars from sucrose into the amyloplast for starch synthesis. The major metabolite absorbed by the amyloplast is 3-phosphoglycerate.

ble 18-kDa protein will be dissociated from the enzyme complex. In the absence of the modifier protein and under the conditions of a damaged cell, the enzyme produces callose. It is interesting to compare the activity of this modifier protein in plants with the animal protein (α-lactalbumin) which has a somewhat similar role in lactose synthesis.

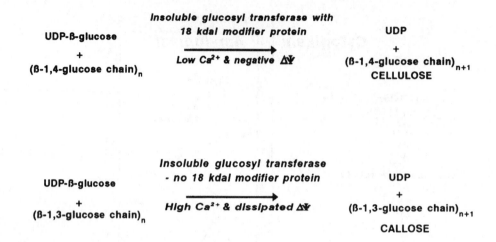

Figure 18.9 Synthesis of cellulose and callose. Normally the conditions favour cellulose synthesis. When the plant is damaged the changes in Ca^{2+} ions and in membrane electrical potential lead to callose synthesis. $\Delta\Psi$ is the electrical potential across the plasma membrane.

19.1 INTRODUCTION

All cells contain lipids, most of which contain fatty acids. Unicellular organisms possess all of the biochemical machinery necessary to synthesize their cellular components and are, therefore, capable of synthesizing the fatty acids and lipids they require. In higher plants and animals, diversification of cells and evolution of tissues have occurred, resulting in the development of tissues with specialized functions. In the context of fatty acid and lipid synthesis this is more apparent in animals than in plants.

In animals, *de novo* fatty acid synthesis occurs mainly in adipose tissue, liver and mammary tissue. Fatty acids may be moved around the body from the site of synthesis in the form of transport lipids, mainly phospholipids and triacylglycerols, but also to some extent as non-esterified fatty acids. Certain tissues are specialized for the production of complex lipids. For example:

- intestinal mucosa transforms the products of lipid digestion into triacylglycerols and phospholipids;
- liver incorporates newly synthesized and absorbed fatty acids mainly into phospholipids which are secreted into the blood as lipoproteins;
- adipose tissue synthesizes predominantly triacylglycerols which are stored in fat droplets and mobilized when required;
- mammary tissue is very active in the synthesis of milk triacylglycerols during lactation.

Most plants do not store large quantities of lipids, with the exception of some oilseeds. Most lipids in plants, therefore, have a structural role as components of membranes and are synthesized *in situ* in each cell. For this reason, plants do not transport fatty acids and complex lipids between their tissues. From an agricultural standpoint the most important plant tissues involved in lipid synthesis are the seeds. In some plant species, seeds produce large quantities of triacylglycerols which are mobilized during germination. A large agricultural and food industry has developed around the extraction and utilization of lipids from oilseeds. Recent developments in biotechnology allow the manipulation of the fatty acid composition of oilseed triacylglycerols. This offers exciting possibilities for the use of genetically manipulated oilseed products for nutritional and industrial purposes.

19.2 TISSUE AND SUBCELLULAR LOCATION OF FATTY ACID SYNTHESIS

In animals, the most important tissues involved in fatty acid synthesis are liver, adipose tissue and mammary tissue. Liver is the site of 90% of fatty acid synthesis in avian species, whereas in ruminants the major site of fatty acid synthesis (90%) is the adipose tissue, and liver makes only a minor contribution to the *de novo* synthesis of fatty acids. In non-ruminant mammals, both the liver and adipose tissue make a contribution to fatty acid synthesis. The relative contributions of the two tissues differ from species to species. In the lactating animal, mammary tissue is a very active site of fatty acid synthesis. For example, a high-yielding dairy cow may secrete as much as 1.25 kg milk fat per day, which represents *de novo* synthesis within the mammary tissue of 400–450 g fatty acid per day.

In plants, most tissues are capable of synthesizing fatty acids. Certain tissues are particularly active in fatty acid synthesis, for example, the seeds of rape, linseed and soyabean. In animals, *de novo* fatty acid synthesis takes place in the cytoplasm; in plants it appears that the chloroplast is the subcellular organelle responsible for the synthesis of fatty acids in mesophyll cells. In other plant cells, particularly those in seeds, the subcellular location of fatty acid synthesis is less clear.

19.3 SOURCE OF THE PRIMARY SUBSTRATE – ACETATE

Acetate is the basic two-carbon unit from which fatty acids are synthesized. Prior to incorporation into fatty acids it must first be converted to acetyl-CoA. As acetyl-CoA is produced in large quantities from pyruvate in the mitochondria, there would appear to be no shortage of this important building block. However, the production of acetyl-CoA from pyruvate and its utilization for fatty acid synthesis take place in different subcellular compartments: production takes place in the mitochondrial matrix, and fatty acid synthesis in the cytoplasm. Because the inner mitochondrial membrane is impermeable to coenzyme A and its derivatives, mitochondrial acetyl-CoA is not readily available for fatty acid synthesis in the cytoplasmic compartment.

In non-ruminant animals, glucose is the main source of the acetate required for fatty acid synthesis. Pyruvate, produced by the oxidation of glucose via the glycolytic pathway, is transported into the mitochondrial matrix. Here pyruvate dehydrogenase converts it to acetyl-CoA which enters the TCA cycle where it condenses with oxaloacetate to produce citrate (Chapter 11). In most tissues citrate is oxidized via the TCA cycle. However, in those tissues where fatty acid synthesis takes place, some of the citrate is diverted to the cytoplasm to serve as a source of acetyl-CoA. In the cytoplasm, citrate is split into oxaloacetate and acetyl-CoA by the enzyme ATP-citrate lyase, sometimes also called the citrate cleavage enzyme. This acetyl-CoA can then be used for fatty acid synthesis.

The oxaloacetate produced by ATP-citrate lyase can be recycled to the mitochondria via pyruvate. This is achieved by the action of a further two enzymes, NAD-dependent malate dehydrogenase, which converts oxaloacetate into malate, and NADP-dependent malate dehydrogenase (the malic enzyme) which catalyses the oxidative decarboxylation of malate to produce pyruvate. This pyruvate can re-enter the mitochondrial matrix. The net effect of this cycle of reactions is the transport of acetyl-CoA from the mitochondrial matrix to the cytoplasm (Figure 19.1).

Another product of the NADP-dependent malate dehydrogenase reaction is NADPH, which is required for fatty acid synthesis. In theory this reaction could provide 50% of the NADPH required for fatty acid synthesis if all the oxaloacetate produced by ATP-citrate lyase were converted to pyruvate. It is likely, however, that between 40 and 50% of the NADPH needed for fatty acid synthesis comes via this route, the remaining 50–60% coming from the oxidation of glucose via the pentose phosphate pathway.

In ruminants, the fermentation of carbohydrate in the rumen produces large quantities of acetate which is transported to the tissues via the bloodstream. This acetate is converted to acetyl-CoA by the action of acetyl-CoA synthetase, which exhibits considerably higher activity in ruminant adipose and mammary tissue than in tissues of non-ruminant animals. Due to the effects of rumen fermentation of dietary carbohydrate, little digestible carbohydrate reaches the small intestine and therefore absorption of glucose from the digestive tract is small. It would, therefore, be wasteful for ruminants to use glucose, produced by gluconeogenesis, for fatty acid synthesis, when there is a plentiful supply of acetate. Although in ruminants citrate can still leave the mitochondria, the conversion of glucose to cytoplasmic acetyl-CoA via citrate is prevented in ruminants by the extremely low activities of ATP-citrate lyase and malic enzyme. Instead of being cleaved to acetyl-CoA and oxaloacetate, this citrate is converted to isocitrate by aconitase, and then acted upon by a cytoplasmic NADP-dependent isocitrate dehydrogenase which produces the NADPH needed for fatty acid synthesis. The α-ketoglutarate also produced can be translocated back into the mitochondrial matrix (Figure 19.2). It is estimated that in ruminant mammary tissue, and probably in adipose tissue, NADP-dependent isocitrate dehydrogenase can produce similar amounts of NADPH to those produced by malic enzyme in non-ruminants.

In plants, acetyl-CoA is produced from glucose via the glycolytic pathway in non-photosynthetic tissues. In photosynthetic tissues, pyruvate can also be produced from glyceraldehyde-3-phosphate synthesized by the Calvin cycle. In addition, there is an active acetyl-CoA synthetase in chloroplasts which is capable of activating any acetate generated outside the chloroplast.

19.4 PRODUCTION OF MALONYL-COA

In addition to acetyl-CoA, malonyl-CoA is also an essential substrate for fatty acid synthesis and is produced by the carboxylation of acetyl-CoA, catalysed by the enzyme acetyl-CoA carboxylase. This is a complex enzyme that occurs in animals, bacteria and plants. Its properties are different in these three different types of organism.

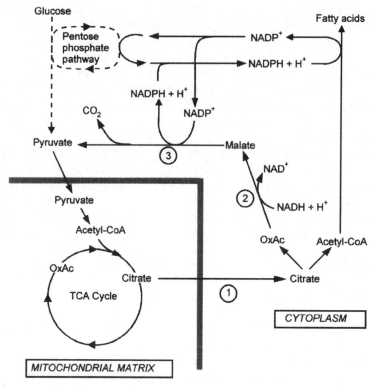

Figure 19.1 Production of NADPH and cytoplasmic acetyl-CoA for fatty acid synthesis in non-ruminant tissues. Enzymes indicated are: 1, ATP-citrate lyase; 2, NAD-dependent malate dehydrogenase; 3, NADP-dependent malate dehydrogenase (the malic enzyme). OxAc = oxaloacetate.

19.4.1 ANIMALS

In animals, the carboxylase is an allosteric enzyme. It exists in two forms, as an inactive monomer and as an active polymer which consists of a linear chain containing approximately 20 monomer units. The monomer unit of acetyl-CoA carboxylase is a 250-kDa multifunctional protein which has three distinct domains. Two of these, biotin carboxylase and carboxyltransferase, have enzymic activities; the third acts as a carrier protein, biotin carboxyl carrier protein (BCCP). Each monomer also contains one molecule of biotin and a regulatory allosteric site.

The production of malonyl-CoA is the sum of two intermediate reactions. Firstly, biotin carboxylase catalyses the carboxylation of biotin attached to BCCP. Secondly, carboxyltransferase catalyses the transfer of the carboxyl group from biotin to acetyl-CoA, producing malonyl-CoA. BCCP carries the biotin molecule. The overall reaction is shown in Figure 19.3.

The control of activity of acetyl-CoA carboxylase is complex. Both acute (short-term) and chronic (long-term) regulation appear to occur, and nutritional and hormonal status have a significant impact on the activity of the enzyme. Polymerization of the inactive monomer units to the active polymer is stimulated by citrate. Citrate acts as an indicator of energy supply and in this way directs excess carbon towards storage as fatty acids. The end products of fatty acid synthesis, fatty acyl-CoAs, cause depolymerization of the enzyme,

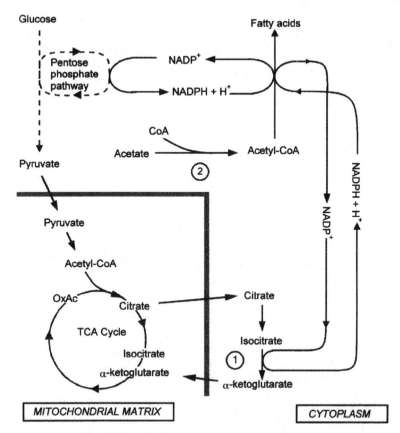

Figure 19.2 Production of NADPH and cytoplasmic acetyl-CoA for fatty acid synthesis in ruminant tissues. Enzymes indicated are: 1, NADP-dependent isocitrate dehydrogenase; 2, acetyl-CoA synthetase. OxAc = oxaloacetate.

thus decreasing its activity and reducing the supply of substrates for fatty acid synthesis.

In addition, the enzyme is regulated by phosphorylation and dephosphorylation caused by glucagon, the catecholamines and insulin. Phosphorylation decreases the activity of the enzyme. The detailed mechanisms of the mode of action of these hormones are still unclear. At present it is thought that glucagon and the catecholamines, which decrease the acetyl-CoA carboxylase activity, cause phosphorylation via an increase in the intracellular concentration of cAMP and the subsequent activation of protein kinases. Insulin, which has lipogenic effects, also promotes phosphorylation: its main effects may be via activation of phosphatases which decrease the phosphorylation state of the enzyme.

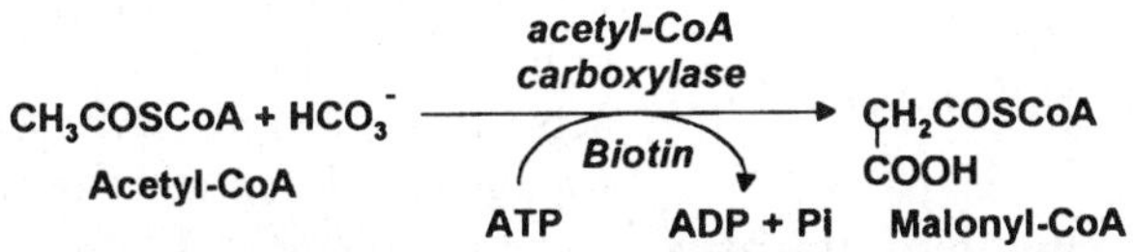

$$CH_3COSCoA + HCO_3^- \xrightarrow[\substack{Biotin \\ ATP \qquad ADP + Pi}]{\substack{\textit{acetyl-CoA} \\ \textit{carboxylase}}} \begin{array}{l} CH_2COSCoA \\ | \\ COOH \end{array}$$

Acetyl-CoA   Malonyl-CoA

Figure 19.3 Overall reaction catalysed by acetyl-CoA carboxylase.

19.4.2 PLANTS

Acetyl-CoA carboxylase in plants is similar in structure to that in other eukaryotic cells. The enzyme is located in the chloroplasts in leaf tissue and in plastids in seeds. Unlike the enzyme in animal tissues, the plant enzyme is not activated by citrate; instead, small changes in stromal pH or Mg^{2+} and K^+ concentration can markedly affect enzyme activity. The enzyme is also regulated by a heat-stable factor found in leaves and is influenced by the ratio of ADP to ATP. High ATP levels activate the enzyme.

19.4.3 BACTERIA

In bacteria, acetyl-CoA carboxylase is active in its monomeric form and there is no evidence that it polymerizes.

The major difference between the enzyme in animals and bacteria is that in bacteria the monomer readily dissociates into its three constituent proteins: biotin carboxylase, biotin carboxyl carrier protein, and carboxyltransferase. Thus the monomer is not a multifunctional protein but consists of individual proteins which come together to form the monomer.

The bacterial enzyme is controlled differently to the animal enzyme. Its activity is reduced by the presence of the nucleosides guanosine-3'-diphosphate, 5'-diphosphate (ppGpp) and guanosine-3'-diphosphate, 5'-triphosphate (pppGpp). These guanosine nucleosides are unique to bacteria and are produced when they are not growing rapidly. In bacteria, the synthesis of fatty acids is mainly to meet the requirement for membrane lipids, so it is logical that fatty acid synthesis should be reduced in conditions when bacteria are not growing and when new membranes are not being synthesized.

19.5 SYNTHESIS OF LONG-CHAIN SATURATED FATTY ACIDS FROM ACETYL-COA AND MALONYL-COA

The synthesis of saturated straight-chain fatty acids from acetyl-CoA and malonyl-CoA takes place on a complex enzyme called fatty acid synthetase. Fatty acid synthetases have been characterized into three groups known as types I, II and III. Types I and II are involved in the *de novo* synthesis of fatty acids; type III is involved in the elongation of existing fatty acids and will be considered later. The main feature that distinguishes the type I and II synthetases is the functional organisation of their constituent proteins.

Type I synthetases are found in animals, yeasts and some bacteria. These enzymes are large multifunctional proteins containing a number of enzyme activities on one polypeptide chain. In addition, they contain a carrier protein called acyl carrier protein or ACP.

Type II synthetases occur in most bacterial and plant tissues. The functional enzyme is a multienzyme complex made up of a number of individual enzymes and ACP loosely bound together. The individual enzymes can be isolated in active form; this cannot be done with type I synthetases.

Despite the differences in the structural organisation of the fatty acid synthetase enzyme in different organisms, the overall sequence of reactions that takes place is similar.

19.5.1 ACYL CARRIER PROTEIN AND ITS FUNCTION

A distinctive feature of fatty acid synthesis is that the acyl intermediates are attached to a low-molecular-weight protein, acyl carrier protein (ACP). This protein, which is an integral part of all fatty acid synthetases, has 77 amino acids. The functional part of the protein is a molecule of 4' phosphopantotheine which is attached to a serine residue at position 36. The 4' phosphopantotheine, which has a terminal SH group to which the acyl intermediates are attached by a thio-ester bond, is identical to that found in CoA. ACP acts as a flexible arm which functions to transfer the fatty acid intermediates from one site of enzyme activity to the next (Figure 19.4).

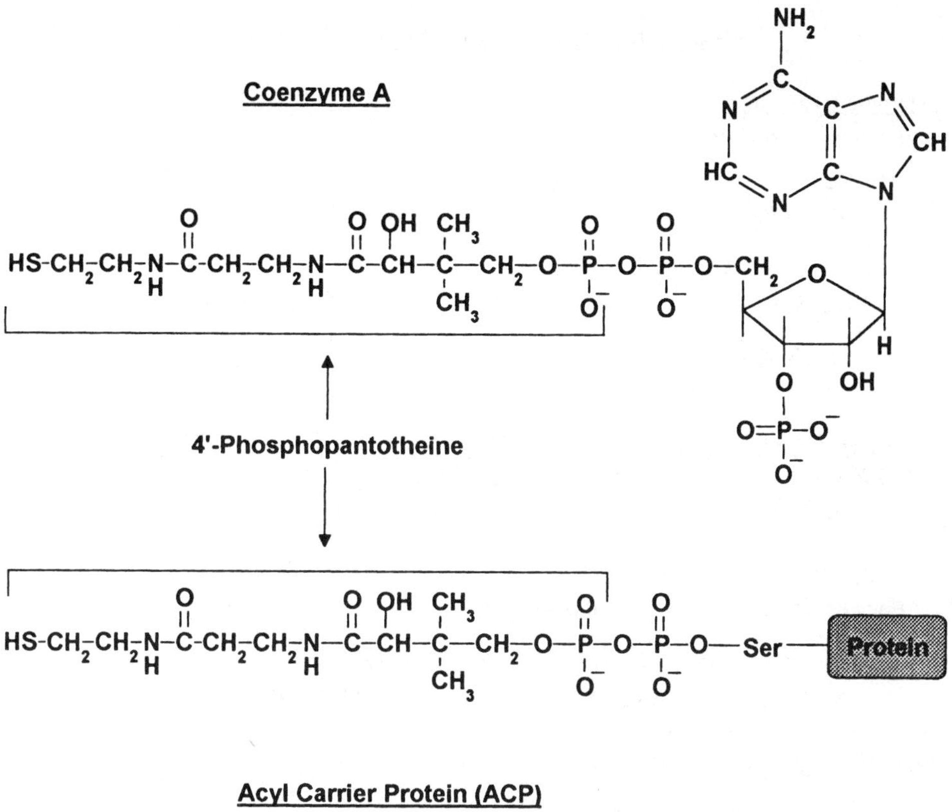

Figure 19.4 Structures of coenzyme A and acyl carrier protein (ACP).

19.5.2 THE REACTIONS OF FATTY ACID SYNTHESIS

The overall scheme of *de novo* fatty acid synthesis is shown in Figure 19.5. The first step in the synthesis of a fatty acid is the attachment of the substrates to the enzyme. Initially acetyl-CoA reacts with the sulphydryl group of ACP to form acetyl-ACP (reaction 1a). This reaction is catalysed by the enzyme acetyl-transacylase, and acetate is termed the primer molecule.

The acetyl residue does not remain on the ACP, however, as ACP is required for the attachment of a malonyl residue transferred from malonyl-CoA. Thus the acetyl residue is transferred from ACP to a sulphydryl group on the enzyme β-ketoacyl ACP synthase (reaction 1b).

Once the acetyl residue has been transferred to β-ketoacyl ACP synthase, the ACP molecule is free to accept a malonyl residue. This reaction is catalysed by the enzyme malonyl transacylase (reaction 2)

With both substrates in position on the enzyme, there then follows a condensation reaction between the malonyl-ACP and the acetyl synthase. This yields the keto intermediate, acetoacetyl-ACP, and is catalysed by the enzyme β-ketoacyl ACP synthase (reaction 3).

During the condensation reaction, malonyl-ACP is decarboxylated and carbon dioxide is released. Isotopic studies have shown that this is the same CO_2 that is added during the acetyl-CoA carboxylase reaction, hence there is no net incorporation of radioisotope into

REACTION SEQUENCE OF FATTY ACID SYNTHESIS

(1) Priming Reaction

(a) Enzyme:- acetyl transacylase

$$CH_3CO\text{-}S\text{-}CoA + ACP\text{-}SH \rightleftharpoons CH_3CO\text{-}S\text{-}ACP + CoA\text{-}SH$$
$$\quad (acetyl\text{-}CoA) \qquad\qquad\qquad (acetyl\text{-}ACP)$$

(b) Enzyme:- β-ketoacyl-ACP synthase

$$CH_3CO\text{-}S\text{-}ACP + Synthase\text{-}SH \rightleftharpoons CH_3CO\text{-}S\text{-}synthase + ACP\text{-}SH$$
$$\qquad\qquad\qquad\qquad\qquad\qquad (acetyl\text{-}synthase)$$

(2) Malonyl Transfer

Enzyme:- malonyl transacylase

$$\underset{\displaystyle COOH}{CH_2CO\text{-}S\text{-}CoA} + ACP\text{-}SH \rightleftharpoons \underset{\displaystyle COOH}{CH_2CO\text{-}S\text{-}ACP} + CoA\text{-}SH$$

$$\quad (malonyl\text{-}CoA) \qquad\qquad\qquad (malonyl\text{-}ACP)$$

(3) Condensation Reaction

Enzyme:- β-ketoacyl-ACP synthase

$$CH_3CO\text{-}S\text{-}synthase + \underset{\displaystyle COOH}{CH_2CO\text{-}S\text{-}ACP} \rightleftharpoons CH_3COCH_2CO\text{-}S\text{-}ACP + CO_2$$

$$\qquad\qquad\qquad\qquad\qquad\qquad\qquad (acetoacetyl\text{-}ACP)$$

(4) First Reduction Reaction

Enzyme:- β-ketoacyl-ACP reductase

$$CH_3COCH_2CO\text{-}S\text{-}ACP + NADPH + H^+ \rightleftharpoons \underset{\displaystyle OH}{CH_3CHCH_2CO\text{-}S\text{-}ACP} + NADP^+$$

$$\qquad\qquad\qquad\qquad\qquad\qquad (D\text{-}3\text{-}hydroxybutyryl\text{-}S\text{-}ACP)$$

(5) Dehydration Reaction

Enzyme:- enoyl-ACP hydratase

$$\underset{\displaystyle OH}{CH_3CHCH_2CO\text{-}S\text{-}ACP} \rightleftharpoons CH_3CH{=}CHCO\text{-}S\text{-}ACP$$

$$\qquad\qquad\qquad\qquad (crotonyl\text{-}ACP)$$

(6) Second Reduction Reaction

Enzyme:- enoyl-ACP reductase

$$CH_3CH{=}CHCO\text{-}S\text{-}ACP + NADPH + H^+ \rightleftharpoons CH_3CH_2CH_2CO\text{-}S\text{-}ACP + NADP^+$$

$$\qquad\qquad\qquad\qquad\qquad\qquad (butyryl\text{-}ACP)$$

Figure 19.5 Reaction sequence of fatty acid synthesis. These reactions are required for the production of the four-carbon fatty acyl intermediate, butyryl-ACP. In the synthesis of palmitoyl-ACP this cycle of reactions is repeated a further six times from priming reaction 1(b). At the start of the second cycle, butyryl-ACP substitutes for acetyl-ACP in reaction 1(b). In subsequent cycles the product of reaction 6 replaces acetyl-ACP in reaction 1(b) until the final product, containing 16 carbons, is produced.

fatty acids when tissue preparations are incubated with [^{14}C]-bicarbonate.

The four-carbon intermediate, acetoacetyl-ACP, is reduced in a reaction catalysed by the enzyme β-ketoacyl-ACP reductase. The hydrogen donor used in this reaction is NADPH and the product is D-3-hydroxybutyrate (reaction 4).

The next step is the dehydration of the hydroxyacyl intermediate resulting in the formation of a Δ^2-*trans* enoyl-ACP. This reaction is catalysed by the enzyme enoyl-ACP dehydratase (reaction 5).

The unsaturated intermediate is then reduced by the addition of hydrogen across the double bond to yield the saturated intermediate butyryl-ACP. This reaction is catalysed by the enzyme enoyl-ACP reductase (reaction 6).

This reaction completes the process by which the original two-carbon unit, acetyl-ACP, is elongated to a four-carbon unit, butyryl-ACP. The butyryl-ACP becomes the starting substrate for the next cycle of reactions. The butyryl residue is transferred to the free SH group of β-ketoacyl-ACP synthase and another malonyl residue is transferred from CoA to ACP. The reaction sequence is repeated to produce a six-carbon intermediate. The process continues until the growing fatty acid has 16 carbons, e.g. palmitoyl-ACP.

In mammary glands, butyrate is also used as a primer molecule, and it has been estimated that butyrate carbon contributes approximately 8% of the total carbon in fatty acids produced by *de novo* synthesis in this tissue.

Odd-numbered fatty acids are produced in small amounts (0.5–1.0% of the total fatty acids) by plant and animal tissues. Ruminant fats contain a small but significant quantity of pentadecanoic (C15:0) and heptadecanoic (C17:0) acids. These fatty acids are produced by the incorporation of the odd-numbered (three-carbon) primer molecule propionyl-CoA, in place of acetyl-CoA. It has been noted that the proportions of these odd-numbered

fatty acids increase when ruminants are fed on diets rich in cereals which increase the molar proportion of propionate in the rumen liquor. In extreme cases, in sheep, this has been shown to lead to unusually high levels of these acids (10–15%) in adipose tissue depots.

19.5.3 CHAIN LENGTH SPECIFICITY OF FATTY ACID SYNTHETASES

Plants and animals

In most species of plants and animals, approximately 90% of the fatty acids produced have 16 carbons, e.g. palmitoyl-ACP. The enzyme responsible for this chain-length specificity is β-ketoacyl-ACP synthase. At the end of each sequence of four reactions, the growing fatty acid is transferred from ACP to the free SH group of β-ketoacyl-ACP synthase. This enzyme will readily accept a fatty acid containing 14 carbons but will hardly ever accept one containing 16 carbons. Thus, the process of fatty acid synthesis stops once 16 carbons have been reached. The newly synthesized fatty acid is released from fatty acid synthetase by acyl-ACP-thioesterase. In animals, thioesterase is an integral part of the synthetase complex, whereas in plants it appears to be a soluble enzyme.

The fatty acid composition of milk fat is characterized by the presence of short- and medium-chain fatty acids (C4:0–C12:0). As these fatty acids are found in only trace quantities in plasma, it is clear that they are synthesized *de novo* in mammary tissue. The exact mechanism that leads to the termination of fatty acid synthesis and release of these fatty acids is not known for all species of mammals. In rat mammary tissue, a second thioesterase has been identified which catalyses the release of medium-chain fatty acids. In the goat, the presence of a transacylase is responsible for the transfer of fatty acids from the synthetase complex into milk triacylglycerols. Other factors may be important in the control of termination of fatty acid synthesis in mammary tissue. For example, the chain length of fatty acids synthesized *in vitro* can be influenced by changing the relative concentrations of acetyl-CoA and malonyl-CoA; the presence of excess acetyl-CoA results in a reduction in the chain length of fatty acids produced. This may be due to competition between the two substrates for binding sites on the fatty acid synthetase enzyme. However, there is no evidence that such a mechanism operates *in vivo.*

Bacteria

The chain length of fatty acids produced by bacterial fatty acid synthetases varies. For example *Escherichia coli* produces both C16 and C18 saturated and unsaturated fatty acids, whereas mycobacteria produce saturated fatty acids of chain length C16 or C24.

19.5.4 SYNTHESIS OF BRANCHED-CHAIN FATTY ACIDS

An interesting feature of fatty acid biosynthesis in bacteria is that most Gram-positive and some Gram-negative organisms produce mainly branched-chain fatty acids of the iso and anteiso series (Chapter 4). These arise as a result of the use of branched-chain primer molecules in place of acetate (Figure 19.6). These primers are produced by the oxidative deamination of branched amino acids. For example, valine is converted to isobutyryl-CoA and isoleucine yields 2-methylbutyryl-CoA.

Fatty acids of the iso and anteiso series are present in the fat depots of ruminant animals. They are found in relatively low proportions (0.5–1.5%) and their presence is due to the digestion and absorption of fatty acids synthesized by rumen bacteria. When sheep and goats are fed diets rich in cereals, the proportion of branched-chain fatty acids in their adipose tissue increases. Analysis of these fatty acids has shown that this is due not only to an increase in the iso and anteiso fatty acids, but also to the presence of unusual methyl branched-chain fatty acids, in which the methyl substituent is located at various places along the fatty acid chain. In some cases di- and tri-methyl substituted fatty acids have been identified. In almost all cases the methyl substituent group is located on even-numbered carbon atoms of the fatty acid chain. This pattern of methyl substitution provided a clue to the mechanism by which these fatty acids were synthesized, and it is now accepted that they arise as a result of the substitution of methylmalonyl-CoA for malonyl-CoA during *de novo* fatty acid synthesis. A mono-methyl-substituted fatty acid is produced when only one molecule of methylmalonyl-CoA replaces malonyl-CoA and the position of the methyl substituent depends upon the stage in fatty acid synthesis at which the replacement occurs. Methylmalonyl-CoA is an intermediate in the conversion of propionate to succinate, a short but important pathway in ruminants which enables them to use the propionate produced in the rumen to synthesize glucose. The feeding of large quantities of cereals which contain readily fermentable starch leads to the rapid production of propionate in excess of the capacity of the liver to metabolize it. This results in elevated levels of methylmalonic acid in the plasma, some of which is excreted in the urine. However, some methylmalonic acid enters the adipose tissue where it is activated and used in fatty acid synthesis. These branched-chain fatty acids have a lower melting point than saturated fatty acids, and accumulation of significant quantities in adipose tissue can result in the production of soft fat.

In water birds a specialized sebaceous gland, the uropygial gland, produces predominantly multi-methyl branched-chain fatty acids, particularly 2,4,6-trimethyldodecanoic acid and 2,4,6,8-tetramethyltetradecanoic acid. This occurs due to the rapid breakdown of malonyl-CoA in the gland by malonyl-CoA decarboxylase and the use of methylmalonyl-CoA as a alternative substrate. These fatty acids may play an important role in the water-proofing of plumage.

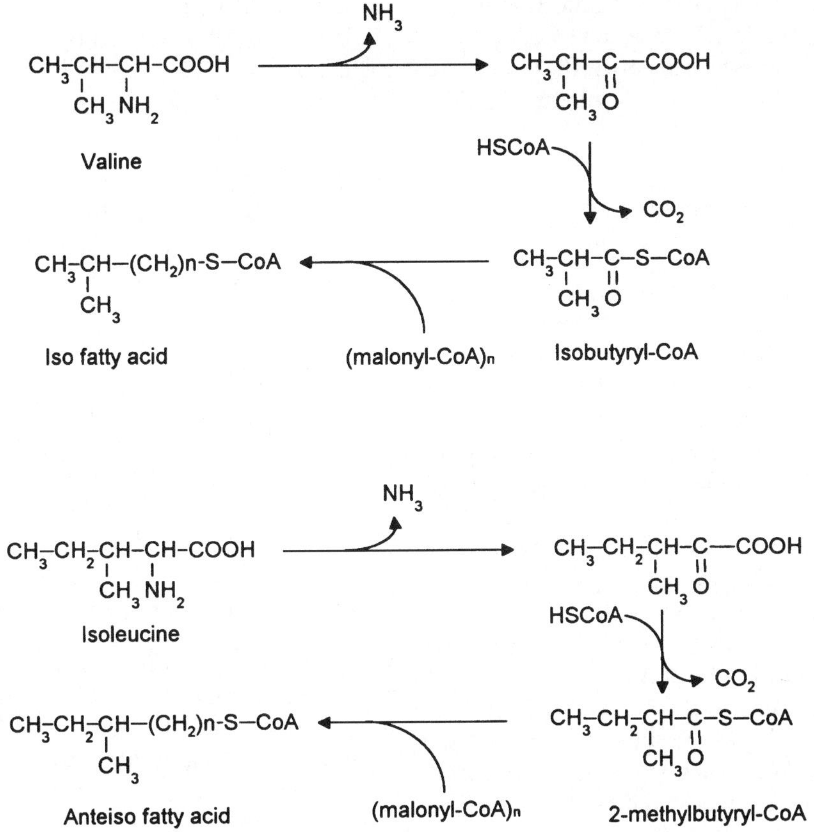

Figure 19.6 Summary of the synthesis of iso- and anteiso-branched-chain fatty acids from valine and isoleucine.

19.5.5 RELEASE OF FATTY ACIDS FROM FATTY ACID SYNTHETASE

Fatty acids may be released from the synthetase in a number of forms. In most animals, fatty acids are released as free fatty acids which may then be modified in several ways. For example, they may be elongated, desaturated and/or incorporated into other lipids such as triacylglycerols and phospholipids. In plants and microorganisms, fatty acids may be released as free acids or may be converted to their CoA esters upon release. In certain microorganisms such as *E. coli* and *Euglena gracilis* the fatty acid is transferred directly from ACP to phosphatidic acid and thereafter into triacylglycerols and phospholipids.

19.6 FATTY ACID ELONGATION

De novo synthesis of fatty acids produces mainly palmitic acid (C16:0). However, analysis of the fatty acids of plant and animal tissues reveals that there are a large number of fatty acids with more than 16 carbon atoms. Some of these may be supplied in the diet in the case of animals, but both plant and animal tissues have the capacity to elongate fatty acids.

The enzyme systems involved in fatty acid elongation are sometimes referred to as type III fatty acid synthetases, and sometimes as fatty acid elongases. The enzymes involved in elongation appear to be membrane-bound and are found in two sites in the cell, in mitochondria and the endoplasmic reticulum. The latter is the main site of fatty acid elongation. The substrates for elongation are fatty acyl-CoAs which are extended by two carbon units which are supplied by malonyl-CoA. Reducing equivalents are provided by NADPH.

19.7 FORMATION OF UNSATURATED FATTY ACIDS

The introduction of double bonds into a fatty acid is called desaturation. There are two processes by which desaturation of fatty acids occurs. In the first, anaerobic desaturation, an unsaturated fatty acid is produced during fatty acid synthesis. This process is of limited importance and is found only in bacteria, particularly Eubacteriales. Typical of this process is the production of vaccenic acid (Δ^{11} *cis*-octadecenoic acid). The synthesis of this fatty acid occurs because the Δ^2 *trans*-decenoyl-ACP, one of the intermediates produced in reaction 5 of Figure 19.5, is isomerized to form Δ^2 *cis*-decenoyl-ACP which is not a substrate for enoyl-ACP reductase. Instead, it acts as the substrate for the condensation reaction (reaction 3) and is elongated to an 18-carbon fatty acid. The second, aerobic desaturation, is found in virtually all organisms. In this process, double bonds are introduced into preformed fatty acids.

19.7.1 DESATURATION OF FATTY ACIDS IN ANIMALS

In animals, the substrates for desaturation are derived from dietary fat or from the products of cytoplasmic fatty acid synthesis and elongation. The process of desaturation takes place on the desaturase complex, which is located on the endoplasmic reticulum and consists of three components: NADH-cytochrome b_5 reductase; cytochrome b_5; and the desaturase enzyme. There is some doubt as to the true number of desaturase enzymes. It is normally assumed that there are four, which are designated Δ^9, Δ^6, Δ^5 or Δ^4 fatty acyl-CoA desaturases. The Δ notation indicates the position in

the fatty acid relative to the carboxyl carbon (carbon no. 1) where the double bond is inserted.

The desaturation of fatty acids in animal tissues is summarized below.

- All desaturases require molecular oxygen and a reduced pyridine nucleotide, and catalyse the desaturation of preformed fatty acids, usually in the form of coenzyme A esters. There is some evidence that the fatty acids in complex lipids can also serve as desaturase substrates.
- For a given desaturase enzyme, the double bond is always introduced into the methylene chain at a fixed position from the carboxyl group and has the *cis* configuration.

e.g. Δ^9 desaturase always introduces a double bond between carbons 9 and 10, not between 8 and 9 or between 10 and 11.

- When the substrate is a saturated fatty acid, the first double bond is inserted between carbons 9 and 10. Unlike plants, animal desaturase systems cannot insert double bonds between carbon 10 and the methyl carbon, for example:

$$e.g.\ C16:0 \rightarrow \Delta^9\,C16:1$$
$$C18:0 \rightarrow \Delta^9\,C18:1$$

- When the substrate is already unsaturated, subsequent double bonds are inserted between the double bond nearest the carboxyl carbon, and the carboxyl carbon itself, in such a way as to (usually) maintain the methylene-interrupted distribution of double bonds (Figure 19.7).

$$CH_3(CH_2)_7CH=CH(CH_2)_7COSCoA$$
$$\Delta^9\text{-oleoylCoA}$$

$$\downarrow\ \Delta^6\ \text{desaturase}$$

$$CH_3(CH_2)_7CH=CHCH_2CH=$$
$$CH(CH_2)_4COSCoA$$
$$\Delta^{9.6}\text{-linoleoyl-CoA}$$

- In a metabolic pathway leading to the formation of polyunsaturated fatty acids

(PUFAs), desaturation usually alternates with elongation.

19.7.2 DESATURATION OF FATTY ACIDS IN PLANTS

The desaturase complex in plants is less well characterized. It appears to be similar to that in animals, with the exception that ferredoxin replaces cytochrome b_5 as the electron carrier. Of particular importance is the fact that in plant tissues, double bonds can be inserted into a fatty acid between the methyl carbon and carbon 10 or a pre-existing double bond. Thus, plants can synthesize linoleic acid ($\Delta^{9,12}$ C18:2) and α-linolenic acid ($\Delta^{9,12,15}$ C18:3), whereas animals cannot.

Desaturation takes place both in the chloroplast stroma and on the endoplasmic reticulum. The initial step in the desaturation process, which occurs in the chloroplast stroma, appears to be elongation of newly synthesized palmitoyl-ACP to give stearoyl-ACP. Stearoyl-ACP is the substrate for a stearoyl-ACP desaturase found in the chloroplast and is converted to oleoyl-ACP. Subsequent fatty acid desaturation can take place by two mechanisms, the eukaryotic pathway and the prokaryotic pathway, and appears to occur once fatty acids have been incorporated into complex lipids (phospholipids and glycolipids). In the eukaryotic pathway, so-called because it involves reactions in both the chloroplast and the endoplasmic reticulum, oleoyl-ACP is converted to oleoyl-CoA in the chloroplast and then transported to the endoplasmic reticulum. Here, it is incorporated into phosphatidylcholine. In this phospholipid form, the oleic acid is desaturated to linoleic acid by Δ^{12} desaturase. The linoleoylphosphatidylcholine is transported back into the chloroplast, where the phosphocholine moiety is removed and the resulting diacylglycerol is converted to monogalactosyldiacylglycerol. At this point linoleic acid is converted to α-linolenic acid by a Δ^{15} desaturase. In the prokaryotic pathway, so named because it

occurs entirely in the chloroplast, the formation of α-linolenic acid occurs from oleate-containing monogalactosyldiacylglycerol as the substrate.

19.7.3 ESSENTIAL FATTY ACIDS

In mammals, PUFAs can be grouped into four distinct families which differ in the number of carbon atoms between the terminal methyl group and the nearest double bond. This pattern arises because mammalian fatty acid desaturases cannot insert a double bond into a fatty acid between carbon 10 and the methyl carbon, referred to in older texts as the ω-carbon. The families of PUFAs are derived from four precursor fatty acids. These fatty acids and the notation used to identify them are shown in Table 19.1.

PUFAs in each of the families are produced from their respective precursor fatty acid by a series of alternating elongation and desaturation reactions. A sequence of reactions for the *n*-9, *n*-6 and *n*-3 family of PUFAs is shown in Table 19.2. Note that for each of the fatty acids within a family, the number of carbon atoms between the methyl carbon and the nearest double bond remains unchanged, regardless of subsequent elongations and desaturations.

Until recently it was assumed that Δ^4 double bonds were inserted by a Δ^4 desaturase. There is now evidence, at least in the case of docosahexaenoic acid (DHA), that an alternative mechanism operates. This involves the elongation of eicosapentaenoic acid (EPA) by four carbon units to produce $\Delta^{9,12,15,18,21}$ C24:5, which is then acted upon by Δ^6 desaturase to give $\Delta^{6,9,12,15,18,21}$ C24:6. This fatty acid undergoes chain shortening by β-oxidation to produce DHA. It remains to be established if this mechanism applies generally to the introduction of Δ^4 double bonds.

Since linoleic and α-linolenic acids cannot be synthesized by mammals, they are termed essential fatty acids (EFAs). PUFAs derived from them are important as components of

Table 19.1 Precursor fatty acids of the four families of polyunsaturated fatty acids

Fatty acid	Structure	Family
Palmitoleic acid	$CH_3(CH_2)_5CH{=}CH{-}(CH_2)_7COOH$	n-9 family
Oleic acid	$CH_3(CH_2)_7{-}CH{=}CH{-}(CH_2)_7COOH$	n-7 family
Linoleic acid	$CH_3(CH_2)_4CH{=}CH{-}CH_2{-}CH{=}CH{-}(CH_2)_7COOH$	n-6 family
Linolenic acid	$CH_3CH_2CH{=}CH{-}CH_2{-}CH{=}CH{-}CH_2{-}CH{=}CH{-}(CH_2)_7COOH$	n-3 family

membrane phospholipids and as precursors of prostaglandins, prostacyclins, thromboxanes and leukotrienes, which are important regulatory compounds.

Table 19.2 shows that the Δ^6 desaturase can introduce a double bond into each of the precursor fatty acids in the n-9, n-6 and n-3 families. When all three precursors are present they compete for the same enzyme; however, the substrate specificity of the Δ^6 desaturase is greatest for α-linolenic acid and lowest for oleic acid, and therefore the relative rates of desaturation of the precursor fatty acids are α-linolenic>linoleic>oleic. When an animal is fed a balanced diet, the intake of linoleic acid is usually greater than that of α-linolenic acid, which ensures an adequate supply of arachidonic acid via the n-6 pathway. In cases of

Table 19.2 Metabolic transformations of the n-9, n-6 and n-3 fatty acid families by desaturation and elongation

	n-9 family	n-6 family	n-3 family
PRECURSOR	Δ^9-C18:1 Oleic acid	$\Delta^{9,12}$-C18:2 Linoleic acid	$\Delta^{9,12,15}$-C18:3 α-Linolenic acid
↓ Desaturation Δ^6 ↓	↓ $\Delta^{6,9}$-C18:2 ↓	↓ $\Delta^{6,9,12}$-C18:3 γ-Linolenic acid ↓	↓ $\Delta^{6,9,,2,15}$-C18:4 ↓
Elongation ↓	$\Delta^{8,11}$-C20:2 ↓	$\Delta^{8,11,14}$-C20:3 ↓	$\Delta^{8,11,14,17}$-C20:4 ↓
Desaturation Δ^5 ↓	$\Delta^{5,8,11}$-C20:3 ↓	$\Delta^{5,8,11,14}$-C20:4 Arachidonic acid ↓	$\Delta^{5,8,11,14,17}$-C20:5 EPA* ↓
Elongation ↓	$\Delta^{7,10,13}$-C22:3 ↓	$\Delta^{7,10,13,16}$-C22:4 ↓	$\Delta^{7,10,13,16,19}$-C22:5 ↓
Desaturation Δ^4	$\Delta^{4,7,10,13}$-C22:4	$\Delta^{4,7,10,13,16}$-C22:5	$\Delta^{4,7,10,13,16,19}$-C22:6 DHA*

Alternative mechanism for the synthesis of DHA

			$\Delta^{7,10,13,16,19}$-C22:5 EPA
Elongation			↓ $\Delta^{9,12,15,18,21}$-C24:5
Desaturation Δ^6			↓ $\Delta^{6,9,12,15,18,21}$-C24:6
β-oxidation			↓ -2C $\Delta^{4,7,10,13,16,19}$-C22:6 DHA*

* EPA, Eicosapentaenoic acid; DHA, docosahexaenoic acid

essential fatty acid deficiency, the availability of oleic acid relative to linoleic acid is increased. The result is an enhanced desaturation of oleic acid which leads to an increase in the tissue content of the *n*-9 fatty acid, $\Delta^{5,8,11}$-20:3, relative to the *n*-6 fatty acid, $\Delta^{5,8,11,14}$-C20:4. Changes in the ratio of these two fatty acids, sometimes referred to as the triene–tetraene ratio, are used as an indicator of essential fatty acid deficiency. Until recently a ratio of greater than 0.4 has been taken as an indicator of an inadequate supply of essential fatty acids in humans; however, recent research has suggested that the ratio should be lower, probably around 0.2.

EFA deficiency is best characterized in the rat, where typical symptoms are markedly reduced growth, scaly paws and tail, reduced fertility, elevated respiratory quotient, high metabolic rate and high skin permeability to water. The fatty acids necessary to prevent deficiency symptoms in the rat are linoleic acid ($\Delta^{9,12}$ C18:2) of plant origin, or arachidonic acid ($\Delta^{5,8,11,14}$ C20:4) which can be formed in animal tissues from linoleic acid. Two other fatty acids related to linoleic acid, in that they have a similar methyl-terminal (*n*-6) chain structure, are also effective: $\Delta^{8,11,14}$ C20:3 and $\Delta^{6,9,12}$ C18:3 (γ-linolenic acid). α-Linolenic acid ($\Delta^{9,12,15}$ C18:3), an *n*-3 fatty acid, is effective in promoting growth but does not prevent the skin deficiency symptoms. Thus, fatty acids related structurally (at the methyl end) to linoleic acid, the *n*-6 family, appear to be the true EFAs.

Grazing ruminants have a diet which is rich in linoleic acid ($\Delta^{9,12}$ C18:2) and α-linolenic acid ($\Delta^{9,12,15}$ C18:3). However, as a result of the activity of rumen microorganisms, most of these fatty acids are biohydrogenated to oleic and stearic acids. Despite this, sufficient PUFAs escape biohydrogenation and pass to the small intestine where they are absorbed and utilized very efficiently. Neonatal ruminants are born with relatively low levels of essential fatty acids. The plasma triene–tetraene ratio in neonatal lambs, kids and calves immediately *post-partum* is considerably higher than 0.4, which suggests that ruminants are born EFA-deficient. The ratio falls markedly within the first few days of birth as the polyunsaturated fatty acids in milk are absorbed.

19.8 SYNTHESIS OF TRIACYLGLYCEROLS

Triacylglycerols are the main storage lipids in animal and plant tissues. In animals, adipose tissue depots, distributed widely around the body and identified by anatomical location (e.g. subcutaneous, perinephric and mesenteric), are the major sites of triacylglycerol storage. In some plants, triacylglycerols may be stored in the seeds. This is particularly true of the seeds of certain plants such as rape, sunflower, soyabean and linseed where the oil (triacylglycerol) content can be as high as 400–500 g kg^{-1}.

The structure of triacylglycerols is discussed in Chapter 4. In animals, the fatty acids contained within triacylglycerols are derived either from dietary lipids following digestion, or by endogenous synthesis of fatty acids as described earlier in this chapter. There are two major pathways for the synthesis of triacylglycerols. The first, the 2-monoacylglycerol pathway, is most active in the intestinal mucosa and derives its substrates from lipid digestion (described in Chapter 28). The second is the glycerol-3-phosphate pathway, which occurs in most tissues.

19.8.1 THE 2-MONOACYLGLYCEROL PATHWAY

The main end products of dietary triacylglycerol digestion in monogastric animals are 2-monoacylglycerols and free fatty acids. These products are incorporated into mixed micelles in the jejunum and ileum. The contents of the micelles are absorbed into the cells of the intestinal mucosa, and it is here that the resynthesis of triacylglycerols takes place. The pathway is a simple two-step process which involves the sequential addition of two fatty acids to 2-monoacylglycerol acceptor molecules catalysed by monoacylglycerol acyltransferase (step 1 in Figure 19.7) and diacylglycerol acyltransferase (step 2 in Figure 19.7).

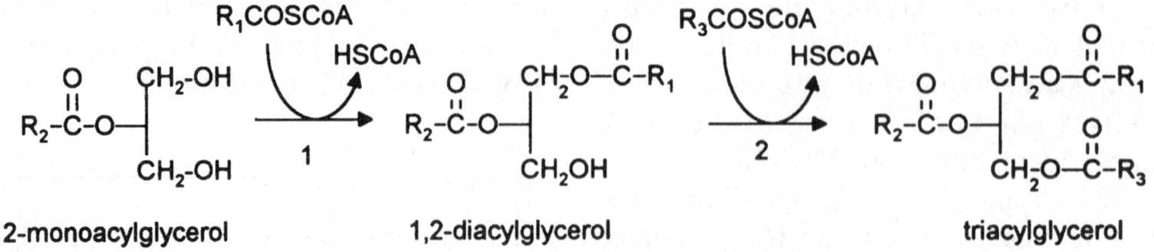

2-monoacylglycerol   **1,2-diacylglycerol**   **triacylglycerol**

Figure 19.7 Synthesis of triacylglycerols by the 2-monoacylglycerol pathway.

The enzymes of this pathway are located on the smooth endoplasmic reticulum. Monoacylglycerol acyltransferase is most active on 2-monoacylglycerol substrates containing short-chain saturated or long-chain unsaturated fatty acids. The enzymes for this pathway are found in other tissues such as liver and adipose tissue; however, as the concentration of 2-monoacylglycerols is very low in these tissues, their contribution to triacylglycerol synthesis is negligible under normal circumstances. It is also important to remember that this pathway is not normally important in ruminant intestinal mucosa because dietary triacylglycerols are hydrolysed to glycerol and free fatty acids in the rumen, and very little lipid escapes the rumen to be digested in the small intestine. However, when protected fat (triacylglycerols coated in formaldehyde-treated casein to prevent hydrolysis of dietary lipid by rumen microorganisms) is fed to ruminants, large quantities of triacylglycerol are digested in the small intestine and the 2-monoacylglycerol pathway becomes more important.

19.8.2 THE GLYCEROL-3-PHOSPHATE PATHWAY

This is the major pathway of triacylglycerol synthesis in tissues other than monogastric intestinal mucosa. The glycerol backbone is provided by glycerol-3-phosphate, most of which is derived from glucose via dihydroxyacetone phosphate (Figure 19.8).

Small amounts of glycerol-3-phosphate are synthesized by the direct phosphorylation of glycerol by the enzyme glycerol kinase (Figure 19.9).

The production of triacylglycerols from glycerol-3-phosphate is shown in Figure 19.10. Initially a fatty acid is added to the 1 position of glycerol-3-phosphate to produce 1-acyl glycerol-3-phosphate. This reaction is catalysed by glycerol-3-phosphate acyltransferase which has a preference for saturated fatty acids. Addition of a fatty acid to the 2 position is then catalysed by 1-acylglycerol-3-phosphate acyltransferase, which has a preference for unsaturated fatty acids. The product is a 1,2-diacylglycerol-3-phosphate, more com-

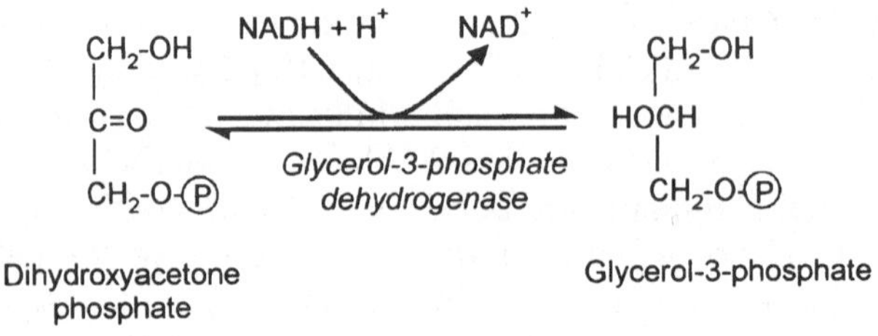

Dihydroxyacetone phosphate   Glycerol-3-phosphate

Figure 19.8 Production of glycerol-3-phosphate from dihydroxyacetone phosphate.

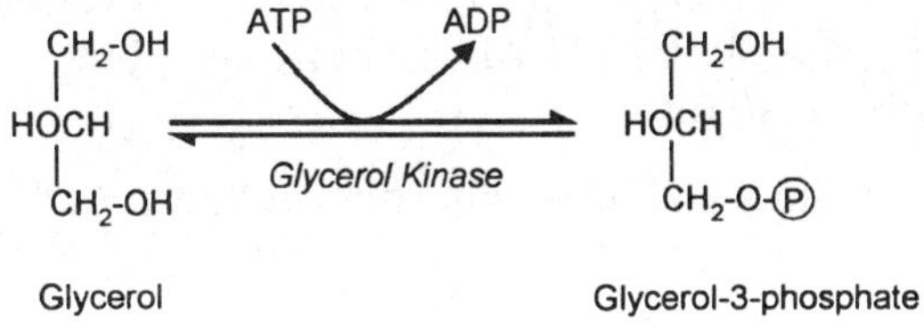

Figure 19.9 Production of glycerol-3-phosphate from glycerol.

reticulum. This dual distribution appears to be a regulatory mechanism, as only the membrane-bound enzyme is active. Factors which favour the synthesis of triacylglycerols, such as energy intake in excess of requirements, increase the proportion of bound enzyme to free enzyme. These effects are probably mediated via changes in hormonal status, e.g. an increase in plasma insulin concentration.

monly known as phosphatidic acid. Phosphatidate phosphohydrolase releases the phosphate from the 3 position of phosphatidic acid to produce a diacylglycerol, which is then further acylated by 1,2-diacylglycerol acyltransferase to yield a triacylglycerol.

Phosphatidic acid is a common precursor of both triacylglycerols and phospholipids. The amount of triacylglycerol formed from phosphatidic acid is determined by the activity of the enzyme phosphatidate phosphohydrolase. This enzyme can exist both free in the cytoplasm and bound on the endoplasmic

19.8.3 TRIACYLGLYCEROL SYNTHESIS IN PLANTS

In plants, triacylglycerol synthesis occurs via phosphatidic acid. Fatty acid synthesis in the plastid produces palmitoyl-ACP, which is converted predominantly to oleoyl-ACP by fatty acid elongation and desaturation. Plastid fatty acyl-ACPs can be converted to fatty acyl-CoAs and exported to the cytoplasm where they may be used for the synthesis of phosphatidic acid on the endoplasmic reticulum. The pattern of fatty acid distri-

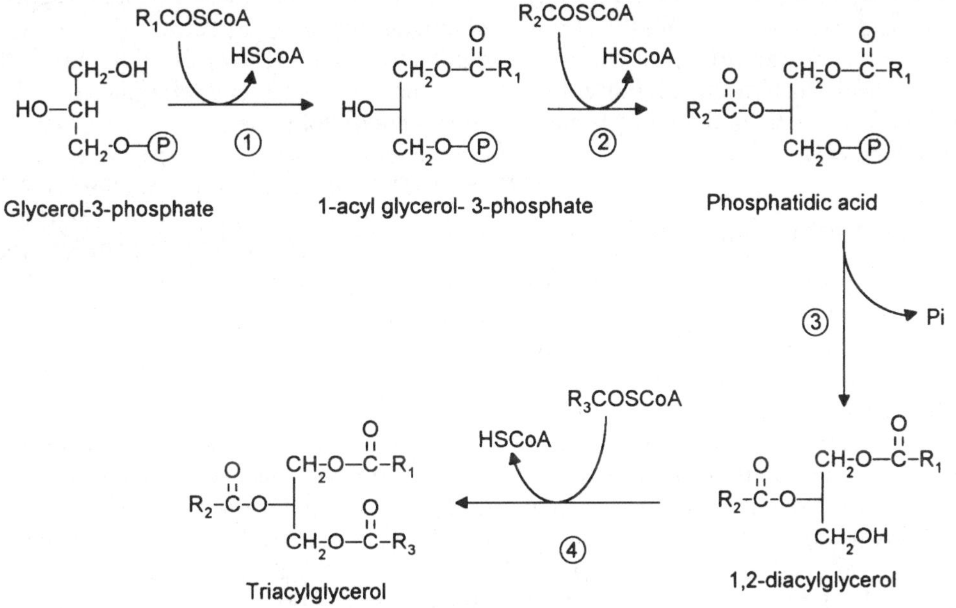

Figure 19.10 Synthesis of triacylglycerols by the glycerol-3-phosphate pathway. Enzymes: 1, glycerol-3-phosphate acyltransferase; 2, 1-acyl glycerol-3-phosphate acyltransferase; 3, phosphatidate phosphohydrolase; 4, 1,2-diacylglycerol acyltransferase.

bution in this phosphatidic acid is usually palmitic acid in position 1 and oleic acid in position 2. Hydrolysis of the phosphate group from position 3 yields a diacylglycerol which is further acylated to triacylglycerol. In many seeds the triacylglycerols are particularly rich in PUFAs (linoleic acid in soyabean, sunflower seeds and safflower seeds; linolenic acid in linseed). This enrichment is apparently achieved by conversion of part of the diacylglycerol pool to phosphatidylcholine, which is subject to the action of desaturases on the endoplasmic reticulum. As this conversion is reversible in many plants, the diacylglycerol pool becomes enriched with PUFAs leading to the production of PUFA-enriched triacylglycerols.

19.9 PHOSPHOLIPIDS

Phosphatidic acid is the basic building block for the synthesis of a wide range of phospholipids. Two strategies are employed to achieve their synthesis. The first strategy leads to the production of phosphatidylserine (PS), phosphatidylinositol (PI), phosphatidylglycerol (PG) and cardiolipin, and involves the synthesis of an activated diacylglycerol, CDP-diacylglycerol.

$$\text{Phosphatidic acid} + \text{CTP} \rightarrow$$
$$\text{CDP-diacylglycerol} + \text{PP}_i$$

PS, PI and PG are produced from CDP-diacylglycerol by exchange of the CDP group with serine, inositol or glycerol-3-phosphate, respectively. Cardiolipin is produced by the condensation of two molecules of PG with the elimination of glycerol (Figure 19.11).

Phosphatidylethanolamine (PE) can be formed by the decarboxylation of PS and subsequently converted to phosphatidylcholine (PC) by three successive transmethylation steps in which the methyl groups are donated by S-adenosylmethionine (SAM). This route of PC synthesis is of only minor importance in plant and animal tissues, but it is the main pathway of PC synthesis in bacteria.

The second strategy involves the activation of the polar head groups, choline or ethanolamine.

$$\text{Choline} + \text{ATP} \rightarrow \text{Phosphocholine} + \text{ADP}$$
$$\text{Phosphocholine} + \text{CTP} \rightarrow$$
$$\text{CDP-choline} + \text{PP}_i$$

Diacylglycerol, produced by the dephosphorylation of PA, displaces CMP from CDP-choline or CDP-ethanolamine to produce PC or PE, respectively. In animals, PS can be produced from PE by head-group exchange, whereas in plants PS synthesis occurs mainly by the CDP–diacylglycerol route plastids (Figure 19.11). In animals, synthesis of phospholipids occurs mainly on the smooth endoplasmic reticulum. However, in plants synthesis of most phospholipids, but not PC and PI, also occurs in chloroplasts and other non-photosynthetic organelles.

19.10 GLYCOLIPID SYNTHESIS

The principal glycolipids of plant tissue are monogalactosyldiacylglycerols (MGDGs) and digalactosyldiacylglycerols (DGDGs) with small amounts of tri- and tetragalactosyl derivatives. These lipids are almost exclusively confined to chloroplast membranes. The synthesis of MGDG requires the production of UDP-galactose. Galactosyltransferases catalyse the transfer of galactose firstly to diacylglycerol (DAG) to yield MGDG, and secondly to MGDG to produce DGDG.

$$\text{DAG} + \text{UDP-Gal} \rightarrow \text{MGDG} + \text{UDP}$$
$$\text{MGDG} + \text{UDP-Gal} \rightarrow \text{DGDG} + \text{UDP}$$

An alternative pathway for synthesis of DGDG and the small quantities of tri- and tetragalactosyldiglycerides involves the transfer of a galactose residue from one molecule of MGDG to the galactose residue of another molecule of MGDG.

$$2\,\text{MGDG} \rightarrow \text{DGDG} + \text{DAG}$$

Analysis of the positional distribution of fatty acids in monogalactosyldiacylglycerols

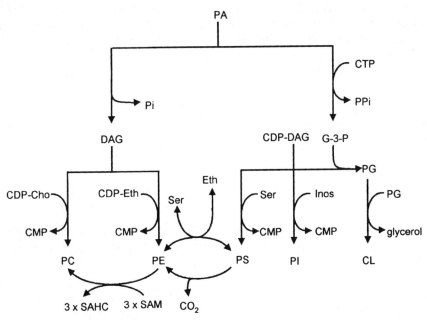

Figure 19.11 Summary of the synthesis of phospholipids. PA, phosphatidyl-choline; PE, phosphatidylethanolamine; PS, phosphatidylserine; PI, phosphatidylinositol; PG, phosphatidylglycerol; DAG, diacylglycerol; CTP, cytidine triphosphate; CDP, cytidine diphosphate; SAM, S-adenosyl methionine; SAHC, S-adenosyl homocysteine; CDP–Cho, CDP–choline; CDP–Eth, CDP–ethanolamine; Eth, ethanolamine; Ser, serine; CL, cardiolipin; Inos, inositol.

from a wide variety of plants has revealed that there are two distinct groupings of plants. One group, typified by the bryophytes, pterido-phytes, gymnosperms, and some angiosperms of the spinach, potato and brassica families, contains an unusual fatty acid, hexadeca-trienoic acid (C16:3), at the 2 position of the MGDG. These plants are referred to as C16:3 galactolipid plants. In more advanced angiosperms such as the Fabaceae, Asteraceae and Poaceae, C16:3 is virtually absent and is replaced by linolenic acid, C18:3. These plants are referred to as C18:3 galactolipid plants.

Synthesis of C16:3 galactolipids takes place entirely in the chloroplasts. This pathway results in the initial production of diacylglyc-erols with C16:0 at the 2 position and C18:1 in the 1 position. Following addition of galactose at the 3 position to produce MGDG, the fatty acids undergo desaturation to produce MGDG containing either C18:3 at position 1 and C16:0 at position 2, or C18:1 at position 1 and C16:3 at position 2.

In contrast, synthesis of the C18:3 galac-tolipids involves both the endoplasmic reticu-lum and the chloroplast. It utilizes the pathway of phosphatidylcholine synthesis. On the endoplasmic reticulum, DAG containing C18:1 in positions 1 and 2 is converted to phos-phatidylcholine, and the fatty acids undergo desaturation to linoleic acid (C18:2). The phos-phatidylcholine is then transferred to the chloroplast where the phosphocholine residue is removed to yield a 1-linoleyl-2-linoleyl dia-cylglycerol from which MGDG is synthesized. The MGDG fatty acids are desaturated to C18:3 prior to transfer of a second galactose residue to produce DGDG.

Interestingly, the sulphur-containing gly-colipid of chloroplasts, sulphoquinovosyl dia-cylglycerol, when synthesized entirely in the chloroplast (the prokaryotic pathway), contains very little C16:3 in the 2 position and usually has C18:3 in the 1 position. When synthesized from the 1-linoleyl,-2-linoleyl diacylglycerol pool originating on the endoplasmic reticulum (the eukaryotic pathway), it contains two C18:3 groups.

19.11 SYNTHESIS OF SPHINGOLIPIDS

Sphingolipids are found in both plants and animals, but are quantitatively more important in the latter. The synthesis of sphingolipids starts with the condensation of palmitoyl-CoA with serine to produce 3-ketosphinganine, which undergoes hydrogenation at the 3-keto group to yield sphinganine (also referred to as dihydrosphingosine).

This is followed by the addition of a fatty acid via the amino group of sphinganine and the conversion of sphinganine to sphingosine by the removal of two hydrogens. This produces the simplest type of sphingolipid, referred to as a ceramide (Figure 19.12).

Ceramides can undergo transformation to the four main types of sphingolipids:

- sphingomyelins
- cerebrosides (also called monoglycosyl ceramides)
- neutral glycosphingolipids (polyglycosylated ceramides)
- gangliosides (polyglycosylated ceramides containing *N*-acetylneuraminic acid).

The conversion to sphingomyelins is brought about by the transfer of phosphocholine from phosphatidylcholine to the free hydroxyl group on carbon 1 of sphingosine. Cerebrosides are formed by the transfer of a

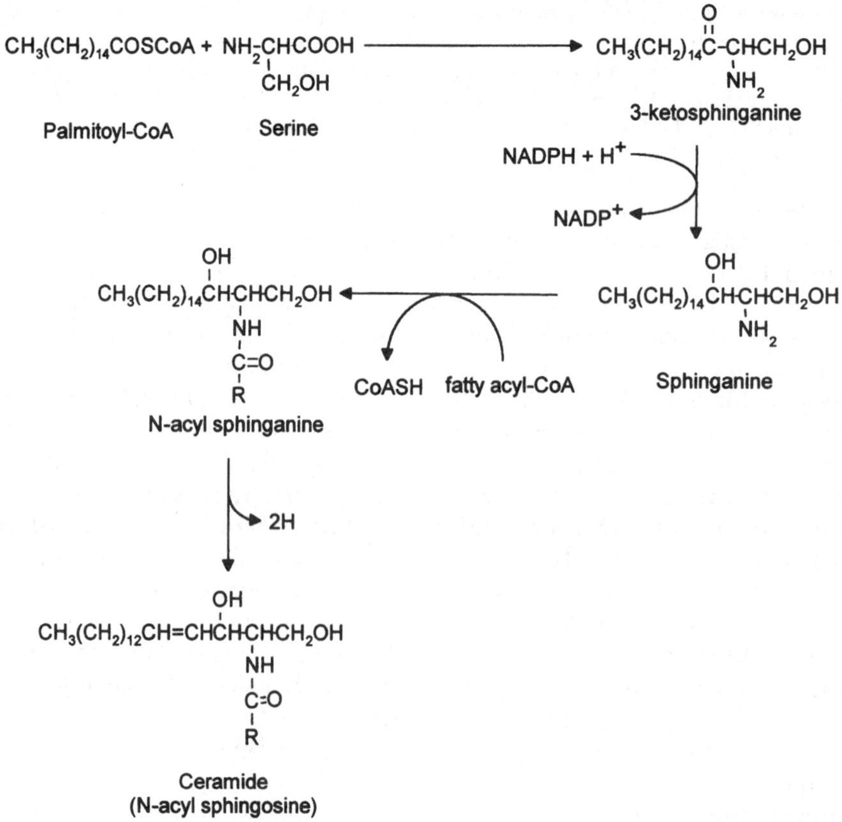

Figure 19.12 Synthesis of ceramides.

sugar residue to the hydroxyl on carbon 1 of sphingosine from a UDP-sugar. The most common sugars found in cerebrosides are galactose and glucose. Neutral glycosphingolipids are synthesized from cerebrosides and may contain six or more sugar units, mainly galactose, glucose or N-acetylgalactosamine. Gangliosides are the most complex form of sphingolipids: they are characterized by the presence of N-acetylneuraminic acid (sialic acid) as the terminal sugar unit (Figure 19.13).

It is now clear that sphingolipids are important in the process of cell recognition. The specific sphingolipid composition of the plasma membrane contributes to the cell's chemical fingerprint, and changes in membrane sphingolipid composition during tissue growth are thought to be related to regulation of cell-to-cell contacts. It has been shown that sphingolipid composition of blood cells is one of the determinants of blood group. Genetic defects in sphingolipid metabolism can have fatal consequences. Two genetically inherited disorders of sphingolipid metabolism, Tay–Sachs disease and Niemann–Pick disease, result in early infant death.

19.12 BIOSYNTHESIS OF TERPENES AND STEROLS

Squalene is a linear triterpene from which plant and animal sterols are derived. The various stages of its synthesis, which occurs in the cytoplasm, involve a number of terpenoid intermediates and are typical of the reactions undergone during the synthesis of many of the higher terpenoids. This section outlines the reactions involved in the production of squalene from acetate and its subsequent conversion to plant and animal sterols.

19.12.1 SYNTHESIS OF MEVALONIC ACID

Experiments have shown that all the carbon atoms in the carbon skeleton of squalene are derived from acetate. The initial stages of squalene synthesis involve the condensation of two molecules of acetyl-CoA to yield acetoacetyl-CoA, followed by the addition of a third molecule of acetyl-CoA to give hydroxymethylglutaryl-CoA (HMG-CoA) (Figure 19.14).

The synthesis of HMG-CoA is also the first stage in the synthesis of ketone bodies from

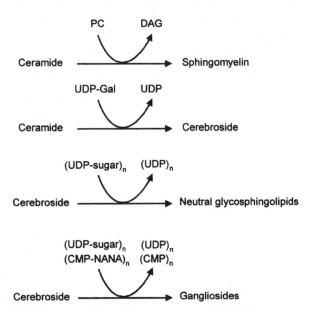

Figure 19.13 Conversion of ceramides to sphingomyelin, cerebrosides, neutral glycosphingolipids and gangliosides.

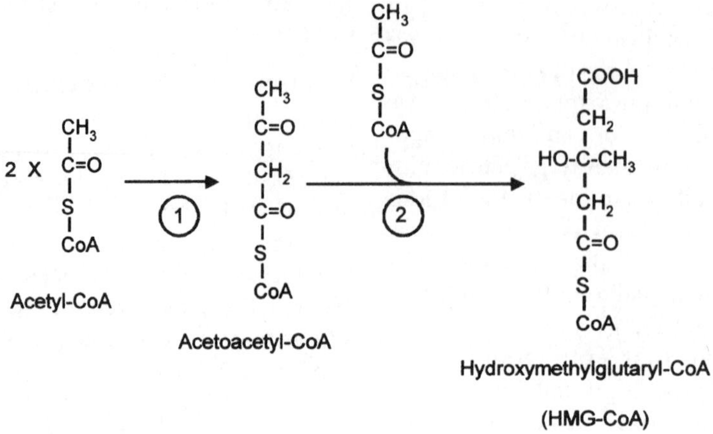

Figure 19.14 Synthesis of hydroxymethylglutaryl-CoA (HMG-CoA).

acetate in liver mitochondria (Chapter 13). HMG-CoA synthetase catalyses step 2 in both processes, and is therefore found in both the mitochondria and the cytoplasm. However, the mitochondrial and cytoplasmic synthetases differ in their catalytic properties and isoelectric points.

From this point the pathways of ketone body and squalene synthesis diverge. In the synthesis of ketone bodies, HMG-CoA is cleaved to yield acetoacetate and acetyl-CoA. The acetoacetyl-CoA may then be metabolized to β-hydroxybutyryl-CoA. In the synthesis of squalene, HMG-CoA undergoes conversion to mevalonic acid, a reaction in which the CoA group is lost, and the resulting carboxyl group is reduced to an alcohol using NADPH as a reducing agent. This reaction is irreversible

and is catalysed by the enzyme HMG-CoA reductase, which is thought to be rate-limiting in cholesterol synthesis in animals (Figure 19.15).

It has been shown that cholesterol synthesis is suppressed by increased consumption of cholesterol. There is evidence, at least in the rat, that HMG-CoA reductase shows changes in activity related to the pattern of food consumption. This is thought to be a response to the loss of free cholesterol and its derivatives such as bile acids. The enzyme has a relatively short half-life (2–4 h) and alterations in its activity are due mainly to changes in its rate of synthesis. In addition, HMG-CoA reductase activity is also regulated by covalent modification via a phosphorylation/dephosphorylation mechanism.

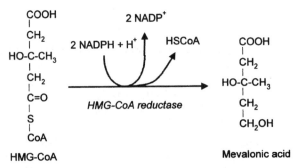

Figure 19.15 Synthesis of mevalonic acid.

19.12.2 CONVERSION OF MEVALONIC ACID TO SQUALENE

Mevalonic acid is first phosphorylated to produce mevalonate-5-phosphate, a reaction requiring ATP and catalysed by mevalonate kinase. This is followed by the addition of a second phosphate group to yield mevalonate-5-pyrophosphate. This reaction is catalysed by phosphomevalonate kinase and also requires ATP. Another molecule of ATP is consumed in a complex reaction catalysed by pyrophosphomevalonate decarboxylase, in which mevalonate-5-pyrophosphate is converted to isopentenyl pyrophosphate. Some isopentenyl pyrophosphate undergoes isomerization to form dimethylallyl pyrophosphate (Figure 19.16).

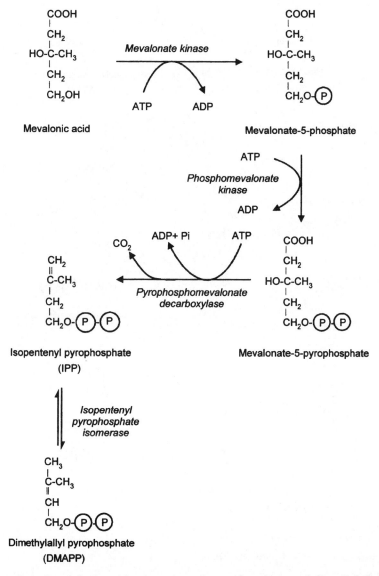

Figure 19.16 Conversion of mevalonic acid to isopentenyl pyrophosphate and dimethylallyl pyrophosphate.

The synthesis of squalene continues with the condensation of isopentenyl pyrophosphate (IPP) and dimethylallyl pyrophosphate (DMAPP) to form geranyl pyrophosphate. This in turn condenses with another molecule of IPP to yield farnesyl pyrophosphate. The enzyme responsible for both of these condensations is prenyl transferase. These reactions are examples of 'head-to-tail' condensations, the nomenclature referring to the methylene or methyl end of IPP and DMAPP as the 'tail' end and the pyrophosphate end as the 'head' end (Figure 19.17).

Squalene is synthesized by the subsequent 'head-to-head' condensation of two molecules of farnesyl pyrophosphate. This reaction occurs

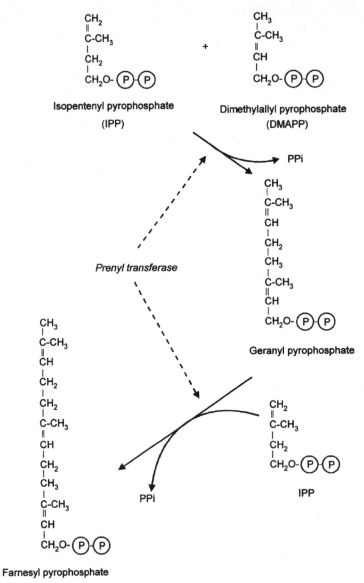

Figure 19.17 Synthesis of farnesyl pyrophosphate.

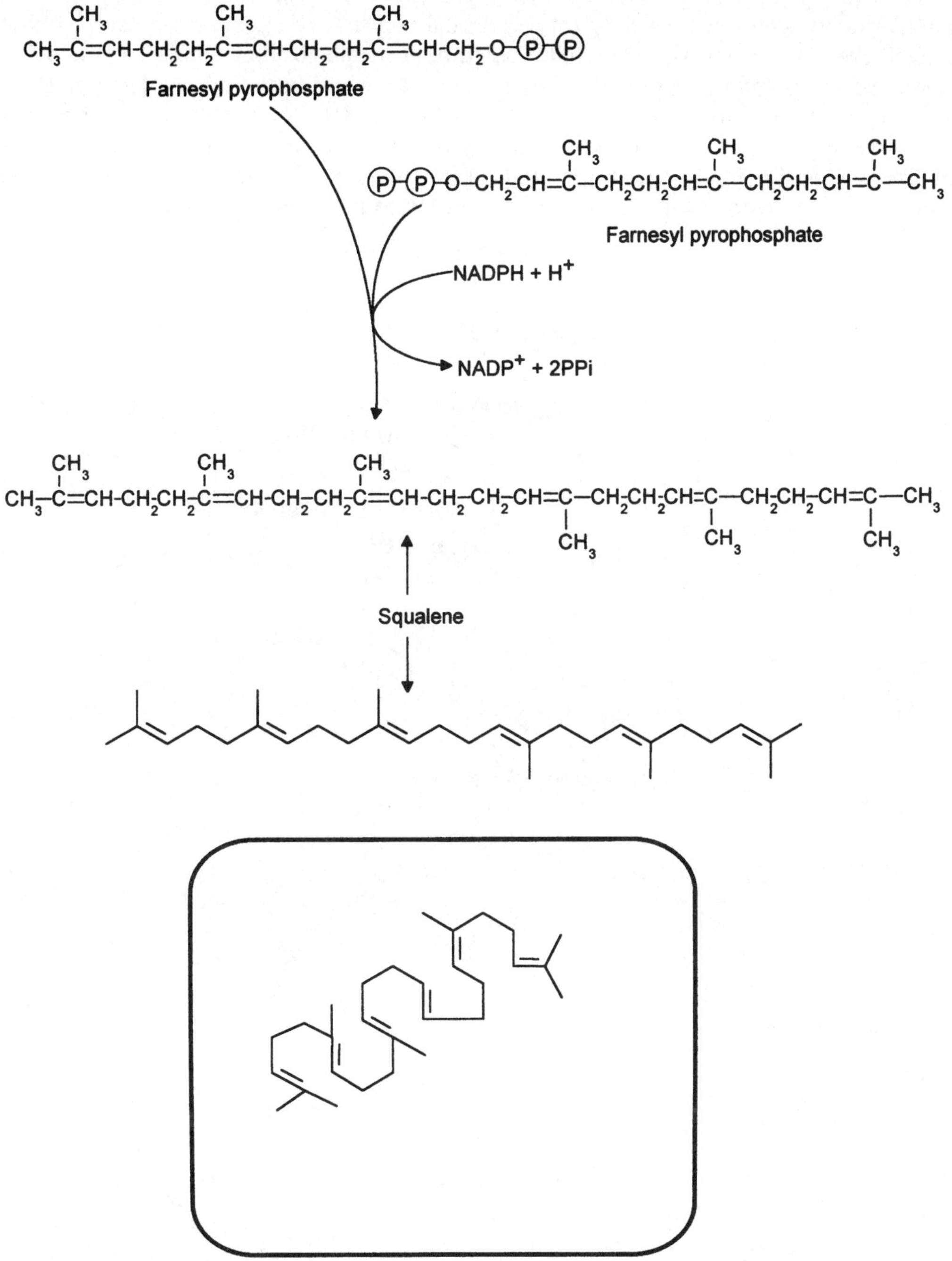

Figure 19.18 Condensation of two molecules of farnesyl pyrophosphate to produce squalene. Inset: alternative structure of squalene showing its similarity to the steroid nucleus.

in two steps, catalysed by the presqualene synthase and squalene synthase (Figure 19.18).

Squalene then undergoes cyclization to form lanosterol in animals and fungi, or cycloartenol in plants and algae. The process involves a two-step reaction in which squalene is first converted to squalene-2,3-epoxide in a reaction catalysed by squalene monooxygenase. Subsequent cyclization to yield lanosterol or cycloartenol is catalysed by cyclase enzymes. Cycloartenol is further metabolized to stigmasterol, and lanosterol to ergosterol and cholesterol (Figure 19.19). The detail of these transformations is still unknown, but that of cholesterol synthesis is thought to involve about 20 reactions located on the endoplasmic reticulum.

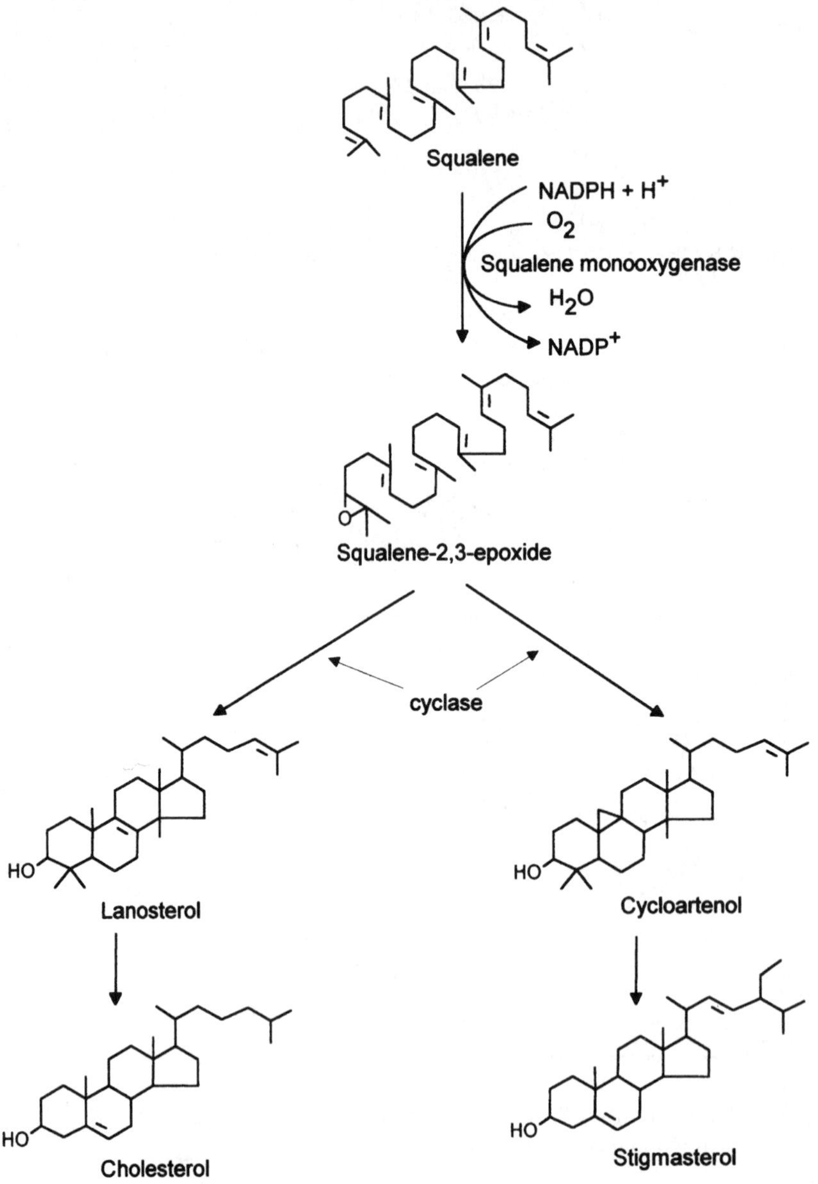

Figure 19.19 Cyclization of squalene to form sterols.

In addition to its function in cell membranes, cholesterol is the precursor of a number of important compounds, including cholecalciferol (vitamin D$_3$) which is involved in the regulation of calcium metabolism and bone growth; the bile acids, essential for efficient digestion and absorption of dietary lipids; and the steroid hormones, which regulate many aspects of growth, development and metabolism.

In plants, geranyl pyrophosphate, farnesyl pyrophosphate and geranylgeranyl pyrophosphate (produced by the condensation of farnesyl pyrophosphate with isopentenyl pyrophosphate) are branch points for the synthesis of a number of non-sterol terpenes. For example the latter provides the carbon backbone for the synthesis of the gibberellins, the carotenoids, the tocopherols and tocotrienols, vitamins K$_1$ and K$_2$ and the phytol side chain of chlorophyll.

20.1 INTRODUCTION

Plants and microorganisms can synthesize amino acids from simple nitrogen compounds such as nitrate or ammonia, or from those released during protein turnover. Animals cannot form amino acids from inorganic nitrogen compounds, but obtain them from protein in the diet or from the breakdown of body proteins.

In addition to the 20 amino acids found in proteins, some plants have pathways which enable them to make other, non-protein amino acids. Some of these are toxic or are of commercial significance (Chapter 5).

Nitrogen is available to plants mainly as nitrate (NO_3^-), ammonia as the ammonium ion (NH_4^+) and nitrogen gas (N_2).

NO_3^- and NH_4^+ are readily available to plants from the soil, but although 78% of the atmosphere consists of nitrogen gas, it is available only after reduction by prokaryotic microorganisms, either free-living or in symbiotic association in root nodules.

Organic nitrogen-containing compounds are rarely available and are not much used by green plants. However, fungi are more dependent on them, and may be able to supply nutrients to higher plants through their mycorrhizal symbiotic associations with roots.

The main sources of nitrogen for non-leguminous plants are NO_3^- or NH_4^+ from the soil. These are mainly derived from fixation of atmospheric nitrogen gas by chemical means or by free-living bacteria and blue-green algae in the soil, or by these organisms in symbiotic associations with higher plants. In most soils, nitrate is more abundant than ammonia because NH_4^+ is readily oxidized to NO_3^- by nitrifying bacteria. In some grassland and forest soils, where nitrification is inhibited by low pH or phenolic compounds in the soil, NH_4^+ may be the major form. Both can be used by the plant but each probably interferes with the utilization of the other, thus the presence of NH_4^+ may inhibit uptake and metabolism of NO_3^-.

20.2 ASSIMILATION OF NITRATE

Any nitrate taken up by plants from the soil must be converted into ammonia before it can be used to make organic nitrogen-containing materials. This process takes place in two stages:

- conversion of nitrate (NO_3^-) to nitrite (NO_2^-) catalysed by the enzyme nitrate reductase;
- conversion of nitrite to ammonia (NH_4^+) catalysed by the enzyme nitrite reductase.

Reduction of NO_3^- and NO_2^- may take place predominantly in the roots or shoots, depending on the species and the availability of nitrogen (see Chapter 26).

Nitrate reductase is a complex enzyme which is located in the cytoplasm. It contains FAD, haem and molybdenum. The conversion of nitrate to nitrite requires a supply of a reducing agent, which is usually NADH but may sometimes be NADPH. The reaction takes place as shown in Figure 20.1a.

Molybdenum forms an essential part of the enzyme. Plants deficient in molybdenum cannot utilize NO_3^- even if it is present in the soil. The heavy metal tungsten can be incorporated into the enzyme in place of molybdenum, but the enzyme formed in this way is inactive.

NO_2^- is very toxic to plants but, because the activity of nitrite reductase is normally higher than that of nitrate reductase, nitrite normally does not accumulate because it is rapidly converted into NH_4^+.

NO_2^- formed by the action of nitrate reductase moves into the chloroplasts in the leaf or the proplastids in the roots, and is reduced in these organelles by nitrite reductase. In chloroplasts, reduced ferredoxin (FdH_2) formed by light-driven electron transport is used as reducing agent. In roots, it appears that NADPH, formed by oxidation of sugars by the pentose phosphate pathway, may be used (Figure 20.1b).

Both nitrate and nitrite reductase are inducible enzymes. Although such enzymes

a

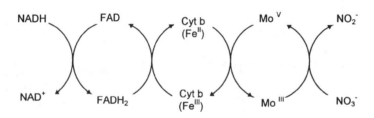

b

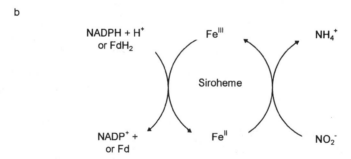

Figure 20.1 Mechanism of (a) nitrate reductase and (b) nitrite reductase. Nitrate reductase normally uses NADH as a reducing agent, but nitrite reductase in different tissues may use NADPH or reduced ferredoxin.

are common in prokaryotic organisms they are unusual in eukaryotes. The activity of nitrate reductase often limits the rate of protein synthesis. Activity of this enzyme is controlled by regulation of the relative rates of its synthesis and degradation. Light and high concentrations of NO_3^- in the cytoplasm both increase activity of the enzyme, whereas its activity is decreased by NH_4^+. The activity of nitrite reductase is also increased under these circumstances.

20.3 NITROGEN FIXATION

Vast quantities of nitrogen are present in the atmosphere in the form of N_2 but this molecule is extremely stable and its nitrogen atoms are available to living organisms only after being reduced to NH_4^+. This reduction is brought about through the process of nitrogen fixation, carried out either by prokaryotic organisms or through the industrial synthesis of fertilizers. Fixation of nitrogen cannot be achieved by eukaryotes functioning alone, although some higher plants obtain reduced nitrogen compounds by entering into symbiotic associations with prokaryotic nitrogen-fixing organisms by forming root nodules. Agriculturally the legumes are by far the most important class of plants of this type, but nitrogen fixation also occurs in a number of pioneer species of trees and shrubs over a wide range of genera. Nitrogen is also fixed by soil- and water-living bacteria, which make a considerable contribution to the nutrition of their environments from which other organisms benefit.

Nitrogen-fixing prokaryotes, whether present in root nodules or free-living, reduce nitrogen to ammonia through action of the enzyme nitrogenase which catalyses the following reaction:

$$N_2 + 16ATP + 8 \text{ electrons} + 8H^+ \rightarrow$$
$$2NH_3 + 16ADP + 16P_i + H_2 \qquad \text{Equation 20.1}$$

Production of hydrogen by reduction of protons appears to be required for binding of nitrogen to the enzyme and is an essential part of the mechanism. The reaction requires a large input of energy and a reducing agent.

Nitrogenase is a large and complex enzyme consisting of two types of protein.

- A small protein containing iron and consisting of two subunits, each of molecular weight about 30 kDa. This dimer contains four atoms of iron associated with four sulphur atoms. The protein binds ATP.
- A large protein containing 28 iron and two molybdenum atoms and consisting of 2α and 2β subunits. α subunits have a molecular weight about 56 kDa and β subunits about 60 kDa.

The Fe protein transfers electrons from the electron donor (reducing agent) to the Fe–Mo protein which actually reduces the nitrogen. The reducing agent is reduced ferredoxin or flavodoxin. These compounds are converted to their reduced forms by NADH or NADPH produced by metabolism of leaf-derived sugars in the nodules (Figure 20.2).

Both the Fe-protein and the Fe–Mo protein are denatured by oxygen, and nitrogenase must therefore be protected from oxygen in the root nodule. This is achieved through the anatomy of the nodule, and by the presence in the nodules of the protein leghaemoglobin. This protein has a high affinity for oxygen and is able to make oxygen available for respiration whilst preventing it from reaching the nitrogenase.

Nitrogen fixation in legumes requires very close integration of the activities of the plant and the bacteria. Thus the reducing agents needed for nitrogenase are derived from sugars formed in the leaves and transported in the phloem. In addition, removal and assimilation of the ammonia into organic nitrogen compounds also requires plant-supplied sugars. Many of the components of the nodule, such as leghaemoglobin, are also products of plant genes. On the other hand, nitrogenase is encoded in bacterial genes.

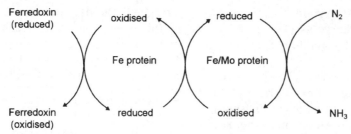

Figure 20.2 Mechanism of nitrogenase. Electrons are transfered from reduced ferredoxin or flavodoxin to an iron and then to an iron/molybdenum protein before reducing nitrogen gas.

20.3.1 MOLECULAR BIOLOGY OF NITROGEN FIXATION

There is a great deal of interest in the molecular biology of nitrogen fixation because of the possibility of increasing the efficiency of the process in legumes and of incorporating genes for nitrogen fixation into non-legumes. This would have the major benefit of reducing fertilizer requirements. Although the process is extremely complex, progress has been made in understanding the means of regulation.

In legumes, fixation takes place in the root nodules, which are formed when soil-living *Rhizobium* bacteria invade the root via root hairs. A particular *Rhizobium* species generally infects only one species of legume, and this specificity is controlled by bacterial *nod* genes, which are required for successful nodulation. A series of flavones, flavanones and isoflavanones, produced by different legumes, activate bacterial *nod* genes and form the basis of the specificity of the nodulation process. Bacterial products cause curling of root hairs and the start of the infection process.

Bacterial *nif* genes code for proteins responsible for the nitrogen fixation process itself. Thus, *nifH* encodes the Fe protein, *nifD* the α subunit, and *nifK* the β subunit of the Fe–Mo component of nitrogenase. Other *nif* genes control synthesis and processing of cofactors and electron donors. Another series of bacterial genes, the *fix* genes, may be associated with ferredoxin production.

Many plant genes are also required for effective nodulation. These include the nodule-specific gene products or nodulins, which are involved in nitrogen and carbon metabolism, nodule structure and regulation of oxygen concentrations (leghaemoglobin).

In addition to ammonia, nitrogenase also produces H_2 as an essential part of its mechanism of action (Equation 20.1). In some species of *Rhizobium* this gas leaks into the soil and the energy associated with it is lost, reducing the efficiency of the process. However in most *Rhizobium* species the H_2 is oxidized to H_2O by an uptake of hydrogenase which produces ATP in the process. Increasing the efficiency of the hydrogenase may thus decrease the energy demands of nitrogen fixation.

20.4 ASSIMILATION OF AMMONIA

Ammonia is very toxic to most cells as it appears to uncouple electron transport from ATP production in mitochondria and chloroplasts. Therefore, however it is produced, it must be rapidly converted into non-toxic organic compounds. In green plants ammonia is produced by nitrogen fixation, nitrite reduction, uptake from the soil and as a result of the operation of photorespiration (Chapter 17). It has been estimated that the quantities of ammonia produced from photorespiration may be as much as 20-fold greater than those produced by uptake of nitrogen from the soil.

The assimilation of ammonia is brought about in a somewhat indirect way through the action of two enzymes called glutamine synthetase (GS) and glutamate synthase

(GOGAT). The reactions are shown in Figure 20.3.

In chloroplasts, GOGAT uses reduced ferredoxin produced during the light reactions as

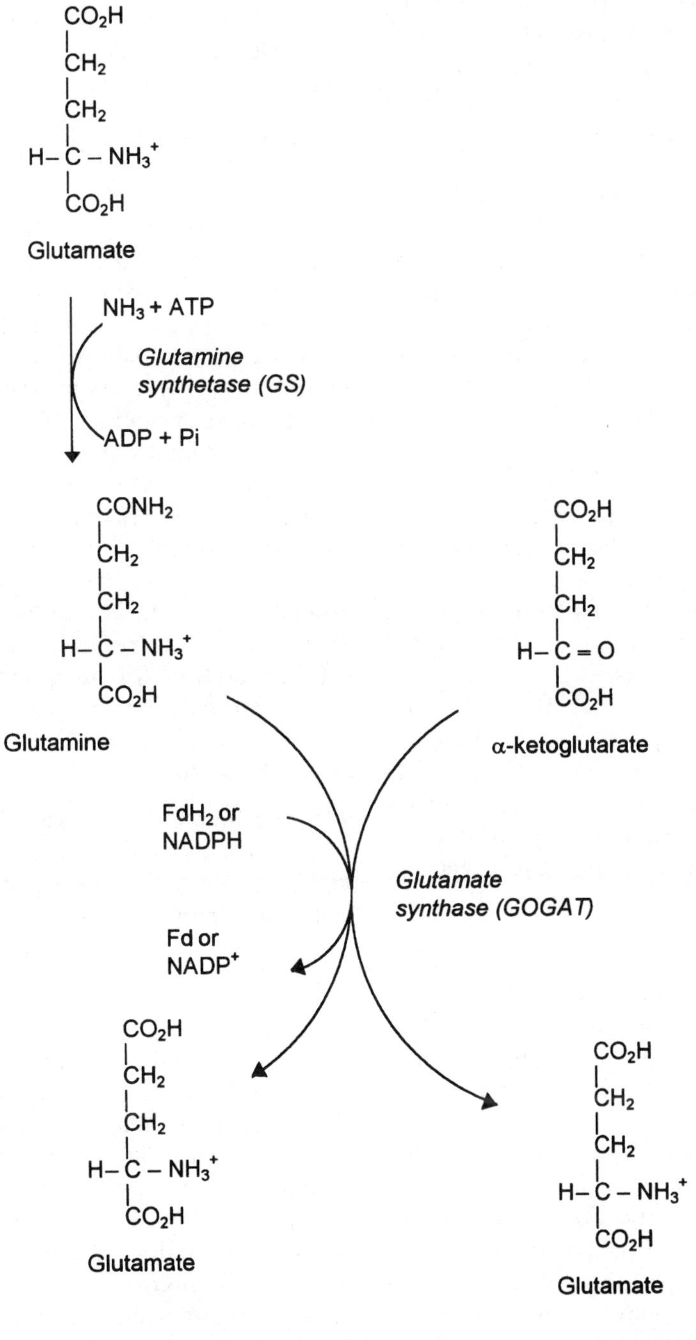

Figure 20.3 Reactions responsible for assimilation of ammonia by glutamine synthetase and glutamate synthase in plants.

the reducing agent, but in non-green tissues NADPH (or NADH) probably replaces FdH_2 as the reducing agent.

Ammonia can also be converted to glutamate by rumen microorganisms. This appears to take place by amination of α-ketoglutarate by ammonia. The process allows ammonia and other non-protein nitrogen sources to be used in the synthesis of microbial protein.

Glutamate may be further transformed into glutamine, asparagine and ureides. These compounds, together with glutamate, serve as the main forms in which nitrogen is transported around most plants. The transport of nitrogen in plants is discussed in more detail in Chapter 26.

20.5 BIOSYNTHESIS OF AMINO ACIDS

Although the formation of organic nitrogen compounds from NO_3^-, NH_4^+ or N_2 takes place only in plants and prokaryotes, pathways for the formation of amino acids from glutamate occur in all organisms, although some cannot synthesize all amino acids.

In general, amino acids are all synthesized in a similar way, by processes which are almost the reverse of the steps which lead to amino acid breakdown. Thus the carbon skeletons are constructed first from compounds such as pyruvate, TCA cycle intermediates or sugar phosphates, giving rise to a series of α-ketoacids, one corresponding to each amino acid. These are then transaminated to form the α-amino acids. The amino group donor is usually glutamate which is converted into α-ketoglutarate and which is therefore a key compound in amino acid metabolism.

The pathways used in the synthesis of the carbon skeletons of amino acids are usually branched so that members of an amino acid family may be formed from common intermediates. Some of the pathways involve only a few steps, but others may require 20 or more reactions. These reactions, being biosynthetic in nature, are usually reductions and require

energy. The pathways by which the amino acids are synthesized are summarized in Figure 20.4.

20.5.1 AROMATIC AMINO ACIDS AND RELATED COMPOUNDS

Aromatic amino acids and other aromatic compounds are synthesized by the shikimic acid pathway from phosphoenolpyruvate and erythrose-4-phosphate. The outline of the pathway is shown in Figure 20.5. Phosphoenolpyruvate (from glycolysis) and erythrose-4-phosphate (from the Calvin cycle or pentose phosphate pathway) react together to form the seven-carbon compound, 3-deoxy-D-arabinoheptulosonic acid-7-phosphate which cyclizes to dehydroquinic acid. Loss of water and reduction lead to shikimic acid. A second molecule of phosphoenolpyruvate reacts with 5-phosphoshikimic acid and serves as the source of the side-chain carbon atoms of the aromatic amino acids, resulting in production of chorismic acid. The aromatic amino acids are synthesized from this compound.

The pathway illustrates many of the general features of pathways by which amino acids are synthesized:

- transamination of α-ketoacids gives rise to amino acids as one of the last steps;
- ATP and NADPH are required;
- the starting materials are small compounds derived from the pathways of intermediary metabolism.

It is a relatively long pathway that is particularly important in plants, which contain many phenolic compounds. The pathway is absent from animals. Phenylalanine and tyrosine or intermediates in the pathway act as precursors of other phenolic compounds.

A particularly important reaction in plants is the deamination of phenylalanine to cinnamic acid (Figure 20.6), catalysed by the enzyme phenylalanine ammonia lyase (PAL). The activity of this enzyme is increased by light, and leads to increased production of

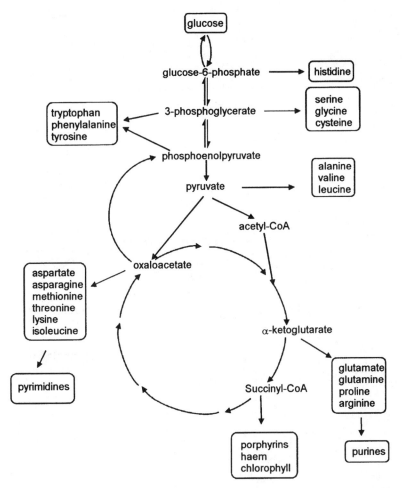

Figure 20.4 An outline of the pathways for the production of carbon skeletons required for amino acid synthesis. The starting materials are mostly intermediates of glycolysis or the TCA cycle.

pigments, lignin and other phenolic compounds. Anthocyanins contain three rings as shown in Figure 5.8. The carbon atoms of the B ring and central ring are formed from the shikimic acid pathway, and those of the A ring come from three molecules of acetyl-CoA.

20.5.2 BRANCHED-CHAIN ALIPHATIC AMINO ACIDS

The aliphatic amino acids, valine, leucine and isoleucine, are synthesized from pyruvate by a branched pathway which is shown in Figure 20.7.

The synthesis of valine begins with condensation of two molecules of pyruvate to form the five-carbon compound α-acetolactate. This is reduced, dehydrated and transaminated to form valine. Leucine arises from α-ketoisovalerate by a branch of this pathway. The synthesis of isoleucine is similar to that of valine, but begins by condensation of pyruvate with a molecule of α-ketobutyrate which contains an extra carbon atom. This then undergoes a series of transformations which are similar to

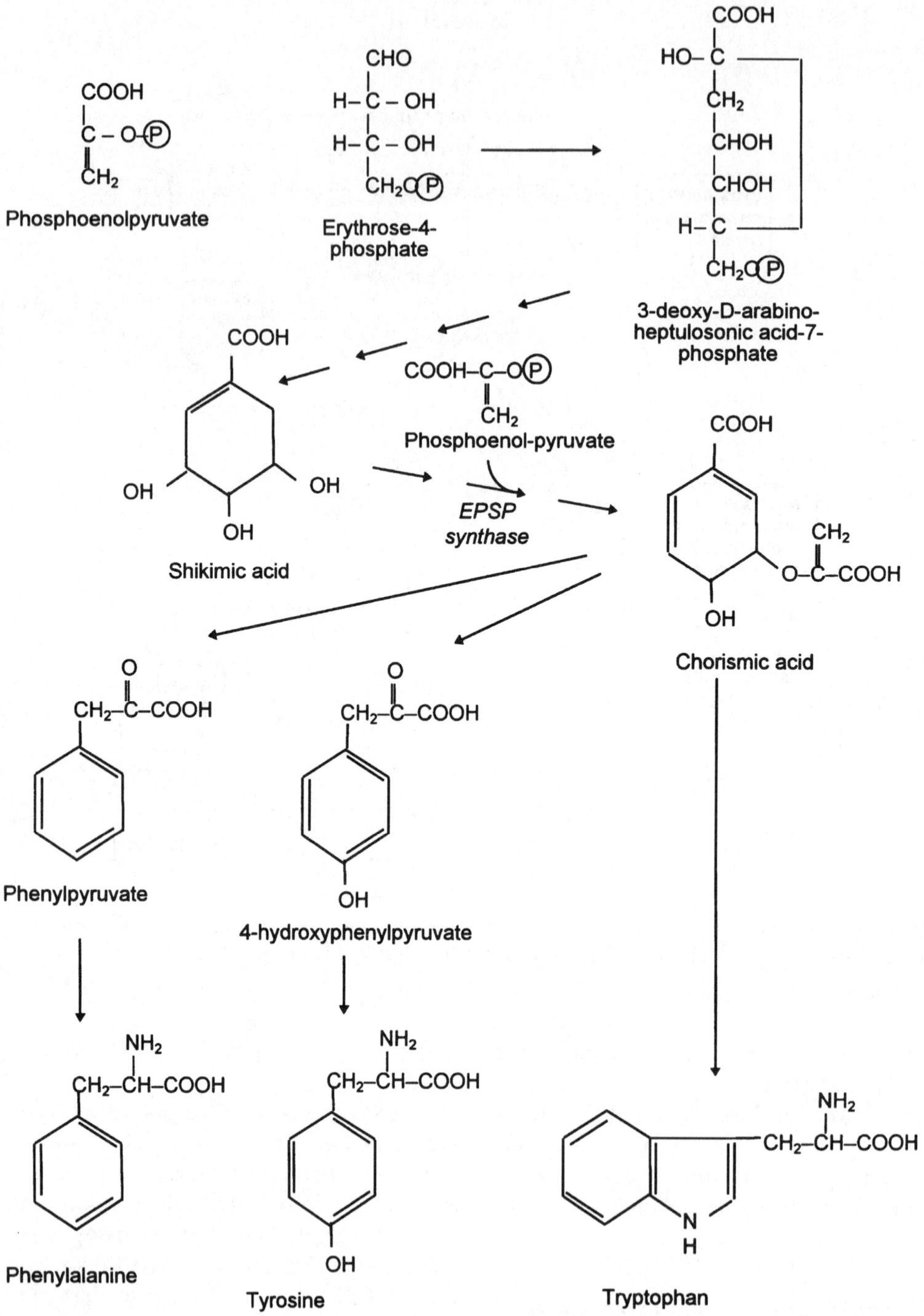

Figure 20.5 Pathway for biosynthesis of aromatic amino acids. The enzyme EPSP synthase is inhibited by the herbicide glyphosate.

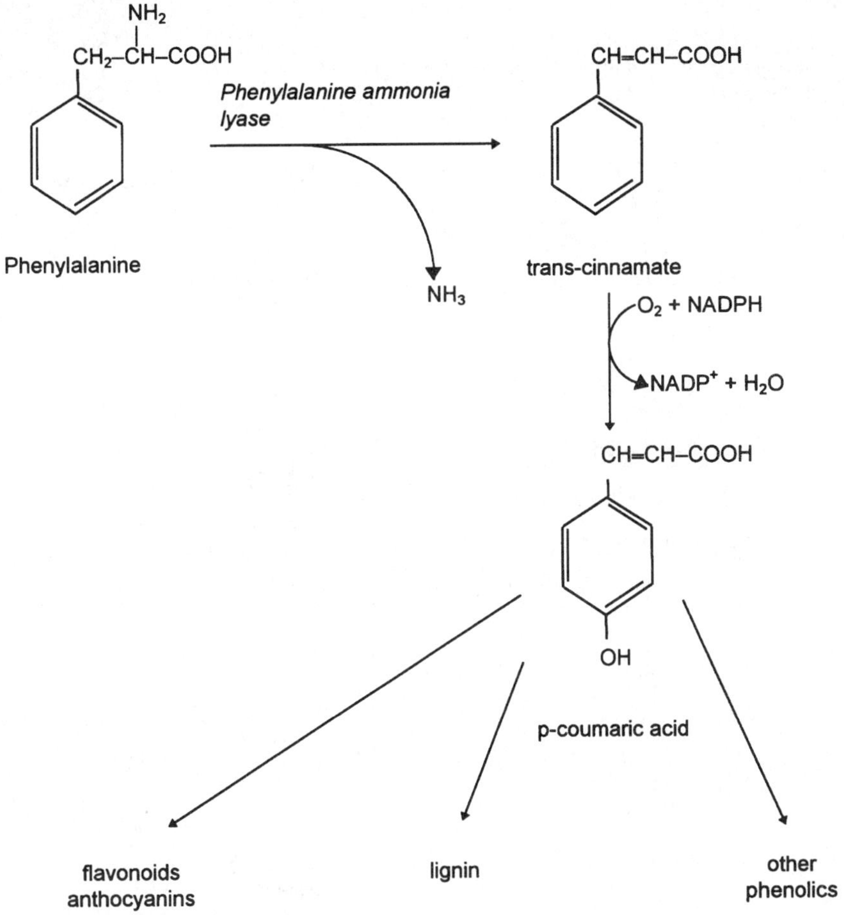

Figure 20.6 The reaction catalysed by the enzyme phenylalanine ammonia lyase (PAL). This is the first step in the pathway for the conversion of phenylalanine into a wide range of phenolic compounds in plants.

those in the formation of valine. Steps in the synthesis of isoleucine and corresponding steps in the synthesis of valine are catalysed by the same enzymes.

20.6 NUTRITIONAL ROLE OF AMINO ACIDS

Organisms differ in their ability to synthesize amino acids. Bacteria and plants have the enzymes needed to make glutamate from inorganic nitrogen compounds such as NH_3 or NO_3^- and a source of carbon. They can also normally use glutamate to make all of the other amino acids. Higher animals cannot make amino acids from inorganic nitrogen compounds and thus require amino acids (usually in the form of protein) in their diets. They can make about half of the 20 amino acids from other amino acids, but cannot make the remainder because they lack the enzymes required for their synthesis. These must therefore be supplied in the diet and are termed essential amino acids (Table 20.1). Often these are the amino acids which have the most complex biosynthetic pathways.

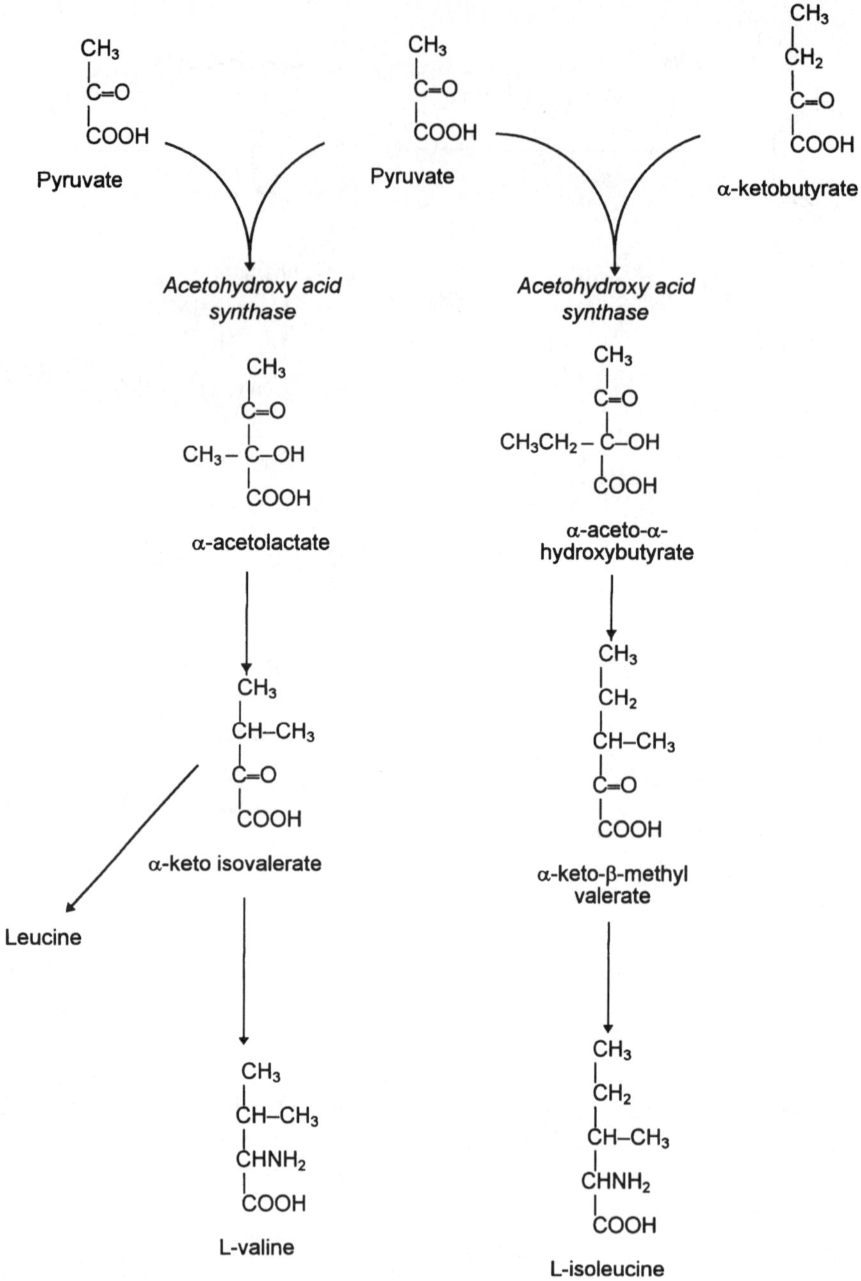

Figure 20.7 The pathways for synthesis of aliphatic amino acids. The enzyme acetohydroxy acid synthase is the site of action of the sulphonylurea herbicides.

A great deal of amino acid interconversion takes place to match the availability of amino acids to the requirements of protein synthesis. This involves:

Table 20.1 Essential and non-essential amino acids in the diet of non-ruminant animals

Essential	Non-essential
Arginine	Alanine
Histidine	Glycine
Isoleucine	Aspartic acid
Valine	Asparagine
Threonine	Glutamic acid
Methionine	Glutamine
Phenylalanine	Proline
Tryptophan	Hydroxyproline
Leucine	Serine
Lysine	Cysteine
	Cystine
	Tyrosine

- transamination of amino acids, conserving their nitrogen as the α-amino group of glutamate, and degrading the carbon skeletons;
- transferring the amino group of glutamate to a new range of α-ketoacids to form a new spectrum of amino acids.

This process takes place in all organisms. It permits them to make best use of the amino acids available in the diet, and it allows them to reprocess amino acids formed by protein turnover to match the changing demands of protein synthesis. In non-ruminant animals, however, the amino acids synthesized as a result of these interconversions are limited to the non-essential ones.

Although bacteria are capable of making all the amino acids, they often do not synthesize a particular amino acid if it is available in the environment. Normally the presence of the amino acid prevents the synthesis of all the enzymes required for its biosynthesis, by switching off their genes. This is an example of the process known as enzyme repression, which prevents wasteful synthesis of materials which are not required. This process is described in more detail in Chapter 21.

Non-ruminant animals must have an adequate supply of all the essential amino acids in the diet. Ruminants, however, contain in their rumen large numbers of microorganisms. These are able to make the whole range of amino acids from nitrogen compounds in the diet. These are then incorporated into the proteins of the microorganism. These amino acids are released, and become available to the animal as the microorganisms are degraded on passing through the digestive system. Thus there is no dietary requirement for essential amino acids for ruminants – provided that they receive an adequate supply of nitrogen-containing compounds in the diet they can obtain all of the amino acids from the microorganisms. Even so, under certain conditions such as high levels of milk production or rapid growth, certain amino acids may not be made sufficiently rapidly and would need to be supplied in the diet. The exact requirements of non-ruminant animals for dietary amino acids are rather complex. They depend on the species, its physiological state and other components of the diet. For example, the need for methionine can be reduced by supplying cysteine, and that for tyrosine by supplying phenylalanine as, in each case, the former can be made from the latter.

The synthesis of proteins requires a supply of all their constituent amino acids in the correct proportions. Rates of protein synthesis will therefore be determined by availability of the limiting amino acid, i.e. the one present in smallest quantity in relation to the demands of the cell. In general the balance of amino acids present in animal-derived foods is closer to the dietary requirements of animals than the balance in most plant-derived foods. Cereal-based diets tend to be deficient in lysine, methionine, threonine and tryptophan, so some or all of these need to be obtained from other components of the diet. In non-ruminant diets, lysine is usually the first limiting amino acid.

20.7 HERBICIDES AND AMINO ACID BIOSYNTHESIS

An interesting feature of the pathways of amino acid biosynthesis in plants is that they are the sites of action for a number of herbicides. A number of very potent herbicides inhibit specific steps in the synthesis of amino acids in plants. As the metabolic pathways affected do not take place in animals, these herbicides should be of intrinsically low animal toxicity. Herbicides which act in this way include:

- the sulphonylureas (e.g. chlorsulphuron) which inhibit acetohydroxy acid synthase, the enzyme which catalyses the initial step in the synthesis of aliphatic amino acids (Figure 20.7);

- glyphosate, which inhibits conversion of shikimic acid into chorismic acid catalysed by the enzyme 5-enolpyruvylshikimic acid-3-phosphate (EPSP) synthase, and therefore prevents synthesis of the aromatic amino acids (Figure 20.5).

These herbicides are non-selective and kill crop plants as well as weeds. Genetic engineers are interested in inserting genes into crops to make them resistant to these herbicides, as this would provide a very effective means of controlling the growth of all weeds in a crop.

THE SYNTHESIS OF NUCLEIC ACIDS AND PROTEINS 21

21.1 INTRODUCTION

DNA is synthesized as the chromosomes replicate prior to cell division. RNA is made continuously, although at varying rates as part of the protein synthesizing machinery of the cell. The synthesis of either type of nucleic acid requires a supply of the purine and pyrimidine nucleotides of which they are composed.

21.2 SYNTHESIS OF PURINE AND PYRIMIDINE NUCLEOTIDES

The structures of the purines and pyrimidines commonly found in nucleic acids are described in Chapter 7. The purines are synthesized as shown in Figure 21.1. Synthesis starts from ribose-5-phosphate which is converted to phosphoribosyl pyrophosphate (PRPP) by attachment of a pyrophosphate group derived from ATP. The purine ring structure is then gradually built up by reaction with glutamine, glycine, 10-formyl tetrahydrofolate, further glutamine, ATP, carbon dioxide and more formyl tetrahydrofolate. The product of the pathway is inosine-5'-monophosphate (IMP) in which the purine ring is already attached to the ribose sugar. Guanosine-5'-monophosphate (GMP) is formed from IMP by oxidation and reaction with glutamine, whilst AMP is produced by reaction with aspartate. The purine nucleoside monophosphates are converted to triphosphates by transfer of phosphates from ATP.

Pyrimidines are synthesized in an entirely different way, the ring being assembled before being linked to the sugar. Synthesis starts with formation of carbamoyl phosphate from ATP, carbon dioxide (as bicarbonate) and glutamine. Reaction with aspartate gives carbamoyl aspartate which cyclizes to form dihydroorotate and then orotate. This reacts with PRPP to form orotidine-5'-phosphate (OMP), which is decarboxylated to UMP (Figure 21.2). Reaction of UMP with ATP forms UTP, from which cytidine 5'-triphosphate (CTP) is synthesized by reaction with ATP and glutamine.

Deoxyribonucleotides required for DNA synthesis are made, in most organisms, by reduction of the corresponding ribonucleotide diphosphates using thioredoxin (a small, sulphur-containing protein) as the reducing agent. Formation of thymidine is brought about by methylation of dUMP, using 5,10-methylenetetrahydrofolate as the source of the methyl group.

21.3 REPLICATION OF DNA

The replication of double-stranded DNA takes place during the S-phase of the cell cycle by a semi-conservative mechanism. In this, the two strands separate and copies of each are made in such a way that each daughter cell inherits one of the original strands and a copy of the other DNA strand (Figure 21.3).

The first stage in DNA synthesis is separation of the DNA strands. Because of the complex manner in which they are twisted around one another, only a part of the chains separates at any one time and forms a replication fork or replicon. This is aided by the presence of unwinding enzymes called helicases.

DNA is synthesized by the enzyme DNA polymerase. There are several forms of this enzyme in both prokaryotes and eukaryotes. Some polymerases appear to be involved in DNA replication and others in DNA repair. DNA polymerase can only make DNA from the 5' to the 3' end. Because DNA is double-stranded and the chains run in opposite directions, it is not possible for both strands to be replicated simultaneously in the same direction. The new strand which needs to be made from the 5' to the 3' end (the leading strand) can be made continuously and in one piece. However the other chain (the lagging strand) must be replicated in short pieces, each of which is made in the 5' to 3' direction, and which are later joined together (Figure 21.4). These pieces are called Okazaki fragments after their discoverer. Okazaki fragments in bacteria may be up to about 2000 bases long, but those found in eukaryotes are usually only about 150 bases long.

DNA polymerase requires a short piece of 'primer' RNA, about 10 nucleotides long, made by an enzyme called primase on which is assembled the DNA. The RNA is later removed and replaced by DNA by another form of DNA polymerase.

Both chains of the DNA molecule are replicated at the same time. In prokaryotes this is

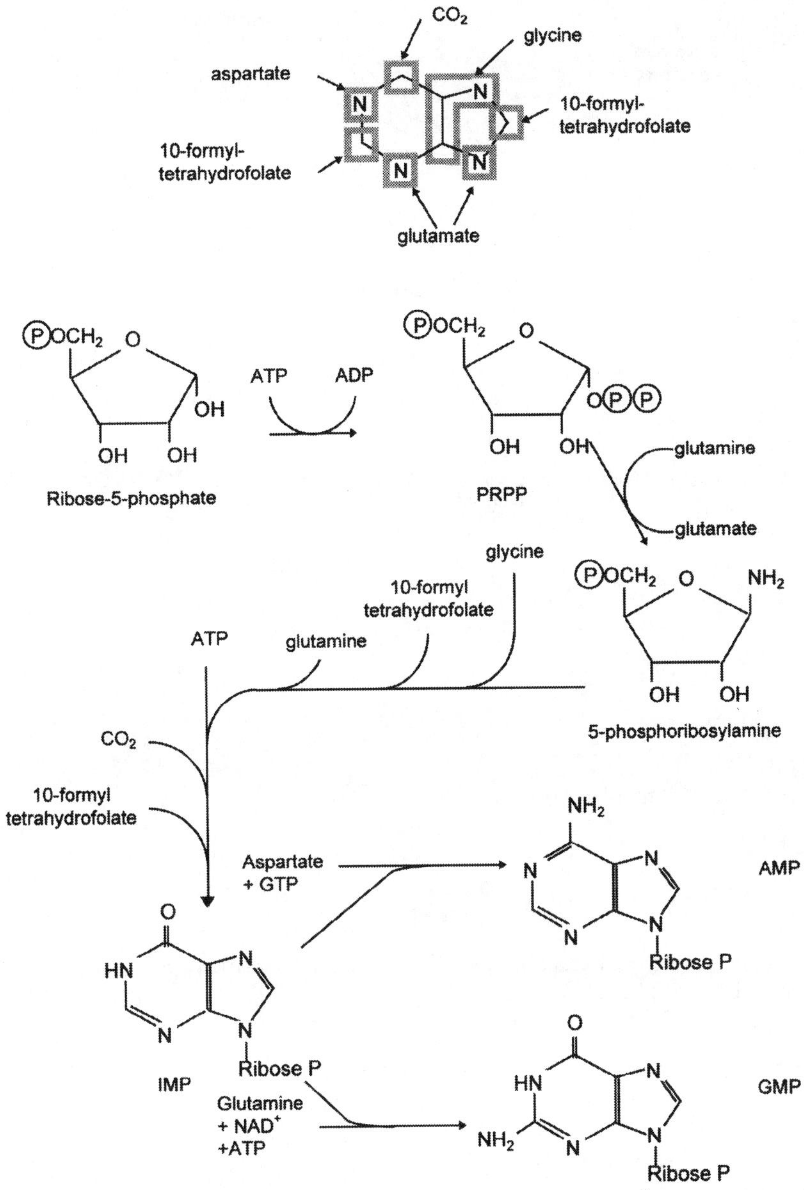

Figure 21.1 Pathway for the biosynthesis of purines. The origin of the atoms in the purine ring is indicated in the upper part of the figure with an outline of the pathway below.

done by a single DNA polymerase molecule with two active sites, but in eukaryotes a separate DNA polymerase catalyses formation of each chain. In prokaryotes the lagging strand may coil around so that, over the section which is being replicated, it actually runs through the enzyme in the same direction as the leading strand.

Once the sections of DNA at each replicon are complete they are joined together through

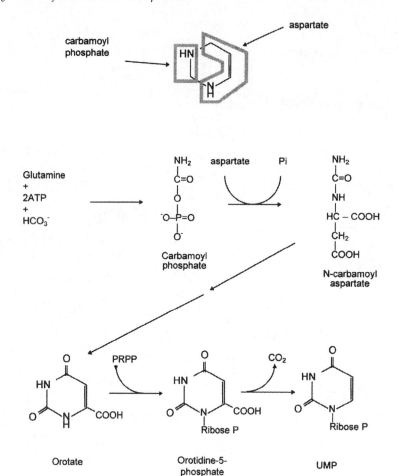

Figure 21.2 Pathway for the biosynthesis of pyrimidines. The atoms of the pyrimidine ring all originate from aspartate and carbamoyl phosphate.

the action of DNA ligase to form continuous DNA molecules.

In prokaryotes such as *Escherichia coli*, replication of DNA starts at one location and progresses in both directions (bi-directional). The origin, where replication begins, may be rich in AT base pairs which are relatively easily separated due to weak hydrogen bonding (Chapter 7). In contrast, in eukaryotes there appear to be multiple origins at which DNA synthesis may begin. There are therefore many replication forks.

DNA polymerase catalyses the following reaction:

$$dNTP + (dNMP)_n \rightarrow (dNMP)_{n+1} + PP_i$$

where dNTP and dNMP represent any nucleoside tri- or monophosphate, respectively.

After synthesis, some modification of the DNA may take place, e.g. some of the cytosine bases may be methylated, and this may play a role in regulation of the expression of some genes.

21.3.1 DNA SYNTHESIS IN VIRUSES

The genetic material in viruses can be single- or double-stranded DNA or RNA. Some virus-

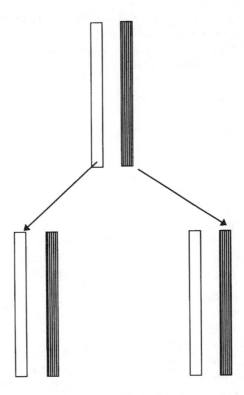

Figure 21.3 Replication of double-stranded DNA. Each daughter cell receives one of the original strands and a newly made strand which is complementary to this.

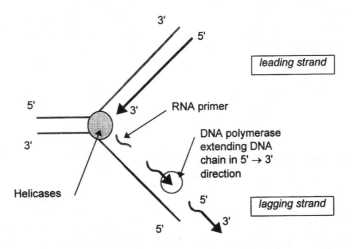

Figure 21.4 Replication of double-stranded DNA catalysed by DNA polymerase. The leading strand is synthesized in a continuous piece, but the lagging strand is made in short pieces which are subsequently joined together.

es carry the genetic information to code for the enzymes needed for replication of their DNA, but others rely on the host's enzymes for this purpose. Some viruses contain RNA as their genetic material, e.g. tobacco mosaic virus, influenza virus, polio virus. In certain types of RNA virus, e.g. some cancer-causing retroviruses and the AIDS virus, an enzyme called reverse transcriptase synthesizes DNA using the RNA as a template. The DNA produced is then incorporated into the host DNA, leading to synthesis of new virus particles. Another virus, hepatitis B, contains DNA as genetic material but uses RNA as an intermediate in its replication.

21.3.2 ACCURACY OF DNA SYNTHESIS

Several mechanisms ensure that the sequence of bases in DNA is replicated extremely faithfully. Some types of DNA polymerase have the capacity not only to add bases to the 3′ end of the growing chain (polymerase activity) but also to remove them (depolymerase activity). If an incorrect base is inserted it is unable to form a base pair with the template, and the depolymerase activity of the enzyme is increased so that the incorrect base is removed before further bases are added. This process is referred to as 'editing' or 'proof-reading'.

In both prokaryotes and eukaryotes, the main function of some DNA polymerases seems to be to repair DNA. In *E. coli*, Pol III appears to be the main enzyme catalysing synthesis of DNA, whilst Pol I seems to be concerned mainly with repair processes. There are also mechanisms for repair of damaged DNA resulting from exposure to UV light, ionizing radiation or chemical mutagens.

21.4 SYNTHESIS OF RNA

The sequence of amino acids in proteins is determined by the sequence of bases in mRNA, which in turn is determined by the sequence of bases in DNA. The information is arranged in such a way that three bases in DNA or mRNA determine the nature of one amino acid. This is therefore called a triplet code.

The flow of information is thus:

$$\text{DNA} \quad \rightarrow \quad \text{RNA} \quad \rightarrow \quad \text{Protein}$$

$$\text{transcription} \qquad \text{translation}$$

The transfer of information from DNA to RNA is known as transcription, and that from RNA to proteins as translation. As the message in DNA and in RNA exists in the same triplet form, it simply has to be 'transcribed' from one to the other. However, during protein synthesis the triplet sequence of bases has to be translated into the sequence of amino acids.

RNA is synthesized by RNA polymerases which catalyse the reaction:

$$\text{NTP} + (\text{NMP})_n \rightarrow (\text{NMP})_{n+1} + \text{PP}_i$$

RNA polymerase can make RNA only by starting at the 5′ end and moving towards the 3′ end. It does this by sequentially inserting bases which are complementary to those in one chain of the DNA. All types of RNA are made by this reaction, although different forms of RNA polymerase may be responsible for the formation of mRNA, tRNA and rRNA.

The RNA which is made by RNA polymerase is called the primary transcript and may undergo further processing before it reaches its final, functioning state. In eukaryotes this may involve removal of introns, addition or removal of other groups, or modification of some of the bases (see section 21.7.2). In bacteria, mRNA does not undergo further processing but precursors of tRNA and rRNA do.

Processing of mRNA

The product of RNA polymerase II is called heterogeneous nuclear RNA (hnRNA) because it contains molecules of a variety of sizes. This must be processed to yield mRNA proper.

Attachment of 3′ and 5′ caps

- 5′-end: the first-made 5′-end is 'capped' by attaching to it a special group. Capping occurs before transcription is complete. The cap is 7-methylguanosine triphosphate and the next residue also has a 2′ methyl group attached to the 2′ OH of its ribose group (Figure 21.5).
- 3′-end: after transcription is complete the enzyme polyA-polymerase adds 100–200 residues of adenylic acid (polyA) to the 3′ end of mRNA molecules. The polyA tail appears to mediate subsequent RNA processing and transport from the nucleus to the cytoplasm. No polyA sequence is added to the mRNA in prokaryotes.

Transcripts produced by RNA polymerase II are generally unstable and short-lived. The stability of mRNA may be related to the length of the polyA tail as older mRNA molecules have shorter tails.

Removal of introns (RNA splicing)

Introns are regions in mRNA which do not code for amino acids and which are removed or excized from the RNA before it is used in protein synthesis. Introns vary in size between about 100 and 10 000 nucleotides or more. There can be very wide differences in the sequence of bases present in different introns, but the few nucleotides at each end are nearly the same in all of them. The sequence has GU at the 5′ end and AG at the 3′ end. This region seems to be complementary to a sequence in the RNA contained in a particle called a small nuclear ribonuclear particle U1 (snRNP). These take part in RNA splicing by forming an RNA–RNA helix with the ends of the introns, so that they are brought together. They can then be cleaved and resealed (Figure 21.5).

Intron removal is not very species-specific. Thus the five introns in bean phaseolin RNA are correctly removed in tobacco. However, although yeast will carry out some intron excision this is often not the same as in higher eukaryotes; only about 5% of the total RNA transcribed leaves the nucleus.

21.5 THE GENETIC CODE

Information required for protein synthesis is carried to the ribosomes, where protein synthesis takes place, by mRNA molecules. A group of three adjacent bases in the mRNA determines the nature of each amino acid in the protein. Given that there are four bases – adenine (A), cytosine (C), guanine (G) and uracil (U) – there are 64 possible different combinations of three of these, each of which could code for one amino acid. For example the sequence AUG in mRNA will result in incorporation of a molecule of methionine into a protein. Not all triplets code for an amino acid, and the code is degenerate as more than one triplet may code for some amino acids. Where this is the case, these degenerate codons normally differ only in the last base (see Chapter 31).

The genetic code shown in Figure 21.6 is almost universal, although minor variations are found in some very simple organisms and in mitochondria.

21.6 PROTEIN SYNTHESIS

Protein synthesis is the assembly of amino acids into a linear sequence using mRNA as a template. The process can be divided into four parts: amino acid activation, initiation, elongation and termination.

21.6.1 AMINO ACID ACTIVATION

Before amino acids can be assembled into proteins, each must be attached to the corresponding tRNA molecule to form an aminoacyl tRNA. This takes place as shown in Figure 21.7. These reactions are catalysed by enzymes called amino acyl tRNA synthetases (or amino acid-activating enzymes). Activation takes place in two stages: first the amino acid reacts with ATP to form an aminoacyl adenylate,

(a)

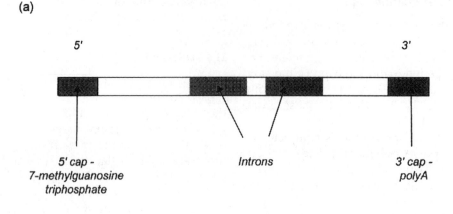

(b)

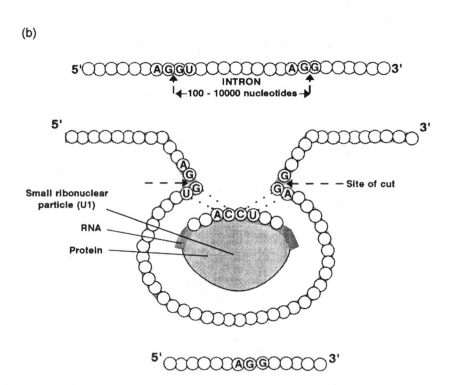

Figure 21.5 (a) The location of caps and introns in eukaryotic hnRNA. (b) Mechanism for removal of introns by snRNP.

which then reacts with the tRNA molecule to form the aminoacyl tRNA. Both reactions are catalysed by the same enzyme. There is at least one tRNA and one enzyme for each amino acid. Each amino acid becomes attached only to its own tRNA molecule (e.g. alanine to tRNA$_{ala}$). This tRNA molecule has an anticodon complementary to the triplet of bases in mRNA

1st position	2nd position				3rd position
	U	C	A	G	
U	Phe	Ser	Tyr	Cys	U
	Phe	Ser	Tyr	Cys	C
	Leu	Ser	STOP	STOP	A
	Leu	Ser	STOP	Trp	G
C	Leu	Pro	His	Arg	U
	Leu	Pro	His	Arg	C
	Leu	Pro	Gln	Arg	A
	Leu	Pro	Gln	Arg	G
A	Ile	Thr	Asn	Ser	U
	Ile	Thr	Asn	Ser	C
	Ile	Thr	Lys	Arg	A
	Met	Thr	Lys	Arg	G
G	Val	Ala	Asp	Gly	U
	Val	Ala	Asp	Gly	C
	Val	Ala	Glu	Gly	A
	Val	Ala	Glu	Gly	G

Figure 21.6 The genetic code. This code is almost universal in all organisms.

which codes for alanine. It is essential that attachment of the amino acids to the tRNA is extremely accurate as any errors here will lead to the insertion of an incorrect amino acid into the protein, with potentially disastrous consequences.

21.6.2 INITIATION

Proteins are synthesized starting from the N-(amino) terminal end (Chapter 5) with a derivative of methionine as the first amino acid in their sequence. In prokaryotes this is N-formyl methionine (f-MET) but in eukaryotes it is methionine itself (MET). In both cases the codon in the mRNA which codes for these amino acids is AUG. In eukaryotic cells, protein synthesis usually begins at the AUG nearest the 5' end of the mRNA. In prokaryotic cells, protein synthesis begins at an AUG sequence which is preceded by a purine-rich region 10 nucleotides away (the Shine–Dalgarno sequence) which binds to the small subunit of the ribosome (Figure 21.8).

Protein synthesis takes place on the ribosomes, which are small subcellular components present either free in the cytoplasm or attached to the membranes of the endoplasmic reticulum. In either case they consist of a small and a large subunit (see Chapter 7). Between the two subunits there is a groove into which a mRNA molecule fits. There are also two sites at which aminoacyl-tRNA molecules can be attached to the ribosomes. One of these is called the A site and the other the P site (Figure 21.9).

In the initial step in protein synthesis, a complex is formed between the mRNA, Met-

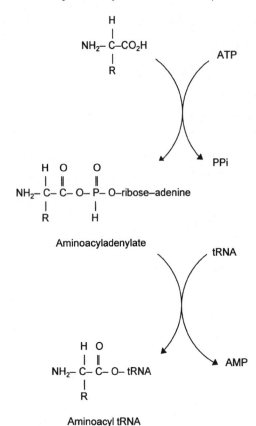

Figure 21.7 Formation of aminoacyl tRNAs. The two-step reaction takes place on the aminoacyl tRNA synthase enzymes and produces an aminoacyladenylate as an intermediate.

tRNA$_{met}$ (or f-Met-tRNA$_{f-met}$), several protein initiation factors, and the small subunit of the ribosome. The large subunit also attaches itself to complete the complex. The anticodon region of the tRNA molecule forms a base pair with the AUG codon of the mRNA and is held at the P site on the ribosome. Formation of this complex requires GTP.

21.6.3 ELONGATION

A second tRNA carrying the amino acid corresponding to the second codon on the mRNA attaches itself to the A site on the ribosome, and GTP is hydrolysed. This tRNA molecule base-pairs to the second codon in the mRNA. The enzyme peptidyl transferase (part of a large ribosomal subunit) now catalyses the formation of a peptide bond between the two amino acids which remain attached to the tRNA at the A site. The ribosome moves one triplet along the mRNA so that the growing peptide and its tRNA are at the P site and the A site is empty. GTP is needed for this process, which is called translocation (Figure 21.9).

21.6.4 TERMINATION

Elongation continues until one of the 'stop' codons is reached on the mRNA. These codons are UAA, UAG or UGA. When the ribosome reaches one of these, the bond holding the growing polypeptide to the tRNA is broken and the polypeptide is released. The tRNA and ribosomal subunits dissociate and are ready to start synthesizing another protein.

Because mRNA molecules are quite long, several ribosomes can synthesize proteins on each mRNA at one time. Under the electron microscope, complexes of this type have the appearance of beads on a string and are called polysomes. There can be one ribosome approximately every 80 bases along the mRNA, so about five ribosomes can operate at once on the mRNA for a protein such as haemoglobin (500 nucleotides). This speeds up protein synthesis considerably.

The equivalent of at least four ATP molecules are used in the synthesis of each peptide bond. In many cells, more energy is used in protein synthesis than in any other process.

21.6.5 POST-TRANSLATIONAL MODIFICATION OF PROTEINS

After release from the ribosome, protein molecules may be modified before they reach their final, 'mature' form. In many proteins, particularly in eukaryotes, some of the N-terminal residues are removed by hydrolysis after synthesis is complete. Thus, although all proteins are synthesized with Met or f-Met as the N-

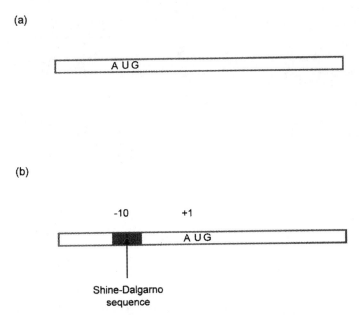

Figure 21.8 mRNA sequences which are involved in initiation of protein synthesis. (a) In eukaryotic cells the AUG nearest the 5′ end of mRNA serves as the start codon for synthesis of the protein. (b) In prokaryotic cells start AUG codons are preceded by the Shine–Dalgarno sequence, which is rich in purines and is complementary to part of the RNA in the ribosomes.

terminal amino acid, many mature proteins do not have these residues at their N-terminus. Sugar or phosphate residues may be added to proteins after they have been synthesized. Addition of sugar residues to form glycoproteins is particularly common in the case of proteins which will be secreted from the cell or incorporated into membranes. In addition, some amino acid side chains may be modified, e.g. proline and lysine may be hydroxylated to form hydroxyproline and hydroxylysine, respectively, as in collagen (see Chapter 30).

Proteins appear to be folded into their native conformation by a sequential mechanism which begins during their synthesis. Regularly repeating structures such as pleated sheet or α-helix are formed first, followed by organization of domains covering larger areas, and then finally combination of these domains to yield the native conformation. As the peptide chain folds, disulphide (–S–S–) bridges are formed between the –SH groups of nearby cys-

teine residues and act as solid 'anchor' points in the structure of proteins, stabilizing their 3-D shape. Inappropriate folding of proteins during synthesis may be prevented by formation of complexes with chaperone proteins.

21.6.6 LOCATION OF PROTEIN SYNTHESIS

The sequence of events described above leads to the release of proteins into the cytoplasm. Enzymes involved in glycolysis, gluconeogenesis, the pentose phosphate pathway, etc. are made in this way. In some cases, however, mechanisms are required to ensure that proteins are targeted to other parts of the cell, or even outside the cell.

Proteins which are to be secreted from the cell or packaged into protein bodies (see Chapter 23) are synthesized with short N-terminal pre-sequences or signal peptides of between 13 and 36 amino acids. The signal peptide is hydrophobic and usually has a basic

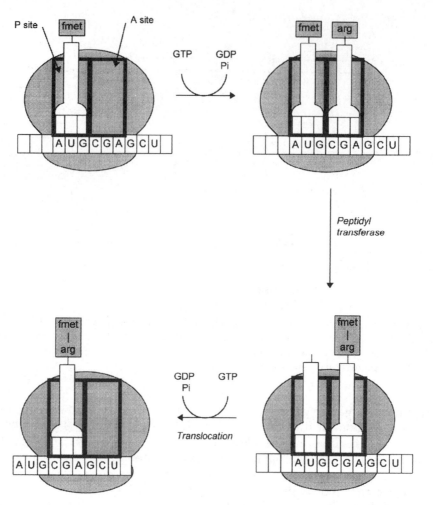

Figure 21.9 The steps concerned in protein synthesis in prokaryotes. In prokaryotes AUG signals the N-terminal amino acid which is f-Met. In eukaryotes, Met is inserted in the N-terminal position.

region near to its N-terminal end. As soon as the signal peptide has been synthesized it forms a complex with a signal recognition particle (SRP) which contains RNA and protein, and the complex binds to the membranes of the endoplasmic reticulum (ER). As the peptide grows it passes through the membranes into the lumen or sac of the reticulum, where it forms a complex with chaperone proteins which prevent it from folding. During transport through the ER membranes, the signal peptide is cleaved by signal peptidases on the lumen side of the membrane. Proteins made in this way are thus released into the lumen of the ER. They are then packaged by the Golgi bodies into vesicles which fuse with the plasma membrane, releasing their contents to the outside of the cell. Many proteins are made and secreted in this way, including those found in blood serum, milk, cell-wall proteins, glycoproteins on the cell surface, digestive enzymes and protein hormones. These are often glycoproteins, i.e. they have sugar residues attached. A further function of the ER membrane is to attach these carbohydrates, most commonly to asparagine residues in the protein.

Some proteins pass, after synthesis, to the mitochondria or chloroplasts. Such proteins are synthesized with pre-sequences called transit peptides at their N-terminal ends, but in this case they are synthesized by ribosomes which are free in the cytoplasm rather than attached to the ER. The transit peptides typically consist of 15–35 residues and are rich in basic and hydroxylated amino acids such as serine and threonine.

The information in these transit peptides is sufficient to target them accurately not just to the correct organelle, but to individual compartments within it, such as the mitochondrial matrix or the space between the chloroplast membranes.

Folded proteins are not easily transported through membranes, and folding is prevented by binding of the peptide to chaperone proteins both before and after transport across the membrane. This ensures that folding takes place only when the entire peptide has crossed the membrane.

All of the information needed to determine the destination of proteins in the cell is contained within the N-terminal peptides. It is thus possible to direct proteins to specific locations simply by adding suitable pre-sequences to their genes.

21.7 REGULATION OF PROTEIN SYNTHESIS

In prokaryotic organisms, protein synthesis is regulated mainly at the level of transcription, i.e. mRNA synthesis is controlled. Because mRNA synthesis and protein synthesis take place in the same cell compartment, the processes occur in rapid succession. Indeed, synthesis of proteins may even begin using incomplete mRNA molecules. There is therefore no time for complex modification of mRNA. In eukaryotes, on the other hand, in addition to transcriptional control, there is the opportunity for control at post-transcriptional

level. This, in combination with the much larger genome and the existence of differentiated cells, results in patterns of regulation which are extremely complex.

21.7.1 REGULATION IN PROKARYOTES

In *E. coli* and other bacteria, a single type of RNA polymerase is responsible for the synthesis of mRNA, tRNA and rRNA. Although made by the same enzyme they are synthesized at different rates, so that they are produced in the correct quantities according to the demands of the cell.

RNA polymerase begins to synthesize RNA at specific locations on the DNA molecules. These are recognized because of the existence of 'promoter sites' which lie upstream of the start of the coding sequence in the DNA. Two sites within this promoter region are important in controlling attachment of the RNA polymerase and initiation of RNA synthesis. These are centred 35 and 10 base pairs before the start of the coding sequence (positions -35 and -10, respectively). They determine where, how strongly, and how frequently RNA polymerase initiates transcription of the structural genes which they control.

The features of promoter sites which are common to different bacteria are summarized in Figure 21.10. At the -35 position the commonest sequence is TTGACA, and at -10 it is TATAAT. The latter is known as the Pribnow or TATA Box. Many promoters differ in one or more positions from these 'consensus' sequences, and generally the better the match the more efficient the promoter.

RNA polymerase consists of 'core' enzyme (itself composed of several subunits) and a protein subunit called σ^{70} which recognises promoter sequences. Variants of σ^{70} may replace it under certain circumstances, and each of these may recognise different promoter sequences, allowing transcription of different sets of genes. Sudden changes in temperature cause the production of different σ subunits which change the specificity of RNA

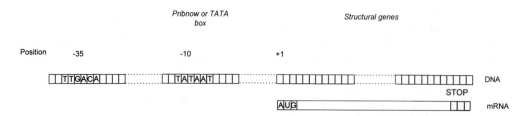

Figure 21.10 RNA polymerase binding sites to prokaryotic DNA. The structural genes beginning at position +1 are preceded by promoter regions around 10 and 35 base pairs upstream.

polymerase, resulting in the production of heat-shock proteins. Changes in σ subunits also occur in response to variation in nitrogen sources, activating genes required for nitrogen assimilation.

The rate at which the basic transcription process occurs can be modified in a great number of ways. Thus there may be 'attenuators', preceding structural genes which cause the synthesis of mRNA to be abandoned after only a short leader piece has been synthesized. These are common, for example, in the genes coding for enzymes involved in bacterial synthesis of amino acids.

Operons

In bacteria, the genes required for a metabolic pathway often occur in the form of an operon. This consists of the genes which code for the enzymes catalysing the process (structural genes) together with genes coding for products which regulate the expression of the structural genes. The lac operon of *E. coli* has been intensively studied and has provided insights into how many processes are regulated in bacteria.

The enzyme β-galactosidase catalyses the hydrolysis of sugars such as lactose, which contain β-galactoside bonds. It converts lactose into glucose and galactose as an essential first step in lactose metabolism. When *E. coli* is grown in the presence of glucose as a carbon source, the levels of this enzyme are very low (about five enzyme molecules per cell) as it is not needed. If the glucose is withdrawn and lactose is provided as a carbon source, β-galac-

tosidase must be made. The amount of the enzyme increases very rapidly so that within 1–2 minutes there may be 5000 molecules per cell. The activity of two other enzymes (β-galactoside permease and β-thiogalactoside transacetylase) also rise at the same time. The manner in which this operon functions is shown in Figure 21.11.

The three enzymes are coded by three structural genes which are adjacent to one another. All are transcribed to form a single piece of mRNA. An operator, which is located next to the structural genes, is able to bind tightly to a protein called the 'repressor protein'. When it does so, it inhibits binding of RNA polymerase and prevents transcription of the three structural genes.

When an inducer (in this case allolactose, a product of lactose metabolism) is present, it binds to the repressor protein and causes a change in conformation. This prevents it from binding to the operator gene, so that transcription of the structural genes can take place.

One operator controls the activity of several structural genes which are adjacent to it. When the operator is unblocked, the single piece of mRNA is produced and all three enzymes are synthesized. Enzymes which are synthesized only when needed, like those described here, are called inducible enzymes.

Other systems which are regulated in a similar manner also occur in bacteria. If *E. coli* is grown in a medium containing NH_4^+ and a carbon source it normally synthesizes the enzymes needed for synthesis of each of the 20 amino acids. However when one of the amino

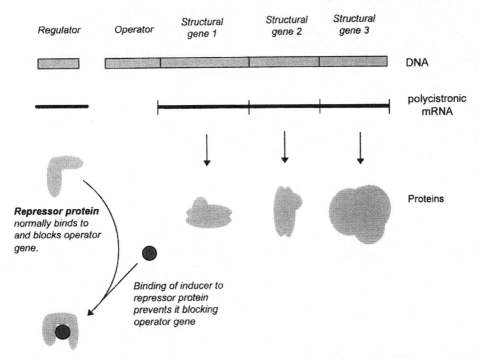

Figure 21.11 The *lac* operon in *Escherichia coli*. Transcription of the structural genes is blocked by binding of a repressor protein to the operator gene. Binding of an inducer to the repressor protein prevents its attachment to the operator, allowing the structural genes to be transcribed.

acids (e.g. histidine) is provided in the nutrient solution, all the enzymes needed for its biosynthesis disappear.

The regulation of this system is very similar to that of the lac operon, but in this case the repressor protein is only able to combine with the operator gene after binding to a co-repressor (histidine or a derivative). When histidine is present it switches off synthesis of all the structural genes in the operon. This is called 'enzyme repression'.

Enzyme induction and enzyme repression are very common in bacteria. These organisms encounter frequent changes in nutrient supply, and mechanisms such as these enable them to make efficient use of resources.

21.7.2 REGULATION IN EUKARYOTES

In eukaryotes, the genes are arranged on a series of chromosomes. Whereas genes for proteins with related functions (e.g. the enzymes catalysing steps in a metabolic pathway) tend to be closely clustered together on the DNA of prokaryotes, those in eukaryotes are often distributed widely amongst the chromosomes. Within chromosomes the DNA is tightly complexed with proteins which modify expression of the genes.

The specialized cell types in eukaryotes, e.g. liver, mammary and brain cells, differ from one another because they contain different proteins. Although all cells of an organism contain identical DNA, only a small part of the total DNA is expressed in any one cell. In different cell types, different parts of the DNA are active. Switching on or off individual genes forms the basis of the process of differentiation or specialization of cells in multicellular organisms.

This complexity means that the manner in which the activity of the eukaryotic genome is regulated is not nearly as well understood as in prokaryotes. However, the following are thought to be important in activating and inactivating genes:

- changes in the distribution and packing of nucleosomes;
- methylation of DNA bases;
- chemical modification (acetylation, methylation and phosphorylation) of histones associated with the DNA;
- changes in patterns of non-histone proteins.

In eukaryotes, RNA synthesis is catalysed by three different types of RNA polymerase. RNA polymerase I forms rRNA; RNA polymerase II forms large molecules of RNA which give rise to mRNA, and RNA polymerase III forms small RNA molecules such as tRNA and the small components of the ribosomes.

The genes in eukaryotes do not occur as operons. Each coding region has its own regulatory sequences. In addition, as discussed below, the coding sequences are often interrupted by non-coding sections which are removed during RNA processing.

There are specific promoter sequences upstream of the start of the coding region of eukaryotic genes. The best characterized of these is the TATA or Hogness box which is about 25 bases upstream of the transcriptional start. The consensus sequence of this is TATAAA. This sequence appears to direct RNA polymerase to the correct AUG codon at which to start transcription (Figure 21.12). Between about 40 and 110 bases upstream of the transcriptional start are found sequences which regulate the frequency of initiation of transcription of the gene. In animals there are commonly two of these, called the GC and CAAT boxes, but in plants the latter is often absent or replaced by alternatives (e.g. CATC box in cereals). Additional enhancer sequences which stimulate the activity of nearby promoters may also be found either up- or downstream of the coding sequences.

21.8 PROTEIN SYNTHESIS IN CHLOROPLASTS AND MITOCHONDRIA

Mitochondria and chloroplasts contain DNA of their own and can carry out protein synthesis. The DNA in both chloroplasts and mitochondria is roughly circular and double-stranded, and there are multiple copies of the DNA molecules. No histones are associated with the DNA in these organelles. In most animals, organelle DNA (mitochondrial) makes up less than 1% of the total DNA of the cell, but in green plant tissues organelle DNA (mitochondrial and chloroplast) can be up to 15% of the total DNA.

Although these organelles are able to synthesize their own proteins, only a few arise in this way. Most organelle proteins are encoded in the nuclear genome, made in the cytoplasm and then transported to the organelle. Details of the processes for which the organelle genomes are responsible are given in Table 21.1.

The genome of mitochondria codes for tRNAs, rRNAs and about 20 proteins, including some of the inner mitochondrial membrane (Table 21.1). Although very variable between species, there is much non-coding DNA. Mitochondrial mRNA has no 5'-caps and no extensive 3'-polyA tails, but it may contain introns. Protein synthesis in mitochondria resembles that in bacteria. Synthesis of proteins starts with f-Met, takes place on 70S ribosomes, and shows similar sensitivity to antibiotics.

The chloroplast genome is much more complex than that of the mitochondria: a large number of proteins are made in this organelle. The DNA exists as double-stranded, circular molecules. The synthesis of RNA in chloroplasts is thought to be very like that occurring in bacteria: there are no 5' caps or 3' polyA sequences, and very little processing of RNA takes place. Protein synthesis is initiated by f-Met and resembles bacterial protein synthesis even more closely than does that taking place in mitochondria. The ribosomes in chloroplasts are of the 70S type and are different from cytoplasmic ribosomes.

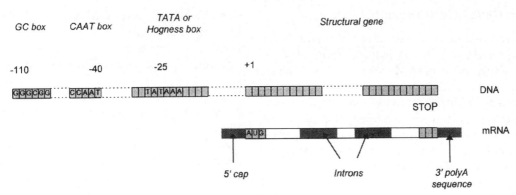

Figure 21.12 Promoter sequences in eukaryotes. Structural genes are preceded by promoter regions 25, 40 and 110 base pairs upstream.

In some cases, some subunits of complex proteins are coded by nuclear and some by chloroplast genes. Rubisco is the best known of such proteins. In this case the small sub-unit is coded for by the nucleus and the large one by the chloroplast, and they are assembled together in the chloroplast to produce active protein.

Table 21.1 Protein synthesis in organelles

Functions	Proteins synthesized
Mitochondrial functions coded by the mitochondrial genome	tRNAs
	rRNAs
	about 20–30 proteins: F_0ATPase (ATP synthesis); subunits of cytochrome c oxidase; cytochrome b; a ribosomal protein product giving rise to cytoplasmic male sterility
Mitochondrial functions coded by the nuclear genome	TCA cycle enzymes
	other proteins involved in electron transport and oxidative phosphorylation
	structural proteins
Chloroplast functions coded by the chloroplast genome	chloroplast tRNAs
	chloroplast rRNAs
	large subunit of Rubisco
	cytochromes f, b-559
	protein synthesis elongation factors
	electron transport coupling factors
	some subunits of the proton-translocating ATPase components of the P_{700}-chlorophyll–protein complex
Chloroplast functions controlled by nuclear genes	chlorophyll synthesis
	chloroplast carotenoids
	aminoacyl-tRNA synthetases
	70S ribosomal proteins
	subunits of the proton translocating ATPase
	small subunit of Rubisco

21.9 GENETIC ENGINEERING

In recent years a wide variety of methods for the investigation and manipulation of gene structure have been developed. Collectively, these have allowed a very rapid increase in understanding of many aspects of gene function and have provided the depth of knowledge necessary to begin to manipulate agricultural systems to increase production and quality. The success of these methods has depended on:

- procedures for identification and sequencing of genes;
- methods of synthesizing DNA of specific sequences;
- methods of producing recombinant DNA by cutting and reassembling the molecules;
- methods of modifying the nucleotide sequence of genes;
- vectors for transferring genes from one organism to another, e.g. viruses and plasmids.

A number of important concepts are outlined below as an introduction to applications of genetic engineering in agriculture.

21.9.1 ENZYMES USED IN DNA MANIPULATION

Restriction enzymes or restriction endonucleases are enzymes which cleave both strands of double-stranded DNA at specific sequences of nucleotides. They are found in many prokaryotes, where their function is to degrade the DNA of invading cells. The enzymes recognise symmetrical sequences of four to eight base pairs in the DNA (Figure 21.13). Restriction enzymes allow DNA molecules to be cleaved at specific sites to produce fragments of various sizes.

DNA ligases join pieces of DNA together into longer molecules, especially where the ends have single-stranded complementary termini, such as are produced by the action of restriction enzymes. DNA ligase, used in con-junction with restriction enzymes, allows genes to be linked to other DNA to produce recombinant products.

Reverse transcriptase catalyses the formation of complementary DNA (cDNA) using RNA as a template. This allows genes to be obtained from corresponding mRNA molecules.

DNA polymerase catalyses polymerization of deoxyribonucleoside triphosphates using a DNA template to produce a complementary copy.

21.9.2 ISOLATION AND SYNTHESIS OF DNA

The structure of genes can be determined by direct sequencing of DNA or indirectly from the sequence of bases in mRNA molecules.

DNA sequencing can be carried out by chemically cleaving DNA molecules at specific bases into fragments of a variety of sizes or, more commonly, by interrupting DNA synthesis, by addition of synthetic nucleotide analogues, to produce fragments of different lengths. Analysis by gel electrophoresis separates the fragments on the basis of their size and provides a simple visual representation of the nucleotide sequence in the DNA.

DNA can be synthesized chemically by condensation of individual deoxyribonucleoside triphosphates. Appropriate protecting groups and 3'-activating groups are required, and the growing chain is normally attached to a solid support to allow easy automation. In this way, synthesis of oligonucleotides up 100 units long is readily achieved. Such molecules are very useful probes which bind to specific complementary sequences within DNA. Binding of radioactively labelled probes is a very sensitive method of identifying genes of known sequence within large amounts of DNA.

cDNA can be synthesized using mRNA templates. RNA is extracted from a tissue which expresses the gene of interest, and mRNA is separated from this by chromatography on columns carrying poly(dT) chains

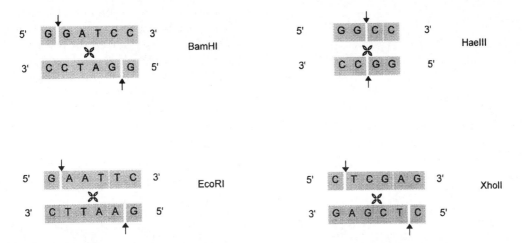

Figure 21.13 The specificity of four restriction endonucleases. Both strands of DNA are cleaved at symmetrical sites. The points of cleavage are indicated by arrows. Where these are not opposite one another, products with single-stranded 'sticky ends' are produced.

which pair with the polyA tails on the mRNA. DNA which is complementary to the RNA is then synthesized on a poly(dT) primer under the action of reverse transcriptase. A complementary copy of this DNA chain can be obtained by treatment with DNA polymerase I from *E. coli*. These processes yield small quantities of cDNA corresponding to all of the types of mRNA present in the cell, and the quantities of DNA can be increased by cloning.

21.9.3 GENE CLONING

Gene cloning can be achieved in vectors or by the polymerase chain reaction (PCR). In the former case, the DNA to be copied is inserted into other, larger pieces of DNA such as those of a virus or plasmid. From there it may be transferred to, and multiplied in, growing cells. For example, bacterial cells may be infected by bacteriophages into which cDNA has been inserted. On division this gives rise to a cell line or clone containing copies of the cDNA. A collection of lines representing cDNA from all the mRNA is called a cDNA library.

Eukaryotic genes may not be correctly expressed in prokaryotes, which cannot remove introns from mRNA. However, cDNA reconstructed from mRNA lacks introns and can generally be expressed correctly in prokaryotic cells.

Genes can also be cloned by PCR. This method uses oligonucleotide primers which are complementary to the DNA on either side of the region to be multiplied (Figure 21.14). New DNA is synthesized by extension of these primers. Double-stranded DNA is denatured at high temperature to separate the strands. The oligonucleotides are then allowed to bind to the complementary DNA chains at low temperature, and extended by use of DNA polymerase. These steps make up a cycle which can be repeated to give rise to exponentially increasing amounts of the required DNA. The process uses heat-stable DNA polymerase enzyme (Taq) from the thermophilic bacterium *Thermus aquaticus*, which survives the thermal denaturation of the DNA. Use of PCR amplification to increase amounts of DNA has allowed analysis and sequencing of very small samples of DNA. Thus the root of a single hair may yield, after amplification by PCR, enough DNA for genetic 'fingerprinting'.

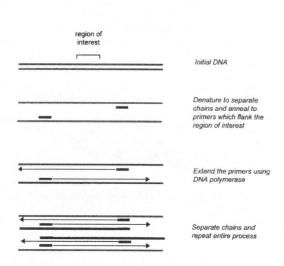

region of
interest

Initial DNA

Denature to separate
chains and anneal to
primers which flank the
region of interest

Extend the primers using
DNA polymerase

Separate chains and
repeat entire process

Figure 21.14 The polymerase chain reaction for cloning of DNA. The initial DNA is denatured to separate chains and annealed to primers which flank the region of interest. Extension of suitable primers using DNA polymerase results in synthesis of copies of the DNA of interest. After separation of the chains these may be used as templates for synthesis of further copies leading to an exponential rise in the yield of DNA.

21.9.4 SCREENING TECHNIQUES

DNA libraries contain large numbers of DNA sequences, only a few of which are likely to be of interest, so some form of screening is often required. This can be achieved by examining binding of radioactive probes which are complementary to the DNA sequences of interest (Southern blotting) or by screening the protein products of the genes with antibodies (Western blotting).

In Southern blotting, DNA is degraded by treatment with restriction enzymes into a large number of fragments, which are separated into bands by electrophoresis and denatured to the single-stranded form. Binding of a radioactive DNA probe to a copy of the gel, made on a nitro-cellulose sheet, allows the restriction fragment containing the required sequence to be identified. Northern blotting is an analogous technique used for analysis of RNA sequences.

21.9.5 RESTRICTION FRAGMENT LENGTH POLYMORPHISM (RFLP)

Because restriction enzymes cleave DNA only at specific sequences of bases, treatment of DNA with these enzymes results in characteristic and repeatable fragmentation patterns. The fragments are easily separated on the basis of size by electrophoresis. Mutations change the fragmentation patterns, which also vary from one individual of a species to another. Thus, analysis of RFLP patterns may be used to identify individuals and to locate genes associated with specific genetic disorders, such as cystic fibrosis.

21.9.6 ANTISENSE RNA

This is RNA which is complementary to mRNA. It forms a duplex with the mRNA and prevents its translation to produce protein. The use of efficient promoter sequences in combination with antisense RNA genes leads to the production of antisense RNA which may prevent expression of specific deleterious genes. Examples of the use of antisense RNA methodology to delay fruit ripening are described in Chapter 27.

21.9.7 VECTORS AND METHODS FOR INSERTION OF DNA INTO CELLS

Vectors have genomes which can be used to introduce DNA into other organisms. Plasmids, bacteriophages and other viruses and bacteria may be used as vectors. Plasmids and λ-phage (a bacteriophage which attacks the bacterium *Escherichia coli*) are most commonly used as vectors for cloning DNA in bacteria. Plasmids are accessory chromosomes found in bacteria and contain circular double-stranded DNA.

The DNA of λ-phage can become incorporated into the bacterial genome and be replicated with it without killing the bacteria. Large parts of the genome of λ are not essential for its action and can be replaced by foreign DNA which is then carried into the bacteria. Specifically modified plasmids and bacteriophages are available which are easier to manipulate and use. For example, they may have modified restriction sites to allow easy insertion of foreign DNA, or carry antibiotic resistance markers simplifying the selection of recombinant organisms.

Uptake of calcium phosphate-precipitated DNA, micro-injection, or genetically modified retroviruses, Baculoviruses or Vaccinia viruses may be used to insert recombinant DNA into animal cells.

The soil bacterium *Agrobacterium tumefaciens* is very commonly used as a vector for insertion of DNA into plant cells. Normally this bacterium causes crown galls in dicotyledonous plants. These arise because of insertion of part (T-DNA, transferred DNA) of a plasmid called the Ti plasmid from *A. tumefaciens* into the genome of the plant cell. Recombinant DNA can be incorporated into the plasmid and thus inserted into the plant genome. Electroporetic insertion of DNA into protoplasts or bombardment with DNA-coated particles may also be used for introducing foreign DNA into plant cells. These methods are especially useful for plants such as monocotyledons, which are not infected by *A. tumefaciens*.

21.9.8 SITE-SPECIFIC MUTAGENESIS (OLIGO-NUCLEOTIDE-DIRECTED MUTAGENESIS)

Site-specific mutagenesis is a potentially very powerful and versatile technique for modifying genes. Using this method a single base can be replaced by any other, leading to replacement of one amino acid by another in the protein for which the gene codes. As an example, serine is coded for by TCT and cysteine by TGT, so that replacement of serine by cysteine requires only replacement of a single C by G in the DNA. To do this, an oligonucleotide about

15 nucleotides long is prepared which is complementary to the region of interest, but in which G replaces C (Figure 21.15). Under particular conditions this hybridizes with the normal gene, as a sufficiently large number of bases are able to form complementary base pairs. The oligonucleotide is then used as a primer for synthesis of new DNA using the original as a template. The copy is complementary with the exception of the one replacement. Replication of both strands yields normal and modified genes which can be separated and cloned. Examples of the possible uses of this method are increasing the lysine content of cereal proteins (see Chapter 23), and changing the specificity of enzymes such as Rubisco (Chapter 17).

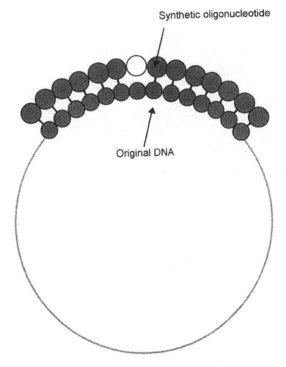

Figure 21.15 Site-specific mutagenesis for modification of base sequences of specific regions of DNA. Synthetic oligonucleotides are made, with all bases except one complementary to the region of interest. One base may be varied as this does not prevent base-pairing to the original. The chain is extended using the original as a template. A complementary copy of this differs from the original only in the nature of one base.

COMPARTMENTS, MEMBRANES AND REGULATION

22

22.1 CELL COMPARTMENTS

The major structural features of cells are described in Chapter 1. It is clear that as organisms have evolved, their architecture has become more complex, as reflected in both the overall shape and size of cells and the diversity of intracellular structures. Prokaryotes have no discernible intracellular membranes and can thus be considered as having a single compartment. Eukaryotic cells contain many different types of membrane-bounded structures. The plasma membrane surrounds the cell, and acts as a barrier between the extracellular environment which may undergo marked changes in composition, and the cell contents which have a relatively constant composition. Within cells are many subcellular organelles. Many of these are surrounded by membranes, e.g. the mitochondria, chloroplasts and nucleus, and have their own composition which may be markedly different from the cytosol.

As more has been learned about the biochemical processes in complex eukaryotic cells, it has become apparent that they have developed metabolic strategies based on their division into numerous compartments within which specific biochemical processes take place, in isolation from (but not totally independent of) processes occurring in other subcellular compartments. Coordination and regulation of biochemical processes in different subcellular compartments requires communication across membranes. This may involve the movement of a substrate or the end product of a pathway from one compartment to another, or more complex regulation requiring the movement of chemical messengers within the cell. Membranes play an important role in this process, acting as selective permeability barriers between the different compartments of the cell, allowing the movement of some molecules but preventing that of others. They also play an integral part in many intracellular signalling pathways and are themselves sites of a number of metabolic processes.

22.2 LIPIDS AND MEMBRANES

The basic structure of all biological membranes is a lipid bilayer composed mainly of polar lipids (phospholipids with smaller amounts of glycolipids and sphingolipids). Cholesterol and cholesterol esters are also present, but tri-acylglycerols are only minor components. The physicochemical properties of polar lipids make them particularly well suited as components of the lipid bilayer. They are amphiphilic molecules containing both fatty acids which are hydrophobic, and polar functional groups which are hydrophilic. In the bilayer the fatty acid moieties of these lipids form a central hydrophobic core with the polar head groups on the surface of each face able to interact with the surrounding aqueous environment.

Chemical analysis of membranes from various sources has shown that all membranes have proteins associated with them. In general, the biological activity of a membrane is reflected in the amount of protein it contains. Thus, the myelin sheath surrounding nerve fibres is composed of approximately 80% lipid and 20% protein. The main function of this membrane is to provide a layer of electrical insulation around the nerve axon. In contrast, bacterial membranes and the inner mitochondrial membrane contain about 25% lipid and 75% protein. These membranes play a role in many metabolic processes. The proteins they contain have many functions, for example, some are carriers involved in membrane transport, others are structural components of the electron transport chain required for the synthesis of ATP, and some are enzymes. In the specialized membranes of the rod cells in the retina, the principal protein is the light-sensitive carotino-protein rhodopsin, which accounts for approximately 90% of membrane proteins. In the plasma membrane of the erythrocyte, a less specialized membrane, about 20 different proteins have been isolated. A highly active membrane such as the inner mitochondrial membrane may contain hundreds of different proteins, some of which are linked together to form complexes required for specific biochemical processes.

Membrane proteins can be classified into two main groups: peripheral and integral pro-

teins. Peripheral (or extrinsic) proteins are relatively loosely attached to the surface of the bilayer by polar interactions between amino acid side chains and the polar head groups of the membrane lipids. They are typical of many cellular proteins that function in an aqueous environment, in that they are globular proteins in which the surface amino acids are mainly polar in nature. Some peripheral proteins are covalently linked to the membrane phospholipid head groups. The second group, integral (or intrinsic) proteins, are embedded in the lipid layer and may span the entire membrane. Very few integral proteins have been analysed for amino acid composition. They are firmly attached to the membrane and are difficult to isolate from other membrane components. That part of the protein which protrudes from the membrane surface and is surrounded by aqueous cell contents has predominantly polar amino acids on the surface of the molecule. However, where the protein passes into the hydrophobic core of the membrane, the amino acids are arranged in an α-helical structure (Chapter 5) in which the non-polar amino acids are found on the surface of the protein. These amino acids may be attached to membrane lipids by covalent linkages.

The most widely accepted model of membrane structure is the fluid mosaic model proposed by Singer and Nicholson in 1972 (Figure 22.1). This proposes that the phospholipid bilayer is fluid and that the phospholipid molecules move in the lateral plane of the membrane. However, there is very little exchange of phospholipids between the two halves of the lipid bilayer ('flip-flop' transitions) as this would involve the movement of the polar head groups of the phospholipid through the hydrophobic membrane core. One of the consequences of the minimal exchange of lipids between the two halves of the lipid bilayer is that it allows asymmetric distribution of phospholipids.

The fluidity of the lipid bilayer is influenced by the type of lipid found in the membrane and, in particular, by the degree of unsatura-

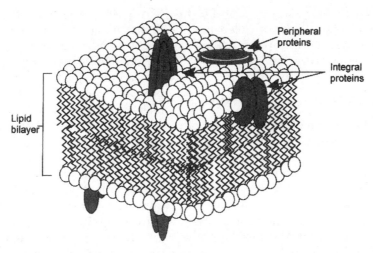

Figure 22.1 Structure of a lipid bilayer and the typical arrangement of peripheral and integral membrane proteins.

tion of the fatty acids in membrane lipids. The presence of a *cis* double bond introduces a kink into fatty acids, thus the presence of PUFAs with *cis* double bonds limits the extent to which the phospholipids in the bilayer can pack together. This increases the fluidity of the membrane, and also has the effect of maintaining the fluidity of the membrane as the temperature falls. The regulation of the fatty acid composition of membranes is not well understood, but it is clear that both diet and changes in environmental temperature influence the ratio of saturated to unsaturated fatty acids in the lipid fraction. The incorporation of sterols and sterol esters also affects membrane fluidity. For example, cholesterol has a dual effect. Its ring structure is rigid and tends to decrease membrane fluidity, but its presence disrupts the packing of the phospholipids and tends to increase fluidity at low temperatures. Together these effects may be useful in maintaining membrane fluidity over a wide range of temperatures.

Integral proteins appear to have a layer or domain of lipid associated with them which may be required to maintain their functional integrity. Extraction of integral proteins from membranes usually results in the co-extraction of a tightly bound lipid fraction. These proteins are able to move in lateral plane of the membrane but are accompanied by their lipid coating. Some membrane proteins are clustered together to form large, functional units such as receptors or transport complexes. Whilst these functional units are able to move within the membrane, the relative positions of individual proteins within the complex remain constant. As with membrane lipids, there is very little movement of proteins in the vertical plane of the membrane. Some integral proteins are glycoproteins, containing carbohydrate side chains of variable length and degree of branching. In the plasma membrane, at least, these glycoproteins are almost always orientated so that the carbohydrate protrudes from the surface of the cell.

22.3 TRANSPORT ACROSS MEMBRANES

Membranes act as selective permeability barriers. With the exception of some hydrophobic molecules which may be able freely to cross the membrane, most important biomolecules are unable to pass through membranes by simple diffusion because they are too large, too polar, or both. However, it is clear that many different types of biomolecules are able to cross membranes by a process called membrane-mediated transport in which the membrane proteins play a very important role.

22.3.1 MEMBRANE TRANSPORT MECHANISMS

A small number of important molecules can pass through membranes by diffusion because they are relatively non-polar, e.g. oxygen, nitrogen and methane. Polar compounds can only pass through membranes by a protein-mediated process known as facilitated transport.

In general, it is assumed that the proteins which aid the movement of molecules across the membrane are integral proteins which span the lipid bilayer. A number of models have been proposed to explain how proteins facilitate transport. In some cases they can form water-filled pores or channels through the lipid bilayer which allow only those molecules or ions small enough to fit into the channel to migrate from one side to the other. While the channel is open, bi-directional movement of molecules or ions can occur by diffusion (Figure 22.2). The net flux of a molecule through the channel is determined by its relative concentration on either side of the membrane, with movement occurring from the area of high concentration to the area of low concentration. If the channel remains open the concentration of any particular molecule will reach equilibrium across the membrane when transport continues to occur at equal rates in both directions. It is unlikely that membrane channels remain permanently open, as this does not allow for any control of the transport process. The conformation of channel proteins can be changed to close the channel, either directly by covalent modification of the protein, e.g. phosphorylation, or indirectly through the concerted effects of adjacent membrane proteins.

Membrane proteins may also act as carriers within the membrane. These have binding sites for molecules undergoing transport and, by analogy with the active site of an enzyme, the carrier-binding sites can exhibit broad or narrow binding specificity. A well characterized example of a membrane carrier system is that for the transport of glucose across the plasma membrane of the human erythrocyte, the so-called glucose permease system. The protein carrier exists in two different conformations. In one conformation the binding site is available on the extracellular surface of the membrane. Attachment of glucose to the carrier-binding site initiates a conformational change in the protein, resulting in a movement of the glucose-filled binding site to the intracellular surface of the membrane where glucose is released (Figure 22.3). Because the erythrocyte metabolizes glucose, its intracellular concentration is usually lower than the extracellular concentration and thus transport of glucose occurs from out to in. However, because the process is freely reversible, glucose cannot be transported into the erythrocyte if the concentration in the surrounding medium falls below that inside the cell.

The carrier exhibits optimum activity towards glucose. The transport constant, K_t, (analogous to the Michaelis constant, K_m, for an enzyme) for glucose is 1.5 mM. Other hexose sugars such as mannose and galactose (which differ from glucose only in the position of one hydroxyl group) can also be carried, but the transport constants for these sugars are much higher, 20 and 30 mM respectively.

The glucose permease system is an example of carrier proteins that transport single molecules, sometimes referred to as uniport systems. There are also examples of membrane carriers which transport two different molecular species, a process known as co-transport. In some cases the different molecules are transported in the same direction, a symport, but they may also be transported in opposite directions, an antiport. The latter is used by the erythrocyte membrane in the transport of waste CO_2 from the tissues to the lungs where it is exhaled. The CO_2 is transported in the blood plasma in the form of bicarbonate ions, HCO_3^-. In the peripheral circulation, CO_2 enters the erythrocyte and is converted to bicarbonate by carbonic anhydrase. The HCO_3^-

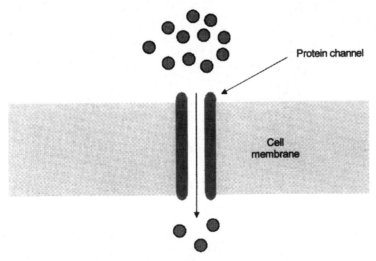

Figure 22.2 Movement of molecules through protein channels in membranes.

ions are rapidly transported out of the ery-throcyte in exchange for chloride ions (Cl⁻), an example of an antiport. Once in the lungs the process is reversed, HCO_3^- is taken up by the erythrocyte and Cl⁻ ions are released. The HCO_3^- is converted to CO_2, released into the plasma and expired (Figure 22.4).

The channel- and carrier-mediated trans-port processes described so far have all involved the movement of molecules from an area of low concentration to an area of high concentration, a process known as passive transport. However, not all molecules cross membranes down a concentration gradient. Movement of molecules from an area of low concentration to an area of high concentra-tion (i.e. against a concentration gradient) is called active transport. This movement can only be achieved by the expenditure of ener-gy. A parallel can be drawn between passive and active transport and the movement of water between two tanks at different heights. If the tanks are connected by a pipe, the hydrostatic pressure will cause the water to flow from the higher tank to the lower tank. This can be compared with passive transport where a concentration gradient acts as the driving force. In order to move water from the low tank to the high tank a pump is need-ed which requires a supply of energy. In active transport, the pump is fuelled by the expenditure of energy provided by the hydrolysis of ATP, oxidation reactions or the co-movement of a counter molecule down an electrical or chemical gradient using a sym-port or an antiport.

A particularly well understood example of ATP-driven active transport is the Na⁺, K⁺-ATPase transport system found in all cell plas-ma membranes. There are large differences in the Na⁺ and K⁺ concentrations of intra- and extracellular fluid (see Table 22.1).

The natural chemical gradients across the plasma membrane tend to force K⁺ out of the cell and Na⁺ into it. Cells maintain the differ-ential concentration of Na⁺ and K⁺ across the plasma membrane by use of the Na⁺, K⁺-ATPase pump. The pump is an antiport which transports two K⁺ ions into the cell in exchange for every three Na⁺ ions carried out. In the process, one molecule of ATP is hydrol-ysed to ADP.

The mechanism of the Na⁺, K⁺-ATPase is shown in Figure 22.5. The ATPase consists of two α and two β subunits. The α subunit con-tains the ATPase activity and the Na⁺ and K⁺ binding sites. The function of the β subunit is not known. The α subunit binds to three Na⁺ ions on the intracellular surface of the plasma membrane. The ATPase then catalyses the

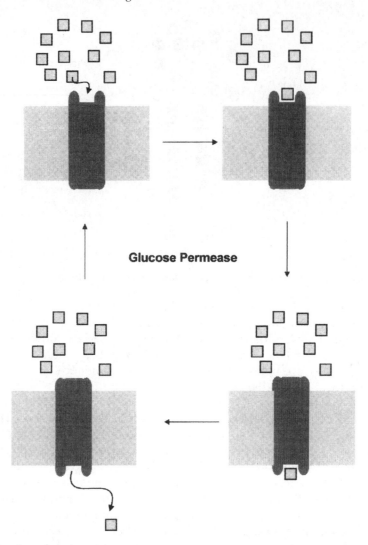

Figure 22.3 Mechanism of glucose permease.

transfer of the terminal phosphate group of ATP to the α subunit which undergoes a conformational change, exposing the Na^+ binding sites on the extracellular surface of the membrane and simultaneously reducing its affinity for Na^+ so that the Na^+ ions are released. At the same time, high-affinity K^+ binding sites, previously inaccessible to the extracellular environment, become available and bind two K^+ ions. The α subunit is subsequently dephosphorylated, allowing the carrier to revert to its original conformation in which K^+ binding sites are exposed on the intracellular surface of the membrane in a low-affinity form, resulting in the release of K^+ into the cell. The Na^+ binding sites are now in the high affinity form on the intracellular side of the membrane, bind to a further three Na^+ ions, and the cycle is repeated.

Maintenance of the correct intracellular Na^+ and K^+ concentrations is vital for cell function and it has been estimated that 25% of

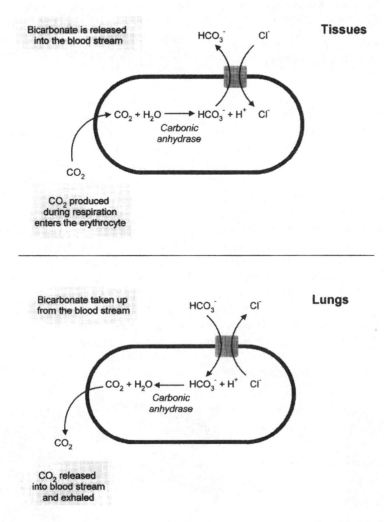

Figure 22.4 Uptake and release of bicarbonate by erythrocytes in exchange for chloride ions, part of the mechanism of CO_2 removal from tissues.

energy expenditure in resting humans can be attributed to the Na^+, K^+-ATPase. This accounts for the consumption of very large quantities of ATP: during a 24-hour period the energy requirements of an adult human account for the hydrolysis of about 200 kg ATP, although at any one time the total amount of ATP in the body is about 50 g. There is, therefore, continual resynthesis of ATP from ADP and P_i. Nonetheless, these fig-

ures suggest that approximately 50 kg of ATP is hydrolysed in a 24-hour period in order to maintain Na^+ and K^+ concentrations in the cell.

Active transport can also be fuelled by the utilization of electrochemical gradients across membranes. This can be illustrated by reference to the transport of lactose into the bacterium *Escherichia coli*. The plasma membrane of *E. coli* contains an electron transport chain which

Table 22.1 Concentration of sodium and potassium in intracellular and extracellular fluid in mammalian cells

Ion	Intracellular concentration (mM)	Extracellular concentration (mM)
Na$^+$	5–15	145–150
K$^+$	140–145	5–10

acts as a proton pump, transporting protons from the intracellular compartment to the surrounding medium. The proton gradient created by this mechanism is an electrochemical gradient referred to as a proton motive force (PMF). The plasma membrane is impermeable to protons, and the PMF is coupled to a transport protein in the membrane which transports both lactose and protons into the cell. By this process, protons return to the cytoplasm down a concentration gradient and provide the energy required to drive the lactose into the cell against a concentration gradient of 1:100. This is an example of co-transport (pro-

tons and lactose) via a symport (Figure 22.6). Proton gradients are not only used to fuel active transport but, as discussed in Chapter 12, they can also be used to provide the energy needed to synthesize ATP in specialized membranes such as the inner mitochondrial membrane. These proton-dependent processes are examples of chemiosmotic coupling.

In the membrane transport described so far, the molecule undergoing transport across the membrane is not modified during the transport process. There is another form of transport in which the transported molecule is chemically modified before release from the membrane. This process is known as group translocation and is typified by the phosphotransferase system (PTS) for sugars, which has been well characterized in several bacteria, particularly *E. coli*, *Salmonella typhimurium* and *Staphylococcus aureus* (Figure 22.7). The transport complex consists of three enzymes and one non-enzymic protein. Enzyme I and HPr are soluble proteins located in the cytoplasm; enzyme II is an integral protein located in the

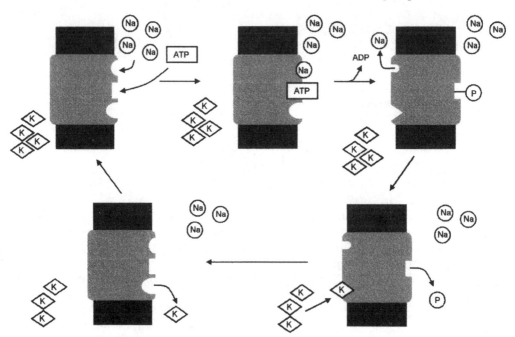

Figure 22.5 Mechanism of Na$^+$, K$^+$-ATPase.

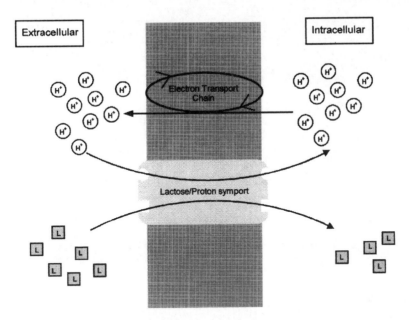

Figure 22.6 Uptake of lactose in *Escherichia coli* by a proton/lactose symport.

cell membrane which acts both as a permease and catalyses the final transfer of phosphate from phosphoenolpyruvate (PEP) to the transported sugar. The latter step also requires the presence of enzyme III which is associated with enzyme II in the membrane.

The specificity for the sugar to be transported is associated with enzymes II and III. PTS can be considered a special form of active transport in that it utilizes PEP as a high-energy phosphate donor. However, it differs from the Na^+, K^+-ATPase previously described because the chemical modification of the transported sugar means that there is no direct accumulation of sugars against a concentration gradient.

22.3.2 OTHER FUNCTIONS OF MEMBRANES

In addition to their role in transport processes, membranes have numerous other important functions. They are involved in the process of signal transduction. Most cellular membranes, but particularly the plasma membrane, contain receptors. These take the form of protein complexes which interact with hormones and other bioactive molecules. Binding of a hormone to its receptor can have a number of consequences. Hormones such as the catecholamines (adrenalin and noradrenalin) interact with plasma membrane receptors and activate the enzyme adenyl cyclase via a receptor-associated G-protein (GTP-binding protein) mechanism. The primary effect of these interactions is to increase the intracellular concentration of 3′, 5′, cyclic AMP (cAMP), one of the so-called 'second messengers', which initiates further metabolic changes in the cell by activation of cAMP-dependent protein kinases. The activation of triacylglycerol lipase (hormone-sensitive lipase) and the regulation of glycogen metabolism are brought about by such cAMP-mediated mechanisms. Other hormones, such as insulin, interact with receptors which have an intracellular tyrosine kinase. This catalyses phosphorylation of tyrosine residues on target proteins which then initiate changes in cellular metabolism. These mechanisms are described in more detail in Chapter 30.

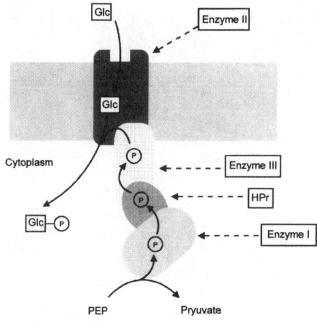

Figure 22.7 Group translocation by the phosphotransferase system.

Proteins on the extracellular surface of plasma membranes are involved in the cell-to-cell contacts and cellular recognition. These proteins are often glycoproteins. In animals, many immune response mechanisms are triggered by the presence of organisms which contain 'foreign' cell-surface proteins. This initiates the activation of the immune system and the production of antibodies, leading to the destruction of the foreign organisms.

22.4 PRINCIPLES OF METABOLIC REGULATION

Part Two describes the metabolism of the main constituents of the cell, carbohydrates, proteins and lipids, and the ways in which their breakdown can provide energy and substrates for biosynthetic processes. For example, sugars and fatty acids can be broken down to supply acetyl-CoA which can be oxidized in the TCA cycle. The reduced cofactors generated (NADH and $FADH_2$) can be used to synthesize ATP by oxidative phosphorylation. TCA-cycle intermediates can also be used as the starting point for the synthesis of amino acids for protein synthesis, of sugars for polysaccharide synthesis and of fatty acids for lipid synthesis. These processes do not occur at random but are coordinated. In many cases they are influenced by the supply of nutrients in the extracellular environment and, in complex multicellular organisms, the relative importance of one metabolic pathway over another is regulated with respect to the stage of growth, reproductive state and tissue specialization.

There are three main ways in which metabolism is coordinated through the regulation of enzymes. Firstly, by allosteric regulation, through changes in the concentration of specific enzyme activators and inhibitors. Secondly, by covalent modification of an enzyme to bring about a change in its activity. Thirdly, by changing the amount of enzyme present through modification of gene expression and protein turnover.

22.4.1 ALLOSTERIC REGULATION

Most metabolic processes involve a series of biochemical transformations in which a substrate is converted to an end product via a number of discrete steps, each step being catalysed by a specific enzyme. This sequence of reactions constitutes a metabolic pathway. Regulation of the flow of metabolites through a metabolic pathway occurs by modifying the activity or amount of one or more of the enzymes catalysing the individual steps. In a relatively simple pathway this regulation may be achieved via only one enzyme, usually the one which catalyses the first reaction in the pathway. This is referred to as the committed step in a metabolic pathway. In more complex pathways, which have many branch points and the potential to convert the substrate to a number of end products, the first enzyme after each branch point is usually regulatory (Figure 22.8). These allosteric enzymes are often complex and are controlled by factors other than the availability of substrate. They are usually rate-limiting enzymes which control the flux of metabolites along a specific pathway. Other enzymes in the pathway are normally present in excess and their activity is determined largely by substrate supply, which in turn depends on the activity of the enzyme catalysing the preceding reaction.

A number of features are typical of allosteric enzymes (see also Chapter 6). They are regulated by the presence of specific signal molecules which are distinct from the substrate and

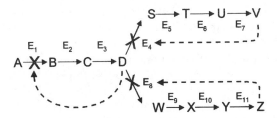

Figure 22.8 Feedback inhibition of allosteric enzymes in a branched metabolic pathway. (E1–E11 represent the enzymes catalysing individual steps; A–D, S–V and W–Z represent the substrates and products of the reactions.)

are often cofactors, such as ATP, NADH and NADPH, the concentration of which reflect energy status or redox potential in a cell. In addition, the end product of a pathway or an intermediate downstream of the allosteric enzyme may act as a regulator. In this latter case the activity of the allosteric enzyme is usually inhibited, and this type of regulation is referred to as feedback or end-product inhibition. This type of mechanism allows the cell to 'sense' when a particular end product is in excess of the requirement. Feedback inhibition slows down the reaction catalysed by the rate-limiting enzyme of the pathway, thereby reducing the rate of production of the end product. If the end product is used by other metabolic processes, its concentration in the cell will fall and the inhibition of the rate-limiting enzyme is relieved, allowing more end product to be formed. This type of mechanism allows acute control of metabolic pathways.

The biosynthetic pathways of many amino acids are branched and are regulated by feedback inhibition.

22.4.2 COVALENT MODIFICATION

This type of modification may alter enzyme activity by changing the conformation of the protein which affects the shape and size of the active site. The most common type of covalent modification is phosphorylation or dephosphorylation of one or more amino acids, most often serine residues. The control of glycogen metabolism provides a good example of this type of enzyme regulation. Glycogen breakdown is catalysed by the enzyme glycogen phosphorylase which exists in two forms, glycogen phosphorylases *a* and *b*. Phosphorylase *a* is the most active form of the enzyme; phosphorylase *b* is relatively inactive. The enzyme consists of two subunits, each of which contains a specific serine residue which undergoes phosphorylation, resulting in the conversion of phosphorylase *b* to phosphorylase *a*. In most cases of covalent modification by phosphorylation the phosphate donor is ATP and the phosphorylation reaction is catalysed by a protein kinase, in this case called phosphorylase kinase. This kinase is itself activated by a non-specific protein kinase. Conversion of phosphorylase *a* back to phosphorylase *b* occurs by removal of the phosphate groups catalysed by another enzyme, phosphorylase phosphatase.

Phosphorylation/dephosphorylation is also used to control the activity of glycogen synthesis through the enzyme glycogen synthetase, which also exists in two forms, *a* and *b*. The active form, glycogen synthetase *a*, is the dephosphorylated enzyme. It is inactivated by phosphorylation, catalysed by protein kinase, and reactivated by removal of phosphate groups by phosphoprotein phosphatase. Thus the regulation of glycogen metabolism is brought about by reciprocal changes in the activities of glycogen synthetase and glycogen phosphorylase. Phosphorylation activates glycogen phosphorylase and inhibits glycogen synthetase, whereas dephosphorylation has the opposite effect (Figure 22.9).

The reciprocal regulation of these enzymes eliminates the possibility of a futile cycle involving breakdown of glycogen to produce glucose-1-phosphate and its reincorporation into glycogen. In fact, in animals, the control of glycogen metabolism is linked to whole-body metabolic needs via fluctuations in the levels of circulating adrenalin and glucagon. Adrenalin acts primarily on glycogen metabolism in muscle. It is secreted in response to nutritional and physiological stress, situations in which glucose supply to muscle tissue may become limiting. It acts on cell membrane adrenergic receptors to increase the intracellular concentration of cAMP. The net effect of this increase is the inhibition of glycogen synthetase, activation of glycogen phosphorylase and an increase in the supply of glucose-1-phosphate. Glucose-1-phosphate is converted to glucose-6-phosphate by phosphoglucomutase and, as muscle cannot release free glucose due to a lack of glucose-6-phosphatase, the glucose-6-phosphate enters the glycolytic pathway and is used to produce the ATP needed to power muscle contraction.

The polypeptide hormone glucagon acts mainly on liver glycogen metabolism through activation of adenyl cyclase. Its effects are similar to adrenalin in that it increases the supply of glucose-6-phosphate. However, as liver has an active glucose-6-phosphatase, there are two major fates for the glucose-6-phosphate: either conversion to glucose and secretion into the blood stream, or entry into the glycolytic pathway.

Often allosteric regulation and covalent modification act in concert on the same enzyme. Glycogen phosphorylase is one such enzyme. In addition to the covalent modification described above, the activity of this enzyme is regulated by the intracellular concentrations of ATP, AMP and glucose, all of which act as allosteric effectors. The enzyme is strongly inhibited by ATP and glucose and is activated by AMP. Thus, the activity of the enzyme is sensitive to the energy status of the cell. When there are adequate supplies of energy in the form of glucose and ATP, the cell has no need to degrade glycogen to glucose. However, high levels of AMP, which reflect reduced availability of ATP, stimulate release of glucose from glucogen and increase the supply of glycolytic substrate for ATP synthesis.

22.4.3 CHANGES IN THE AMOUNT OF ENZYME

A third way in which regulation is achieved is by controlling the synthesis of enzymes. Whilst allosteric regulation and covalent modification can be considered acute responses to changes in metabolic state, control of the rate of synthesis of enzymes is a longer term effect. It involves the regulation of gene expression and is discussed in more detail in Chapter 21. This type of regulation is typified by the lac operon in bacteria which controls the production of β-galactosidase in response to nutrient supply. Repression of enzymes required for amino acid synthesis in bacteria is controlled in a similar manner. Changes in the amount of enzyme also occur in plants and animals as a

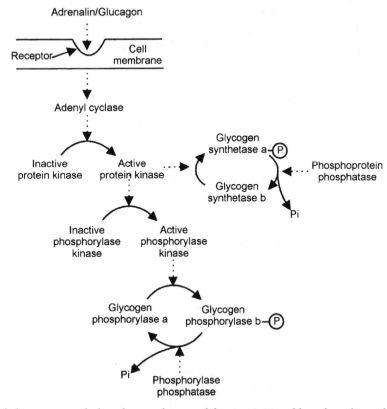

Figure 22.9 Control of glycogen metabolism by covalent modification initiated by adrenalin and glucagon.

regulatory mechanism. However, the ways in which this is brought about differ between prokaryotes and eukaryotes.

22.4.4 COORDINATED REGULATION OF PATHWAYS

The direction of metabolism is controlled by the coordinated regulation of interconnecting pathways. This regulation is brought about both by allosteric regulation of enzyme activity and by covalent modification of enzymes in response to changes in hormonal status. This can be illustrated by examining the way in which glycogen metabolism, glycolysis, the TCA cycle and oxidative phosphorylation are coordinated. The allosteric effects of substrate and coenzyme concentrations on the enzymes

in these pathways are illustrated in Figure 22.10.

The main function of glycogen stored, for example, in muscle tissue is to provide glucose-6-phosphate for ATP production. *In vivo*, in tightly coupled mitochondria, the flux of electrons through the electron transport chain and the synthesis of ATP are regulated by the energy requirements of the cell. Utilization of ATP for metabolic processes is the trigger which initiates oxidative metabolism. The energy status of the cell is reflected in the ratio of [ATP]/([ADP] + [P_i]). This ratio is normally high, i.e. high cellular concentrations of ATP and low cellular concentrations of ADP. Under these conditions, the supply of ADP for ATP synthesis is limiting and the flux of protons through the F_0F_1-ATPase is low. The energy

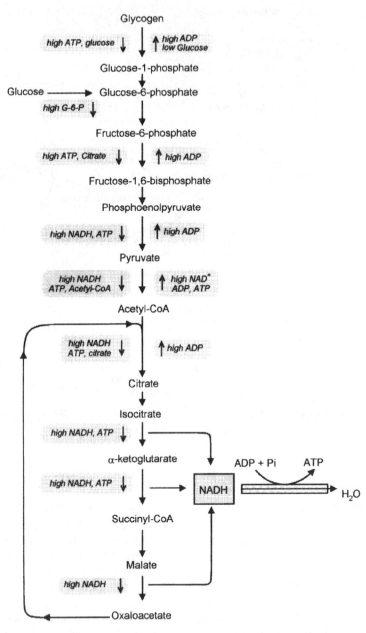

Figure 22.10 A summary of glycolysis, the TCA cycle and electron transport showing the main points where the metabolism of glucose and glycogen is regulated by changes in the concentration of intermediates: NAD^+, NADH, ATP and ADP.

required for the electron transport chain to pump electrons across the inner mitochondrial membrane increases as the magnitude of the proton gradient increases; thus when the proton gradient is high, electron transport slows down and the rate at which NADH and $FADH_2$ are reoxidized decreases. This leads to a more reduced environment in the cell because of the high NADH/NAD$^+$ ratio. The combined effect of a high ATP concentration and the high NADH/NAD$^+$ ratio reduces the rate of oxidative metabolism. This is brought about by:

- the direct effect of ATP on key regulatory enzymes in the TCA cycle, glycolysis and glycogenolysis, e.g. α-ketoglutarate dehydrogenase, citrate synthase, pyruvate dehydrogenase, pyruvate kinase, phosphofructokinase-1 and glycogen phosphorylase;
- the lack of NAD$^+$ and FAD which reduces the activity of dehydrogenases of the TCA cycle and glycolysis;
- the consequent build-up in concentration of intermediates which cause feedback inhibition of the enzymes which produce them, e.g. succinate inhibits α-ketoglutarate dehydrogenase, citrate inhibits citrate synthase, acetyl-CoA inhibits pyruvate dehydrogenase, and glucose-6-phosphate inhibits hexokinase.

This concerted regulation ensures that glycogen and glucose are not wastefully metabolized when the energy requirements of the cell are low. Depletion of ATP reserves rapidly releases the ATP inhibition of oxidative metabolism, increases the rate of electron transport and reduces the NADH/NAD$^+$ ratio in the cell, allowing oxidative metabolism to increase until a new energy equilibrium is established.

Superimposed on this level of control is a further tier of regulation, the covalent modification of a number of key enzymes. This type of regulation modifies the metabolic activity of cells and tissues in response to changes in whole body requirements. The control of carbohydrate metabolism via the pathways considered above again provides good examples. In animals, the availability of food is reflected in changes in the concentration of glucose in the blood. The endocrine response to an increase in blood glucose levels is secretion of insulin from the pancreas, which stimulates tissue uptake of glucose for glycogen, fatty acid and amino acid synthesis. In contrast, when blood glucose falls the secretion of insulin decreases and that of glucagon increases.

Glucagon not only stimulates the mobilization of glycogen reserves (section 22.4.2) but also stimulates gluconeogenesis and inhibits glycolysis. The balance between these two pathways is controlled by a relatively newly discovered compound, fructose-2,6-bisphosphate (F-2,6-P). This is synthesized by a second type of phosphofructokinase referred to as PFK2 (to distinguish it from PFK1 which catalyses the mainstream reaction of glycolysis). F-2,6-P is broken down to fructose-6-phosphate (F-6-P) by the action of fructose-2,6-bisphosphatase. The unique feature of the metabolism of F-2,6-P is that the enzymes responsible for its synthesis and degradation are located on the same protein, which acts as a bifunctional enzyme. The nature of the enzyme activity expressed by this protein is determined by its phosphorylation state. In the dephosphorylated state it functions as PFK2, but when phosphorylated it acts as F-2,6-bisphosphatase. F-2,6-P is synthesized from F-6-P and is a potent activator of PFK1. Thus when the F-6-P concentration is high, F-2,6-P is produced, PFK1 activity is increased, and the flux of glucose through the glycolytic pathway increases. The primary regulator of PFK2 and F-2,6-bisphosphatase activity is glucagon, which acts via cAMP and protein kinase to inhibit PFK2 and activate F-2,6-bisphosphatase. The net effect of glucagon is to reduce the concentration of F-2,6-P and shift the balance between glycolysis and gluconeogenesis in favour of the latter. Thus, in liver, glucagon promotes the production of glucose by stimulating both glycogenolysis and gluconeogenesis. Glucagon also inhibits glycolysis via pyruvate kinase. The activity of this enzyme is reduced by phosphorylation initiated by glucagon and mediated via a cAMP dependent kinase (Figure 22.11).

Allosteric regulation and covalent modification are found in other areas of metabolism. For example, in fatty acid metabolism the pres-

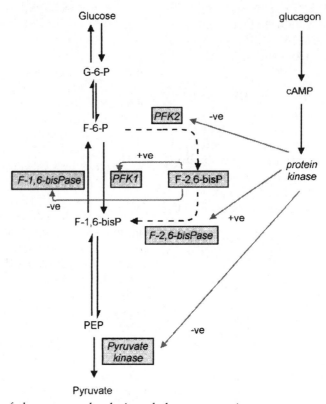

Figure 22.11 The effect of glucagon on glycolysis and gluconeogenesis.

ence of citrate, an intracellular indicator of high energy status, stimulates acetyl-CoA carboxylase which catalyses the production of malonyl-CoA, a substrate for fatty acid synthetase. This has the effect of channelling excess energy to fat storage. Insulin, which acts as a sensor of high blood glucose levels in animals, promotes the activation of pyruvate dehydrogenase and ATP-citrate lyase via a phosphorylation mechanism. Both of these enzymes catalyse reactions leading to the production of acetyl-CoA which can be used for fatty acid synthesis. To ensure that newly synthesized fatty acids are not immediately oxidized, malonyl-CoA inhibits carnitine acyltransferase I, thus preventing fatty acids from being transported into the mitochondria where β-oxidation takes place.

When energy supply is limiting it would be wasteful to convert glucose to fatty acids, and this is prevented in several ways. Firstly, a fall in blood glucose stimulates lipolysis. The presence of high levels of acyl-CoAs in tissues where fatty acid synthesis takes place inhibits acetyl-CoA carboxylase and limits substrate availability for the synthesis of fatty acids. Secondly, glucagon and adrenalin concentrations increase in response to reduced energy supply. Both of these hormones cause deactivation of acetyl-CoA carboxylase via a phosphorylation mechanism. These relationships are summarized in Figure 22.12.

22.4.5 COORDINATION OF METABOLISM IN DIFFERENT TISSUES

In mammals, where tissues have developed with specialized biochemical functions, the endocrine system serves as a means of integrating tissue metabolic activity through

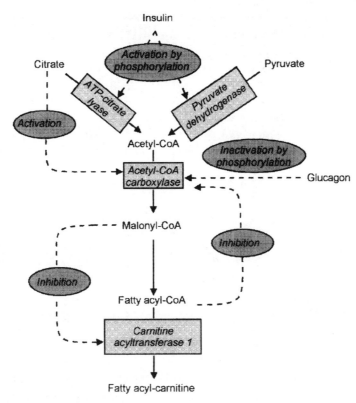

Figure 22.12 The control of lipid metabolism is coordinated through allosteric regulation and covalent modification of enzymes.

changes in hormonal status. This can be illustrated by considering the relationship between carbohydrate and lipid metabolism in liver and adipose tissue, and the way in which particular physiological states, such as pregnancy and lactation in sheep and cows, change the balance and direction of their metabolic activity.

A well nourished animal fed above its maintenance level is in positive energy balance, and therefore glucose synthesis in the liver is adequate to meet the needs of growth and allow synthesis of glycogen and deposition of fat. In contrast, in late pregnancy in sheep carrying multiple foetuses, or in early lactation in the high-yielding dairy cow, the glucose demand for the foetuses or the mammary gland is exceptionally high and feed intake cannot match metabolic requirements. As a result

such animals are in a state of negative energy balance. The drain on blood glucose causes a decrease in the plasma concentration of insulin and a corresponding increase in that of glucagon. Plasma catecholamine levels also increase. These hormonal changes modify the metabolic activity of liver and adipose tissue. The increase in glucagon levels stimulates hepatic gluconeogenesis and channels all available glucogenic substrates towards the synthesis of glucose. At the same time, the hormonal shift has profound effects on adipose tissue metabolism (Figure 30.11). Firstly, the decrease in plasma insulin concentration reduces the uptake of glucose by adipocytes and inhibits fatty acid synthesis. This limits the supply of substrates (glycerol-3-phosphate and fatty acids) for triacylglycerol synthesis and reduces lipogenesis. Secondly, the

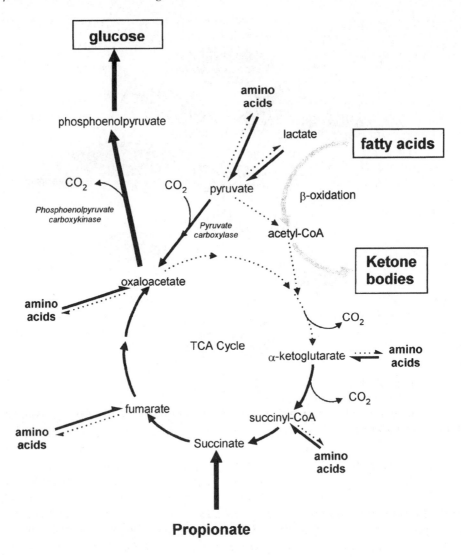

Figure 22.13 Negative energy balance in animals leads to the production of ketone bodies. The solid arrows indicate the direction of flow through metabolic pathways.

increase in glucagon and catecholamine concentrations stimulates the activity of triacylglycerol lipase and thus increases lipolysis. In addition, these hormones inhibit fatty acid synthesis. This lipolytic effect may be further enhanced by release of the inhibitory effect of insulin on lipolysis. The net effect of these changes in adipose tissue is the mobilization of fatty acids.

These fatty acids can be utilized for energy production by some other tissues in the body and therefore have a sparing effect on glucose utilization. Liver is an important site of fatty acid oxidation which produces acetyl-CoA. Under normal circumstances this would be further oxidized in the TCA cycle after its condensation with oxaloacetate to form citrate. However, during negative energy balance,

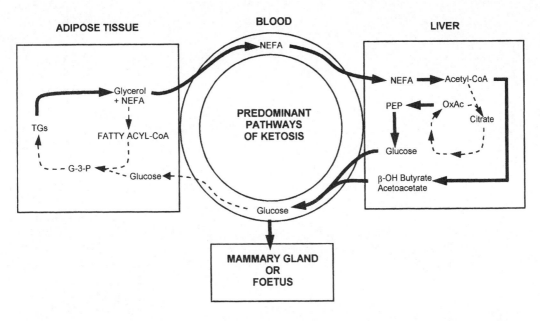

Figure 22.14 The relationship between metabolism in adipose tissue and liver during ketosis.

oxaloacetate is in short supply as any that is produced (for example from propionate or amino acids) is rapidly syphoned off for the synthesis of glucose. In this situation, acetyl-CoA cannot be oxidized and is instead converted to ketone bodies (Figure 22.13). These can be exported from the liver to other tissues which use them as substrates for energy production. The synthesis of ketone bodies is described in Chapter 13.

The presence of excessive concentrations of ketone bodies in the blood results in the development of the clinical condition known as ketosis. The two major ketone bodies are acids (β-hydroxybutyric acid and acetoacetic acid), which, if present in excess in the blood, can cause acidiosis (a decrease in blood pH), which in some cases is fatal.

It is not uncommon for the pregnant ewe and lactating cow to suffer from a less severe, subclinical form of ketosis. When the lambs are born or the level of lactation falls, these animals will eventually return to positive energy balance. When this occurs the requirement for hepatic gluconeogenesis is reduced and, in adipose tissue, lipogenesis takes over from lipolysis. The relationship between metabolism in adipose tissue and liver during ketosis is summarized in Figure 22.14. The changes described above are an extreme case of a normal metabolic response to changes in energy intake. Less dramatic shifts in the metabolism of liver and adipose tissue occur continually in animals and humans. For example, short periods of fasting, even the overnight fast between an evening meal and breakfast, will cause an increase in glycogenolysis, gluconeogenesis and lipolysis.

PART THREE

STRATEGIES FOR PROCESSING OF NUTRIENTS IN PLANTS

SEEDS AND GERMINATION 23

23.1 SEEDS AND PLANT DEVELOPMENT

The seed is the means by which a plant reproduces and multiplies, and which allows it to persist through adverse conditions. Inside a protective seed coat, it consists of an embryo together with nutrient reserves, and which support the growth of the embryo until the seedling is able to carry out photosynthesis and obtain nutrients from the soil via the root system.

During the phase of seed development and maturation on the mother plant, the embryo develops from the fertilized ovule and the nutrient reserves may accumulate in either the endosperm or the cotyledons, which form part of the embryo itself.

Wherever the nutrients are stored, their function is to provide the materials which are needed to support the growth of the developing seedling. They must thus act as a source of energy, carbon, nitrogen, sulphur and other major and trace elements.

23.2 SEEDS AS FOOD AND AGRICULTURAL COMMODITIES

Seeds are major world food crops. The principal seeds which are of agricultural importance are cereals, legumes and oilseeds. The cereal staples (wheat, rice and maize) together provide more than 50% of the world's food supply, and a further 13% is provided by legume seeds (peas, chickpeas, beans, lentils, soyabeans and cowpeas). Cereal seeds are particularly important because, in addition to supplying energy, they are the major global source of protein in human and animal diets. Most cereal proteins are deficient in some essential amino acids. Where they provide most of the protein intake, it is likely that the diet as a whole will be deficient in these amino acids. Thus attempts have been made to improve the quality of cereal proteins as this would improve human nutrition, particularly in developing countries where they are the major source of nitrogen in the diet. This could reduce the need for high-protein additives such as soyabean, fish or pure amino acids in the nutrition of man or other monogastric animals.

In addition to nutritional uses, seeds also have important commercial and industrial applications. Thus, cereal grains may be malted to provide raw materials for the brewing and distilling industries, and starch and oils may be extracted from seeds and processed to provide important industrial raw materials.

23.3 SEED COMPOSITION

Although seeds of different species have the same functions, their composition varies a great deal. Typical compositions of some different types of seeds are shown in Table 23.1. In many seeds the energy reserves consist of carbohydrate, principally starch. In others starch may be almost completely absent and instead large amounts of lipid may be present. Because of its higher energy content, the use of lipid, instead of starch, as an energy store allows seeds to be smaller, lighter and more easily dispersed (Chapter 9).

Cereal grains contain large amounts of starch. Protein levels are moderate but, as discussed elsewhere, the quality is poor for feeding to non-ruminants. The lipid content is low, but it is higher in oats and maize than in most other cereals.

The pea is typical of many legumes and contains less carbohydrate and more protein than cereals. The protein is also of higher quality than cereal protein. There is very little fat in most legumes but soyabeans are rather different, containing moderate amounts of both fat and carbohydrate, and being intermediate in composition between legumes and oilseeds such as rapeseed, linseed and sunflower.

In oilseeds the main storage reserve is in the form of triacylglycerols, and these crops are grown principally as a source of oil for the food industry. They contain very little carbohydrate, but the meal which remains after extraction of the fat is a source of good quality protein which can be used in animal feeds.

23.3.1 SEED CARBOHYDRATES

Seed carbohydrates consist principally of starch, which is found as granules in the endosperm. Starch consists of a mixture of amylose (straight-chain structure) and amylopectin (branched-chain structure) and the proportions of these two forms are largely genetically controlled. In barley, wheat, oats, maize and rye there is between 27 and 29% amylose in the starch whereas in rice there can be up to 37% amylose. In some mutants the proportions may be quite different, e.g. in 'waxy' maize there is virtually no amylose. The proportion of amylose present in the starch increases during development of the grain. As high-amylose starch has the ability to form films and fibres, it has considerable industrial potential. The composition affects the physical properties of the grain so that high-amylose starch leads to a 'floury' endosperm, whilst low-amylose starch causes the endosperm to be 'flinty'. In granules the starch exists in an organised, semi-crystalline form which is almost entirely due to the amylopectin fraction. The amylose and amylopectin molecules are probably held together through the formation of hydrogen bonds and there is also some tightly bound water which is necessary for the ordered structure of the starch granule. The crystalline structure of starch in the granules has to be broken down before complete enzymic breakdown can take place. Physical processing of starchy feeds, e.g. dry heat or hydrothermal processing, leads to gelatinization of the starch and may have a substantial effect on the nutritive value of feeds. Cereal starch granules tend not to burst on gelatinization, whilst those of some other plants (e.g. potato tubers) do.

Table 23.1 The approximate composition of various types of seeds

Seed	Water	Carbohydrate	Protein	Fat
		(g kg fresh wt^{-1})		
Wheat	140	660	132	20
Barley	106	836	79	17
Oats	89	728	124	87
Dried peas	133	500	216	13
Soyabean	70	235	368	235
Rapeseed	70	250	212	486

Starch from some sources is an important industrial raw material. It may be phosphorylated which alters its physical properties and potential uses.

23.3.2 SEED LIPIDS

Plants in general do not contain large amounts of lipid: for example, the content in cereal grains varies from 10–30 g kg^{-1} dry matter in wheat, barley and rye to 40–60 g kg^{-1} dry matter in oats and maize. This lipid is concentrated in the embryo (rice embryos contain a particularly high concentration: 350 g kg^{-1} dry matter within the small amount of embryo). The lipid is in the form of triacylglycerols and is quite unsaturated, containing much linoleic and oleic acid, and it therefore tends to oxidize readily. The presence of these unsaturated lipids in the diet may lead to the production of unacceptably soft fat in pigs.

Some commercially important crops accumulate large amounts of oils in their seeds (Table 4.4). This is often accompanied by useful quantities of protein, enhancing the value of the crop still further. Important oilseed crops include sunflower, rapeseed, soyabean, castor bean and groundnut.

Chemically, plant seed oils are triacylglycerols. These contain fatty acids of varying chain lengths (Table 4.3). A number of plant oils have important industrial uses, including manufacture of cooking fats, lubricants, soaps, pharmaceuticals, fuels and other chemicals.

Some oilseeds have toxic constituents. For example, rape and mustard both contain erucic acid (Chapter 4) and glucosinolates (Chapter 3). Because of the nutritional importance of rapeseed, varieties with low levels of both constituents have been developed.

23.3.3 SEED PROTEINS

The nitrogen in most seeds is found as a protein and a non-protein nitrogen (NPN) fraction. The proteins consist of specialized storage and non-storage proteins. The latter are mainly enzymes needed in the developing or germinating seed. The storage proteins act as reserves of nitrogen for the germinating seed. Their properties, like their functions, are rather different from most other proteins. They have a relatively high nitrogen content to allow them to store the nitrogen efficiently, they are stable to desiccation and rehydration, and are readily broken down by proteolytic enzymes during germination.

The NPN fraction makes up 2–12% of the total seed nitrogen. It consists principally of free amino acids, nitrate, ammonia and small peptides. Although it is quite a small proportion of the total nitrogen under most circumstances, it may rise as high as 35% when plants are subjected to environmental stresses.

Cereal proteins

The protein content of cereal grains is normally between 80 and 120 g kg^{-1} dry matter, and it is present in highest concentrations in the embryo and aleurone layer.

The nutritional quality (amino acid composition) of most cereal proteins is poor in comparison with those from other sources (Table 23.2). The general levels of essential amino acids are low, but the limiting amino acids vary in different cereals. Thus lysine is the first limiting amino acid in maize, sorghum, barley, wheat and triticale. The second to become limiting in maize is tryptophan; in barley and sorghum it is threonine. Diets in which these cereals predominate tend to be deficient in these amino acids, unless the diet also contains protein from other sources in which they are more abundant.

Cereal proteins are subdivided into albumins, globulins, prolamins and glutelins on the basis of their solubility (Chapter 5). Although this classification is rather crude, the classes correlate to a considerable degree with the function and amino acid composition of the proteins. In most cereals the storage proteins are mainly prolamins or glutelins, whilst non-storage proteins are albumins and globu-

Table 23.2 Essential amino acid content of cereals and other foods

Amino acid	Barley	Rice	Pea	Cabbage	Beef	Cod
			(g amino acid kg food[-1])			
ile	3	2.6	2.5	1.0	10.4	9.2
leu	5.7	5.6	4.0	1.8	16.3	14.7
lys	2.2	2.5	4.3	1.0	18.5	17.0
met	1.4	1.4	0.6	0.3	5.5	5.0
phe	4.3	3.3	2.7	1.0	9.1	7.2
thr	2.8	2.3	2.3	1.2	9.4	8.3
trp	1.4	0.9	0.6	0.3	2.6	2.0
val	4.2	3.9	2.7	1.4	10.7	10.0
arg	4.1	5.1	5.4	2.8	13.7	11.1
his	1.8	1.6	1.3	0.9	7.5	5.0

Data from Amino Acids and Fatty Acids (1st Supplement to The Composition of Foods, 4th edition) are reproduced with the permission of The Royal Society of Chemistry and the Controller of Her Majesty's Stationery Office.

lins. As the distinction between albumins and globulins is not clear-cut, they are often grouped together. In most cereals there is also an insoluble residue which contains structural proteins, such as those of the cell wall.

In barley, wheat, maize, rye and sorghum, the prolamin fraction makes up most of the protein (Table 23.3). In rice, the most abundant types of proteins are glutelins, whilst in oats they are globulins. In cereal processing the term gluten is often used to refer collectively to the endosperm proteins.

When the compositions of the different fractions are compared (Table 23.4) it is clear that the low content of essential amino acids in most cereals (of which barley is typical) is determined mainly by the composition of the prolamin fraction, as this is most abundant, accounting for 30–60% of the total grain nitrogen.

The prolamins from each species are given trivial names which are often used and which are shown in Table 23.3. The amino acid composition of the prolamins differs from species to species, but when compared to typical proteins they are all rich in glutamine and proline, and deficient in lysine and most other charged amino acids. Tryptophan is absent from zein but present in reasonable quantities in hordein and gliadin. The glutelin fraction contains insoluble enzymes, ribosomal proteins and some membrane proteins.

Table 23.3 Percentage composition of the proteins from cereal seeds

Cereal	NPN+ albumin +globulin	Prolamin	Glutelin	Insoluble	Name of prolamin
Barley	27.2	45.2	18.0	5.2	Hordein
Wheat	33.1	60.7	---6.2---		Gliadin
Maize	6.8	55.4	22.9	–	Zein
Rye	26.4	40.5	24.1	4.3	Secalin
Sorghum	13.2	54.5	25.5	–	Kafirin
Rice	15.7	6.7	61.5	15.4	–
Oats	67	9	23	–	Avenin

Modified with permission from Bright, S.W.J. and Shewry, P.R. (1983) Improvement of protein quality in cereals., *CRC Critical Reviews in Plant Science*, 1, 49–93. Copyright CRC Press, Boca Raton, Florida. © 1983.

Table 23.4 Essential amino acid composition of protein fractions from barley grain

Amino acid	Globulin	Albumin	Prolamin	Glutelin
	(g amino acid kg protein⁻¹)			
ile	33	62	54	52
leu	68	86	69	87
lys	53	67	7	40
met	15	24	13	19
phe	28	51	30	36
thr	33	46	26	42
trp	8	15	8	13
val	55	78	47	66
arg	110	65	30	60
his	18	25	13	25

Modified with permission from Folkes, B.F. and Yemm, E.W. (1965) The amino acid content of the proteins of barley grains. *Biochemical Journal*, **62**, 4–11.

Unusually, in rice the major storage protein is also a glutelin.

As the major storage proteins of oats and rice (a globulin and a glutelin, respectively) are relatively rich in lysine and threonine, their protein is of nutritionally higher quality than that of most other cereals.

The non-storage proteins, generally albumins and globulins, constitute the basic structural and metabolic machinery of the cell. The amino acid composition of these proteins is usually much better than that of the storage proteins. In both fractions, lysine and threonine make up over 3% of the proteins (Table 23.4). The corresponding fractions from maize and wheat have compositions similar to those of barley.

Many attempts have been made to improve the nutritional quality of cereal proteins. The strategies employed have included:

- reducing the proportions of nutritionally poor storage proteins – some spontaneous high-lysine mutants of maize, such as opaque-2, have a reduced prolamin content together with increased levels of lysine-rich proteins;
- increasing the proportion of lysine-rich albumins and globulins – the high-lysine barley line, Hiproly, has increased levels of metabolic proteins;
- increasing the free amino acid content;
- modifying the structural genes coding for the storage proteins by site-specific mutagenesis to increase their lysine content.

Although the last approach is attractive, it is difficult to achieve. A number of factors contribute to this:

- the proteins are encoded by complex multigene families;
- genetic manipulation of cereals (as with all monocotyledons), especially gene transfer and regeneration, are very difficult to achieve;
- modification must not prevent the normal functions of the seed.

Legume proteins

Legume proteins are of higher quality than those of cereals, mainly because of their higher lysine content (Table 23.2). The proteins, however, have a low content of the sulphur-containing amino acids, cysteine and methionine. Some legume seeds also contain toxic compounds which are discussed in more detail in Chapter 5.

Most proteins contain only 1–2% methionine, but some plants contain sulphur-rich proteins in their seeds. Brazil nuts and sunflower, for example, have proteins of molecular weight 10 000–12 000 which contain 15–20% methionine and 8% cysteine. Introduction of the Brazil nut genes into soyabean or other legumes would be one way of increasing the sulphur content of the seed. Maize also contains a small sulphur-rich protein, but it is present in only very small amounts. It may also be possible to increase expression of the gene for this protein to increase the sulphur amino acid content of maize.

23.3.4 SEED MINERALS

Phytic acid (inositol hexaphosphate) is an important component of cereal seeds (Figure 23.1). It serves as a store of phosphate for the seedling, but its phosphate groups carry negative charges which bind calcium, iron and zinc ions. This reduces the availability of these minerals to animals and can result in deficiency symptoms even when the diet appears adequate. The phosphate groups are removed from phytic acid through the action of an enzyme, phytase, which is produced by germinating seeds, fungi and microorganisms. Thus, phytate in the diet of ruminants is broken down in the rumen and the phosphate is made available. In non-ruminant animals, however, up to two-thirds of the phosphate in the diet may be in the form of phytate and be unavailable. This is not only nutritionally important but also leads to excretion of phosphate and pollution of the environment.

23.4 GERMINATION

When seeds germinate it is essential that the stored reserves are degraded to provide nutrients at the correct time during embryo growth. This is achieved through hormonal control mechanisms which increase the activity of proteolytic enzymes, starch-degrading enzymes, nucleases, phytase, etc. at the appropriate time.

Some of these enzymes are synthesized *de novo* during germination but others are formed during seed development and are stored, in an inert form, during seed maturation. The latter type only require activation or release during germination.

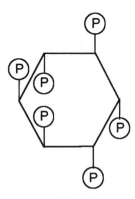

Figure 23.1 Structure of phytic acid.

23.4.1 STARCH BREAKDOWN

In most seeds starch forms the main energy reserve. It is degraded through the action of a variety of amylolytic enzymes to form simple sugars which may be converted into sucrose.

A number of enzymes are potentially capable of contributing to starch breakdown.

- α-Amylase hydrolyses α-1,4 bonds of amylose at random to produce a mixture of glucose and maltose. (Note that the names α- and β-amylase do not refer to the types of glycosidic bonds hydrolysed.) α-Amylase does not degrade maltose to glucose. It also attacks amylopectin but cannot hydrolyse α-1,6 bonds at branch points and does not act well on bonds near to branch points or near to the end of the chain. It exists as groups of isoenzymes with different isoelectric points, some of which may be able to attack native starch grains.

- β-Amylase is not a very widespread enzyme, being restricted to cereal seeds and a few other plants. It hydrolyses α-1,4 bonds of amylose at the penultimate position releasing maltose. It also attacks amylopectin, but stops two to three glucose units from branch point and cannot hydrolyse α-1,6 bonds at branch points. It eventually degrades amylopectin to limit dextrin, during which 50–60 % of the glucose is released. It cannot attack starch granules, and in many cases its presence does not appear to be essential for starch breakdown to take place.

- α-Glucosidase hydrolyses α-1,4 bonds of maltose and oligosaccharides.

- oligo-1,6-Glucosidase hydrolyses α-1,6 bonds in isomaltose and dextrins. It is present in germinating seeds and, together with α and β-amylases, allows complete degradation of starch by attacking the branch points in dextrins released from amylopectin.

- Starch phosphorylase is a widespread enzyme in leaves and many storage organs. Although present in some seeds it does not appear to be of major importance in mobilizing their reserves. The product of action of the enzyme is glucose-1-phosphate.

In most seeds, including cereals, α-amylase and β-amylase appear to be more important than phosphorylase, but starch breakdown in pea seeds may be brought about by starch phosphorylase.

Breakdown of starch reserves in wheat and barley has been extensively studied. It appears to be regulated by the plant hormone gibberellic acid (GA). This is produced by the embryo as it starts to grow, is released and passes to the aleurone layer, where it increases the activity of many hydrolytic enzymes. These are then released into the starchy endosperm where the reserves are broken down. The presence of GA activates the genes for α-amylase in the aleurone layer, leading to increased synthesis and release. There are two isoenzyme groups of α-amylase, one with a low isoelectric point (pI 4.5–5.0) and one with a high pI (5.9–6.5). These differ in their capacity to attack native starch grains and in the time at which they are produced during germination. Similarly α-glucosidase and limit dextrinase also increase in activity in response to GA. In contrast, β-amylase is present in the endosperm of the dry seed in an inactive form. During development of the seed it becomes covalently linked to proteins in the protein bodies by formation of disulphide bridges. In this form it lacks enzyme activity, but the active form is released during germination when GA-induced proteolytic enzymes break down the proteins to which it is linked.

23.4.2 BEER AND WHISKY PRODUCTION

About 20% of the barley grown in Britain is used for malting and hence for alcohol production. Alcohol production depends on the fermentation of simple sugars which are released from stored starch when cereal grains germinate. The initial stage is called malting: barley grains are allowed to germinate for a limited time, during which amylases are produced and partial breakdown of starch into simple sugars occurs. This process is stopped by gentle heating of the grain (kilning) to produce malt. Peat may be added to the fuel to impart characteristic flavours, for example in malt whisky production. The rootlets of the germinated grain (culms) are removed and may be used as cattle feed. Malt is ground and extracted by heating in water at about 65° C for several hours (mashing). This process allows starch to be degraded by enzymes to form sugars and it gives rise to wort. If beer is being produced, hops are added and the wort is boiled to denature enzymes, precipitate proteins and sterilize the liquid. After cooling, brewer's yeast is added which ferments the sugars to produce ethanol. The biochemistry of this process is described in Chapter 16.

Malt whisky is produced using barley only as the source of both enzymes and fermentable sugars. During this process no hops are added and the wort is not boiled. After fermentation the wash is fractionated by distillation through several stills and the product is stored in oak casks for at least 3 years, usually longer. Grain whisky is made by adding cooked maize to malted barley as a source of additional starch. The barley serves mainly as a source of starch-degrading enzymes.

For making malt whisky, very stringent quality standards are set for the barley. The grain acts as a source of both starch and amylases which must be present in the correct proportions. In general, the starch and protein content of grain are inversely related to one another. Thus grain with a high protein content will have a high amylase activity but a low starch content and will be unacceptable for malt production. Maltsters normally accept only barley with a nitrogen content of less than about 1.7% nitrogen (on a dry matter basis). The presence of large amounts of nitrogenous constituents in malting barley results in unacceptable flavours and may produce a cloudiness or haze when used in beer production, particularly when the beer is chilled. High activities of β-glucanase enzymes, which attack glucans in cell walls, are also desirable. Failure to break down the walls slows down starch degradation during malting and leaves residues which may hinder filtration steps in brewing.

In addition to its chemical composition, maltsters also require grain with a high and uniform germination. This minimizes losses resulting from some grains developing too far and using their sugars as an energy source to support their growth. Thus two-row barleys are preferred because of their more uniform grain size. The speed and uniformity of germination may be improved by addition of gibberellic acid during malting.

23.4.3 PROTEIN BREAKDOWN

Plants contain many proteases which are required not only for germination, but also for protein turnover, which takes place in all plant tissues to a greater or lesser degree. In seeds, proteases are associated with the protein bodies, which function as specialized vacuoles. These enzymes are mainly acid endopeptidases and carboxypeptidases with acidic pH optima (3–6).

The production of proteases in cereals appears to be induced by GA. In addition to increasing protein breakdown, these enzymes may also release other enzymes, e.g. β-amylase, from their association with proteins.

Hydrolysis of stored proteins releases a mixture of amino acids. The balance of amino acids demanded by the developing tissues usually does not match that of the amino acids released from stored protein, so a considerable amount of interconversion by transamination, breakdown and resynthesis of carbon skeletons takes place.

23.4.4 LIPID BREAKDOWN

Hydrolysis of stored lipids produces fatty acids and glycerol. The glycerol is converted into dihydroxyacetone phosphate in the cytoplasm and then used mostly for synthesis of sucrose. Fatty acids are degraded by the β-oxidation pathway (Chapter 13) to form acetyl-CoA which is used to synthesize carbohydrates through use of the glyoxylate cycle (described in Chapter 11).

VEGETATIVE GROWTH OF PLANTS

24.1 INTRODUCTION

Once the root system has developed and photosynthesis is well established, seedlings are no longer dependent on nutrients supplied by the seed. However the functions of the root and the shoot are closely integrated and each is dependent on the other for its survival.

The root system takes up water and mineral elements from the soil. Some elements, such as nitrogen, may also be converted into organic compounds in the root. The uptake and assimilation of materials from the soil requires energy and carbon skeletons which are largely provided from the shoot in the form of sucrose via the phloem. Some of these sugars are oxidized in respiration, whilst others are used to make the carbon skeletons themselves. Thus the root system relies on shoot-produced sugars. It may also be dependent on shoot-produced hormones to stimulate and regulate its growth. The shoot, on the other hand, depends on the root system for a supply of water and nutri-ents and for root-produced hormones such as gibberellins and cytokinins, which in turn stimulate shoot development.

An important metabolic function of leaves is to carry out photosynthesis, providing a source of energy for the entire plant. However, leaves also control water loss, are involved in nitrogen metabolism and storage, and undergo senescence. In addition to carrying out photosynthesis, chloroplasts also assimilate ammonia produced in photorespiration, synthesize amino acids, and reduce sulphate and nitrite. Leaves serve as a temporary store for sugars produced in photosynthesis (in the form of starch) and of protein (mainly in the form of Rubisco) which may later be mobilized and contribute to the development of storage organs.

The stem transports nutrients and water and may carry out significant amounts of photosynthesis, and store nutrients. In some plants, particularly at early stages of their development, the shoot consists mainly of leaf tissues. In others, however, stem tissue may make up a considerable proportion of the shoot, and some of this may become lignified and take on a structural role. This is reflected in the high fibre content of such shoots. This consists principally of the thickened walls of cells which provide support and allow transport through the shoot to take place.

The components of the shoots of some plants are of great commercial significance. Leaves and stems are eaten by man in the form of green vegetables and salads, and by animals in the form of herbage. They may be used in the form of timber and fibre products for construction purposes, and as a source of raw materials for fermentation and other industrial processes.

During vegetative growth some plants may produce storage organs such as tubers or roots which store carbohydrates and which are important agricultural products.

24.2 COMPOSITION OF SHOOTS

24.2.1 PLANT CELL WALLS

One of the characteristic features of plant cells is the presence of a cell wall surrounding the plasma membrane. The properties of these walls are important for a number of reasons.

- Plants have no separate skeleton to provide support, and the cell wall provides structural rigidity through the development of turgor.
- The presence of a cell wall limits the growth of the cell. Before plant cells can grow, the cell wall must be weakened. One of the important functions of auxins and possibly also of other growth-stimulating plant hormones is to initiate this weakening.
- Interaction with the cell wall is one of the first events in the attack of pathogens on plant cells. The properties of the walls are therefore of major importance in determining resistance or susceptibility of plants to disease. As pathogens attack the cell wall, fragments of the wall components are released and these may induce the production of phytoalexins – a sort of primitive immune system.
- The composition of the cell wall has very large effects on the ability of both ruminants and non-ruminants to digest foodstuffs. Changes in the wall occurring during development change the digestibility and palatability (fruit ripening, etc.) of both animal and human foods.

At least 90% of the structural material of cell walls of all higher plants is made up of the polysaccharides cellulose, hemicellulose and pectin, described in Chapter 3. The remaining 10% is made up of a glycoprotein called extensin. This protein is rich in the unusual amino acid hydroxyproline and also in alanine, serine and threonine. It is very insoluble and difficult to extract from the wall. Most of the hydroxyproline residues are glycosylated with arabinose-containing oligosaccharides. Lignins (Chapter 5) are also present in some cell walls.

Linkage of cell wall components

The cell wall of the sycamore maple (*Acer pseudoplatanus*) has been very thoroughly studied and much is known about the detail of its structure. The structure of the cell wall of other dicotyledons is thought to be similar.

The wall consists of cellulose microfibrils running through a matrix made up of the other polysaccharides (Figure 24.1). The microfibrils tend to run parallel to one another and are cross-linked by the other sugars. The following are important in maintaining this three-dimensional structure:

- xyloglucan (hemicellulose) is hydrogen-bonded to cellulose fibrils via its glucan backbone [1];
- xyloglucan is linked through the reducing ends to the arabinogalactan (pectin) [2];
- arabinogalactans are covalently linked to rhamnogalacturonans (pectin) via rhamnose residues [3].

The rhamnogalacturonan therefore cross-links cellulose chains via arabinogalactans and xyloglucans.

In grasses and possibly other monocotyledons, single units of ferulic acid and *p*-coumaric acids are esterified to the arabinose side chains of the arabinoxylan fractions of the hemicellulose. These feruloyl esters may be involved in cross-link formation catalysed by peroxidase.

It is thought that cell-wall protein may be cross-linked by the peroxidase-catalysed oxidative coupling of tyrosyl residues to form isodityrosine. This may tighten the wall structure and limit cell expansion.

Lignin

The name lignin does not refer to a single chemical structure, but rather to a range of structures of high molecular weight formed by random polymerization of phenolic monomers.

The polymerization of *p*-coumaryl, coniferyl and sinapyl alcohols to form lignin is initiated through the action of cell-wall peroxidases which convert these compounds to aroxyl radicals (Figure 24.2). The unpaired electron within these moves around the radical (mesomerism). Covalent bond formation between unpaired electrons in two radicals leads to the formation of dimers. The nature of the bond formed depends on the exact position of the unpaired electrons at the moment of reaction. The products can themselves generate further radicals so that complex polymers are formed, containing structures such as those shown in Figure 5.6. The formation of this network enmeshes the other wall components, but may also result in formation of covalent bonds between lignin and polysaccharide molecules. Covalent bonding to the

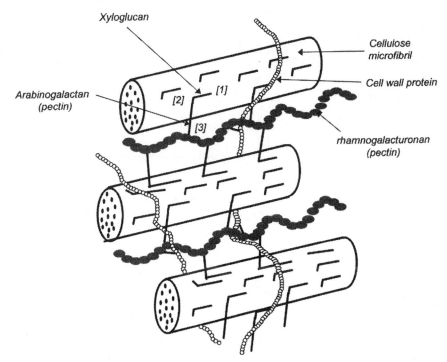

Figure 24.1 Linkage of the components of a plant cell wall. The cellulose microfibrils are cross-linked by xyloglucans. These are hydrogen bonded to the microfibrils [1] and covalently linked to arabinogalactans [2]. The arabinogalactans are also covalently linked to rhamnogalacturonans [3].

arabinoxylans and glucuronides of the hemi-celluloses is common.

Role of lignin in digestion of plant cell walls

Lignin is the main factor limiting the digestibility of forages. The bonds in lignin are extremely resistant to enzymic and chemical attack. The presence of lignin also reduces the ability of microorganisms to degrade cell-wall proteins and carbohydrates. In some types of forage, lignification renders the material more indigestible than in others. The lignins from grasses are very much more soluble in alkali than those from wood or non-grass forages. This seems to be because of the high content of esters and the low methoxyl group content of grasses. In contrast, ether-type linkages to hemicellulose are not restricted to the grasses and these are more stable in alkaline conditions.

Role of cell wall in control of cell expansion

Auxin treatment causes cells to expand. It acts by promoting pumping of protons from the cell into the cell wall, thus decreasing the wall pH. This acidification weakens bonds between components of the hemicellulose fraction and, because the wall is subjected to turgor pressure, allows the cellulose microfibrils to move away from one another. This mechanism

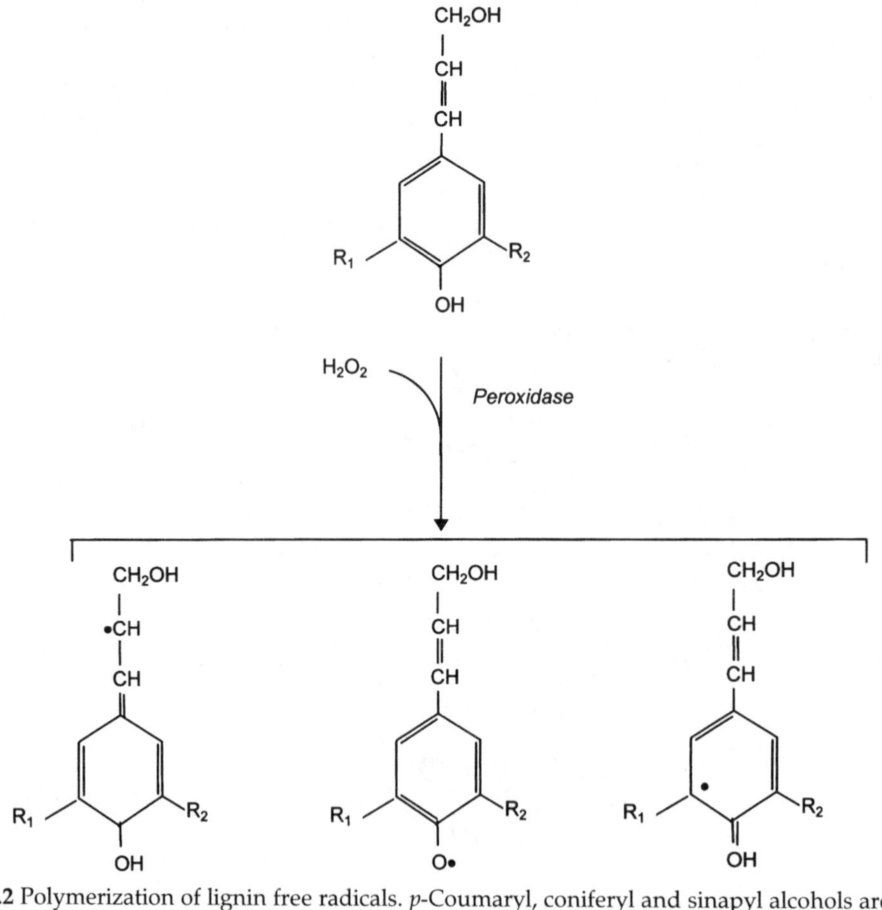

Figure 24.2 Polymerization of lignin free radicals. *p*-Coumaryl, coniferyl and sinapyl alcohols are converted to free radicals by the enzyme peroxidase. The unpaired electron moves rapidly around the free radical and may form bonds with other free radicals or polysaccharides.

allows cells to expand perpendicular to the orientation of the microfibrils but not parallel to them, so that the direction of cell expansion can be regulated by the orientation of microfibrils in the wall.

24.3 IMPORTANT SHOOT CROPS AND THEIR PRODUCTS

24.3.1 TEMPERATE GRASSES

Grasses and other pasture plants are the main components of the diet of herbivores. The overall composition of grasses and green vegetables are given in Figures 9.7 and 9.8, where they are compared with the composition of cereal grains.

Carbohydrates

In comparison with cereal grains, temperate grasses have a much higher content of cellulose, hemicellulose and pectin associated with cell walls and they usually contain little starch. Instead the main storage carbohydrates are the fructans, but there may also be appreciable amounts of sucrose, fructose, glucose and other small sugars. The chemical nature of fructans is described in Chapter 3. They are very soluble in water and tend to be found mainly in the stem. Their content in grasses may vary, being promoted by high light intensity, high photosynthetic rates and low temperature. Under favourable conditions they may constitute to 30% of the dry matter. There is marked diurnal variation in the levels.

The fructan content of grasses is very important in determining both palatability and suitability of the grass for silage making. They serve as the main substrates for fermentation occurring during ensiling. Both the fructose liberated by hydrolysis of fructans, and pentoses released by hydrolysis of hemicellulose, may be fermented to lactic acid during ensiling (Chapter 16).

The cellulose and hemicellulose content of grasses is variable, increasing as the plants mature. In addition, the cell-wall components also become increasingly lignified towards maturity. This decreases the digestibility of the other nutrients, except for the soluble carbohydrates which are readily utilized.

Mature forages are of lower quality than young ones. This results mainly from the increase in the proportion of the forage consisting of stems and the rapid decrease in digestibility of this fraction with age. Although in many grasses the leaves assist the stems in providing support and may become lignified with age, they usually lose their quality less quickly than stems. In most forage legumes the stems support the plant and the leaves serve as metabolic organs, so the leaves do not become highly lignified and do not lose quality. In a few grasses, e.g. timothy grass (*Phleum pratense*) and sugarcane, the stem may serve as a storage organ so that it may have higher quality than the leaves.

Proteins

The amino acid composition of proteins from different grasses varies very little. Up to 50% of the protein is made up of Rubisco, the enzyme responsible for fixation of CO_2 in photosynthesis. Grass proteins are of higher biological value than seed proteins: they are rich in arginine and contain appreciable quantities of lysine. The first limiting amino acid is methionine and the second is isoleucine. There is also some non-protein nitrogen present, principally in the form of free amino acids and their amides (e.g. glutamine and asparagine). Nitrate may also be present under some circumstances and may be toxic because of its reduction to nitrite in the rumen.

Lipids

The lipid content of grasses is usually low (less than 60 g kg^{-1} dry matter). There is very little triacylglycerol, most (60%) consists of glycolipid which forms part of the membranes within the leaf. Between 60 and 75% of the total fatty acids are in the form of linolenic acid.

24.3.2 TROPICAL GRASSES

In tropical grasses, starch is the main storage carbohydrate and accumulates mainly in the leaves.

High temperature appears to be a major factor which increases the rate of lignification. In this context it is interesting that C_4 grasses are usually of lower nutritional value than C_3 ones (tropical forages in general are of lower nutritional value). Tropical grasses have a low protein content which appears to be an intrinsic part of C_4 metabolism.

Sugarcane is an example of a tropical grass which stores sucrose in the stem. It provides about 65% of the world sugar supply and, as a result of breeding programmes, is very high yielding. It is the most efficient of crop plants in converting solar energy into food.

24.3.3 FORAGE LEGUMES

Legumes also form valuable forage crops. In pastures, clovers may be common. Clovers are nutritionally superior to grasses in their protein content, and in some minerals (calcium, phosphorus, magnesium, cobalt and copper), and their nutritive value falls little with age. Their inclusion in the diet also increases palatability and feed intake. Sucrose and starch are the main carbohydrates; fructans are virtually absent.

A number of nutritional disorders may result from ingestion of legumes. Bloat, resulting from sheep and cattle grazing on clover- or lucerne-dominated pastures, is caused by inadequate eructation of fermentation gasses as a result of foam formation caused by soluble legume leaf proteins. Oestrogenic compounds in some legumes lead to infertility in sheep. Some may contain toxic amino acids or vitamin K antagonists which cause sweet clover disease.

24.3.4 BRASSICAS

Different varieties of brassicas are grown for human consumption or as forage crops. The versatility of some species is seen in the variety of parts of the plant which may be eaten. Thus cabbages are apical buds, Brussels sprouts are axillary buds, cauliflower and broccoli are flowers, etc. In the human diet they are mainly used to add variety and as a source of fibre, vitamins (especially C and E) and minerals (e.g. calcium). The presence of goitrogens and S-methylcysteine sulphoxide (Chapters 3 and 5) may affect animals eating large quantities of brassicas.

24.3.5 STRAWS

Straws consist of the stems and leaves of cereals or legumes which remain after collection of the seed. They all have high fibre and lignin contents and are of low nutritive value. Their use is restricted to ruminants. The dietary intake of straws is low unless they are improved by addition of nitrogen in the form of protein or urea. Legume straws have a higher protein and mineral content than cereal straws.

The digestibility of straws may be improved by treatment with alkalis. This breaks ester bonds between lignin and cell-wall polysaccharides, making the carbohydrates more available. Sodium hydroxide or ammonia are usually used, and in the latter case the added nitrogen also increases the crude protein content.

24.3.6 HAY

Hay-making provides a means of storing herbage by reducing its moisture content. Fresh herbage typically has a moisture content of 650–850 g kg^{-1} but this must be reduced to less than about 150 g kg^{-1} if the product is to be stored successfully. This low moisture content inhibits respiration and microbial growth and so minimizes loss of nutrients from the hay during storage.

Some loss of nutrients may take place during drying. Thus, plant enzymes may degrade fructans to fructose and some sugars may be respired before the moisture content has been reduced sufficiently. In addition, microbial spoilage, oxidation and leaching of soluble materials may also occur if the drying process is prolonged. The protein and mineral content of legume hays are higher than those of cereal hays. Overheating during storage of hays may lead to loss of protein as a result of cross-linking of amino acids, especially lysine, to cell-wall carbohydrates. Treatment of hays with preservatives, such as propionic acid, reduces fungal growth and allows them to be stored at moisture contents of up to 500 g kg^{-1}.

24.3.7 SILAGE

Silage is produced by controlled fermentation of a green crop of high moisture content. The aim of ensiling is to encourage the growth of microorganisms which bring about fermentation of sugars, converting them to organic acids. This lowers the pH and so allows storage without spoiling. During the fermentation process anaerobic conditions must be maintained and the pH must be reduced sufficiently to minimize growth of undesirable microorganisms. The fermentation pathways which function during ensiling are described in Chapter 16.

The principal bacteria active in conditions in the silo are lactic acid bacteria. These are facultative anaerobes which can survive under aerobic or anaerobic conditions. Under anaerobic conditions they ferment water-soluble carbohydrates to organic acids, mainly lactic acid. Some degradation of hemicellulose occurs, yielding pentose sugars which may also be fermented to lactic or acetic acid. They grow rapidly under acidic conditions. Strictly anaerobic *Clostridia* may also be present, and some species ferment sugars to butyric acid, or ferment amino acids to acetic and butyric acids, ammonia and amines. Enterobacteria ferment soluble sugars to ethanol, acetic acid and hydrogen. They are facultative anaerobes and therefore compete with lactic acid bacteria. However their growth is optimal at pH 7 and their activity is significant only at the start of the fermentation process.

The exact course of the fermentation depends on the nature of the crop, its moisture content, exclusion of air, etc. Well-made silage has a pH around 4 and a high lactic acid content. Good silage also usually contains acetic acid, smaller amounts of propionic and butyric acids and some ethanol. Most of the protein of the original crop is hydrolysed to amino acids but these are not extensively deaminated, so the ammonia content is low. If anaerobic conditions are maintained little loss of energy or dry matter takes place during ensiling. Badly preserved silage, in which fermentation by *Clostridia* or *Enterobacter* predominates, often results when the moisture content is too high. It is usually only weakly acidic (pH 5–7) and the predominant acids are acetic or butyric. Substantial degradation of amino acids to form ammonia or toxic amines may occur.

The quality of the fermentation process may be improved by the use of inoculants of lactic acid bacteria. Selective inhibition of fermentation by other types of bacteria can be achieved through addition of acids such as formic acid, possibly together with formaldehyde.

24.3.8 WOOD

Wood is made up of cells with heavily thickened and lignified walls and with very little cell content. In softwoods the cells are mainly tracheids, which provide both conduction and support. Hardwoods contain a mixture of vessels, which carry out conduction, and fibres, which provide support. Because hardwoods have fewer fibre-like cells and these tend to be shorter than in softwoods, they yield weaker papers. They may be used for high quality writing papers and have attractive decorative properties. Wood at the centre of the stem (heartwood) is often darker and harder than that at the outside (sapwood). The cells in the heartwood are dead and contain extractives, some of which are polyphenolics, and which penetrate the walls and lumen of the cells. These give the heartwood its darker colour and make it aromatic as well as making the cells more resistant to attack by fungi and insects.

Wood consists primarily of cell walls and thus contains cellulose, hemicellulose and lignin. Typical values for the gross composition of softwoods and hardwoods are given in Table 24.1. The differences are relatively small, although hardwoods tend to contain slightly more cellulose and less lignin. The structure of hemicellulose molecules varies widely between species. However, in general, the main constituent of softwood hemicellulose is mannose, followed by xylose, glucose, galactose and arabinose. In hardwoods xylose is by far the most abundant monomer, followed by mannose, glucose and galactose with small amounts of arabinose and rhamnose.

As noted in Chapter 5, lignin of softwood trees is formed by polymerization of coniferyl alcohol units whilst that of hardwoods is formed from both coniferyl and sinapyl alcohol groups (see Table 5.1). Hardwood lignin thus has a higher methoxy (-OCH$_3$) group content (18–22%) than softwood lignin (12–16%). Hardwood lignins are also of lower molecular weight, possibly because syringyl groups cannot cross link at the 5 position because of the extra methoxy group which they contain. Carbon–carbon (–C–C–) bonds are much more difficult to break than ether (–C–O–) bonds. The lower proportion of –C–C– bonds in hardwood lignin and its lower molecular weight makes it pulp more rapidly than softwood lignin.

Wood extractives

These can be extracted with non-polar and polar solvents. The term resins is used to refer to those extracted by organic solvents. Many different compounds may be present. Some resins, such as frankincense, are produced commercially.

In different species of softwoods, resins may contain resin acids (which are present in turpentine), fatty acids, sterols and polyphenolics. Hardwood extractives include mono- and triterpenoids, triacylglycerols and polyphenolics. These compounds may be responsible for the durability and colour of the wood, and may be of value in the tanning industry.

Paper making

This depends on the reduction of wood to its constituent fibres (pulping). Paper is made from pulp by depositing the suspension of fibres on a moving screen to form a web, in which the fibres are randomly aligned and interwoven. The web is then pressed and dried to produce paper, in which the fibres are held together by hydrogen bonds.

Pulp can be prepared by mechanical or chemical methods. Mechanical pulps are made by compression and grinding of wood chips, combined with heat treatment. They are chemically unaltered and thus have a high lignin content. They tend to be bulky and porous and cannot be bleached to yield a very white product. They also tend to yellow when exposed to light. This process produces paper which is well suited for printing of newspapers, etc.

Chemical pulping aims selectively to remove lignin but there is also some loss of carbohydrates, especially hemicellulose, so that yields of pulp are lower than those obtained by mechanical pulping. Chemical pulping may be achieved by sulphite pulping or, more commonly, by the Kraft process. In sulphite pulping, chips are heated with solutions of sulphur dioxide at various pH values. This cleaves ether bonds in the lignin and sulphonates the products to give rise to small, soluble compounds. Some of these lignosulphonates are used as pellet binders in animal feedstuffs. In Kraft processing, solutions of sodium hydroxide and sodium sulphide are

Table 24.1 Gross composition of wood from hardwood and softwood trees

Substance	Softwoods (%)	Hardwoods (%)
Cellulose	42	45
Hemicellulose	27	30
Lignin	28	20
Extractives	3	5

Reproduced with permission from: Walker, J.C.F. (1993) *Primary Wood Processing*, Chapman & Hall, London.

used. These also cleave ether bonds and degrade the lignin, but they are less selective than sulphite so there is greater loss of carbohydrates. Chemical pulps produced by these processes can be bleached to very high whiteness. Bleaching removes further lignin and may be achieved by treatment with chlorine, oxygen or chlorine dioxide. Use of chlorine as a bleaching agent has declined because of the need to minimize the environmental impact of organochlorine compounds produced.

24.4 IMPORTANT ROOT AND TUBER CROPS

In addition to their roles in absorbing nutrients and anchoring the plant to the ground, some roots have developed into storage organs. This is particularly the case with biennials and herbaceous plants, allowing them to overwinter successfully. Grass roots are also important in storing reserves to supply the needs of the growing shoot in spring. The stored reserves in the roots of weeds such as bracken, docks and dandelions are central to their survival and to the difficulty of eradicating them. Tubers, such as those of the potato, are underground storage organs which are formed by swelling of specialized underground stems called stolons and they therefore exhibit many of the characteristics of stems. Roots and tubers are important crops and are used as both human and animal foods.

24.4.1 ROOT CROPS

True roots make a small contribution to the human diet but are widely used in animal feeds. Thus carrots, parsnips, radish and beetroot are used as human food, and various root crops including turnips and swedes (*Brassica* spp.) and mangels and fodder beet (forms of *Beta vulgaris*) provide animal feed. Sugar beet provides 35% of the world sugar supply, and also produces by-products which are used as animal feeds. The storage organ arises partly from swelling of the tap root and partly from the hypocotyl. In some instances (e.g. carrots and parsnips) the hypocotyl is relatively unim-

portant but in others (e.g. radish) most of the growth occurs in the hypocotyl and is above ground.

Root crops generally have a low protein and dry matter content. A considerable part of the dry matter consists of sugars. In swedes and turnips the main sugar is glucose, whereas in mangels, fodder beet and sugar beet it is sucrose. Sugar beet is grown principally for sugar production and the residue after extraction of the sucrose (sugar beet pulp) is used for feeding to ruminants.

24.4.2 TUBERS

Important tubers include potatoes, cassava, sweet potatoes and yams. Tubers generally have a higher dry matter and a lower fibre content than root crops. They also differ in having starch as their main storage reserve instead of free sugars.

Potato

The potato is an important crop in Europe, where 75% of world production takes place. Cool nights and warm days promote development of its tubers. Typically 70% of the dry matter of the tuber consists of starch. Some phosphorus is associated with potato starch, which makes it more viscous than most starches and gives it specific industrial uses. About 6% of the dry matter is protein, and there are substantial quantities of low-molecular-weight nitrogen compounds, including free amino acids. The quality of the protein is relatively high, being quite rich in lysine but it is limited by its methionine and cysteine content. Potatoes contain several alkaloids, of which the most important are derivatives of solanidine (see Chapter 5). These compounds are toxic when they occur at levels in excess of about 0.15 mg g^{-1} and environmental factors such as exposure to light and sprouting may promote their accumulation.

Because of their low fibre content potatoes are suitable for feeding to pigs and poultry, however the protein in uncooked tubers has a very low digestibility because of the presence of a protease inhibitor which prevents digestion. As the inhibitor is denatured by heat, it is normal to cook potatoes before feeding to these animals. This is not necessary for ruminants because the inhibitor is destroyed in the rumen.

Browning reactions

Plant products are subject to several types of 'browning reactions'. These occur in potatoes and are also common in other plants. Brown colours may arise in plant products in several ways. In uncooked products the enzyme phenolase may act on tyrosine to produce quinones which undergo further, non-enzymic oxidations, to form coloured products such as melanin. A range of non-enzymic browning reactions may also occur. In cooked potatoes, for example, the phenolic compound, chlorogenic acid, may react with iron to produce grey-coloured products. Differences in the susceptibility of different varieties are related to the amount of chlorogenic acid which they contain.

The Maillard reaction may take place at high temperatures between carbonyl groups of sugars and free amino groups of amino acids or proteins. The amino acid lysine is particularly susceptible to such reactions. The products are brown in colour and are mainly responsible for the coloration which accompanies frying, roasting and other cooking carried out at high temperature. In some instances the colour may be unacceptable and may also significantly reduce the availability of amino acids in the food.

A considerable proportion of the potato crop is processed into crisps, chips and dried products. The free sugar content of potato tubers is important in determining their suitability for high-temperature cooking. The reducing sugars react with amino acids by the Maillard reaction and may result in development of an unacceptable degree of browning. Potatoes to be used for making crisps must contain no more than 2.5–3 mg reducing sugar g^{-1} fresh weight. Unfavourable storage conditions, especially storage of tubers at temperatures below about $10°$ C, may substantially increase the concentrations of both sucrose and reducing sugars. This process, referred to as low-temperature sweetening, results from changes in the relative proportions of starch and free sugars because of differences in sensitivity to temperature of some of the enzymes involved in sugar metabolism. A particularly important factor is the disproportionately great loss of activity of phosphofructokinase (PFK) as the temperature is reduced, because of weakened interaction between the subunits of this allosteric enzyme. This inhibits glycolysis so that free sugars, produced by starch degradation, accumulate and are converted into sucrose.

Cassava (manioc)

Cassava is an important tropical perennial, tuber-bearing plant. It has a higher dry matter but lower protein content than potato tubers. The tubers can be used to make tapioca. The tubers of some varieties contain cyanogenic glucosides, including linamarin, which are broken down by hydroxy nitrile lyase, to form HCN (see Chapter 3). However, traditional methods of preparation of foods from these tubers minimize the breakdown or wash the glucosides from the tubers.

Sweet potato

Like cassava, sweet potatoes have a higher dry matter but lower protein content than potato tubers. They also have a high content of vitamins A, B and C.

25.1 FLOWERING

Most plants flower at some stage in their life. Flowers in the form of cut flowers or flowering pot plants are of importance both in the horticultural industry and as sources of perfumes; some, such as cauliflower and broccoli, may be eaten – cauliflower curd has a high protein and carbohydrate content. However the major importance of flowers is in the fruit or seed formation which follows.

In many plants, flowering takes place when a terminal vegetative bud, which normally gives rise to leaves and stem, is converted into a flower bud. The mechanisms which control this transition are complex and poorly understood. Environmental conditions may induce flowering in many plants, e.g. exposure to particular temperatures or day lengths, but the response to these stimuli is not uniform between species. Flowering behaviour may also be influenced by nutrition; the application of large amounts of nitrogen often favours vegetative growth at the expense of flower initiation. It is believed that flower induction is mediated by plant hormones whose concentration or distribution is influenced by environmental and other stimuli. This is discussed further in Chapter 27.

Flower initiation may be thought of as a switch which activates a pre-determined programme of development leading automatically to flower, seed and fruit production. Dramatic changes in metabolism and redirection of nutrients occur as flowers, seeds and their surrounding structures begin to develop.

25.2 FRUIT DEVELOPMENT AND COMPOSITION

Fruits develop from the tissues of the ovary. They vary considerably in shape and in the nature of the tissues which make up most of the fruit – in many the pericarp predominates and may develop in several layers, whilst in others the receptacle is the major structure. The composition of a range of fruits is given in Table 25.1.

Most fruits have a relatively low dry matter, protein and lipid content. Often between 60 and 80% of the dry matter is made up of carbohydrate which consists of both polysaccharides, such as starch, and simple sugars such as glucose and sucrose. The relative proportions of these may change during ripening and may contribute to changes in flavour.

Many fruits also contain large quantities of fruit acids and some vitamins, especially vita-

Table 25.1 Composition of typical fruits in the ripe state

Fruit	Water (g kg fruit^{-1})	Carbohydrate (g kg fruit^{-1})	Protein (g kg fruit^{-1})	Fat (g kg fruit^{-1})	Starch (ripe)/starch (unripe) (%)	Sugars (ripe)/sugars (unripe) (%)
Orange*	861	85	11	1	–	–
Tomato	931	31	7	3	–	–
Banana*	751	232	12	3	6	2000
Apple*	845	118	4	1	5	99

*Based on flesh only
Water, carbohydrate, protein and fat content from *The Composition of Foods*, 5th edn. are reproduced with the permission of The Royal Society of Chemistry and the Controller of Her Majesty's Stationery Office.

min C. The nature of the fruit acids differs from one type of fruit to another. Malate predominates in apples, bananas, cherries, plums and pears, but citrate is the main acid in fruits such as citrus, figs, raspberries and strawberries. Tomatoes and gooseberries contain approximately equal amounts of malate and citrate, and in grapes the principal acid is tartrate (Figure 25.1). These acids are stored in the cellular vacuoles in the flesh and contribute to the acidic taste of many fruits, particularly in their unripe state.

25.3 FRUIT RIPENING

As fruits ripen they undergo extensive changes in composition. Frequently there is a decrease in starch and an increase in free sugars, a loss of chlorophyll and an increase in yellow and red pigments. A loss of cell-wall components, particularly pectin, leads to softening of the cells and changes in texture. In apples loss of starch occurs without an increase in free sugars, whereas in bananas starch breakdown results in a dramatic increase in free sugar levels during ripening (Table 25.1).

25.3.1 CHANGES IN COLOUR

Ripening of fruits is often accompanied by a change in colour from green to predominantly red or orange hues. This is due to degradation of chlorophyll and increased synthesis and accumulation of carotenoids (Chapter 8) in the plastids, and of anthocyanins (Chapter 5) in the vacuole. However this process does not occur in all fruits – pears and kiwi fruit, for example, remain the same colour as they ripen. The colour changes may be restricted to the outer coat, as in apples, or occur throughout the fruit, as in tomatoes.

25.3.2 CHANGES IN TEXTURE

Ripe fruits are usually much softer in texture than unripe fruits. This is a result principally

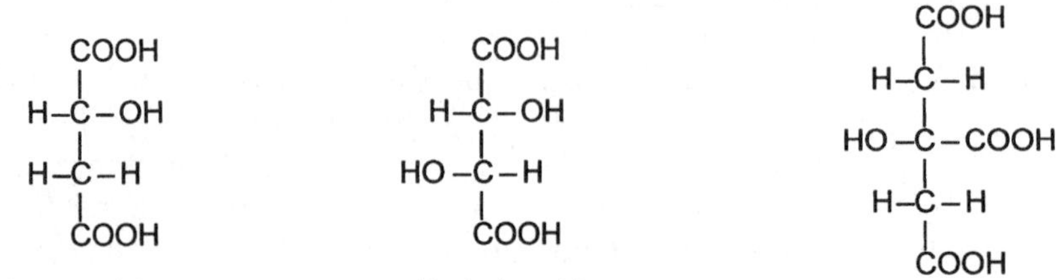

Figure 25.1 Structures of some acids which commonly occur in fruits.

of changes in the structure of the cell walls. Often during ripening there is a decrease in the polygalacturonic acid content of the middle lamella of the cell walls. This is the region which is shared by adjacent cells and which cements them together. It consists mainly of pectin made up of polygalacturonic acid residues. Some of the free CO_2^- groups may be linked to one another by formation of cross-links via Ca^{2+} ions, and others may be esterified by methyl groups. During ripening, the pectin may be degraded by polygalacturonase, thus weakening the wall. This enzyme acts on demethylated pectin. Demethylation of the pectin during ripening, through the action of the enzyme pectin methylesterase, allows attack by polygalacturonase. The presence of calcium in fruit delays softening and prolongs storage life, probably by preventing degradation of pectin polymers, particularly in the middle lamella. Physiological disorders of apple, such as bitter pit, appear to be associated with low calcium levels in the fruit and occur because of poor transport of calcium into the fruit. They can be alleviated by direct application of calcium to the fruit, but not usually by application to the soil.

The flesh of many fruits contains stone cells which have hard, highly lignified cell walls and which contribute to the texture of fruit such as pears. They also often contain tannins which are bitter and which contribute the astringent flavour to fruits such as persimmon.

25.3.3 CHANGES IN FLAVOUR

Generally, as fruits ripen there is a decrease in acidity and an increase in sweetness. The decreased acidity often results from reduction in fruit acids, although in fruit such as bananas there is actually an increase in acid content during ripening. At the same time there is usually an increase in the concentration of simple sugars such as sucrose and, less frequently, glucose. In many cases these sugars are produced by metabolism of starch which is often present in considerable quantities in unripe fruit. In others there may be little or no carbohydrate reserve in the unripe state, and these must rely on sugars which are imported into the fruit during the ripening process itself. This is the case in melons and grapes which cannot sweeten after detachment of the fruit from the plant. These different patterns of development have obvious implications for fruit handling and storage.

Whilst the content of complex sugars, fruit acids etc. generally decreases during ripening, that of proteins usually increases. This is in part because of the need to synthesize additional enzymes which are required for the ripening process itself. Thus the breakdown of starch, chlorophyll and cell-wall components, and the synthesis of new pigments, all require the production of additional enzymes.

In many fruits ripening is accompanied by the development of very characteristic aromas and flavours. These are generally due to the presence of specific combinations of volatile compounds such as esters. Their synthesis requires the activation of specific metabolic pathways.

25.3.4 RESPIRATION IN RIPENING FRUIT

Biosynthetic activities in ripening fruit require energy, which is provided through respiration occurring in the fruit. Many fruits show very characteristic changes in respiration during ripening. In some, respiration shows a climacteric during which there is a rapid and short-lived increase in the rate of respiration at the beginning of the ripening process (Figure 25.2). In climacteric fruits the magnitude of the increase in respiration rates varies widely. It is about five-fold in avocado and three-fold in tomatoes and bananas, but no increase is detectable in citrus fruits, pineapples, grapes etc. Ethylene appears to be produced and to influence ripening in climacteric fruits, but to play no part in the normal ripening of non-climacteric fruits. The ripening of many fruits can be slowed dramatically by removal of ethylene from the storage environment. The role of ethylene in control of ripening is discussed further in Chapter 27.

It is not clear why some fruits show a climacteric pattern of ripening whilst others do not. Climacteric and non-climacteric fruits do not fall into groups of related plants, and no obvious biochemical characteristics have been correlated with such patterns of development. Indeed in the case of tomato (which is normally a climacteric fruit) there are non-climacteric mutants which show no autocatalytic ethylene production, but which have otherwise identical compositions.

Understanding of the processes occurring during ripening has been helped by the existence of mutants. These lack specific enzymes which are thought to be involved in the ripening process and they thus allow the importance of the enzymes to be assessed. Some, such as the *nor* and *rin* mutants of tomato, lack the capacity to synthesize ethylene and are deficient in lycopene (red colour) and polygalacturonase.

There is considerable interest in carrying out genetic manipulation to modify patterns of fruit ripening, and this is discussed in Chapter 27.

25.4 SEED DEVELOPMENT

Growing seeds are powerful sinks to which nutrients move from photosynthetic and other tissues. As discussed in Chapter 23, the principal storage compounds in most seeds are starch and proteins. In certain specific types of seeds quantities of oils also accumulate.

25.4.1 STARCH BIOSYNTHESIS

Starch consists of a mixture of the straight-chain molecule amylose, and the branched-chain molecule amylopectin (see Chapter 3). It is found within plastids in the form of starch grains or granules. Amylose normally makes up 15–30% and amylopectin 70–85% of the total, and in general the proportion of amylose

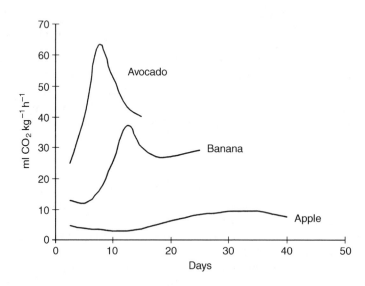

Figure 25.2 Changes in respiration rates during ripening of climacteric fruits. The intensity and duration of the increase in respiratory activity vary widely from species to species. In other fruit such as citrus, pineapples and grapes no climacteric is seen. (Redrawn from Biale, J.B. (1950) Postharvest physiology and biochemistry of fruits. *Annual Review of Plant Physiology,* **1,** 183–206.)

present increases with increasing maturity of the grain.

Starch is synthesized from sucrose which may initially accumulate in vacuoles. Some invertase is present in starch-synthesizing tissues but its activity is rather low, and the enzyme sucrose synthase is probably the main enzyme responsible for sucrose breakdown, as the reaction which it catalyses is easily reversible. The synthesis of starch takes place in the amyloplast and appears to be catalysed by starch synthase firmly bound to the starch granules. ADP-glucose acts as the glucose donor. Glucose and fructose resulting from sucrose breakdown probably enter the amyloplast after conversion to triose phosphate.

Starch synthase catalyses the transfer of glucose residues from ADP-glucose to the non-reducing end of pre-existing primer molecules in the presence of K^+ to produce α-1,4 bonds, but it is not known how the primers themselves arise. It is possible that starch phosphorylase, which normally catalyses starch breakdown, may be used to synthesize short straight-chain oligomers of glucose. The action of starch synthase produces only linear, α-1,4-linked molecules. Branches are introduced into such straight-chain molecules by branching enzyme. This hydrolyses α-1,4 bonds and transfers the resulting short oligosaccharides to a primary -OH group, thus creating an α-1,6-linked branch. The action of branching enzyme therefore increases the number of non-reducing chain ends where starch synthase can act. Both starch synthase and branching enzyme exist in multiple forms, but the significance of this is unknown.

25.4.2 PROTEIN SYNTHESIS

Seed storage proteins are synthesized from sucrose and a source of nitrogen such as glutamine or asparagine. Both of these are brought to the developing seed in the phloem. In many plants, much of the nitrogen is formed by re-mobilization of protein previously stored in the leaves (Chapter 26). The pathways for biosynthesis of the amino acids which are required to make proteins have been described in Chapter 20; they are very active in developing seeds.

Seed storage proteins accumulate in protein bodies, which are membrane-surrounded organelles, often spherical, with a diameter of between 0.1 and 25 μm.

The genes coding for storage proteins are complex and the regulation of their expression is imperfectly understood. Transcription of the nuclear genes gives rise to mRNA from which introns may be excised. The mRNA leaves the nucleus and is translated on the rough endoplasmic reticulum, as the proteins produced are targeted to the protein bodies. In the lumen of the endoplasmic reticulum the proteins are processed into their mature forms. In the case of legume storage proteins the modifications are complex, involving hydrolysis of the polypeptide chain at specific points, the formation of disulphide bridges, and attachment of sugar groups to some amino acid side chains (glycosylation). Some of these processes are brought about by the Golgi bodies. Processing of cereal protein pre-cursors is usually simpler, not normally involving peptide hydrolysis or glycosylation. The prolamins are deposited in different protein bodies from the other storage proteins. In this case they may pass directly from the endoplasmic reticulum to the protein bodies, whilst other proteins pass through the Golgi bodies.

25.4.3 BIOSYNTHESIS OF FATS

Oils are synthesized from sucrose transported in the phloem from photosynthetic tissues. The glycerol and acetyl-CoA needed for lipid synthesis can be made from sucrose via the glycolytic pathway. The conversion of acetyl-CoA into oleic acid takes place in the proplastids, as described in Chapter 19. Modification of oleic acid and subsequent synthesis of triacylglycerols occurs on the endoplasmic reticulum membranes.

PLANT NUTRITION 26

26.1 INTRODUCTION

Plant growth depends on the presence of a considerable number of elements. Some of these are required in large amounts, others in only very small quantities. The presence of impurities in growth media may make it very difficult to establish whether some elements are truly essential or not. It is usual to consider an element to be essential if a plant cannot complete its life cycle (i.e. produce viable seeds) without it, or if it has been shown to form part of a compound which is essential to the plant. On this basis 17 essential elements, listed in Table 26.1, have been identified.

Those elements which are present in the plant at concentrations of $1\,g\,kg^{-1}$ or more are called the major elements or macronutrients, whereas those present at $100\,mg\,kg^{-1}$ or less are the trace elements or micronutrients.

26.2 BIOCHEMICAL FUNCTIONS OF MAJOR PLANT NUTRIENTS

26.2.1 NITROGEN

Nitrogen is an extremely important element. It forms a part of the structure of amino acids, nucleic acids, alkaloids, etc. The pathways by which inorganic nitrogen is incorporated into organic compounds are described in Chapter 20, and the way in which it may be remobilized during filling of storage organs is discussed later in this chapter.

26.2.2 SULPHUR

Sulphur occurs mainly in the form of the amino acids cysteine and methionine, which are constituents of many proteins. Cysteine is also found in glutathione, a tripeptide (γ-glutamyl-cysteinyl-glycine), which serves as a

Table 26.1 Levels of essential minerals in plant tissues

Element	Concentration(mg kg^{-1} dry matter)
Hydrogen	60000
Carbon	450000
Oxygen	450000
Nitrogen	15000
Potassium	10000
Calcium	5000
Magnesium	2000
Phosphorus	2000
Sulphur	1000
Chlorine	100
Iron	100
Boron	20
Manganese	50
Zinc	20
Copper	6
Nickel	–
Molybdenum	0.1

Modified from: Stout, P.R. (1961) *Proceedings of the Ninth Annual California Fertiliser Conference*, pp. 21–23.

redox buffer, keeping –SH groups in their reduced state.

Sulphur also forms part of several vitamins and coenzymes e.g. CoA, lipoic acid, biotin and thiamin (Chapter 8). Iron–sulphur proteins, such as ferredoxin and those involved in electron transport, contain inorganic sulphur.

Of particular interest to agriculturalists is the existence of volatile or toxic sulphur compounds in some plants. The characteristic flavours and odours of onion, garlic and some brassicas is due to the mercaptans, sulphides and sulphoxides which they contain. Similar compounds contribute to the taint which develops in milk exposed close to swedes and turnips. Toxic sulphur-containing compounds include glucosinolates which are found in plants such as oilseed rape. These may be converted into a variety of toxic thiocyanates, isothiocyanates

and goitrin which cause goitre, and liver and kidney damage, in non-ruminants.

Assimilation of sulphur

Sulphur is available to the plant mainly as sulphate (SO_4^{2-}). In plants and animals it is present mainly in its reduced (–SH) form so that assimilation requires reduction. This reduction cannot be carried out by animals, which are therefore dependent on plants and microorganisms for a supply of reduced sulphur in the form of cysteine and methionine.

Sulphate is absorbed from soil by the roots. A small amount is reduced in the roots but most is transported to the shoot, where it is reduced in the chloroplasts. The first step in the reduction involves the reaction of sulphate with ATP to produce adenosine-5′-phosphosulphate (APS) and pyrophosphate in a reaction catalysed by ATP sulphurylase (Figure 26.1). The sulphate group of APS is then transferred to the sulphur atom of another compound (possibly glutathione) before it is reduced using electrons supplied by reduced ferredoxin. Sulphide results (possibly still bound to glutathione) and this is rapidly converted into cysteine by reaction with O-acetylserine (see Figure 26.1). Methionine and other sulphur-containing compounds are made from cysteine.

Sulphur deficiency

Sulphur deficiency is quite unusual, as most soils contain adequate sulphate. It results in inhibition of protein synthesis and accumulation of amino acids which do not contain sulphur. Growth rates, especially of the shoot, are reduced. Chlorosis occurs first in the young leaves because there is very little remobilization of sulphur from old leaves. Young leaves thus depend on the roots for sulphur uptake.

Sulphur deficiency could become more common because of lowered SO_2 emissions into the air and reduced use of sulphur-con-

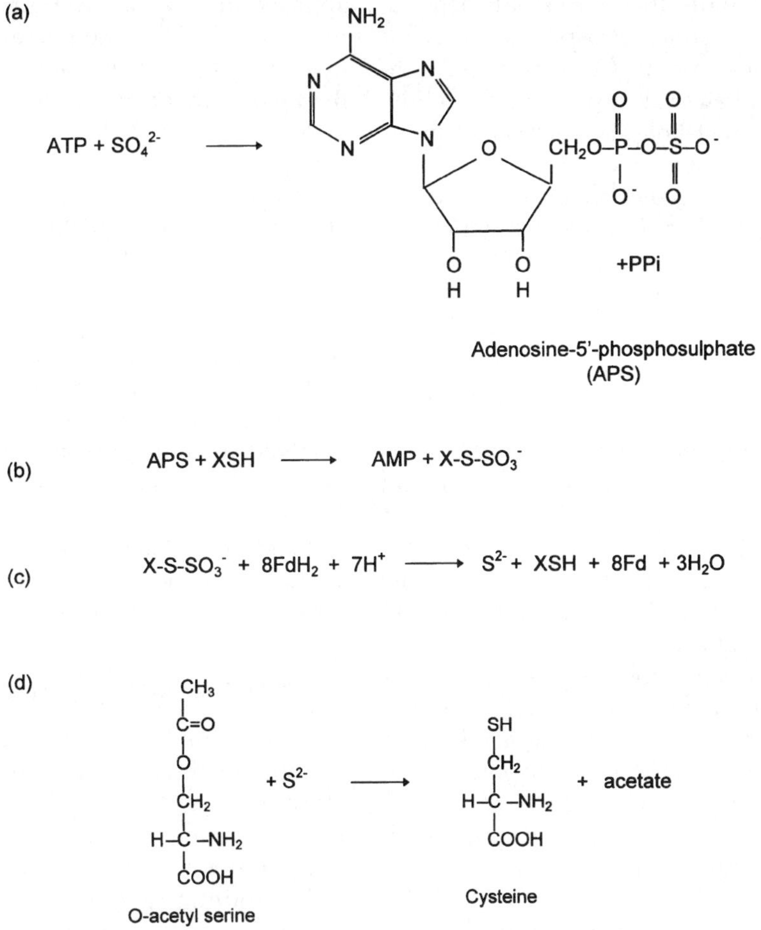
(a)
ATP + SO₄²⁻
NH₂
N
N
N
N
O
CH₂O–P–O–S–O⁻
O
O
O⁻
O
O
H
O
H
+PPi
Adenosine-5'-phosphosulphate
(APS)
(b)
APS + XSH → AMP + X-S-SO₃⁻
(c)
X-S-SO₃⁻ + 8FdH₂ + 7H⁺ → S²⁻ + XSH + 8Fd + 3H₂O
(d)
CH₃
C=O
O
CH₂
H–C –NH₂
COOH
+ S²⁻
SH
CH₂
H–C –NH₂
COOH
+ acetate
O-acetyl serine
Cysteine

Figure 26.1 Reactions for the assimilation of sulphate into organic compounds. (a) APS is formed by reaction of ATP with sulphate, catalysed by ATP sulphurylase; (b) the sulphate group of APS is transferred to X, which may be glutathione; (c) the sulphate group is reduced to sulphide (S^{2-}); (d) sulphide reacts with O-acetyl serine to form cysteine, from which other sulphur-containing compounds arise.

taining fertilizers, e.g. ammonium sulphate or superphosphate.

26.2.3 PHOSPHORUS

In soils, phosphorus is present almost exclusively as phosphate. Most is in the form of inorganic orthophosphate (P_i) but some occurs as organic phosphate esters. P_i is released from these by the action of phosphatases, either from plant roots or microorganisms.

Uptake into the plant occurs against a very strong concentration gradient and therefore depends on energy provided by respiration. Phosphorus is taken up only in the form of the ion $H_2PO_4^-$ and hence rates of uptake are very pH-dependent.

In plants, phosphorus exists as P_i or organic phosphates such as glucose-6-phosphate, ATP, phospholipids, DNA, RNA and phytic acid. Normally P_i is much more abundant than organic phosphorus. In plants which are deficient in phosphorus, the reserve of P_i decreases but the organic phosphorus content remains fairly constant. Conversion of P_i to organic phosphates takes place very rapidly. Phosphorus is metabolically very active and is readily remobilized from older tissues.

Phosphorus is needed for export of sugars from the chloroplast. Phosphorus deficiency results in reduced growth, reduced tillering in cereals, and particularly in reduced fruit and seed development.

26.2.4 POTASSIUM

Potassium is quantitatively the most important cation in plants, and is the major ion which maintains turgor. Changes in turgor pressure resulting from movements of K^+ are responsible for leaf movements and for the opening and closing of stomata in response to changes in water availability.

Potassium is very mobile in the plant and is readily redistributed from older to younger tissues. It is the only essential mineral cation which is taken up by active transport. It activates many enzymes e.g. starch synthetase, and promotes the synthesis of enzymes such as RUBISCO, thus enhancing the rate of photosynthesis.

Deficiency of potassium causes chlorosis of leaves and weakens the stems of cereals. It also leads to low cell pH and results in the production of toxic amines such as putrescine.

26.2.5 CALCIUM

Most soils contain adequate calcium, which is absorbed as Ca^{2+}. Concentrations of calcium in the cytoplasm of cells are very low and most of the calcium in the plant is present in vacuoles and cell walls. Some calcium may be bound to calmodulin and may act as a second messenger within cells. Calcium is essential for cell elongation and division, and for maintenance of membrane permeability and of cell-wall structure through cross-linking of pectin (Chapter 24). Redistribution of calcium appears to play an important part in the response of cells of the root and stem to gravity (gravitropism). Calcium accumulates on the lower side of horizontal roots and on the upper side of stems, and in both cases inhibits elongation of that side of the organ.

Symptoms of deficiency of calcium are most obvious in young, meristematic tissues.

26.2.6 MAGNESIUM

Magnesium is a component of chlorophyll and deficiency leads to interveinal chlorosis, which is first seen in older leaves. It is involved in transfer of phosphate groups as the true substrate of many enzymes using ATP is a magnesium complex of ATP (Chapter 8). In addition it activates other enzymes such as RUBISCO and may play a role in control of carbohydrate synthesis in chloroplasts.

26.3 TRACE ELEMENTS – MICRONUTRIENTS

Trace elements or micronutrients are those essential elements which are present in the plant at concentrations of $100\ mg\ kg^{-1}$ or less. Their main functions are given in Table 26.2.

In addition to the elements listed in the table, sodium and silicon appear to be required by certain species, and may prove to be needed by all. Other elements such as cobalt may be needed by symbiotic bacteria.

26.4 TOXIC EFFECTS OF MINERALS

At concentrations above those adequate to meet the needs of plants, a number of minerals may be toxic. Many non-essential elements are also toxic to plants. These include the metals lead, cadmium, silver, mercury and tin, which often result from pollution, and aluminium, which is naturally abundant and becomes available under acidic conditions. Increased availability of aluminium is a major factor which reduces plant growth in acidified soils resulting from acid rain. Some plants have become resistant to toxic metals, appar-

Table 26.2 Functions of trace elements in plants

Element	Function
Fe	Forms part of the structure of haem, cytochromes and non-haem iron proteins, also needed for synthesis of chlorophyll
Mn	Resembles Mg^{2+} and can replace it in some functions; component of indole acetic acid oxidase and needed for photolysis of water by PSII; deficiency leads to interveinal necrosis seen in young leaves
Cu	Component of plastocyanin, cytochrome oxidase, ascorbic acid oxidase and polyphenol oxidase, needed for the synthesis of tryptophan and therefore of indole acetic acid; also needed for desaturation of fatty acids and for superoxide dismutase
Zn	Component of carbonic anhydrase which catalyses conversion of bicarbonate into CO_2, and of many dehydrogenase enzymes e.g. glutamic dehydrogenase, lactic dehydrogenase, alcohol dehydrogenase; needed for superoxide dismutase activity
Mo	Component of nitrogenase and nitrate reductase, essential for nitrogen assimilation
B	Needed for synthesis of uracil and production of UDP sugars and nucleic acids; cell division in meristems is particularly affected by its deficiency; transport and synthesis of sucrose (which requires UDPG) also reduced
Cl	Needed for evolution of oxygen by PSII in photosynthesis; Cl^- acts as a negatively charged counterion to K^+
Ni	Needed for urea breakdown in tropical legumes

ently by production of chelating agents called phytochelatins, which are small peptides containing the sulphur amino acid cysteine. Binding of the metals to these sulphur atoms renders the metal non-toxic. Other ions such as selenium, although not essential and without effects on the plants, may be accumulated by some plants which may then be toxic to animals which eat them.

26.5 INTERACTION BETWEEN CARBON AND NITROGEN METABOLISM

Individual tissues or organs of the plant have specialized functions but they must work together throughout the life of the plant to ensure its survival and efficient growth. Carbon compounds derived from photosynthesis taking place in the leaves and other green parts of the shoot are distributed to other plant organs. The root system, on the other hand, is responsible for the uptake and distribution of sufficient soil-derived nutrients and water for the entire plant. The assimilation of nitrogen and other nutrients in the roots depends on the shoot for a supply of car-

bon compounds. These serve as a source of energy and carbon skeletons, to which nitrogen-containing groups may be attached. The growth of storage organs depends on resources supplied by both shoots and roots. Nutritional demands of individual parts of the plant, which vary from time to time, are met by a flow of nutrients from site to site. These flows may change in the short term, allowing temporary stores of starch formed in leaves in the light to be redistributed to other organs during darkness. Over longer time scales, stores of both nitrogen and carbon built up during vegetative growth may later be redistributed to developing storage organs.

To explain how these nutrient flows are controlled, the concept of 'sources' and 'sinks' within plants has been developed. Sources are those parts of the plant which can supply assimilates and sinks are those to which they move. In most plants the leaves are the most obvious source as a result of their photosynthetic capabilities, but storage organs supporting re-growth, and cotyledons and endosperm supporting seedling growth, also act as sources. Organs such as developing seeds, tubers, roots, stems, flowers and young leaves, in which growth or storage are taking place, may be regarded as sinks. Tissues may act as sources or sinks over short or long periods of time and many may act in both capacities at different times during growth.

26.5.1 CARBON ASSIMILATION

Carbon assimilation takes place principally in the leaves through photosynthesis. Rates of photosynthesis in leaves depend on the leaf area, as this affects their ability to absorb light. Newly emerged leaves have a small surface area but a rapid growth rate. They are therefore not able to produce all the assimilates required to support their own growth and must import materials from larger leaves or from seed storage reserves. This situation persists until the leaf reaches about one-third of its maximum size, when rates of assimilate production approximate to the demands of the leaf. Leaves above one-third of maximum size generally become net exporters of assimilates which may be used to support the growth of nearby, newly formed and expanding leaves, or may contribute to the filling of seeds or other storage organs.

26.5.2 NITROGEN ASSIMILATION

The biochemical processes which are responsible for the assimilation of inorganic nitrogen into organic compounds are described in Chapter 20. Nitrogen comes from the soil in the form of nitrate or ammonia or, in the case of legumes, from nitrogen gas in the air.

Nitrate reductase and nitrite reductase may be located mainly in the roots, mainly in the leaves, or divided between the two. The nature of nitrogen-containing compounds found in the xylem of plants of different species thus covers a very wide spectrum, reflecting these metabolic differences (Figure 26.2).

Cocklebur (*Xanthium pennsylvanicum*) represents one extreme of the spectrum. In this plant over 95% of nitrogen in the xylem con-

sists of nitrate as the roots have a very low capacity to assimilate nitrogen. However most plants have some nitrate reductase activity in the roots and thus some organic nitrogen-containing compounds are found in the xylem. In wheat and maize, up to approximately 80% of the nitrate entering the plant passes unaltered through the roots. In barley about 50% is converted to organic nitrogen in the roots, and in most legumes even more is assimilated. Usually one nitrogen-rich compound, which in most non-legumes is glutamine, predominates in the xylem or phloem.

In legumes, although glutamine is the initial product of nitrogen fixation, little is normally found in the xylem of plants which are actively fixing nitrogen. Instead it is normally converted mainly to the ureides, allantoin or allantoic acid (in tropical legumes such as cowpea or soyabean) or to asparagine (in temperate legumes such as peas, clovers or lupins). When legumes assimilate nitrate, some may pass unaltered into the xylem and some may be reduced to asparagine or glutamine before entering the xylem. Thus in tropical legumes ureides only predominate in the xylem whilst the plants are assimilating nitrogen principally by nitrogen fixation. Both temperate and tropical legumes transport a mixture of asparagine and glutamine in the phloem.

Nitrate reductase is unusual because it is one of the few well-characterized inducible enzymes in higher plants. Its activity is low when nitrate is not available but increases when nitrate is supplied. Roots have a finite capacity to reduce nitrate, and as more nitrate is applied increasing proportions pass unaltered into the xylem.

Wherever it occurs, the incorporation of inorganic nitrogen into organic compounds requires a reducing agent (such as NADH, NADPH or reduced ferredoxin) and α-ketoglutarate. NADH, NADPH and α-ketoglutarate are produced by metabolism of sugars or by diversion of photosynthetic products from the production of sugars. Thus the metabolism of nitrogen is a drain on the carbohydrate metab-

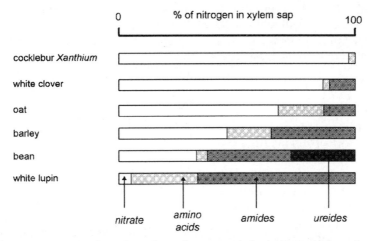

Figure 26.2 Nitrogen compounds in the xylem of a range of plants. In cocklebur almost all nitrate passes through the roots unchanged. Cereals convert up to about half of the nitrate taken up to amino acids and particularly to glutamine. In most legumes, in addition to reduction of nitrate, ureides may arise from the products of nitrogen fixation. (Redrawn from Pate, J.S. (1973) Uptake, assimilation and transport of nitrogen compounds by plants. *Soil Biology and Biochemistry*, **5**, 109–119.)

olism of the plant, and uptake of large amounts of nitrogen may deplete sugar reserves.

Where nitrogen is converted into organic compounds in the roots, they must have a supply of sugars available. This is achieved by the transport of sucrose from the leaves to the roots in the phloem. In the roots sucrose is hydrolysed to glucose and fructose, which may then be oxidized to provide NADH, NADPH and α-ketoglutarate. Where most nitrogen assimilation occurs in the leaves, less sucrose has to be transported to the root system.

Assimilation of nitrogen through the action of nitrogenase also requires large amounts of energy and carbon skeletons which must both be provided by catabolism of sugars.

Because the metabolism of one depends on the other there is often a strong inverse correlation between the carbohydrate and nitrogen content of plants. For example, late application of nitrogen fertilizers to barley increases the protein and decreases the starch content. Both of these are undesirable in barley intended for malting – the former because of the production of nitrogen compounds which impart off-flavours to the product, and the latter because

it reduces the amount of fermentable sugar available. Similarly, late application of nitrogen to sugar beet depletes the sugar content which reduces quality.

There is a constant flow and recycling of nitrogen around the plant. Nitrogen taken up and assimilated by the roots passes in the xylem to the shoot, where it may be temporarily stored as protein. This may be degraded to provide amino acids which are re-exported in the phloem to support growth of other parts of the plant. In non-legumes nitrogen is exported from the leaves mainly as glutamine, but in legumes there is often a mixture of glutamine and asparagine. Glutamine and asparagine may be used to make amino acids by transamination and the amino acids may be converted into protein. Proteins undergo constant turnover and are eventually broken down to re-form amino acids. These are transaminated to form keto acids which are subsequently oxidized. At the same time α-ketoglutarate is converted to glutamate (see Figure 14.1) which is available for synthesis of new amino acids or for export to other parts of the plant. The amino acid composition of leaves and the storage proteins in seeds, roots or bark differ considerably. Thus extensive amino acid interconversion accompanies the mobilization of nitrogen reserves.

26.5.3 SENESCENCE AND NUTRIENT CYCLING

Protein turnover results in continuous synthesis of proteins from amino acids and their subsequent breakdown. Some proteins turn over faster than others, in general enzymes turn over faster than structural proteins. Half-lives vary from a few hours to several days. In growing leaves, proteins are synthesized faster than they are broken down so that the net nitrogen content of the leaves increases, but when the leaves become senescent there is a net loss of nitrogen as rates of breakdown exceed those of synthesis. This nitrogen may be transported and stored in other parts of the plant. Much of the nitrogen found in seeds, for example, is taken up from the soil before flowering, and may have been cycled several times and into several different organs before reaching the seeds. This point is illustrated by Figure 26.3 which shows the nitrogen content of bean plants during their growth.

The nitrogen content of the leaves reaches a maximum in the early stages of seed development and then declines as the seeds fill. Most of the nitrogen in leaves is in the form of Rubisco and this enzyme is extensively degraded as seeds fill. The photosynthetic capacity of the leaves therefore declines. It has been estimated that up to 40% of seed nitrogen is derived from leaves and a further 20% from pods and endosperm. The remaining nitrogen is assimilated during seed fill.

Such patterns of nitrogen flux are commonly found in many monocarpic annual plants including cereals. The redistribution of nitrogen in cereals may be even greater than in legumes, with up to 90% of the nitrogen found in the mature plant being taken up from the soil by the time the plant is half-grown, and 85% of the nitrogen in wheat leaves being transported to the developing grain. The photosynthetic tissues closest to the ear, including the glumes and flag leaf, normally provide much of the nitrogen.

In many perennial plants, nitrogen required for seed development is obtained from leaves without correlated senescence occurring. In perennials, much of the nitrogen made available when the leaves senesce in the autumn is transported and stored in the crown and roots (herbaceous perennials) or in the bark (woody perennials). In the latter case arginine-rich proteins are found in the bark. Stored protein is used as a source of amino acids needed to support new growth in spring.

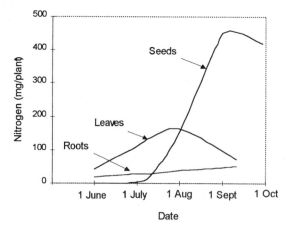

Figure 26.3 Nitrogen content of broad bean plants during growth. Nitrogen accumulates in the leaves during vegetative growth; it is then mobilized and transported to the seeds where it is used to synthesize storage proteins. (Modified from Emmerling, A. (1880) *Ladw. Versuchsshtat*, **24**, 113)

REGULATION OF PLANT GROWTH AND DEVELOPMENT 27

27.1 INTRODUCTION

Plants must regulate and integrate the activities of all their parts. They must also respond to changes in their environment and do so in a bewildering variety of ways. To achieve this they have complex mechanisms for perception of external events and for controlling their growth and development. Some aspects of these control mechanisms, including the role of plant hormones, are considered in this chapter.

27.2 RESPONSES TO LIGHT

The intensity, duration and quality of light to which plants are exposed varies greatly at different locations, depending on latitude in a predictable manner, and less predictably on factors such as weather. The responses of plants to light are extremely complex but can be classified into the groups described below on the basis of the action spectra and the intensities of light required.

27.2.1 EFFECTS ON PHOTOSYNTHESIS

The action spectrum for photosynthesis is similar to the absorption spectra of the chlorophylls, with peaks in the blue and red regions. For most C_3 plants at normal carbon dioxide concentrations, as light intensity is increased photosynthetic rates increase until a limiting value is reached. This is called light saturation, and for many C_3 crops is reached at an intensity between a half and a quarter of normal sun-

light. In C_4 plants, rates of photosynthesis increase almost linearly right up to full sunlight intensity. For many plants growing at high light intensities, the amount of Rubisco appears to be rate-limiting and is only just sufficient to account for observed rates of carbon fixation.

C_4 plants have the highest photosynthetic rates and CAM plants the lowest. Alpine and arctic plants must grow rapidly in the short growing seasons, and have rates of photosynthesis which greatly exceed those of respiration. Leaves of plants which grow in shade direct more of their biochemical activity to trapping light and less to carbon dioxide fixation. They have high chlorophyll contents, large, irregularly orientated grana stacks, and low levels of Rubisco and other photosynthetic enzymes.

For crops in the field, the light intensity at the lower leaves may be much lower than at the upper leaves, so most crops are never saturated by sunlight when grown at normal plant densities.

27.2.2 PHYTOCHROME-MEDIATED RESPONSES

Plant responses which can be brought about by red light but reversed by far-red light, or *vice versa*, often involve phytochrome. Such responses include photoperiodism, control of stem extension, apical dominance, seed germination and synthesis of phenolics.

Phytochrome is a pigment–protein complex which exists in two forms. One of these (Pr) absorbs red light whilst the other (Pfr) absorbs mainly far-red light. Absorption of red light by Pr converts it to Pfr, whilst absorption of far-red light by Pfr converts it to Pr. Pr is blue and shows an absorption maximum at about 666 nm and Pfr is olive green and absorbs at about 730 nm. Spectra of the two forms are shown in Figure 27.1a. As both Pr and Pfr absorb red light, this light establishes a steady state in which about 75–80% of the total phytochrome exists as Pfr. Far-red light, on the other hand, is absorbed only by Pfr and therefore causes almost complete conversion of phytochrome to Pr.

Light from most sources contains both red and far-red light and establishes an equilibrium between the different forms of phytochrome. In normal daylight the ratio of red to far-red light (R:FR) is about 1.15. In comparison, white fluorescent light tubes give out light with R:FR ratio of between 3 and 6, and incandescent bulbs a ratio of about 0.7–0.8. At dawn or dusk, when the sun is less than 10° above the horizon, the light is enriched in far-red light as a result of passage of light through a longer air path which preferentially scatters shorter wavelengths. Leaves absorb mainly blue and red light and allow green and far-red to pass. Thus light in the shade of leaves is considerably enriched in far-red light compared with sunlight. In the shade of opaque objects, most light comes from the blue sky and so it is slightly enriched in blue.

Phytochrome is a dimer of two identical polypeptide chains each of about 120 kDa. Each chain carries a chromophore group attached via a cysteine S atom. When light is absorbed the chromophore changes from the *cis* to the *trans* form, or *vice versa*, and this in

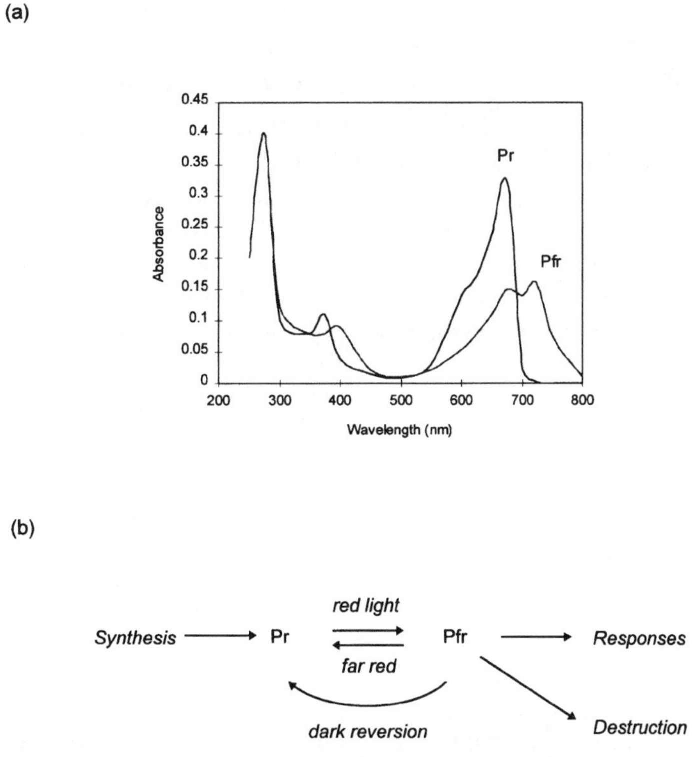

(a)

(b)

Figure 27.1 (a) The absorption spectra of the red- and far-red-absorbing forms of phytochrome (Pr and Pfr, respectively). Red light (λ = approx. 660 nm) is absorbed by both forms, but far-red (λ = approx. 730 nm) is absorbed only by the far-red form. (b) The synthesis, interconversion and destruction of phytochrome as it occurs in most dicotyledons.

turn causes subtle changes in the conformation of the protein. It appears that Pfr is the active form of phytochrome which triggers positive or negative developmental responses. In most dicotyledons and gymnosperms, Pfr may also be converted into Pr by the process of 'dark reversion'. This is a relatively slow, pH-dependent process which takes place over a period of hours and which does not require light (Figure 27.1b). It occurs only very slowly in monocotyledons.

27.2.3 OTHER RESPONSES TO LIGHT

Some responses of plants show action spectra with peaks in the blue or long wavelength ultraviolet (UV-A) regions of the spectrum. Phototropism is one such response. The optimum wavelength is often around 450 nm and is thought to result from absorption of light by a flavoprotein.

Some responses of plants require much higher (but still less than the intensity of daylight) light intensities than those in which phytochrome is involved. These are known as 'high irradiance responses' (HIR). They may involve light in the UV-A, red and far-red regions of the spectrum. In many cases there may be cooperation between phytochrome and blue-absorbing pigments. Other responses, such as the inhibition of mesocotyl elongation in dark-grown oats, require even lower light intensities than phytochrome, the equivalent of 1 second of full moonlight may be enough. Such responses are triggered by red light but not reversed by far-red.

27.3 RESPONSES TO TEMPERATURE

27.3.1 PHOTOSYNTHESIS

Optimum temperatures for photosynthesis are often similar to the daytime temperature at which plants normally grow. C_4 plants usually have higher optimum temperatures than C_3 plants because of their lower rates of photorespiration. At temperatures of 10–25° C the efficiencies of C_3 and C_4 plants are about the

same. Both need about 15 photons to fix one molecule of carbon dioxide. At lower temperatures, C_3 plants require only about 12 photons whilst C_4 plants need 14 because extra ATP is used in carbon dioxide fixation. Above about 30° C, C_3 plants become less efficient because of the operation of photorespiration.

27.3.2 VERNALIZATION

Winter cereals require exposure to low-temperature (vernalization) to promote flowering. Most biennials also require exposure to cold (several days to several weeks at temperatures just above freezing) without which they will not flower. It is the bud which is the site of response. When buds from vernalized plants are transferred to non-vernalized plants they flower. The vernalization effect can also be carried over in cuttings, suggesting that the stimulus is chemical in nature.

27.4 RESPONSES TO ATMOSPHERE

Photosynthesis and respiration work in opposition to one another and, although some respiration is required to provide energy which 'drives' the plant's metabolism, beyond a certain level it reduces the productivity of the crop.

The carbon dioxide content of the atmosphere is about 0.035% or 350 µmol mol^{-1}. After remaining stable at about 280 µmol mol^{-1} for many centuries, it has risen dramatically since about 1850 as a result of the burning of fossil fuels. Approximately 13% of the atmospheric carbon dioxide is used in photosynthesis each year, and about the same amount exchanges with carbon dioxide dissolved in the oceans. Carbon dioxide, and other greenhouse gases such as methane, absorb long-wavelength radiation. When radiation from the sun reaches the earth it is re-radiated at longer wavelength, and accumulation of greenhouse gases in the atmosphere prevents loss of heat, thus increasing the energy retained. Increased cloud cover and dust reflect back solar radiation and cause a cooling effect. How these opposing factors might affect specific locations or the entire planet is of great interest to agriculturalists and is the subject of much speculation.

Photosynthesis in C_4 plants is usually saturated by carbon dioxide levels close to or just above those in the atmosphere (350 μmol mol^{-1}). In contrast, raising the carbon dioxide concentration of the atmosphere in which C_3 plants are growing decreases the rate of photorespiration by competing with oxygen for Rubisco, leading to faster net photosynthetic rates. In greenhouses, especially in winter, supplementation with carbon dioxide at concentrations of up to about 1000 μmol mol^{-1} may be used. Above this concentration stomata close and photosynthesis is inhibited.

If carbon dioxide levels are decreased below atmospheric levels, net rates of photosynthesis decrease and become zero at the carbon dioxide compensation point where the rates of photosynthesis and photorespiration are equal. For C_4 plants this occurs when the carbon dioxide concentration is between 0 and 5 μmol mol^{-1} and for C_3 plants between 35 and 45 μmol mol^{-1}. (Figure 27.2).

Oxygen is required for respiration. Normal dark respiration, carried out by most tissues, uses the mitochondrial electron transport chain and cytochrome oxidase as its last step. This enzyme has very high affinity for oxygen and is saturated at concentrations well below the normal atmospheric level, so that changes in atmospheric oxygen concentration have virtually no effect on rates of respiration. However, rates of photorespiration are reduced by lowering oxygen concentrations as Rubisco has a low affinity for oxygen (Chapter 17).

27.5 RESPONSES TO STRESS

27.5.1 TEMPERATURE STRESS

Low temperature is probably the most important factor which limits plant distribution. Many plants which normally grow in hot climates may be injured below a critical temperature of between about 0 and 20° C (chilling

injury). Others, typically from colder environments, may only be damaged when frost occurs (freezing injury). Plants may also be damaged when exposed to high temperatures.

Freezing injury

During natural freezing, ice crystals form in the extracellular spaces and grow by addition of water from within the cells, which therefore become desiccated. This causes the cells to contract and the concentration of solutes inside them to increase, reducing the freezing point. In hardened plants there is usually no damage to cell membranes or organelles, and when they thaw water is taken back into the tissues. In cold-susceptible plants the membranes are often damaged and become leaky as the cells shrink, and they are not then able to take up water after thawing. When rapid freezing is induced experimentally the cell contents may freeze.

Tolerance to extracellular ice formation may depend on accumulation of non-damaging solutes within the cell (cryoprotectants or compatible solutes) which decrease the freezing point and limit water loss. Production of cold-stress proteins or changes in membrane structure also increase resistance.

One of the most crucial factors which may cause development of ice in plants is the presence of ice-nucleation sites. These may be within the tissues or may be provided by bacteria on the surface. As bacterial species differ in their capacity to act as nucleation sites, frost damage may be reduced by spray application of cultures of bacteria with low nucleation potential, which replace the naturally occurring ones.

Chilling injury

Some plants may be damaged by exposure to lower than critical temperatures, even though it does not result in freezing. The symptoms of such injury seem to arise mostly from changes which take place in the cellular membranes at

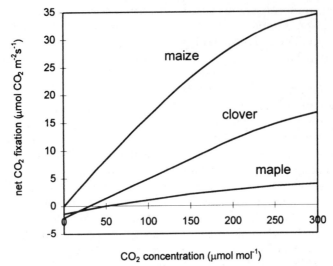

Figure 27.2 The effect of carbon dioxide concentration on net rates of photosynthesis. Photosynthesis in C_4 plants is almost saturated by atmospheric levels of CO_2 (350 μmol mol^{-1}) but rates of photosynthesis carried out by C_3 plants generally increase up to about 1000 μmol mol^{-1}. At low CO_2 levels, net rates of photosynthesis become zero for C_4 plants between 0 and 5 μmol mol^{-1} CO_2 and for C_3 plants between 35 and 45 μmol mol^{-1}.

the critical temperature. Above the critical temperature the membrane is in a fluid state, but below this temperature it becomes semi-crystalline. The critical temperature is influenced by the ratio of unsaturated to saturated fatty acids and by the sterol content. The changes affect the permeability of the membranes and the activity of enzymes embedded in them, and lead to redistribution of metabolites and damage to membrane-dependent processes such as photosynthesis and ATP production. Chilling-resistant or hardened plants have increased levels of unsaturated fatty acids which could account for their lower critical temperatures.

High-temperature stress

Plants are adversely affected by high temperature. Some effects may result from differences in the temperature sensitivity of individual metabolic pathways. Thus in C_3 plants, photorespiration becomes increasingly important in relation to photosynthesis as the temperature rises, reducing the plants' capacity for net carbohydrate synthesis. High temperatures may result in decreased protein synthesis, increased cytoplasmic viscosity and loss of membrane semi-permeability. Membrane-dependent processes such as photosynthesis are particularly sensitive, especially photosystem II. Photosynthetic ATP production is also inhibited because of the release of free fatty acids which act as uncouplers of the electron transport process.

Although the production of many proteins is decreased at high temperatures, that of a specific set of proteins, the heat-shock proteins, is increased in response to temperatures 10–15° C above the optimum growth temperatures. The production of heat-shock proteins also occurs in most other organisms. A 70-kDa protein seems to be common to most organisms, but plants may also produce smaller (15–27-kDa) ones. Some heat-shock proteins are associated with particular organelles (the nucleus, chloroplast etc.) whilst others form cytoplasmic aggregates. These proteins may protect proteins and nucleic acids from damage, possibly by stabilizing their folding, and may also bind ions released as membranes change their permeability at high temperature.

27.5.2 WATER STRESS

Many plants are exposed to significant degrees of water stress and suffer considerable water loss from their tissues. As the water content falls the concentration of solutes in tissues increases to the point where damage to enzymes might occur. In some plants, specific, non-damaging solutes accumulate to high levels and this helps water retention. Such compounds include sucrose, glycerol, mannitol, proline and betaine. Proline levels may increase to between 10 and 100 times their normal values. Other plants simply lose large amounts (up to 70%) of their water but still survive. Even moderate drought may considerably reduce yields of agricultural crops, especially if it coincides with particularly sensitive phases of the plants' development such as germination, seedling growth or flowering. Rates of photosynthesis may not recover for several days after stress is relieved, and old leaves may be shed and growth delayed.

Rates of photosynthesis, respiration, protein synthesis and nucleic acid synthesis are decreased by water stress: cell growth and cell wall synthesis are particularly sensitive. Water stress reduces protein synthesis, acting at the level of translation, but it induces the synthesis of a small group of stress-specific proteins and some hydrolytic enzymes such as α-amylase and ribonuclease. A general effect of many types of stress, including water stress, is to increase concentrations of free amino acids.

ABA levels increase, particularly in leaves, as water stress develops. This causes K^+ to leave the guard cells and the stomata to close, reducing both rates of water loss and stem growth. Resistant varieties of some plants may have increased levels of ABA. Levels of ABA also increase in response to nutrient deficiencies or excesses, chilling, waterlogging and salinity. Ethylene may also play a role in

responses of plants to stresses such as flooding, mechanical stress and pollutants.

In barley and maize, synthesis of 60–70-kDa proteins is induced by water stress and in many plants osmotin (26 kDa) is produced in response to water stress as well as salt stress (Section 27.5.3).

27.5.3 SALT STRESS

In hot, dry environments, salt often accumulates because of high rates of evaporation, and many areas of irrigated land have become saline for this reason. Soils close to the sea are also often highly saline. High salinity makes it difficult for plants to obtain water from the soil and to grow, because of the negative osmotic potential of the soil. Salinity also requires the plant to deal with toxic amounts of certain ions, particularly sodium, chloride and carbonate. Some crops, e.g. beet, tomato and rye, are salt tolerant but others, such as peas and onions, are not.

Plants growing in saline soils may have difficulty acquiring potassium, as sodium reduces its uptake. Salt-adapted plants show enhanced discrimination between potassium and sodium, and the discrimination is increased by calcium.

Some salt-tolerant plants, e.g. mangroves and salt-tolerant wheat, exclude sodium and chloride from their tissues. Others may take up salt but exude it again. Many accumulate non-toxic compatible solutes such as proline, betaine, sugars, galactosyl glycerol, polyols or malate which help to maintain osmotic balance with the soil. Some adapted plants have weakened cell walls, allowing faster growth for the same osmotic potential, whilst in others some enzymes are unusually stable at high salt concentrations.

Salt-stressed plants contain reduced quantities of most proteins (because of reduced protein synthesis and increased proteolysis), however, the production of certain specific proteins may be increased. Cultured tobacco cells synthesize new proteins in response to increased salt levels. One of them, a 26-kDa protein called osmotin, may make up more than 10% of the total protein in stressed cells, and may facilitate solute accumulation or make cells more resistant to osmotic stress. Maize, rice and citrus also produce 25–30-kDa proteins in response to salt stress.

27.6 NATURE OF PLANT HORMONES

In differentiated organisms, including higher plants, the activities of individual tissues and organs must be regulated so that they operate in a coherent manner. They must also be able to respond to environmental stimuli. Plant-growth substances or plant hormones form one element of such control processes.

Plant hormones are normally defined as naturally occurring compounds which influence biochemical processes in plants at concentrations far below those at which nutrients or vitamins are active. In comparison with animal hormones, there are fewer plant hormones and each produces a very wide range of effects and acts on many different tissues. Some appear to act in tissues where they are synthesized as well as moving around the plant. There are also very pronounced patterns of interaction between the different groups of hormones. To emphasize these differences, plant hormones are sometimes known as plant-growth substances.

Hormone concentrations might be influenced by rates of hormone synthesis, by transport and uptake into target tissues, and by metabolism in these tissues to form inactive or bound forms. The sensitivity of tissues to hormones may also vary as a result of changes in the numbers or types of receptors present in target tissues. Thus hormone action in plants is extremely complex and, on the whole, poorly understood.

The main groups of plant hormones are the auxins, gibberellins, cytokinins, abscisic acid and ethylene. In addition, a number of other phytochemicals such as the brassinosteroids, dihydroconiferol, triacontanol, methyl jas-

monate and polyamines, when applied to plants, have pronounced effects on their growth. As it is not at present clear whether these have hormonal functions, they are not discussed further in this chapter.

A considerable number of synthetic compounds have been synthesized and shown to modify the growth of plants. Such compounds are commonly known as plant-growth regulators (PGRs) and many are commercially important in crop production. Many of them are synthetic analogues of hormones which mimic their effects or interfere with their production or action.

27.7 AUXINS

Indole-3-acetic acid (IAA) is the main endogenous auxin. Some other endogenous compounds also show auxin activity, e.g. indoleacetonitrile and indoleacetaldehyde, but this is because they are converted into IAA by the plant.

Since the discovery of IAA a large number of synthetic organic compounds which also have auxin activity have been synthesized and used as herbicides and plant-growth regulators.

27.7.1 BIOCHEMISTRY OF AUXINS

IAA can be synthesized by plants from the amino acid tryptophan. Although alternatives exist, in most plants the route via indole-3-pyruvic acid is probably preferred (Figure 27.3). In tissues infected by *Agrobacterium tumefaciens*, the bacterium causing crown gall, tryptophan is converted to IAA via the intermediate indoleacetamide (Section 27.9.3).

IAA can be oxidized to form methylene-oxindole (catalysed by the well characterized peroxidase enzyme, IAA oxidase) or to oxindole-3-acetic acid. IAA may also be converted into a number of conjugates, collectively known as 'bound auxins', by conjugation with inositol, glucose or amino acids such as aspartic acid (Figure 27.4). Bound auxins are inactive but may release IAA on hydrolysis and

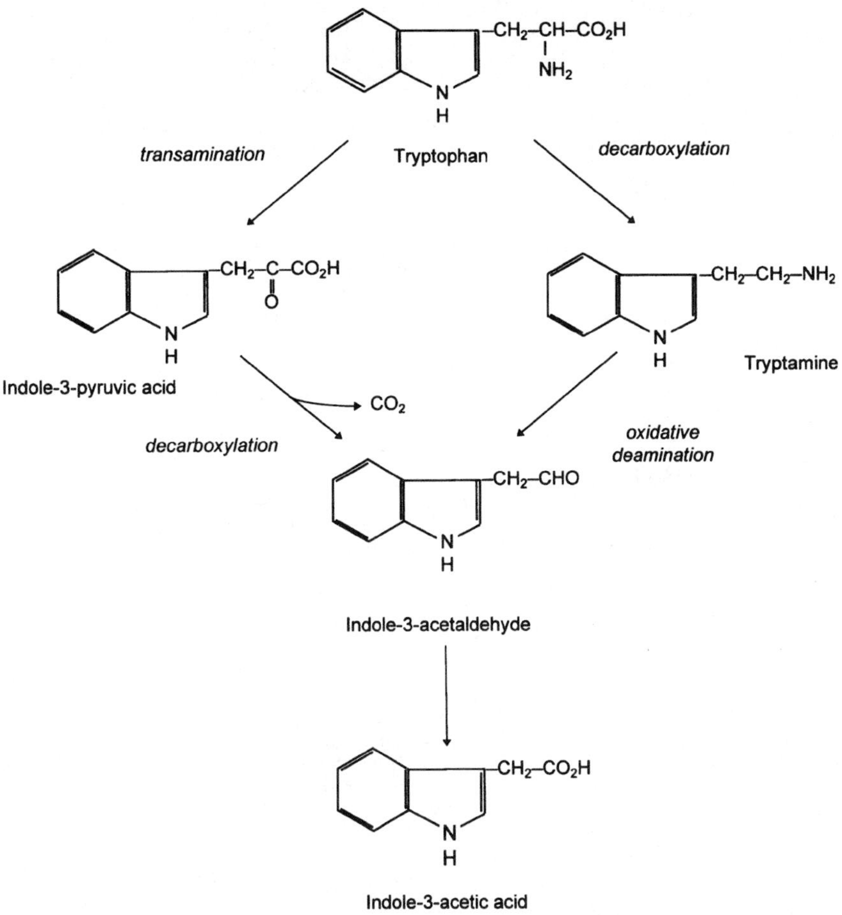

Figure 27.3 Alternative pathways for the biosynthesis of auxins in higher plants. The route via indole-3-pyruvic acid is most common.

may be forms in which IAA is transported in germinating seeds.

27.7.2 SYNTHETIC AUXINS

A large number of synthetic compounds (Figure 27.5) have been shown to have auxin activity. Compounds such as indole-butyric acid (IBA), α-naphthalene acetic acid (NAA) and 2,4-dichlorophenoxyacetic acid (2,4-D) are widely used as plant-growth regulators. These compounds are more often used than IAA as they are cheaper and more persistent in the plant because they are not attacked by IAA oxidase.

When applied at higher concentrations, a wide range of auxins, including those shown in Figure 27.5, act as herbicides. Symptoms of auxin herbicide treatment are distorted growth, epinastically curved leaves and split stems, many of which are probably a result of stimulation of ethylene production.

A major advantage of auxin herbicides is that they are selective in their action. Cereals and other monocotyledons are quite resistant to them, but dicotyledons are usually very sensitive. This selectivity depends mainly on differences in the nature and orientation of monocot and dicot leaves, so that the sprays may be retained and penetrate to different

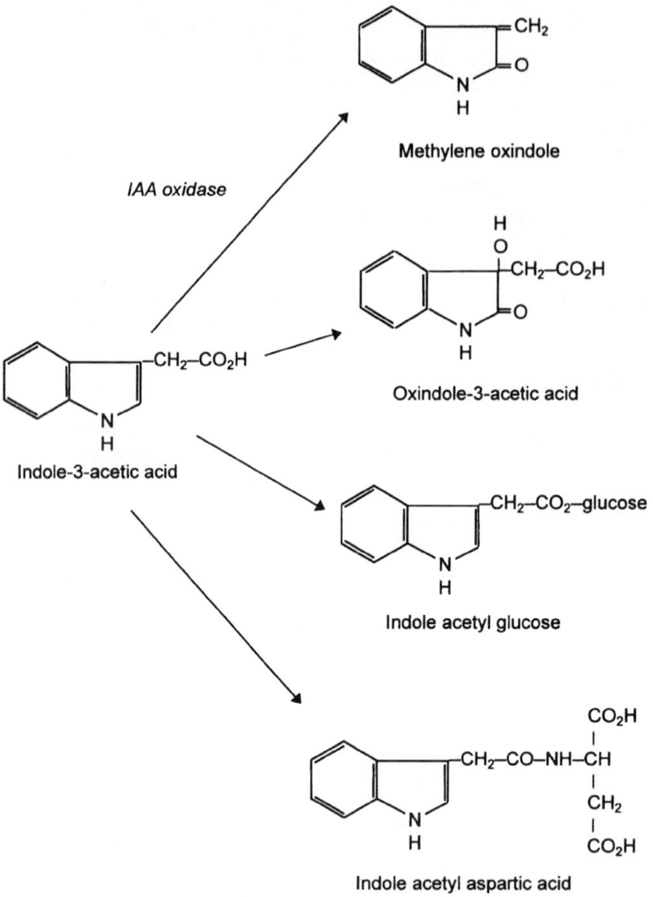

Figure 27.4 Auxin metabolites. Oxidation leads to irreversible loss of auxin activity, but conjugation with sugars or amino acids produces bound auxins from which free auxin may be regenerated by hydrolysis.

extents. In addition, differences in the metabolism of auxins between different species exist and contribute to their selectivity. The phenoxybutyric acids, e.g. MCPB and 2,4-DB, do not themselves have auxin activity but in some species they are degraded into the corresponding phenoxyacetic acids, which are active, by the β-oxidation pathway which removes pairs of carbon atoms from the fatty acid side chains (Chapter 13). Most legumes, including peas and clover, are unable to bring about this conversion and are resistant, allowing weeds to be selectively killed in these crops.

27.7.3 SITES OF SYNTHESIS AND TRANSPORT OF AUXINS

Meristems, enlarging tissues including the shoot apex, enlarging leaves and opening buds, and developing fruits and seeds are major sites of auxin production. Longitudinal movement of auxins is strongly polar and requires ATP. In both the root and the shoot, movement is generally downwards and quite slow, (about 1 cm h^{-1}). Lateral movement of auxin occurs in both roots and shoots and gives rise to phototropic and gravitropic responses.

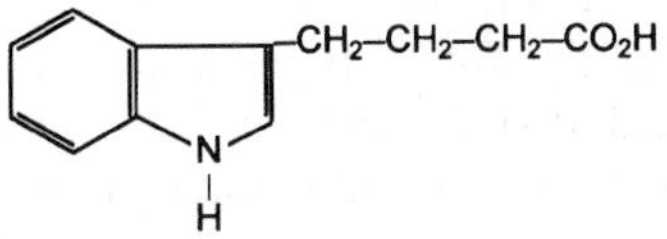

Indolebutyric acid
(IBA)

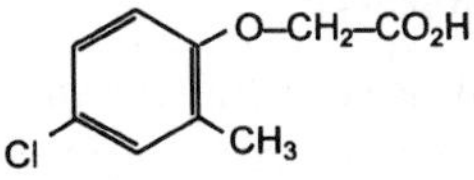

(2-methyl, 4-chlorophenoxy)acetic acid

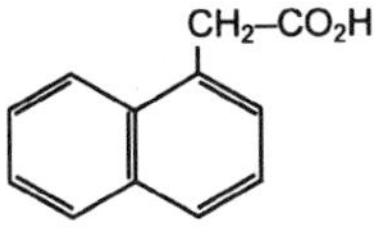

α Naphthaleneacetic acid
NAA

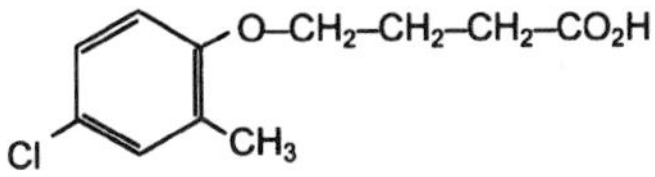

4-(2-methyl, 4-chlorophenoxy)butyric acid

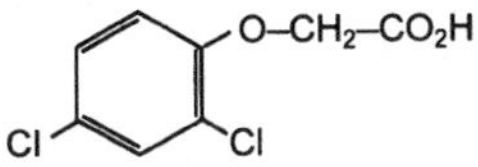

(2,4-dichlorophenoxy)acetic acid
2,4-D

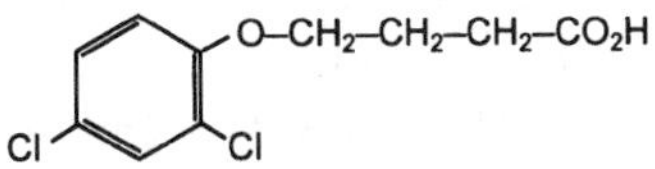

4-(2,4-dichlorophenoxy)butyric acid
2,4-DB

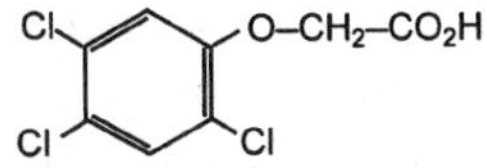

(2,4,5-trichlorophenoxy)acetic acid
2,4,5-T

Figure 27.5 Structures of a selection of synthetic auxins. IBA and NAA are commonly used as plant-growth regulators whilst the phenoxy derivatives are mainly used as herbicides.

27.7.4 PHYSIOLOGICAL ACTIVITIES AND APPLICATIONS OF AUXINS

Effects on stem elongation

Auxins stimulate the elongation of stem sections, but not usually of intact stems which receive sufficient auxin from the shoot apex to promote optimal growth.

Binding of auxins to the plasma membrane of stem cells activates a proton pump, causing secretion of protons from the cell into the wall. This activates enzymes which hydrolyse wall polysaccharides, or weakens hydrogen bonds between wall components and allows the wall to be stretched.

Unilateral illumination of shoots causes auxin to become asymmetrically distributed and results in curvature towards the light (phototropism). Light perception in the apex results in reduced rates of transport from the apex on the illuminated side, so that the concentration of auxin on the shaded side is increased relative to that on the illuminated side, thus promoting its elongation.

Effects on cambial development

Both auxins and gibberellins stimulate cambial cell division, but their effect when applied together is much greater than either applied alone (synergism). In addition, auxin is essential for further differentiation of cells to form xylem, whilst gibberellins are needed for phloem formation.

Role in apical dominance

The presence of the shoot apex normally inhibits growth of lateral buds (apical dominance). In many plants removal of the shoot apex allows the lateral buds to grow, but this can be prevented by application of auxin to the cut stem. Thus auxins originating in the shoot apex are thought to inhibit the development of lateral buds. In contrast, root-produced cytokinins promote such development, as application of cytokinins to some inhibited buds can stimulate their growth.

Effects on root development and differentiation of callus

Auxins stimulate the production of lateral roots, thus promoting root branching. Both endogenous and exogenously applied auxins stimulate the development of adventitious root primordia from the stem, causing rooting of cuttings. Synthetic auxins such as IBA (indole butyric acid), NAA (naphthalene acetic acid) or naphthalene acetamide are widely used in commercial rooting powders, IBA and NAA often being used together. Although auxins promote initiation of roots, they generally inhibit root elongation as the concentration of auxin required for maximal root growth is very low (approximately 10^{-10} M).

Auxin accumulates on the lower side of both horizontal roots and shoots, but whereas it promotes elongation of that side of the shoot, it probably inhibits that of the root. Thus roots bend downwards whilst shoots bend upwards. The gravitational stimulus is detected by statoliths containing heavy starch grains which settle to the lower side of the statolith. This is believed to cause an asymmetrical distribution of Ca^{2+} ions which in turn causes redistribution of IAA.

Callus may differentiate to form roots or shoots when exposed to auxins or cytokinins. High auxin/cytokinin ratios promote root formation, whilst low ratios promote shoot formation.

Effects on flowering

Application of 2,4-D or NAA is used commercially to induce and increase the uniformity of flowering of pineapple. Auxins generally do not promote, and may even inhibit, flowering in most other species.

Auxins favour female flower production in a number of plants, including cucumbers. The effects of auxins on flowering may be due to their induction of ethylene production, as ethephon or ethylene produce similar responses (Section 27.11.4).

Effects on fruit development

Auxins in pollen ensure that a rapid burst of ovary growth, accompanied by abscission of the stamens and petals, usually follows pollination (fruit set).

When flowers are not fertilized abscission of the flower usually occurs, but in some species this can be prevented by auxin application. Many synthetic auxins increase fruit-set in fruit containing many ovules, e.g. fig, strawberry, squash, tomato and aubergine. This may be used to overcome pollination difficulties in indoor tomatoes, or depressed fruit set due to low temperatures in the field.

Auxins may also cause abscission of flowers of fruit trees. Several auxins, including NAA or naphthaleneacetamide, can be used to remove some flowers and prevent development of excessive numbers of fruit, which tend to be small.

Stimulation of fruit development by auxin in pollen lasts only a short time and continued development of the fruit depends on hormones, especially auxin, produced by the developing seed.

Mature apples and pears often fall off the tree just before harvesting because of low levels of auxin in the fruit, but application of NAA or 2,4,5-trichlorophenoxypropionic acid may be used to prevent this.

27.8 GIBBERELLINS

The gibberellins (GAs) contain 19 or 20 carbon atoms and are based on the *ent*-gibberellane ring structure (Figure 27.6). They have been given numbers such as gibberellin A_1 (GA_1) etc., approximately in order of their identification; about 100 are now known.

Most higher plants contain a number of GAs although not all have intrinsic biological activity. In addition, GAs are also found in fungi: GA_3 and a mixture of GA_4 and GA_7 from these sources are available commercially.

27.8.1 BIOCHEMISTRY OF GIBBERELLINS

GAs are diterpenoid derivatives and are synthesized by the condensation of five-carbon (C-5) units head-to-tail. Successive linking of these units results in formation of geranyl pyrophosphate (C-10), farnesyl pyrophosphate (C-15) and geranylgeranyl pyrophosphate (C-20) (Figure 27.7). Farnesyl pyrophosphate may also serve as a precursor for the synthesis of ABA (section 27.10.1) and steroids. Geranylgeranyl pyrophosphate undergoes cyclization to form *ent*-kaurene via copalyl pyrophosphate. Steps up to this point are catalysed by soluble enzymes.

The next steps take place on the endoplasmic reticulum and result in formation of GA_{12} aldehyde, from which other GAs arise by steps which vary from genus to genus.

The existence of mutants which are defective in GA synthesis has allowed the biosynthesis of GA in shoots of pea and maize to be studied very thoroughly. GA_{12}-aldehyde is converted to GA_1 by a pathway called the early 13-hydroxylation pathway, and GA_1 has been identified as the form which actually promotes shoot growth in these plants.

GAs may be metabolized into a number of derivatives such as glucosides and protein–GA conjugates, which may function as storage forms.

A number of plant-growth retardants inhibit specific steps in the pathway leading to the biosynthesis of GA_{12}-aldehyde (section 27.8.4).

27.8.2 SITES OF SYNTHESIS AND TRANSPORT OF GIBBERELLINS

GAs appear to be synthesized in leaf primordia in the shoot apex, in root tips, and in developing fruits and seeds.

There are no clear patterns of movement of GA in the plant although they generally move towards young, growing tissues in both xylem and phloem.

27.8.3 PHYSIOLOGICAL ACTIVITIES AND APPLICATIONS OF GIBBERELLINS

Effects on vegetative growth

Application of GAs to intact plants often results in spectacular stimulation of shoot elongation, but they generally have little effect on elongation of isolated stem sections. They appear to stimulate cell division in the shoot apex and to increase cell wall plasticity in stem cells. However the effects on wall plasticity do not seem to result from acidification of the wall, and synergistic interactions of gibberellins and auxins in controlling elongation of cells are seen.

Many dwarf plants have abnormalities in gibberellin biosynthesis or responsiveness. The former types have been particularly useful in elucidating the pathways of gibberellin biosynthesis whilst the latter form the basis of many commercially important high yielding, semi-dwarf cereal varieties.

GA can be used to overcome some environmental requirements for growth. Thus, GA reduces the chilling requirement of rhubarb

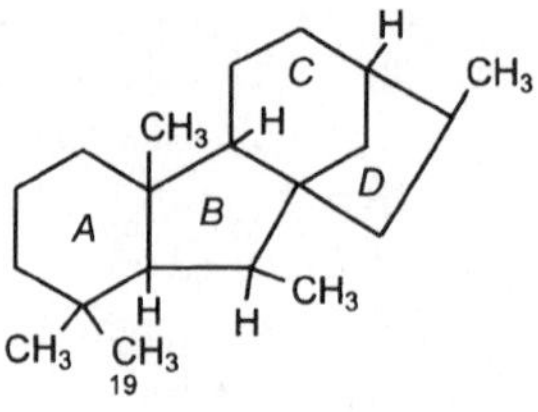

ent-gibberellane

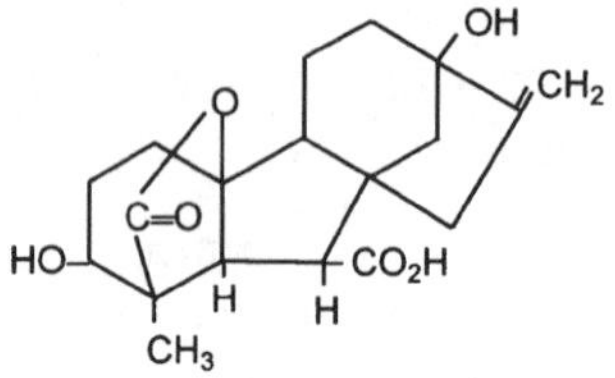

GA$_1$

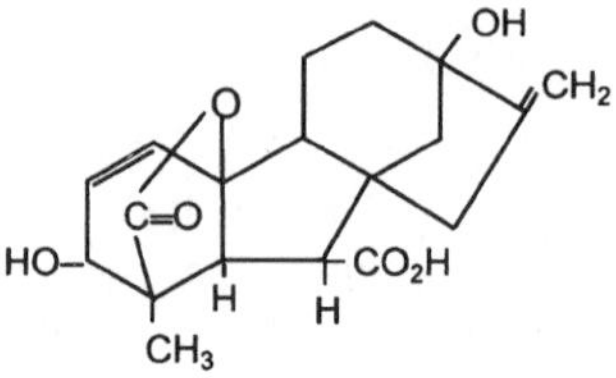

GA$_3$

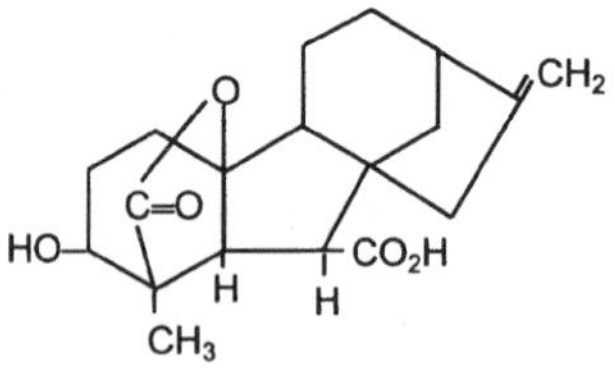

GA$_4$

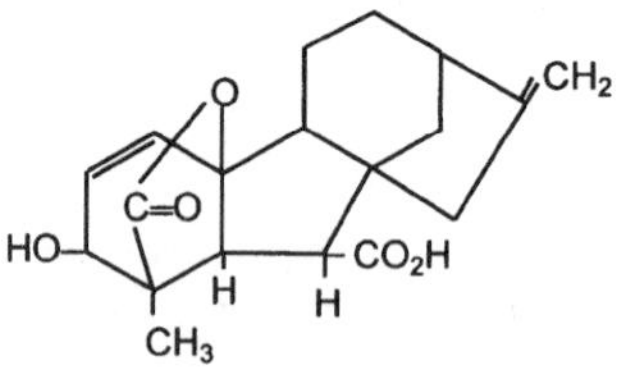

GA$_7$

Figure 27.6 The structure of the gibberellins is based on the *ent*-gibberellane ring. GA$_1$ is the biologically active form in pea and maize shoots, and GA$_3$, GA$_4$ and GA$_7$ are commonly used as growth regulators.

and increases the yield, replacing the need for forcing. It can also be used to overcome dormancy in potato tubers.

Effects on flowering

Young plants may show a juvenile growth phase in which they will not flower, however favourable the environmental conditions. Application of GA may be used to reduce the length of this unproductive juvenile phase of growth.

GA enhances flowering in many plants and in some, such as the globe artichoke, this may increase yield. It may replace the need for long days to induce flowering in rosette plants. GA may also replace the need for exposure to low temperatures (vernalization) to induce flowering in some biennial or spring-flowering perennials. GA has a number of indirect effects on flowering: for instance, it stimulates production of male flowers in cucumbers, leading to better pollination, and stimulates bolting in many rosette or biennial plants such as lettuce and carrots, enhancing seed production. It has also proved very useful in the control of flowering in conifers as part of breeding programmes, reducing the time needed to produce seed.

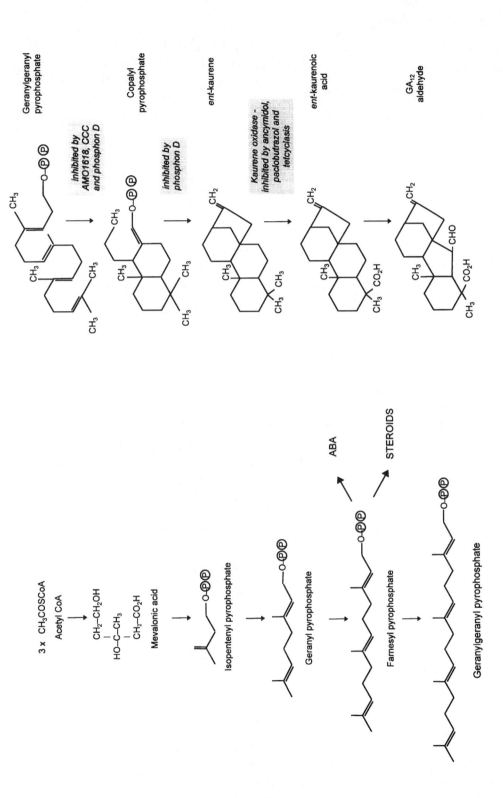

Figure 27.7 The biosynthesis of gibberellins from acetyl-CoA. Condensation of five-carbon units head-to-tail leads to production of geranylgeranyl pyrophosphate. Cyclization and oxidation result in production of GA_{12} aldehyde from which other GAs arise. Enzyme-catalysed reactions in the second part of the pathway are inhibited by growth retardants.

Effects on fruit development

Application of GA can overcome ineffective pollination in fruit trees such as plum, peach, cherry, apple and pear, which flower early in the year and are particularly susceptible to frost damage. GA may also be used to promote fruit set in tomatoes and mandarin oranges. In some plants, combinations of GA with auxin or GA with auxin and cytokinin may be more effective than GA alone.

GA is often applied pre-bloom to open clusters of grapes, giving greater separation. This ensures that the berries dry more rapidly and that the fruit do not damage one another, releasing juice and increasing susceptibility to fungal infections.

Effects on seed dormancy and germination

Treatment of seeds with GA can lead to more uniform germination and hence better seedling establishment, so that mechanical harvesting of the crop becomes more effective. It is widely used in malting as it stimulates germination and α-amylase production in the grain, so that conversion of starch to fermentable sugars is both accelerated and made more complete.

Effects on cambial development

GA stimulates cambial growth and differentiation in conjunction with auxins (Section 27.7.4).

27.8.4 GROWTH RETARDANTS

Growth retardants reduce the growth of plants by inhibiting the biosynthesis of GA. They reduce the rate of stem elongation without reducing leaf number or leaf area. The structures of some commonly used growth retardants are shown in Figure 27.8.

More recently introduced growth retardants such as paclobutrazol and tetcyclasis are effective at approximately one thousandth the concentration at which older retardants, such as CCC, are normally used.

Growth retardants inhibit the biosynthesis of GA and many of their effects can be reversed by simultaneous application of GA. AMO1618, CCC and phosphon D inhibit synthesis of kaurene, whereas paclobutrazol, ancymidol and tetcyclasis inhibit kaurene oxidase (Figure 27.7). Although the general effects of all the growth retardants are similar, in that they reduce internode length in stems, particular retardants have advantages for specific applications.

Dwarfing of pot plants

Application of growth retardants to pot plants such as chrysanthemums, azaleas, poinsettias, carnations and geraniums produces bushier plants with more flowers. They are very rapidly translocated and may be applied as a foliar spray (except phosphonium chloride which may scorch leaves), as a soil drench, or incorporated into the soil.

Lodging control in cereals

Lodging of cereals occurs in response to wind and rain, particularly in fertile soils and when plants are weakened by eyespot infection. Application of CCC is widely used to reduce lodging of wheat and oats even at high fertilizer application rates.

Correlative effects

Individual parts of a plant compete with one another for assimilates. Thus, yields of roots or fruits or seeds may often be increased by restricting vegetative growth by application of growth retardants. In addition, reduced growth of parts of no commercial value may result in compensatory growth of more valuable parts of the plant. Effects of this type are known as 'correlative effects'.

Application of daminozide, CCC or paclobutrazol restricts the growth of apple and pear trees, producing more compact plants which flower earlier. In young, vigor-

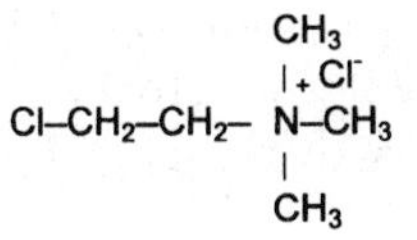

CCC
Chlorocholine chloride
Chlormequat chloride
Cycocel

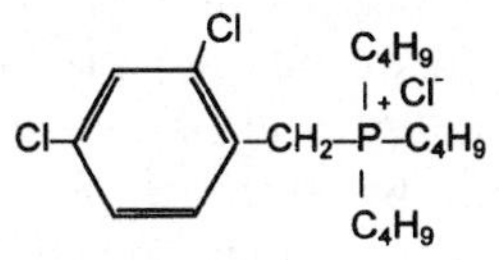

Phosphon
Chlorphonium chloride

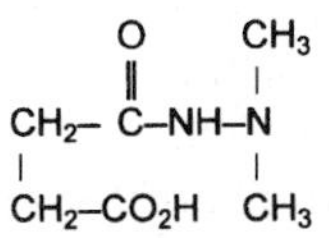

Daminozide
Alar
B9

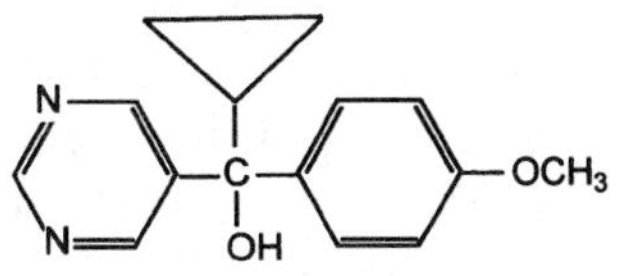

Ancymidol
A-Rest

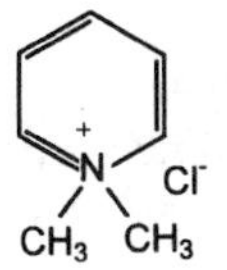

Mepiquat chloride

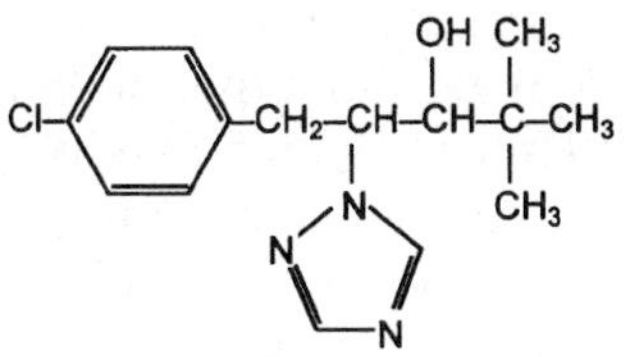

Paclobutrazol

Figure 27.8 Structures of some commonly used growth retardants. Quaternary ammonium and phosphonium derivatives such as CCC and phosphon were amongst the first to be discovered. More recently introduced compounds include paclobutrazol which is related to the triazole fungicides.

ous fruit trees, fruit drop may occur because of competition for assimilates between the large number of rapidly growing fruits and the vegetative growing points. This can be greatly reduced by pinching out the shoot tips or by spraying with daminozide, which inhibits shoot growth and diverts nutrients to the fruitlets.

Early application of growth retardants to cereals may reduce growth of the shoot apex, allowing development of additional tiller buds and stimulating root development. This gives rise to better established plants with more fertile tillers, with more heads and with the potential to give higher yields.

Fruit quality

Daminozide (Alar) increases anthocyanin production (and hence the red colour in apples), decreases pre-harvest drop, and reduces softening in cold conditions. It has therefore been extensively used on red apples, especially in the USA. In 1989 there was considerable concern because the compound was reported to be weakly carcinogenic; despite this not being substantiated by an independent enquiry, the compound was withdrawn from use for food crops worldwide because of public reaction.

27.9 CYTOKININS

27.9.1 BIOCHEMISTRY OF CYTOKININS

Naturally occurring cytokinins are substituted adenine derivatives. Most are derivatives of zeatin or of N^6-isopentenyladenine (Figure 27.9). These compounds exist in the free form as bases, their 5'-ribosides (zeatin riboside and isopentenyl adenosine, respectively), and 5'-ribotides (zeatin ribotide and isopentenyl AMP).

Cytokinins may be converted into glucosides or amino acid conjugates. These are inactive but some may serve as a store from which active hormone can be regenerated by hydrolysis.

Cytokinins also occur as components of about 10% of tRNA molecules. When present, they always occur at the position immediately next to the 3' end of the anticodon, and the tRNA molecules containing them all bind to mRNA codons beginning with U.

Cytokinins are synthesized by transfer of an isopentenyl group from dimethylallyl pyrophosphate to the N^6-amino group of free AMP or to an adenosine residue at the appropriate position in RNA. The enzyme which catalyses this reaction has been characterized (section 27.9.3). Zeatin derivatives are formed by hydroxylation of the isopentenyl side chain. Some free cytokinins could arise from breakdown of cytokinin-containing tRNA but this does not appear to be an important source. The physiological significance of cytokinins in RNA and their relationship to free cytokinins is unclear.

Some cytokinins can be oxidized by the enzyme cytokinin oxidase, which catalyses removal of the side chain from the N^6-amino group of the purine ring. The products of

cleavage of isopentenyl adenine are adenine and 3-methyl-2-butenal [$CHO-CH=C(CH_3)_2$] and all biological activity is lost, so this is a means of inactivating cytokinins.

27.9.2 SITES OF SYNTHESIS AND TRANSPORT OF CYTOKININS

In vegetative plants, the root tip seems to be the major site of cytokinin biosynthesis. These compounds are also found in relatively large amounts in young fruits, seeds and leaves and may be synthesized in some of these.

Cytokinins produced in the roots are transported in the xylem to the shoot. In other respects their transport is very restricted, and when applied exogenously they tend to remain at their sites of application.

27.9.3 PHYSIOLOGICAL ACTIVITIES AND APPLICATIONS OF CYTOKININS

Role in cell division and differentiation of callus

Callus cells replicate their chromosomes but are unable to divide unless cytokinins are added to the growth media. Stimulation of cell division is the most characteristic effect of these hormones. The effect of cytokinins on differentiation of callus is discussed in section 27.7.4.

Most plant cells growing in tissue culture require added hormones to grow rapidly, but when cells become infected with the crown gall bacterium *Agrobacterium tumefaciens* they become hormone-autotrophic. The Ti (tumour inducing) plasmid which becomes incorporated into the host-cell genome carries genes coding for the biosynthesis of both auxin and cytokinins. As crown gall tissues contain much higher levels of hormones than normal tissues, they have been able to provide a great deal of information about auxin and cytokinin synthesis.

Auxin biosynthesis in these galls requires two enzymes – tryptophan-2-monooxygenase and indoleacetamide hydrolase, which catal-

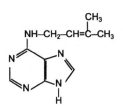

Zeatin

N⁶-Isopentenyl adenine

Zeatin riboside

N⁶-Isopentenyl adenosine

Zeatin ribotide

N⁶-Isopentenyl AMP

Figure 27.9 Structures of cytokinins. Naturally occurring cytokinins are mostly derivatives of zeatin and isopentenyladenine. They exist in the free form as bases, 5′-ribosides and 5′-ribotides.

yse the steps shown in Figure 27.10a. These are coded for by the genes *tms1* and *tms2*, respectively.

Two genes in *A. tumefaciens* code for enzymes which catalyse cytokinin synthesis. One of these, *tmr* (also called *ipt*), codes for the enzyme DMAPP:AMP transferase which catalyses the reaction of dimethylallyl pyrophos-

phate with AMP to form isopentenyl-AMP (Figure 27.10b).

Inactivation of *tms* genes coding for auxin biosynthetic enzymes results in greatly increased cytokinin/auxin ratios, leading to crown galls which produce shoots, whilst inactivation of genes coding for cytokinin biosynthetic enzymes (*tmr*) results in decreased

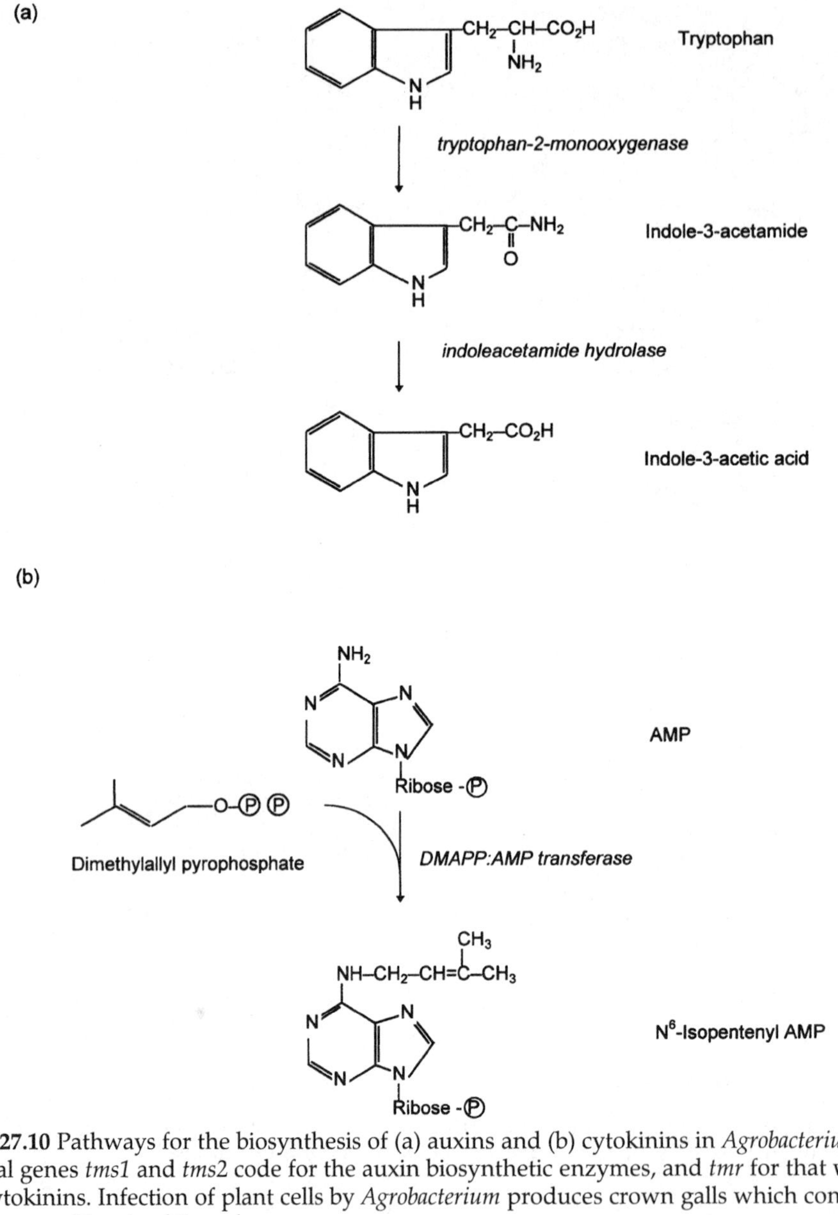

(a)

(b)

Figure 27.10 Pathways for the biosynthesis of (a) auxins and (b) cytokinins in *Agrobacterium tumefaciens*. Bacterial genes *tms1* and *tms2* code for the auxin biosynthetic enzymes, and *tmr* for that which synthesizes cytokinins. Infection of plant cells by *Agrobacterium* produces crown galls which contain much higher than normal levels of these hormones.

cytokinin/auxin ratios and crown galls which produce roots.

Leaf development and senescence

Cytokinins promote the growth of leaves and delay senescence. There is considerable evi-dence that a decrease in protein synthesis occurs during senescence. Although leaves normally senesce quickly when detached from the plant, this is prevented if they produce roots or if a drop of cytokinin solution is applied to the leaf. In the latter case the region in contact with the cytokinin remains as a 'green island' as the remainder of the leaf yellows. It is thus generally considered that root-produced cytokinins delay leaf senescence by maintaining protein synthesis in leaves.

Apical dominance

Cytokinins promote development of lateral buds in some species (section 27.7.4).

Applications

The principal synthetic cytokinins available are benzyl adenine (BA) and kinetin. These are used in combination with GA to improve fruit set in grapes and figs, and to increase lateral bud growth and fruit quality in apples. Although cytokinins have potential uses in delaying senescence of green vegetables, mushrooms and cut flowers they are not widely used in this way.

27.10 ABSCISIC ACID (ABA)

The structure of abscisic acid is shown in Figure 27.11. Biological activity appears to be associated with just this single compound. It is a sesquiterpenoid containing 15 carbon atoms. It is optically active, existing as +/- isomers at the 1'-position, and there are also *cis*/*trans* isomers about the 2-double bond. The naturally occurring compound is (+)-2-*cis* ABA. It occurs widely in angiosperms, gymnosperms, ferns and mosses but a different compound, lunularic acid, may fulfil similar functions in liverworts and some algae.

27.10.1 BIOCHEMISTRY OF ABSCISIC ACID

ABA is synthesized from mevalonic acid. The steps as far as the 15-carbon compound farnesyl pyrophosphate (FPP) are the same as those described for GA biosynthesis (Figure 27.7). The conversion of FPP into ABA occurs via another C-15 compound, xanthoxin. In higher plants it is likely that carotenoids are intermediates in the conversion of FPP to xanthoxin.

ABA is very rapidly and extensively metabolised in higher plants to form mainly ABA-glucose ester and phaseic and dihydrophaseic acids (Figure 27.11). Phaseic and dihydrophaseic acids show little biological activity and may be formed as a way of inactivating ABA.

27.10.2 SITES OF SYNTHESIS AND TRANSPORT OF ABSCISIC ACID

Abscisic acid is synthesized in mature green leaves, especially in response to water stress. It is also present in large quantities in seeds. ABA appears to move quite readily between the leaves, shoots and roots.

27.10.3 PHYSIOLOGICAL ACTIVITIES AND APPLICATIONS OF ABSCISIC ACID

Role in response to stress

When leaves are subjected to water stress the rates of biosynthesis of ABA are greatly increased. This is triggered by changes in turgor pressure which are detected by the plasma membrane. The elevated levels of ABA in the leaves inhibit K^+ accumulation in guard cells, reducing their turgor and causing the stomata to close, thus reducing water loss. When plants are re-watered, ABA concentrations drop to pre-stress levels within 4–8 hours, in this case due to decreased biosynthesis and rapid metabolism to phaseic and dihydrophaseic acids.

The potential use of ABA as an antitranspirant is of considerable interest. Partial closing of the stomata can decrease transpiration rates without affecting photosynthesis, and can enable plants to make more efficient use of water. ABA itself is expensive and has short-lasting effects. Derivatives such as the methyl ester of ABA, however, can control stomatal opening for 9–10 days. Such compounds would be of particular value for plants introduced into climatically unsuitable, marginal areas especially at times when they are particularly susceptible to water stress, such as during reproduction.

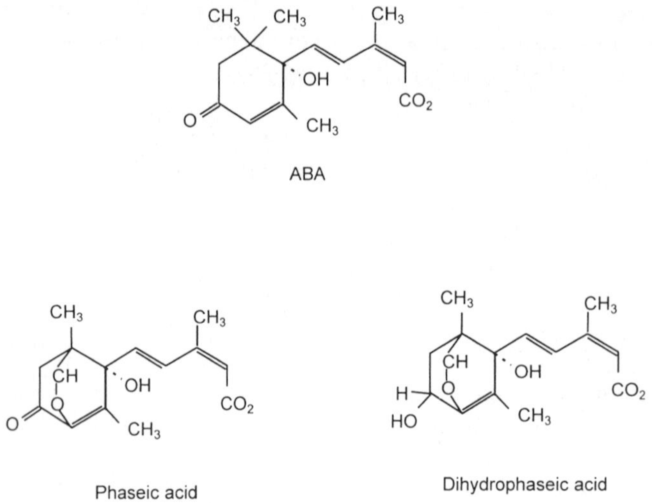

ABA

Phaseic acid

Dihydrophaseic acid

Figure 27.11 The structures of abscisic acid (ABA) and metabolites. ABA is rapidly metabolized to form phaseic acid which is then converted to dihydrophaseic acid.

Salinity and low temperature may also cause increased synthesis of ABA and this may help the plant resist these stressful conditions (section 27.5) .

Role in control of assimilate movement

ABA, auxins and gibberellins appear to increase assimilate transport from leaves into developing seeds. ABA also activates genes coding for specific storage proteins during seed development and prevents premature germination of the seed before they desiccate and mature.

Role in inhibition of growth and in dormancy of buds and seeds

ABA commonly inhibits the growth of cells and often counteracts the growth-promoting activity of auxins, gibberellins and cytokinins.

In some trees, levels of ABA in leaves and buds increase in late summer when the buds become dormant, and levels decline as buds are chilled and dormancy is broken. Application of ABA can cause some buds to become dormant. In many species, however, there is no correlation between ABA levels and depth of bud dormancy.

Although ABA inhibits germination of the seed of many species, and levels of endogenous ABA fall when dormancy is broken in some seeds, there is no universal correlation between seed dormancy and ABA levels.

27.11 ETHYLENE

Ethylene is unique amongst the hormones in being a gas and having a very simple structure (Figure 27.12a). A number of synthetic compounds decompose (either spontaneously or through the action of enzymes) to produce ethylene, and some other unsaturated hydrocarbons may mimic its effects.

27.11.1 BIOCHEMISTRY OF ETHYLENE

Ethylene is derived from carbon atoms 3 and 4 of methionine. Biosynthesis takes place via amino cyclopropane carboxylic acid (ACC) as shown in Figure 27.12.

ACC synthase [1] catalyses conversion of S-adenosyl methionine to ACC, which is converted to ethylene by the action of ACC oxidase [2], formerly known as ethylene-forming enzyme (EFE). As ACC oxidase requires oxygen, ACC accumulates under anaerobic conditions. The activity of ACC oxidase increases under some conditions which increase ethylene production such as stress, in fruit ripening or in response to ethylene itself (autocatalysis).

27.11.2 SITES OF SYNTHESIS AND TRANSPORT OF ETHYLENE

Most tissues can produce ethylene: ripening fruit, the shoot apex, nodes of dicotyledonous seedlings, and flowers and leaves undergoing senescence are particularly rich sources. Ethylene production in most tissues is increased in response to mechanical stresses such as rubbing, pressure or damage or by treatment with high concentrations of auxins.

27.11.3 METHODS OF MODULATING THE EFFECTS OF ETHYLENE ON PLANTS

Ethylene-induced responses can be manipulated in several ways.

Several small, unsaturated hydrocarbons and carbon monoxide have similar, but much weaker biological activity than ethylene (Table 27.1). These compounds bind to the ethylene receptor which has relatively low specificity for ethylene. The receptor contains copper and its function is inhibited by copper-chelating agents such as ethylene diamine acetic acid (EDTA), and by Co^{2+} and Ag^+. CO_2 acts an antagonist and competes with ethylene for binding to the receptor.

Compounds such as ethephon (Etherel) and etacelasil (Figure 27.12) decompose in the plant to produce ethylene. Silane derivatives such as etacelasil decompose more rapidly than ethephon and produce slightly different responses.

When ACC is applied to plants it may be converted into ethylene by the action of ACC oxidase in an oxygen-requiring reaction (see Figure 27.12).

High auxin concentrations usually inhibit plant growth, because they induce the production of ethylene in the plant by increasing transcription of the gene coding for ACC synthase.

The biosynthesis of ethylene is reduced by aminoethoxyvinylglycine (AVG) and aminooxyacetic acid (AOA), which inhibit ACC synthase by blocking the action of pyridoxal phosphate. A single application of AVG may prevent flowering in bromeliads for several months. AOA has similar activity to AVG and is considerably cheaper.

Ag^+ and carbon dioxide are widely used to reduce ethylene activity and to delay flower senescence and fruit ripening, as discussed elsewhere in this chapter.

27.11.4 PHYSIOLOGICAL ACTIVITIES AND APPLICATIONS OF ETHYLENE

Effects on fruit ripening

There is much evidence that the ripening of fruit is controlled principally by ethylene. In most fleshy fruits the onset of ripening is preceded by a burst of ethylene production which stimulates respiration. Ethylene production is an autocatalytic process; trace amounts lead to production of increasing quantities.

Ethylene may change membrane permeability leading to the release of the enzymes needed for ripening. It may also increase synthesis of proteins, including those required for hydrolysis of cell components and for the

Table 27.1 Relative activity of ethylene analogues in inducing ethylene-like responses

Structure	Name	Relative effectiveness
$CH_2=CH_2$	Ethylene	1
CH_3-$CH=CH_2$	Propylene	1/100
$CH\equiv CH$	Acetylene	1/2800
$CH_2=C=CH_2$	Allene	1/29000
$C=O$	Carbon monoxide	1/2700

Data reproduced with permission from Burg, S.P. and Burg, E.A. (1967) Molecular requirements for the biological activity of ethylene. *Plant Physiology*, 42, 144–152.

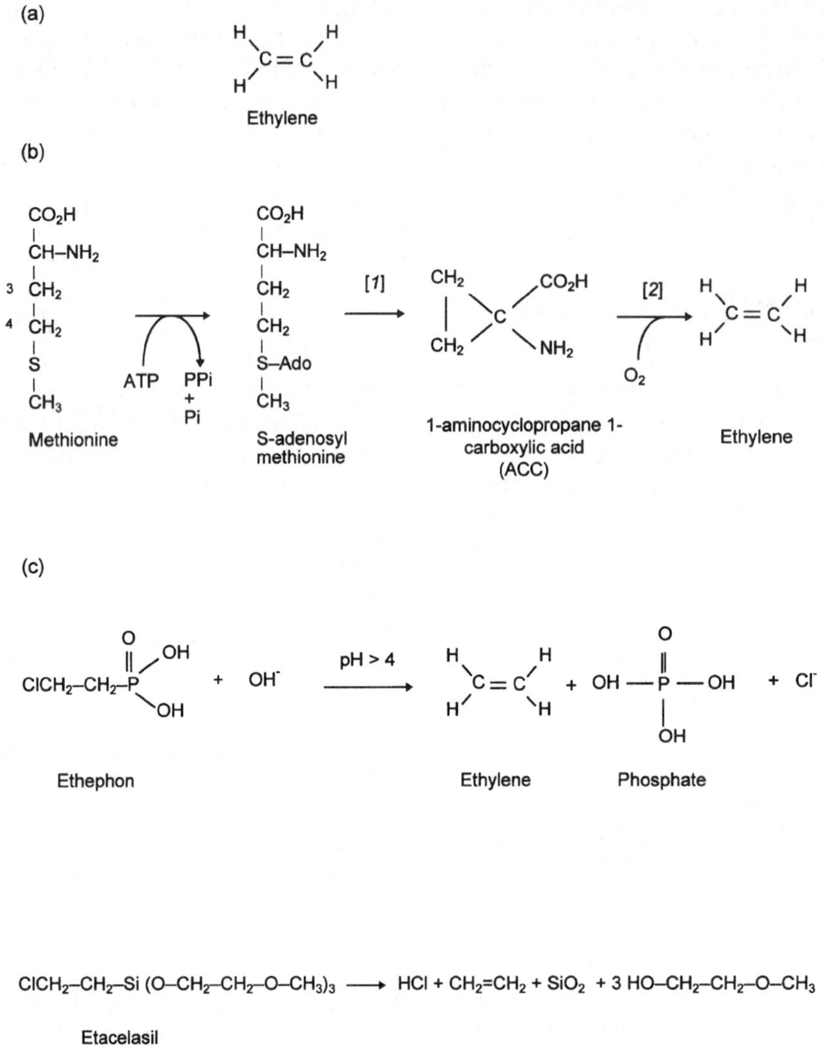

(a)

Ethylene

(b)

(c)

Etacelasil

Figure 27.12 The chemistry of ethylene and related compounds. (a) The structure of ethylene. (b) Pathway for the biosynthesis of ethylene from methionine: ACC synthase catalyses reaction [1] and ACC oxidase catalyses reaction [2]. (c) Reactions for the breakdown of ethephon and etacelasil to release ethylene.

increase in respiration. Both the biosynthesis of ethylene and the increased respiration triggered by ethylene require oxygen.

Storage of fruit in air with a low oxygen content, a low ethylene content or a high carbon dioxide content delays ripening. Low temperature prevents ethylene production and reduces respiration rates. Conversely ripening can be promoted, either on the plant or after harvesting, by applying ethephon to fruits such as apple, tomato, banana, melon, pepper and coffee.

The molecular biology of the ripening process has been studied extensively because of the potential commercial value of fruits which ripen more slowly and have a longer shelf life. Ripening may be delayed by reducing ethylene synthesis through the use of antisense genes for ethylene biosynthetic enzymes, or through the use of antisense genes for enzymes, such as polygalacturonase (PG) and pectinmethylesterase (PME), which are involved in ripening itself. The use of transgenic plants containing antisense genes for ACC synthase or ACC oxidase has allowed ethylene synthesis to be almost completely eliminated. Ethylene synthesis has also been reduced in transgenic plants containing the gene for ACC deaminase, which rapidly degrades ACC, reducing its availability for ethylene synthesis. Insertion of antisense genes for PG into tomatoes reduces the activity of this enzyme and somewhat delays the softening which accompanies ripening. Transgenic fruit of this type are marketed by Calgene under the trade name FlavrSavr®.

Promotion of abscission of leaves and fruits

Ethylene promotes abscission of both leaves and fruits. Abscission occurs when cells in a separation layer synthesize pectinase and specific isozyme forms of cellulase in response to ethylene. These enzymes degrade cell walls and cause the cells to part.

Where mechanical harvesting is employed, ethylene-generating compounds may be used to remove leaves in order to facilitate harvesting of seeds or fruits such as cotton and some types of beans. Fruit harvesting may also be facilitated by treatment with a 'harvest-aid' which reduces the force required to remove the fruit, thus reducing mechanical damage to both fruit and tree. Ethephon is extremely effective when used on apples, cherries, citrus fruits, olives, pears and grapes. However other ethylene-generating compounds such as etacelasil, which is commonly used in olives, cause less leaf abscission.

Effects on stem elongation and leaf development

Exposure of pea seedlings to ethylene reduces stem elongation and causes a downward curvature of the leaves called epinasty. Ethylene promotes elongation of cells on the upper side of the petiole but not those on the lower side, resulting in curvature. Symptoms such as these may be seen in plants exposed to ethylene or high concentrations of auxins, which stimulate ethylene production, or in plants subjected to stress.

Cells cannot extend parallel to the orientation of the cellulose microfibrils in their cell walls, but are able to do so perpendicular to them. In many stem and root cells, microfibrils are laid down predominantly with a horizontal orientation, allowing the cells to extend vertically. Exposure to ethylene changes the orientation, reducing vertical and promoting horizontal growth. This is important in the emergence of seedlings where thickening of the stem and root gives them greater strength and enables them to push through compacted soils.

Ethylene synthesis can be stimulated by stresses such as rubbing or cutting stems and leaves, attack by pathogens or exposure to drought or flooding. In some plants, ethylene produced on wounding initiates secretory processes. A commercially important example is the stimulation, by application of ethephon, of latex flow in rubber trees following tapping. A 10% solution of ethephon in palm oil applied to the bark of the tree just below the normal cut increases latex and dry rubber production by up to 100%. Ethephon appears to prevent the vessels from becoming blocked.

Ethylene diffusion through water is slow and it cannot readily escape into waterlogged soils. In addition, roots become anaerobic, reducing ethylene formation from ACC (section 27.11.1). Under these conditions ACC moves to the shoot where, in the aerobic environment, it is rapidly converted into ethylene and is responsible for many of the characteristic symptoms of waterlogging. Some types of rice are unusual because ethylene promotes elongation of their internodes.

Effects on flowering and flower senescence

Ethylene-producing chemicals are used to promote flowering in mangoes and bromeliads (section 27.7.4).

Ethylene production, particularly by the stigma and style, decreases the storage life of cut flowers by promoting senescence. Auxin in pollen may also stimulate ethylene production after fertilization. Some species such as carnations are particularly susceptible to ethylene.

The storage life of cut flowers can be increased by a process called pulsing, in which they are immersed in solutions containing a mixture of sucrose, a biocide, a weak acid, an anti-ethylene agent and sometimes a growth regulator. Sucrose replaces carbohydrates which are depleted by respiration occurring during storage. It also decreases ethylene production and increases the ability of plant tissues to retain water. Biocides such as silver nitrate, 8-hydroxy quinoline and aluminium sulphate prevent growth of bacteria which may plug the stem or produce toxic metabolites and ethylene. Silver nitrate is an anti-ethylene agent as well as an effective bacteriocide. Silver thiosulphate is less effective as a biocide but is very effective indeed in inhibiting both the production and action of ethylene. It may double the life of those flowers, such as carnations, which are particularly sensitive to ethylene. Treatment of cut flowers with cytokinins improves keeping qualities but is rarely more effective than silver thiosulphate treatment.

The manipulation of the atmosphere in which flowers are stored can be beneficial to their longevity in much the same way as for fruits. Thus lowering temperature and raising CO_2 concentration, reducing O_2 concentration and removal of ethylene from the atmosphere are all beneficial.

27.12 MISCELLANEOUS PLANT GROWTH REGULATORS

27.12.1 MORPHACTINS

The morphactins such as chlorflurecol are a group of compounds which are derivatives of fluorene-9-carboxylic acid. They disrupt coordinated growth and reduce apical dominance, possibly by affecting basipetal auxin transport. Chlorflurecol methyl is used in combination with maleic hydrazide to delay growth of grasses and broad-leaved weeds on roadside verges.

27.12.2 MALEIC HYDRAZIDE

Maleic hydrazide has been used for over 40 years and was, at one time, the most commonly used plant-growth regulator. It is a growth inhibitor, restricting cell division in both the apex and sub-apical region. It therefore reduces not only stem elongation, but also the initiation of leaves and flowers. It is used to control growth of suckers in tobacco, and also at the bases of amenity trees, to prevent sprout growth in potato tubers and onions, and to reduce the growth of hedges and grasses on verges.

27.12.3 GLYPHOSINE

Glyphosine is used as a sugarcane-ripening agent. It inhibits the late stages of sugarcane growth, causing sucrose to be stored in the stem instead of being used for fibre production, and leads to increases in yield of sucrose of 10–15%.

PART FOUR

STRATEGIES FOR PROCESSING OF NUTRIENTS IN ANIMALS

DIGESTION AND ABSORPTION IN RUMINANTS AND NON-RUMINANTS

28

28.1 INTRODUCTION

Animals consume food consisting of a complex matrix containing both simple molecules (free sugars and amino acids, etc.), and macromolecules such as structural carbohydrates, protein and lipids. For the most part, simple molecules require little modification before they can be absorbed from the digestive tract and utilized by the tissues. However, macromolecules almost always require dismantling into their simpler building blocks before they can be absorbed across the intestinal wall. Digestion is the process by which these dietary macromolecules are broken down. This is followed by absorption of digestive end products across the wall of the different compartments of the digestive tract, and their distribution to body tissues via the blood stream. Digestion is a physicochemical process in which enzymes, secreted both by the digestive tract and by microorganisms present within it, are utilized to catalyse the breakdown of complex molecules. The end products of digestion vary from species to species. The nature of these products is influenced by the extent to which food is degraded by microbial fermentation or endogenously synthesized enzymes.

28.2 STRUCTURE OF THE DIGESTIVE TRACT

Animals have developed a wide variety of digestive strategies to cope with the great variation in the nature and digestibility of the feedstuffs they consume. This is reflected in the diversity of digestive anatomy seen in the animal kingdom. Among the monogastric species there are considerable differences in the size of the digestive tract, and in the contribution made by each of the three main compartments of the digestive tract, the stomach, small intestine and hindgut (caecum and colon), to the process of digestion.

In carnivorous animals such as feline and canine species, the digestive tract is relatively short and simple. Digestion occurs mainly within the stomach and small intestine through the action of endogenous enzymes. The hindgut, where fermentation takes place, is poorly developed. In contrast, monogastric herbivores such as the rabbit and horse are dependent not only on enzymic digestion in the stomach and small intestine, but also on the subsequent microbial fermentation of the structural carbohydrates from forages in the caecum and/or colon. Omnivores such as the pig fall between these two extremes.

Ruminants differ from monogastric herbivores in digestive anatomy. They are reliant on pregastric fermentation of dietary constituents in the rumen. Fermented food passes from this large fermentation chamber through the omasum and into the abomasum, which is the true glandular stomach of ruminant species. The importance of the rumen is reflected in the fact that it represents between 60 and 70% of the volume of the digestive tract. The relative sizes of the digestive compartments in monogastric and ruminant species are shown in Table 28.1.

28.3 CARBOHYDRATE DIGESTION IN MONOGASTRIC ANIMALS

In monogastric animals, little digestion of dietary carbohydrate occurs before it reaches the small intestine. Saliva contains an amylase which is capable of cleaving the α-1,4 glycosidic bonds between the glucose residues of amylose, amylopectin and glycogen. This enzyme (which has a pH optimum of 7.0) is identical to pancreatic amylase, but normally has little effect on dietary carbohydrate because of the relatively short time it is in contact with food before being inactivated by the acidic conditions of the stomach.

Pancreatic α-amylase and β-amylase are the two enzymes mainly responsible for degradation of α-linked carbohydrates in the small intestine. They are secreted in the pancreatic juices and mixed with digesta in the proximal duodenum. α-Amylase is an endoglycosidase which catalyses the random hydrolysis of α-1,4 glycosidic bonds between glucose residues (Figure 28.1). β-Amylase is an exoglycosidase which removes glucose residues from the non-reducing ends of glucose chains. The linear chains of amylose are degraded to a mixture of glucose and the disaccharide maltose. Amylopectin and glycogen are branched structures (Chapter 3) in which glucose residues at the branch points are linked by α-1,6 glycosidic bonds that are not hydrolysed by amylases. The end product of amylopectin and glycogen degradation by amylase is a highly branched structure known as limit dextrin. This is degraded by the action of debranching enzyme to produce numerous short, linear chains of glucose that are then further degraded to maltose and glucose by amylases. Maltose is hydrolysed to glucose by maltase, one of a number of disaccharidases associated with the brush border of the small intestine. Other disaccharidases are sucrase, which converts sucrose to glucose and fructose; and lactase, which hydrolyses lactose to glucose and galactose.

Non-starch polysaccharides in roughages and cereals contain a variety of sugars (Chapter 3). Cellulose is a homopolymer of glucose residues linked by β-1,4 glycosidic bonds which are not hydrolysed by enzymes secreted into the small intestine. Hemicelluloses contain a mixture of arabinose, xylose, mannose, galactose and glucuronic acid, linked by a variety of

Table 28.1 Volume of main compartments of digestive tract expressed as a percentage of the total volume of the tract

| | *Compartment of digestive tract* | | | | |
Animal	*Rumen*	*Stomach*	*Small intestine*	*Caecum*	*Colon*
Dog	–	62	23	–	14
Pig	–	29	33	8	30
Rabbit	–	15	12	23	50
Horse	–	9	21	16	54
Sheep	60	8	19	3	10
Cow	67	5	18	3	7

glycosidic bonds which are resistant to amylase hydrolysis. Typical hemicelluloses, xylans, contain a core polymer of β-1,4 xylose and a number of branches with a variety of other sugars. Glucose-containing hemicelluloses, β-glucans, are typical of plant seeds. These hemicelluloses are known as the cereal gums. They constitute a major proportion (70–75%) of the carbohydrate content of the endosperm cell wall, but overall are a minor component of the grain.

These non-starch polysaccharides pass to the large intestine where they undergo degradation by microbial fermentation. The pathways of fermentation are discussed in Chapter 16. The end products of the fermentation process are methane, carbon dioxide, and the volatile fatty acids: acetate, propionate and butyrate. In monogastric herbivores the hindgut is greatly enlarged to allow time for microbial fermentation of these carbohydrates. The volatile fatty acids are absorbed across the intestinal wall and are a valuable source of energy. Measurements in the pig suggest that volatile fatty acids absorbed from the hindgut can contribute as much as 1.2 MJ kg$^{-0.75}$ live weight compared with 2.5–3.3 MJ kg$^{0.75}$ from VFA absorbed from the rumen in cattle and sheep. (See Chapter 30 for an explanation of metabolic body weight, kg$^{0.75}$.)

There is considerable interest in the use of microbial enzymes, now produced on a commercial scale, to improve the digestibility of

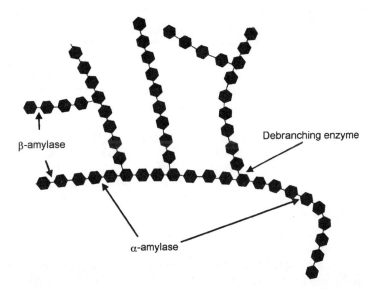

Figure 28.1 Site of action of α-amylase, β-amylase and debranching enzyme on starch and glycogen.

cereal non-starch polysaccharides in pigs and poultry. These carbohydrates increase the viscosity of the digesta in the small and large intestine, resulting in depressed digestibilities of other nutrients and reduced growth performance. The use of carbohydrase enzyme supplements has been shown to reduce digesta viscosity, improve mixing of digesta and increase growth performance, presumably through enhanced nutrient absorption.

28.4 CARBOHYDRATE DIGESTION IN RUMINANTS

The digestion of carbohydrates in ruminants occurs mainly in the rumen. Rumen microorganisms produce a variety of enzymes capable of degrading starch, cellulose and other non-starch polysaccharides. Starch is rapidly degraded to glucose by the action of bacterial amylases. Cellulose is more slowly degraded, by the action of β-1,4-glucosidases, to cellobiose which is then split into glucose. Hemicelluloses are degraded to xylose-rich oligosaccharides and then to monosaccharides. The free sugars, mainly glucose, are rapidly metabolized to pyruvate by the glycolytic pathway and are then converted to the volatile fatty acids acetate, propionate and butyrate by the pathways described in Chapter 16. The nature of the carbohydrate in the diet has a marked influence on the molar proportions of the volatile fatty acids produced. Forage diets produce a predominantly acetate fermentation. The introduction of cereals into the diet results in a shift towards the production of larger quantities of propionate (Table 28.2).

Some starch may escape fermentation in the rumen and pass into the small intestine where it is digested to glucose by amylases. The type and degree of processing of cereals influences the amount of starch which may reach the small intestine. When whole cereal grains are consumed, the presence of the intact pericarp may prevent access of the microbial enzymes to starch granules which pass undegraded to the small intestine. In con-trast, when finely ground cereals are eaten, the small starch particles may pass rapidly through the rumen before fermentation of the starch is complete. For this reason cereals are often rolled, bruised or flaked to rupture the pericarp and expose the starch granules without markedly reducing particle size.

Non-starch polysaccharides, not degraded in the rumen, pass through the small intestine and undergo further fermentation in the hindgut; however, this makes only a small contribution to the total VFA production in the gut.

28.5 ABSORPTION AND UTILIZATION OF GLUCOSE

The absorption of glucose from the small intestine occurs via active transport. The apical membrane of the mucosal cells lining the lumen of the digestive tract is highly specialized for the absorption of nutrients The surface area of this membrane is extremely large due to the presence of numerous microvilli. The disaccharidases which act upon maltose, sucrose and lactose are intimately associated with the extracellular surface of these microvilli. The monosaccharides produced are passed to a Na^+-dependent carrier protein that transports the sugars into the cell by utilizing the difference in the Na^+ concentration of the digesta and mucosal cell cytoplasm. Na^+ is transported into the mucosal cells down a concentration gradient and 'pulls' sugars into the mucosal cell against a concentration gradient. A low intracellular Na^+ concentration is maintained by a Na^+/K^+ ATPase in the basal membrane of the cell, which removes excess Na^+ to the extracellular fluid in exchange for K^+ (Figure 28.2).

Absorbed sugars enter the portal circulation and pass through the liver, which is able to metabolize a variety of them for metabolic purposes. For instance, galactose and fructose feed into the glycolytic pathway or can be converted to glucose. Glucose may be incorporated into glycogen or metabolized to provide

Table 28.2 Molar proportions of VFA in rumen fluid from sheep fed differing proportions of hay and concentrates

	Molar proportions of volatile fatty acids		
Hay:concentrate ratio	Acetate	Propionate	Butyrate
0.8:0.2	0.61	0.22	0.09
0.6:0.4	0.61	0.25	0.11
0.4:0.6	0.52	0.34	0.12
0.2:0.8	0.40	0.40	0.15

Adapted from: McDonald, P., Edwards, R.A., Greenhalgh, J.F.D. and Morgan, C.A. (1995) *Animal Nutrition*, 5th edn, Longman Scientific and Technical, Harlow, Essex.

energy and the precursors for the synthesis of amino acids and lipids. Glucose in excess of hepatic requirements passes to the peripheral tissues. In muscle, the main fate of glucose is oxidation to provide energy or storage as glycogen. In adipose tissue, glucose may be used to provide acetyl-CoA for fatty acid synthesis and glycerol-3-phosphate for triacylglycerol synthesis. The mammary gland has a very high glucose demand for energy and for lactose, fatty acid and triacylglycerol synthesis (see Chapter 31). The brain and nervous tissue are largely dependent on glucose as a supply of energy, and this is the main reason for the very tight regulation of plasma glucose concentrations in animals. In man, the nervous system accounts for about 70–80% of the total body glucose requirement in the resting state. The nervous system is a much smaller proportion of body weight in farm animals, and therefore places a lower demand on the available glucose. In the sheep, which is adapted to lower plasma glucose concentrations, this may account for only 10–15% of total body glucose.

Plasma glucose concentrations are regulated by the hormones insulin and glucagon. When plasma glucose concentrations rise, for example after ingestion of a starch-rich diet, the pancreas secretes insulin which increases the uptake of glucose by the tissues and promotes glycogen synthesis. Glucagon secretion by the pancreas is stimulated by a fall in plasma glucose, and promotes the breakdown of glycogen in tissue reserves and the synthesis of glucose via gluconeogenesis. The plasma concentrations of insulin and glucagon change in a reciprocal manner. Changes in the ratio of their plasma concentrations reflect changes in the plasma glucose concentration.

28.6 DIGESTION OF LIPIDS IN MONOGASTRIC ANIMALS

The main types of lipids found in feedstuffs are triacylglycerols, the storage lipids of plants and animal tissues; glycolipids and phospholipids, the structural lipids of plant and animal membranes; cholesterol and cholesterol esters. The relative proportions of these types of lipids vary with the nature of the food consumed.

28.6.1 DIGESTION IN THE STOMACH

The main site of lipid digestion is the small intestine; however, minor modifications to dietary lipid take place before the food reaches the intestine. A lipase has been detected in saliva and in the stomach, and these two enzymes appear to be identical. They have a pH optimum of 4 and thus retain activity in the acidic gastric environment. In adult animals they are of limited importance, but in suckling animals they may make a significant contribution to the digestion of milk fat. The limited information available on these enzymes suggests that they preferentially hydrolyse the fatty acid esteri-

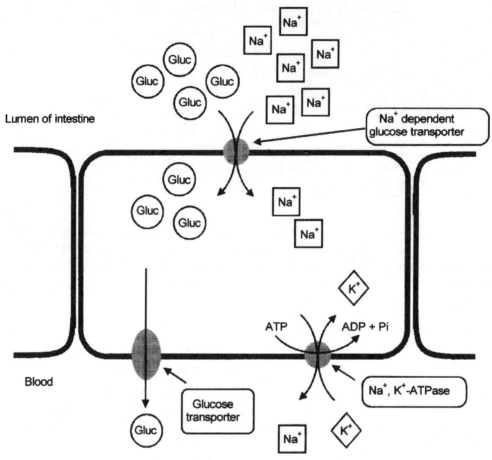

Figure 28.2 Mechanism of glucose absorption from the small intestine.

fied to the 3 position of the triacylglycerol and exhibit specificity for short- and medium-chain fatty acids. This further supports the suggestion that they are most important in suckling mammals as milk fat is unusual in containing relatively high proportions of short- and medium-chain fatty acids. The stomach also contributes to lipid digestion by denaturation of the protein component of ingested lipoproteins, and by the production of an oil-in-water emulsion due to the vigorous mixing of the stomach contents.

28.6.2 DIGESTION IN THE SMALL INTESTINE

In the small intestine, efficient lipid digestion requires the presence of two main compo-

nents: bile, which contains bile salts, is required for the emulsification of lipid contained in digesta leaving the stomach; and pancreatic juice which contributes the enzymes pancreatic lipase, phospholipase (A1 and A2) and cholesterol esterase, and the non-enzymic protein co-lipase. On entering the small intestine, digesta is mixed with pancreatic juice which contains bicarbonate. This neutralizes the acidic digesta leaving the stomach. Typical pH values for the gut contents of the pig are 2.4 (stomach); 6.1 (proximal duodenum); 6.8 (distal duodenum); 7.4 (jejunum) and 7.5 (ileum).

Pancreatic lipase is responsible for the release of fatty acids from triacylglycerols. The enzyme is most active between pH 6 and 7, and preferentially hydrolyses the fatty acids esterified to the 1 and 3 positions of the tri-acylglycerol molecule. 2-Monoacylglycerols and free fatty acids are the main end products of lipase action. A small amount of the 2-monoacylglycerols may be further digested by the action of isomerases, which catalyse migration of the fatty acid from the 2 position to the 1 position to produce 1-monoacylglycerols, which are promptly hydrolysed by pancreatic lipase. Between a quarter to a third of dietary triacylglycerols are completely degraded to glycerol and free fatty acids (Figure 28.3).

Pancreatic lipase acts at the surface of fat droplets in the digesta. The rate of lipolysis is dependent on the surface area of the droplets. Bile salts have a detergent action, breaking the large fat droplets into numerous smaller ones, thereby increasing the surface area available for attachment of pancreatic lipase. The main types of bile salts found in bile are tauro-cholate, glycocholate, taurodeoxycholate, gly-codeoxycholate, taurochendeoxycholate and glycochendeoxycholate. The ratio of taurine-esterified bile salts to glycine-esterified bile salts is about 1:3 in monogastrics and 2.5:1 in ruminants. The detergent effects of bile salts stem from their amphiphilic properties. The hydrocarbon ring structure of these molecules is planar. The polar functional groups are located on one surface of the molecule while the opposite surface is non-polar; thus they are able to coat the surface of fat droplets and provide each droplet with a polar surface which prevents the droplets from coalescing.

The bile salt layer surrounding the fat droplets is a barrier preventing the direct attachment of pancreatic lipase. However, co-lipase, a small protein secreted by the pancreas, is able to penetrate the bile salt layer and provides anchor points on the droplet to which the lipase can attach (Figure 28.4).

As digestion progresses, the products of tri-acylglycerol breakdown are transferred from the bulk fat phase to structures called micelles.

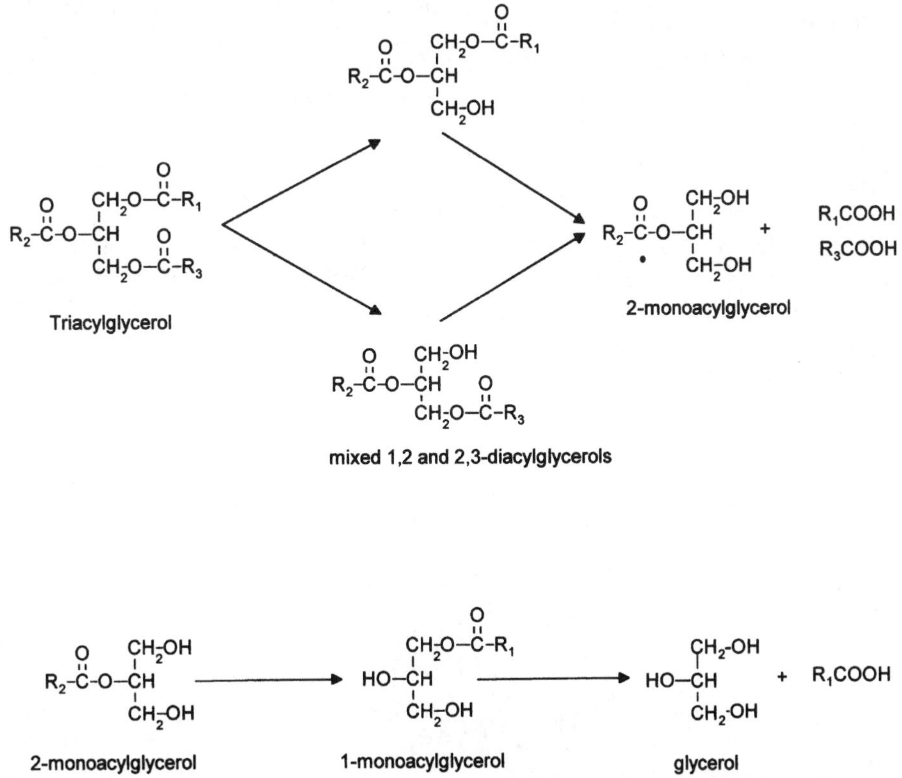

Figure 28.3 Breakdown of triacylglycerols in the small intestine.

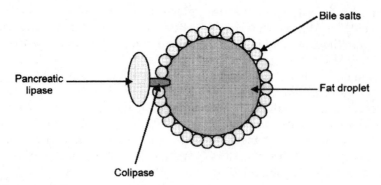

Figure 28.4 Attachment of pancreatic lipase to fat droplets in the small intestine.

These structures are roughly spherical in shape and consist of an outer coating of bile salts and phospholipid (both undigested phospholipid and lysophospholipids) surrounding a core of undigested triacylglycerols, together with the polar digestion products fatty acids and 2-monoacylglycerols, which are orientated with their polar functional groups towards the surface of the micelle (Figure 28.5).

Because these micelles contain a mixture of the products of digestion (tri-, di-, and 2-monoacylglycerols and fatty acids) they are referred to as mixed micelles.

Phospholipid is digested by the action of phospholipases A1 and A2 (see Figure 13.13) yielding 2-acyl lysophospholipids and 1-acyl lysophospholipids, respectively. The distribution of polar groups in these molecules gives them amphiphilic properties which enables them to work in concert with the bile salts at the surface of the fat droplets and enhances solubilization of the digesta lipid phase.

28.7 ABSORPTION OF LIPID FROM THE SMALL INTESTINE

The end products of lipid digestion are absorbed across the intestinal wall in the jejunum and proximal ileum. The mechanism of transfer of micellar contents from mixed micelles is poorly understood, but it is clear that bile salts do not enter the intestinal mucosa in this region of the small intestine. They are absorbed in more distal regions of the ileum and recycled via the liver. This constitutes the enterohepatic circulation of bile salts.

28.7.1 GLYCERIDES

The detailed mechanism of uptake of fatty acids and 2-monoacylglycerols into the intestinal mucosa is unknown. It has been generally accepted that uptake is by passive diffusion, but recent work suggests that membrane-mediated processes involving fatty acid-bind-

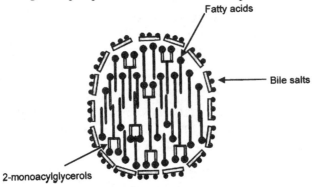

Figure 28.5 Representation of a mixed micelle.

ing proteins may be involved. In the intestinal mucosa a process of resynthesis of lipid occurs. Both triacylglycerols and phospholipids are synthesized. The synthesis of triacylglycerols occurs mainly by the 2-monoacylglycerol pathway, a pathway unique to the intestinal mucosa (Chapter 19). Absorbed fatty acids are activated to their coenzyme A derivatives and are then esterified to 2-monoacylglycerols by the sequential action of monoacylglycerol acyltransferase and diacylglycerol acyltransferase. There may also be limited synthesis of triacylglycerols by the glycerol-3-phosphate pathway (Chapter 19). The role of fatty acid-binding proteins in the process of resynthesis remains unclear. These proteins are present in significant quantities in the mucosal cells, but it is not known whether they merely act as a vehicle for the transport of absorbed fatty acids, or whether they play a more important role in the regulation of lipid synthesis.

28.7.2 PHOSPHOLIPIDS

The major substrates for phospholipid synthesis are the lysophospholipids absorbed from the intestinal lumen. These molecules undergo re-acylation by lysophosphotidate acyltransferase on the mucosal endoplasmic reticulum. Some *de novo* synthesis of phospholipid also occurs in the mucosa via the glycerol-3-phosphate pathway. The main end product of this pathway is phosphatidylcholine.

28.7.3 CHOLESTEROL

Cholesterol absorption from the intestinal lumen is an inefficient process. Much of the absorbed cholesterol is re-esterified during passage through the mucosal cell, probably by a reversal of the cholesterol esterase reaction.

28.7.4 CHYLOMICRON FORMATION

Resynthesized lipids are incorporated into chylomicrons for transport to the blood stream. This type of lipoprotein is associated exclusively with the intestinal mucosa. They are relatively large, between 50 and 200 μm in diameter, and consist of a hydrophobic lipid core containing triacylglycerols, cholesterol and cholesterol esters surrounded by a protein and phospholipid coat. Their average composition is 85% triacylglycerol, 9% phospholipid, 4% cholesterol and cholesterol esters and 2% protein, mainly apolipoproteins B-48, C-II and C-III. The synthesis of chylomicrons occurs on the mucosal cell Golgi bodies. The nascent chylomicrons are transported to the basal membrane of the mucosal cells and are then released into the lymphatic system.

28.8 UPTAKE OF ABSORBED LIPID BY BODY TISSUE

The lymphatic system draining the digestive tract empties via the thoracic duct into the subclavian vein. Chylomicrons entering the blood stream are rapidly transported around the body, where the lipid contained within them can be utilized by the tissues. The process of lipid uptake by tissues is mediated by the enzyme lipoprotein lipase, which is widely distributed throughout the body and is associated with the endothelial cells of the capillaries. The presence of chylomicrons containing apolipoprotein C-II activates the enzyme which catalyses the hydrolysis of triacylglycerols to glycerol and free fatty acids in the lumen of the capillaries. The released fatty acids are absorbed across the capillary wall into the underlying tissue (Figure 28.6).

The fate of absorbed fatty acids is determined by the function of the tissue into which they are absorbed. In adipose tissue, they are reincorporated into triacylglycerols and stored for future use. In muscle and other tissues, they may undergo oxidation to provide energy or be used for structural lipids. In liver, they may be oxidized, converted to structural lipids, or incorporated into triacylglycerols and phospholipids prior to inclusion in very low density lipoproteins (VLDLs) which are secreted into the blood stream for use by other tissues in the body. In mammary tissue, lipoprotein lipase provides the mechanism for

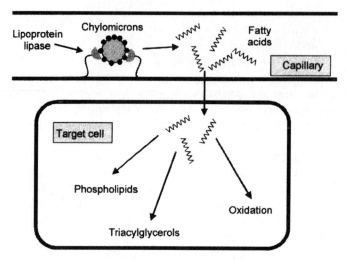

Figure 28.6 Tissue uptake of fatty acid via the action of lipoprotein lipase.

the uptake of circulating fatty acid of both dietary and endogenous origin for use in the synthesis of milk fat.

28.9 DIGESTION OF LIPIDS IN RUMINANT ANIMALS

The digestion and absorption of fatty acids is different in ruminants from that in non-ruminants, because the rumen microflora has a profound effect on the nature of the lipid reaching the small intestine. The main types of lipids present in a ruminant diet are galactolipids (the main dietary lipid of a grazing ruminant), triacylglycerols and phospholipids. Rumen microorganisms secrete lipases which very rapidly hydrolyse dietary lipids to free fatty acids, glycerol, sugars and phospholipid bases. The glycerol and sugars are quickly fermented to volatile fatty acids. Fatty acids are further modified by biohydrogenation in which most of the unsaturated fatty acids are converted to their respective saturated derivatives (Chapter 16). The consequence of this metabolism is that lipid leaving the rumen is composed almost entirely of saturated fatty acids and smaller proportions of dietary unsaturated fatty acids which have escaped complete biohydrogena-

tion. Also present are geometrical (*trans*) and positional isomers of unsaturated fatty acids (intermediates in the biohydrogenation process). Fatty acids passing out of the rumen are bound to the surface of feed particles. It is somewhat paradoxical then that in the grazing ruminant, whose diet is particularly rich in polyunsaturated fatty acids ($\Delta^{9,12}$-C18:2 and $\Delta^{9,12,15}$-C18:3), the fatty acids entering the small intestine are predominantly saturated in nature.

Lipid added to ruminant diets can be protected against the effects of rumen microorganisms. The most effective form of protection is achieved by coating fine droplets of lipid with casein which is treated with formaldehyde to cross-link the protein molecules. This renders the protein undegradable by the rumen microflora and prevents enzymic breakdown of the encapsulated lipid. When the protected lipid enters the abomasum (the gastric stomach), the acidic conditions cause denaturation of the casein coating and the lipid is released to be digested in the small intestine. This type of protection offers several benefits. Firstly, it prevents biohydrogenation, allowing unsaturated fatty acids to pass to the small intestine unaltered and thus offering the potential to manipulate the fatty acid composition of ruminant tissues and products. Secondly, it protects the rumen microflora from the adverse effects of high levels of lipid in the diet. Under normal conditions dietary lipid should not exceed 5–7% of the dry matter content of the diet, as above this level dietary lipid, particularly if rich in unsaturated fatty acids, depresses rumen fibre digestion. Protection of dietary lipids prevents these adverse affects and allows the energy density of ruminant diets to be increased. This is particularly important in high-yielding dairy cows (see Chapter 31). Alternative strategies to the use of casein-protected fats have been developed, such as the feeding of the calcium salts of fatty acids and high-melting-point fats, which are insoluble in the rumen and therefore do not adversely affect fibre digestion. However, biohydrogenation is not prevented completely by the feeding of these insoluble fat sources.

Enzymic lipid digestion in the small intestine of a ruminant fed unprotected fat is minimal as the bulk of the dietary lipid has already been digested to free fatty acids in the rumen. The most important function of the small intestine is release of fatty acids from the surface of feed particles and their incorporation into micelles prior to absorption. This is achieved by the secretion of bile containing bile salts and phospholipids (principally phosphatidylcholine). Ruminant bile is particularly rich in taurine-conjugated bile salts, which are soluble and partially ionized at pH 2.5 (compared with the glycine-conjugated bile salts, the largest constituent of monogastric bile salts, which are insoluble below pH 4.5). This difference is important in the solubilization of feed particle-bound fatty acids in ruminants, as the pancreatic juice contains a relatively low concentration of bicarbonate and intestinal pH rises more slowly in the ruminant than in the pig. Typical pH values for the abomasum and small intestine of the sheep are 2.0 (abomasum); 2.5 (proximal duodenum); 3.5 (distal duodenum); 3.9 (proximal jejunum);

4.7–7.6 (distal jejunum) and 8.0 (ileum) (see comparable figures for the pig in section 28.6.2). The rate of transfer of fatty acids from feed particles to micelles increases as digesta passes from the duodenum into the jejunum. This process is aided by the presence of bile phospholipids. The amphiphilic nature of the phospholipids, and the lysophospholipids produced from them by pancreatic phospholipase A1 and A2, enhances the solubilizing effect of the bile salts. Absorption of fatty acids from micelles occurs in the jejunum and is virtually complete by the time the digesta reaches the ileum.

Pancreatic lipase is present in ruminant pancreatic juice, but is effectively redundant in a grazing ruminant. In intensively fed ruminants, where the proportion of triacylglycerol in the diet increases, small amounts of lipid may escape digestion in the rumen and provide a substrate for the enzyme. In animals fed protected fat, dietary triacylglycerol enters the duodenum unaltered and is digested by the action of pancreatic lipase in a similar manner to that in monogastric animals. Under these circumstances the end products of digestion are a mixture free fatty acids and 2-monoacylglycerols.

It is assumed that the mechanism of uptake of fatty acids from the ruminant digestive tract into the intestinal mucosa is similar to that in monogastric animals. However, the pathway of triacylglycerol resynthesis in the mucosal cells of animals fed unprotected fat differs from that in monogastric animals, as there is a virtual absence of 2-monoacylglycerol. Instead, triacylglycerol synthesis occurs by the glycerol-3-phosphate pathway (Chapter 19). The formation of phospholipid in mucosal cells is predominantly by the reacylation of 1-lysophospholipids absorbed from the intestinal lumen.

The newly synthesized lipids are incorporated into lipoproteins. In the ruminant both chylomicrons and VLDL are synthesized. Nascent lipoproteins are secreted into the lymphatic system, enter the blood stream, and are absorbed by similar mechanisms to those described for monogastric animals.

28.10 LIPID DIGESTION IN POULTRY

In poultry, lipid digestion follows a similar pattern to that in monogastric mammals. The main difference that has been observed is in the mechanism of delivery of absorbed lipid to the blood stream. Poultry have a poorly developed digestive lymphatic system, and as a result the chylomicrons are released from the intestinal mucosa directly into the portal blood stream. These chylomicrons are sometimes referred to as portomicrons. A major consequence of this different absorption mechanism is that dietary lipid passes through the liver before being circulated to the peripheral tissues.

28.11 DIGESTION OF PROTEINS IN MONOGASTRIC ANIMALS

28.11.1 DIGESTION IN THE STOMACH

The stomach is the first major site of protein digestion in monogastric species. Two factors affect the nature of proteins in the stomach: firstly the acidic environment, pH 2.0–3.0 (corresponding to [H^+] concentration of approximately 0.1–0.01 M) denatures ingested proteins; and secondly the action of the protease pepsin, which partially degrades the dietary protein into smaller peptides.

The acid environment of the stomach is due to the secretion of hydrochloric acid by the parietal or oxyntic cells of the stomach mucosa. This is achieved by the mechanism shown in Figure 28.7. Carbon dioxide absorbed from the blood stream and produced in the cell by oxidative metabolism combines with water to form carbonic acid. At physiological pH (pH 7.4) carbonic acid dissociates to produce hydrogen and bicarbonate ions. The hydrogen ions are transported across the apical membrane of the parietal cells in exchange for potassium ions present in the stomach contents. Both of these ions are transported

against their respective concentration gradients, and thus ATP is consumed by the membrane transporter, the H^+, K^+-ATPase. The chloride ions present in the stomach are derived from the plasma. They enter through the basal membrane against a concentration gradient via two types of membrane transporter. The first, a bicarbonate/chloride antiport, transports bicarbonate from the cytoplasm to the plasma in exchange for chloride. The second is a sodium/bicarbonate symport, which co-transports sodium and chloride into the cell. The driving force for the accumulation of chloride ions in the cytoplasm is the combined effect of the movement of sodium and bicarbonate out of the cell down their respective concentration gradients. The surge of bicarbonate into the plasma during rapid HCl secretion causes the blood plasma to become slightly alkaline, a phenomenon known as the 'alkaline tide'. Chloride and potassium ions move from the cytoplasm to the stomach contents down a concentration gradient via a passive membrane symport. The high intracellular concentration of potassium ions and low intracellular concentration of sodium is maintained by the Na^+, K^+-ATPase. The net result of the concerted action of these transport mechanisms is a hydrogen ion concentration in the stomach contents which may be as much as one million times greater than that in the plasma (pH 1.0 or [H^+] = 10^{-1} M as opposed to pH 7.0 or [H^+] = 10^{-7} M).

Pepsin is secreted by the chief cells of the stomach mucosa. The enzyme is secreted in an inactive zymogen form, pepsinogen, and is converted to active pepsin by the action of pepsin already present in the stomach. The activation involves the removal of 42 amino acid residues from the N-terminal end of the zymogen polypeptide. The secretion of proteolytic enzymes as inactive zymogens is a protective mechanism which prevents proteolytic attack within the secretory cells. Pepsin has a pH optimum of around 2.0 and exhibits a specificity for peptide bonds on the carboxyl side of the aromatic amino acids phenylala-

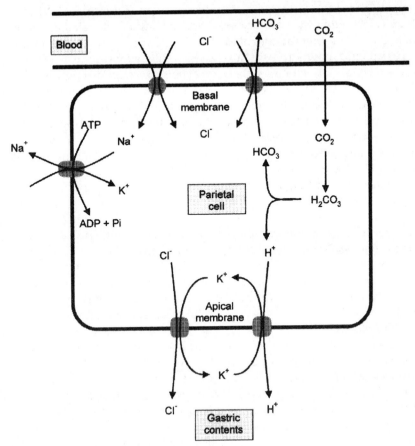

Figure 28.7 Transport mechanisms involved in the secretion of HCl by parietal cells of the stomach.

nine, tryptophan and tyrosine. The end product of pepsin digestion of proteins is a mixture of shorter-chain polypeptides with predominantly N-terminal aromatic acids.

The secretion of both hydrochloric acid and pepsinogen is regulated hormonally. Gastrin production by the gastric epithelium is stimulated by the presence of protein in the stomach, by distension of the stomach, and by the autonomic nervous system. Gastrin acts upon the parietal and oxyntic cells to increase acid and enzyme release. Secretin, produced in the duodenum in response to increased acidity of duodenal contents, inhibits acid and pepsinogen release. Two other hormones, cholecystokinin and gastric inhibitory factor, are released in response to the presence of fatty acids and other lipids in the small intestine. They also inhibit acid and pepsinogen release.

28.11.2 DIGESTION IN THE SMALL INTESTINE

Partially degraded protein leaves the stomach and enters the duodenum, where the digesta is mixed with pancreatic secretions that contain the proteolytic zymogens, trypsinogen, chymotrypsinogen, procarboxypeptidases A and B, and proelastase. All of these zymogens are activated by the removal of an amino acid sequence from the N-terminal end of the protein by a common activator, trypsin. This results in the coordinated activation of these pancreatic enzymes and ensures that dietary protein is efficiently degraded by a battery of enzymes which have differing amino acid specificities. Trypsin and chymotrypsin are endopeptidases which act at various points along the polypeptide chain. Trypsin cleaves peptide bonds formed by lysine or arginine residues; chymotrypsin has a preference for peptide bonds on the carboxyl side of phenylalanine, tryptophan and tyrosine. The carboxypeptidases and elastase are exopeptidases which remove amino acids from the ends of the polypeptide chain. Carboxypeptidases A and B remove amino acids from C- and N-terminal ends of peptides, respectively. Carboxypeptidase A is specific for phenylalanine, tryptophan and tyrosine, whereas carboxypeptidase B is specific for lysine or arginine residues. Elastase exhibits specificity for the neutral amino acids leucine, isoleucine and valine.

Pancreatic proteolytic enzymes are secreted as zymogens to ensure that there can be no premature activation of the zymogens before they reach the intestinal lumen. The pancreas also produces an inhibitory protein, pancreatic trypsin inhibitor, which binds very tightly to the active site of any enzyme activated prior to release from the exocrine cells.

The sequence of activation of the pancreatic zymogens in the intestinal lumen is initiated by a specialized proteolytic enzyme, enteropeptidase (previously called enterokinase), produced by the intestinal mucosa. This enzyme removes a six-amino acid sequence from the N-terminal end of trypsinogen, thereby converting it to active trypsin. Only a small amount of trypsinogen need be activated in this way as trypsin itself will then catalyse the activation of trypsinogen and the other pancreatic proteolytic zymogens. This regulatory mechanism, the activation of enzymes by proteolytic cleavage, has been studied in detail for chymotrypsinogen. In the zymogen form the protein has a sequence of 245 amino acids. A 15-amino-acid sequence at the N-terminal end of the protein appears to block the active site of the enzyme, preventing it from binding

to protein substrates. When this short sequence of amino acids is removed the protein undergoes a conformational change which fully exposes the active site.

The short peptides produced by the action of the pancreatic proteases are further degraded in the small intestine by aminopeptidases produced by the intestinal mucosa. These enzymes, which act both intracellularly and in the intestinal lumen, catalyse the sequential removal of amino acids from the N-terminal end of peptides to produce a mixture of short peptides and amino acids. In addition, the intestinal mucosa contains dipeptidases which act on dipeptides absorbed from the digestive tract.

Absorption of small peptides and amino acids from the intestinal lumen occurs via membrane transport proteins (symports) which are specific for small groups of structurally related amino acids and also co-transport Na^+ from the lumen of the intestine into the intestinal mucosa. Na^+ is usually present in higher concentration in the digesta than in the mucosal cell, and therefore moves into these cells via passive transport, dragging the amino acids with it. Once in the mucosal cell, Na^+ is actively transported into the extracellular fluid by the action of the Na^+, K^+-ATPase in the basolateral membrane of the cell. This maintains a concentration gradient between the mucosal cell cytoplasm and the intestinal lumen. This process concentrates amino acids in the cytoplasm of mucosal cells, and allows them to move across the basal membrane and into the portal blood stream by passive transport down their concentration gradient. The intracellular hydrolysis of dipeptides to free amino acids by dipeptidase also contributes to the amino acid concentration gradient. Although ATP is not directly involved in the action of the amino acid transporters, it provides the driving force for amino acid absorption by creating a low intracellular Na^+ concentration through the action of the Na^+, K^+-ATPase (Figure 28.8). Absorbed amino acids are transported to the liver where they may undergo a number of metabolic fates described in Chapter 14.

28.12 DIGESTION OF PROTEIN IN RUMINANTS

Microbial proteases degrade proteins to peptides and amino acids in the rumen. These amino acids are utilized in two ways: they may be directly incorporated into microbial protein; or they may be further metabolized by transamination and deamination to organic acids, ammonia and carbon dioxide. In this way they can contribute to the volatile fatty acid pool in the rumen. The branched-chain amino acids valine, leucine and isoleucine are degraded in this way to produce the branched-chain volatile fatty acids isobutyric acid, isovaleric acid and 2-methylbutyric acid, respectively.

Ammonia may be used as a convenient nitrogen source for microbial synthesis of amino acids which are thereafter incorporated into microbial protein. When ammonia is produced in excess of microbial requirements, it is absorbed across the rumen wall, transported to the liver and converted to urea. A proportion of this urea is recycled to the rumen, either directly from the blood or via the saliva, and may be used as a non-protein source of nitrogen for microbial amino acid synthesis. Excess urea is excreted in the urine. The quantity of

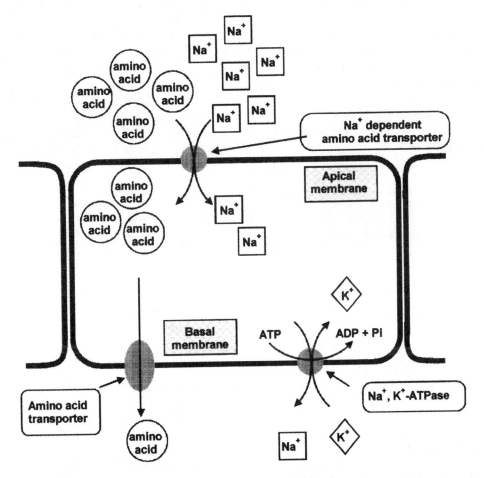

Figure 28.8 Mechanism of amino acid absorption from the small intestine.

ammonia produced in the rumen and of urea synthesized in the liver is dependent on the quantity and quality of dietary protein. Large quantities of highly degradable protein lead to the rapid production of ammonia in excess of the capacity of the rumen microorganisms to use it. Under these circumstances urea excretion increases. When poor quality, low-protein diets are fed, rumen ammonia production is low and the quantity of urea nitrogen recycled to the rumen by direct transfer from the blood and via saliva may be greater than the ammonia nitrogen absorbed from the rumen. The additional urea nitrogen is derived from the catabolism of amino acids by the liver.

The degradability of dietary proteins in the rumen varies considerably. Some, such as casein, are efficiently degraded in the rumen, whereas others such as albumin or the complex mixture of proteins in fish meal are poorly degraded. The degradability of a protein is determined by a combination of factors such as physical structure, solubility characteristics, and amino acid composition and sequence in relation to the substrate specificity of the microbial proteases.

Further digestion of undegraded dietary and microbial protein in the abomasum and small intestine is similar to that in monogastric animals. Abomasal and pancreatic secretions are similar to those in non-ruminants; however, the pattern of digesta flow dictates that these secretions are produced in a continuous manner in contrast to the more intermittent secretions of monogastric species. A summary of protein metabolism in the rumen is shown in Figure 28.9.

28.13 INHIBITORS OF DIGESTIVE ENZYMES

Some plants contain proteins which inhibit the action of digestive enzymes and which may result in poor animal performance when present in the diet. Legume seeds, for example, contain proteins known as protease inhibitors which can inhibit trypsin and chymotrypsin. Their presence results in depression of growth when raw or unheated legume products are fed to non-ruminant animals. The inhibitors are inactivated in the rumen, so their presence in ruminant feeds does not cause the same problems. Some bean seeds contain inhibitors

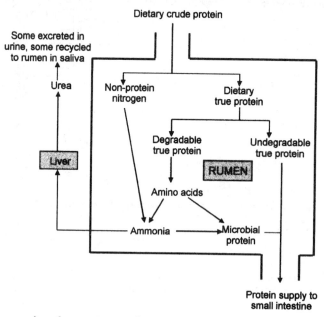

Figure 28.9 Summary of crude protein metabolism in the rumen.

of amylases which may impair starch digestion. Because these inhibitors are proteins they are inactivated by heating, and it is common practice to subject feeds containing bean products to heat treatment prior to feeding.

Inhibitors such as these often act as deterrents to insects feeding on them the plants.

Their occurrence is not limited to seeds, and protease inhibitors may also be present in fruits, leaves and tubers. In some cases they appear to be produced in response to damage caused by insects.

MAINTENANCE

29.1 INTRODUCTION

In the short term, if an animal is not fed it starts to lose weight as body tissues are utilized to provide for its metabolic needs. In the longer term it will die. On the other hand, if an animal is very well fed its body weight will increase by the processes of growth. At a dietary level in between these two extremes there is a point at which the animal's body weight remains constant. This is known as the maintenance level of feeding. (Some authors define maintenance as the point at which the body's energy content is constant; similarly, it can be defined as the point of constant body protein composition. The agricultural definition of constant body weight provides a compromise which is very useful in farming practice.) Even though body weight may be constant, the animal will in fact change its body composition, usually losing fat and adding water and protein.

In agriculture there are circumstances when a maintenance level of feeding may be desirable, for instance breeding males may need to be kept at a near-constant weight over a period of many months or even years. Maintenance represents the cost, in nutritional terms, of keeping the animal alive. In growing animals, this cost has to be met before any production can be seen, because only when nutritional levels are above maintenance will there be any accretion of tissues. The maintenance level of nutrition is usually expressed in terms of energy, and is the energy requirement of a non-productive animal (i.e. one that is not growing, producing milk or laying eggs, etc.) at the point where the dietary energy intake exactly equals energy output in various forms:

- heat production
- losses in faeces
- losses in urine
- losses in hair and skin.

Of these losses the largest are the first two. The losses in faeces will depend largely upon the nature of the animal's diet, although there is some contribution to the faeces that comes from the animal's body by the passage of sloughed cells of the gut.

29.2 BODY TEMPERATURE AND HEAT PRODUCTION

All of the animals commonly used in agriculture are homoeothermic – they maintain their body temperatures within precisely defined limits. The actual mean temperatures recorded vary from species to species: man (37° C) has a slightly lower body temperature than most livestock such as cows and sheep (38.5 and 39.5° C, respectively) and much below the hen (41° C). Temperatures can change with physiological state and in response to the environment. This is very clearly seen in the camel, which can vary its temperature over a range of about 5° C during the daily fluctuation in ambient temperature. In most animal habitats and under most agricultural systems, the animal's body temperature is below that of the environment. Even most tropical climates have a mean ambient temperature below body temperature. Heat production is therefore a very important function of the body and one which has to rely upon biochemical reactions.

In contrast, there are situations in which the heat produced by an animal under hot conditions can exceed the animal's ability to lose heat to the environment. The animal must invoke extra cooling mechanisms that are themselves dependent upon the expenditure of biochemical energy.

The heat produced by an animal which has no dietary input is known as its fasting heat production. A beef bull of 650 kg body weight produces about 41 MJ day^{-1} – roughly the same output of heat as a single bar (500 W) electric heater running continually. Most of this energy comes from the hydrolysis of ATP (molecular weight = 414) which produces a free energy change of about 30 kJ for every mole. To produce 41 MJ the bull has to hydrolyse about 1344 moles of ATP, equivalent to 556 kg of ATP – nearly its own body weight. Clearly, this can only happen if the ATP is continually being synthesized and hydrolysed.

Figure 29.1 gives a generalized impression of the heat production by an animal at various environmental temperatures. The two important figures for any animal are its lower and upper critical temperatures. If the environmental temperature drops below the lower critical temperature, then heat production will rise. Similarly, above the upper critical temperature

the demands of sweating will further increase the heat load on the animal. The figures for the upper and lower values will vary depending on a number of factors including species, breed, acclimatization, level of production, etc. For example, for a temperate dairy cow the lower temperature may be as low as –15° C, whereas for a tropical breed it may be +10° C. The upper critical temperature for a temperate dairy cow is about 20° C and for a tropical animal it may be in the high twenties.

Like many other metabolic processes, heat production is proportional to the animal's metabolic body weight, as can be seen from the interspecies comparisons in Figure 29.2. The slope of the line is approximately 0.67, which is slightly different from the power of 0.75 often used for calculations of metabolic body weight.

29.3 BIOCHEMICAL PRODUCTION OF HEAT

Heat production can be divided into two areas:

- heat produced as a by-product of metabolism;
- heat produced in response to a perceived need of the body to raise its temperature.

Under many conditions the heat produced as a by-product of metabolism will be sufficient to meet or exceed the needs of the animal for heat. Where the animal is at or above the upper critical point, energy has to be expended to disperse the excess heat. In an animal that is kept in an environment below its critical temperature, extra heat is required to maintain body temperature. This requires some system that can sense body temperature and another that can communicate to the tissues that they should produce more heat. The measuring system appears to be located in the hypothalamus, from which 'messages' are sent via the sympathetic nervous system and probably through hormones such as those of the thyroid. The detailed control mechanisms involve the integration of the efforts of many different physiological systems, and are beyond the scope of

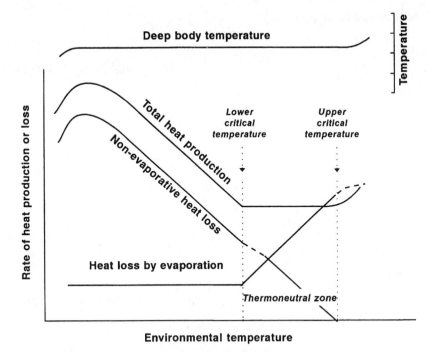

Figure 29.1 Heat production of cattle at different environmental temperatures: heat production rises when animals are kept in environments either hotter or colder than their thermoneutral zone. (After Blaxter, K.L. (1989) Metabolic effects of the physical environment, in *Energy Metabolism in Animals and Man*, Cambridge University Press, Cambridge, UK.)

this book (for further information see Ruckebusch *et al.*, 1991. *Physiology of Small and Large Mammals*. Marcel Dekker, Philadelphia).

29.3.1 BACKGROUND HEAT PRODUCTION FROM THE MAINTENANCE OF ION GRADIENTS IN CELLS

Na+,K+- ATPase pumps

Within animal cells the sodium ion concentration is held at a lower value than that of the surrounding fluid, whereas the potassium ion concentration is much higher (Table 8.3). The ionic gradient leads to the development of an electrical charge such that the contents of the cell are at a negative potential in relation to their surroundings. This charge is known as the membrane potential; in a resting nerve cell, for instance, it is about 60 mV. Gradients of this type do not just happen by chance, they have to be actively created and continually maintained by pumping sodium ions out of and potassium ions into the cell. This requires energy in the form of ATP. For example, the Na+,K+ ATPase found in cell membranes uses the energy released from the hydrolysis of one molecule of ATP to transfer three Na+ ions out of the cell in exchange for two K+ ions (Figure 22.5). The requirement for energy is particularly high in nerve cells where the transmission of an impulse leads to the loss of the membrane potential. Estimates vary as to the precise overall importance of sodium pumps in basal heat production; figures of 5–50% of total resting body heat production have been quoted by different researchers.

Other ATP-dependent ion pumps

It is paradoxical that exposure of animals to high temperatures increases their metabolic

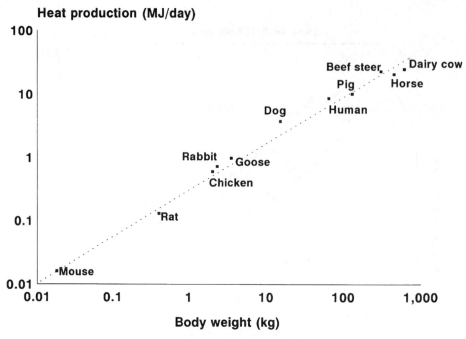

Figure 29.2 Heat production in a range of mammalian species. Heat production is logarithmically related to body size.

needs and consequently increases their heat production. The reason for this is to be found in the large amount of water that animals need to evaporate in order to keep their body temperature within the physiological limits. In hot climates, a mature sheep can evaporate over 5 litres of water per day in the form of sweat and of fluid from the surfaces of the lungs.

The maintenance of ionic balance is only one function of membrane pumps; other metabolites such as glucose have to be pumped through membranes in exchange for Na^+. These pumps too are dependent on the hydrolysis of ATP and consume large amounts of energy.

Muscle is maintained in its resting state by the concentration of Ca^{2+} ions in vesicles within the sarcoplasmic reticulum (see Chapter 32). This requires the continual operation of a Ca^{2+} ATPase.

29.3.2 BACKGROUND HEAT PRODUCTION FROM PROTEIN TURNOVER

Protein synthesis

A growing heifer synthesizes about 1.5 kg of protein per day; of this the majority is broken down again into amino acids, and at most only about 0.2 kg or so may be added to the animal's body (assuming that she is growing at 1 kg day^{-1}). A mature cow at maintenance (not lactating or growing and not pregnant) has a protein synthesis rate in the region of 2.0 kg day^{-1}. By far the largest part (about 40%) takes place in the gut, and of this about 40% is lost to the animal in the faeces. Muscle, skin and other tissues account for approximately equal shares of the remainder, with a smaller amount (about 7%) being made and broken down in the liver.

A very rapidly growing beef animal may exceptionally gain 2.0 kg per day of body weight but a more normal figure would be 1.0 kg. Of this, only about 20% will be protein which indicates that, at the most, a rapidly growing animal will only add 200–400 g protein per day, figures which are very small in comparison to the overall protein turnover rates of the animal. In the whole animal, protein synthesis accounts for between a sixth and a quarter of total energy expenditure. Calculations using whole animals have suggested that there is an energy requirement of about 7.5 kJ for every gram of protein synthesized. This suggests that every amino acid incorporated requires the expenditure of about 7 moles of ATP, which is greater than is accounted for in the pathway of protein synthesis (see Chapter 21). This extra energy must either be used in steps ancillary to protein synthesis itself, such as interchanges between non-essential amino acids, or it is due to the energy needed for protein breakdown.

Protein breakdown

Protein breakdown in cells takes place by one of two mechanisms. One uses the range of proteolytic enzymes located in the lysosome; the other is the ubiquitin pathway which is ATP-dependent. Ubiquitin is itself a small protein which can attach to lysine residues in a protein that is earmarked for destruction. This protein–ubiquitin complex is then degraded to small peptides and free amino acids by large multi-enzyme complexes called proteasomes.

The large amounts of synthesis and breakdown at first appear to be rather pointless in terms of the overall metabolic economy of the animal, and the question has to be asked, why?". There are several reasons. Some proteins are actually made with a deliberately short life span, others are designed to last much longer. The life expectancy of a protein molecule is marked in its N-terminal amino acid. Figure 29.3 shows the longevity in yeast cells of proteins with different N-terminal residues.

Table 29.1 Effect of different N-terminal amino acids on the half-life of proteins in yeast cells

Amino acid	Half-life of protein
Arginine	2 min
Asparagine	3 min
Aspartate	3 min
Histidine	3 min
Leucine	3 min
Lysine	3 min
Phenylalanine	3 min
Tryptophan	3 min
Glutamine	10 min
Tyrosine	10 min
Glutamate	30 min
Isoleucine	30 min
Alanine	>20 h
Cysteine	>20 h
Glycine	>20 h
Methionine	>20 h
Proline	>20 h
Serine	>20 h
Threonine	>20 h
Valine	>20 h

From: Varshavsky, A. (1992) The N-end rule. *Cell*, **69**, 725–735.

It can be seen that there is an enormous difference in the life span of proteins depending upon the identity of one single amino acid (Table 29.1). Many of the proteins that are marked for rapid destruction are enzymes that have a regulatory function controlled by messengers outside the cell. The rate of synthesis of the enzyme is modified by the external stimulus such as a steroid hormone, and so long as the enzyme is rapidly destroyed its activity at any one time is determined by the strength of the stimulus.

Some of the amino acids that are produced by the breakdown of tissue proteins are re-used for synthesis. Others are broken down to urea, in which form some nitrogen passes to the urine whilst other nitrogen is recycled through blood to the salivary glands. The urea in saliva then provides a source of simple nitrogen to the microbes of the rumen.

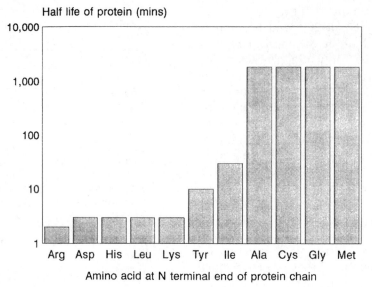

Half life of protein (mins)

Figure 29.3 The effect of N-terminal amino acids on the half-life of polypeptides in yeast. Polypeptides with N-terminal amino acids to the left of isoleucine (Ile) have half-lives of 0–60 minutes, whereas those to the right of isoleucine have half-lives of many hours.

The overall loss of nitrogen at maintenance is the sum of nitrogen lost in the urine and faeces. The daily extent of this total loss of nitrogen in ruminants at maintenance is estimated at 350 mg per $kg^{0.75}$. (The metabolic rate of animals correlates well with their body weight raised to the power of 0.75 ($W^{0.75}$). On this scale the metabolic rates of mice and elephants are quite similar. The same exponent can be used to compare the metabolic activity of animals of the same species but of differing body weight.)

29.3.3 BACKGROUND HEAT PRODUCTION FROM OTHER METABOLIC EVENTS

The activities of the sodium pump and of protein turnover are the two largest single contributors to heat production, but even at their highest they can only account for about a half of the maintenance heat production. The rest must come from a myriad of smaller contributions. There are small, negative, free energy changes associated with many biochemical reactions; much of the energy that this represents will be coupled to other reactions so that they can progress even when the free energy changes are not necessarily favourable. Some energy from these reactions will be lost as heat.

Another possible source of heat comes from the so-called futile cycles. Some of these have already been described in relation to gluconeogenesis (Chapter 18); another, larger cycle can be put together by looking at the synthesis and hydrolysis of triacylglycerols, and a single turn of this cycle leads to the hydrolysis of seven molecules of ATP (Figure 29.4). From a metabolic point of view, these reactions must be carefully controlled to ensure that they do not waste ATP by hydrolysing it back to ADP as soon as it is formed. However, they also provide a method of producing heat within cells.

29.3.4 HEAT PRODUCTION FROM MUSCULAR ACTIVITY

Movement requires physical energy and in turn this must be supplied from chemical energy by the hydrolysis of ATP. Not all of the energy obtained from ATP hydrolysis is converted to perform physical work – this excess energy is lost as heat. It is tempting to think of this heat as being wasted, but in many situations the principal role of muscle may be its ability to produce heat rather than physical effort.

Shivering

It is a common experience that the human response to cold is to shiver, a behaviour pattern that we share with the rest of the mammals. This activity leads to the production of large amounts of heat from the hydrolysis of ATP, both by myosin and in the Ca^{2+} ATPase (Chapter 32). In animals kept in environments below their lower critical temperature, this provides a rapid method for the conversion of the chemical potential energy of tissue stores into thermal energy released by ATP hydrolysis. The mechanical energy that is produced is small in relation to the direct thermal output.

Possible non-coordinated muscle activity

It has been shown that resting thigh muscles from rats respond to administration of the hormones noradrenaline and adrenaline by increasing their oxygen consumption and hence their heat production. The effect on the muscle is an indirect one: it cannot be demonstrated with excised muscles in an isolated system. It has been calculated that the heat production from resting muscles in rats could in this way provide a large proportion of the thermal energy output of the animal. The hormones cause increases in the blood flow through muscles, but their effects on oxygen consumption are much greater. It is tempting to suggest that a combination of the ATPase activity of myosin and the Ca^{2+} ATPase within muscle form a convenient mechanism for heat production in animals.

29.3.5 HEAT PRODUCTION FROM UNCOUPLED PHOSPHORYLATION

The young of many mammals, and the adults of some (such as humans) have brown adipose tissue (BAT), this is darker in colour than the normal white fat deposits. It appears to have a very specific function in relation to

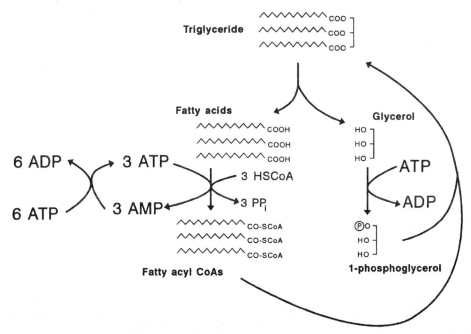

Figure 29.4 The triacylglycerol futile cycle which operates in adipose tissue. This leads to the hydrolysis of seven molecules of ATP for each turn of the cycle.

heat production. Its dark colour is related to an unusually high density of mitochondria. Brown adipose tissue is unique in having a protein, thermogenin or uncoupling protein (UCP) which removes the link between mitochondrial proton transfer and ATP generation (see Chapter 12). The activity of this protein is controlled in two ways. In the short term, the activity of pre-existing protein can be switched on and off. After long-term stimulus, the level of uncoupling protein in BAT increases. In both short- and long-term regulation the message to the cell comes via the neurotransmitter, noradrenaline. The noradrenaline itself is produced within the sympathetic nervous system and binds to special binding sites (termed β_3 adrenoceptors) which are unique to BAT.

The main substrate for BAT is, not surprisingly, the fatty acids liberated from the triglycerides stored in the tissue. The BAT therefore represents a combined heater and fuel-storage unit.

30.1 INTRODUCTION

Growth is one of the fundamental processes by which cheap agricultural inputs are converted into expensive products, generally meat. It is associated very closely with another process, that of development. A mature animal is not just an expanded version of a new-born one: the various tissues of the body change in their importance during development. To take a simple example, a new-born calf has very little need for teats, a male will never develop a need for teats, but they will become central to the existence of a dairy cow in adult life. For this reason, most of the development of the teats does not take place until adult life. In general, animals have waves of growth for different tissues, with a hierarchy of growth of brain, bone, muscle and fat. This hides some gross approximations: some bones have to develop much earlier than others, for example, as the brain is developing the skull has to grow to accommodate it. Looking at the whole animal there are also waves of growth that change its proportions during development. In general, growth moves from head to tail – the front limbs and their associated tissues mature earlier than the rear ones. In agricultural terms, this has implications for the planning of production, because the expensive cuts of meat tend to be those produced at the rear of the animal.

The rate of growth of an animal can be changed by dietary manipulation, altering the availability of the raw materials needed for body tissues; however the effect on the pattern of development is much smaller. No matter how well an animal is fed, its tissues will tend to grow in the same order: brain, bones, muscle and fat.

30.1.1 GENERAL PRINCIPLES

Tissue accretion is the process by which new tissue is laid down, and which defines the rate at which the animal grows. The rate of accretion can be thought of in terms of the overall growth rate of tissue, or as the rate for individual chemical components such as protein, lipid or minerals. Accretion is always calculated as the difference between synthesis and breakdown. Even the largest accretion rate is very small in relation to overall turnover. This suggests ways in which animal growth can be made more efficient: either a small increase in the rate of synthesis or a small decrease in the rate of degradation will lead to a larger increase in growth rate.

30.2 RATES AND PATTERNS OF GROWTH

The term, 'rate of growth' is usually employed describe the speed at which animals increase their overall live weight, although it may at times be more important to think in terms of the speed at which the carcass of the animal grows. (The carcass is the body minus the gastro-intestinal tract, head, feet, skin, blood, etc.) Rate of growth depends upon many factors, both internal and external to the animal. The most important internal factors are the ani-

mal's age, health and genetic composition. External factors include climate, season, the availability of food and water, and the necessity or otherwise of having to perform strenuous physical work.

The rates of growth are often controlled by the producer in order to ensure that the animal's productivity is well adapted to the farming system. For instance, broiler chicks are encouraged to grow at a very fast rate so as to achieve marketable weight in about 6 weeks. On the other hand, chicks that are destined to become egg layers are fed so that they do not reach mature weight until they are ready to start laying at about 22 weeks old. Similarly, dairy heifers are fed so as to gain a little over 0.5 kg per day, which means that they will be at the right weight for breeding at around 15 months of age. If they are fed too well and reach the appropriate weight much before this, the mammary gland will not be sufficiently well developed to support high-yielding lactations. By contrast, intensively fed beef animals which are produced for their meat may grow at much faster rates.

Metabolic body weight ($W^{0.75}$) can be used to study the relative potential of an animal for growth. Figure 30.1 shows that as an animal grows, its rate of growth relative to metabolic body weight declines. The rate of deposition of protein also decreases, following closely the pattern of the rate of change in body weight. However, the relative rate of fat deposition stays almost constant in proportion to metabolic weight. Thus a heavier animal lays down fat at a much faster rate than a light one.

The composition of the animal's body changes during development. To achieve this change, differences in the composition of what is added to the body during growth (known as the composition of gain) must be even more extreme.

Figure 30.2 shows the compositions of each extra kilogram of gain in a growing steer from birth at about 50 kg to slaughter at about 500 kg. The increase in fat and decrease in water during growth are demonstrated very clearly.

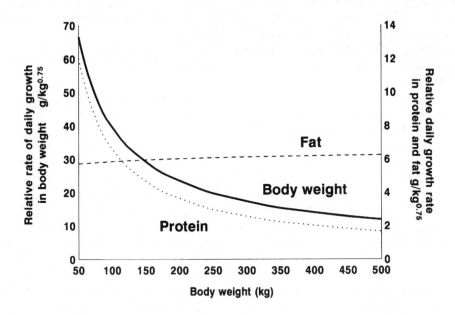

Figure 30.1 The relationship between body weight, relative rate of growth of body weight, protein and fat expressed on a metabolic body weight basis (g kg$^{-0.75}$) in cattle.

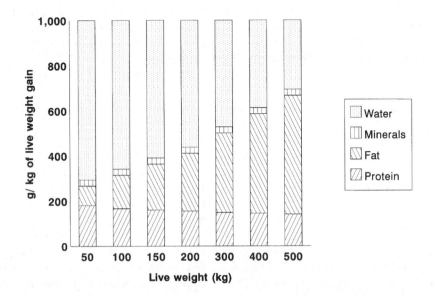

Figure 30.2 Composition of weight gain in a growing steer from 50–500 kg live weight.

The rate of increase in the amount of protein depends very much upon the quantity of dietary amino acids available to the animal.

The question then arises as to what the animal does with quantities of nutrients supplied in the diet which exceed its ability to deposit protein. Amino acids that are supplied in excessive amounts are deaminated (Chapter 14) and the resulting α-ketoacids are used as substrates for energy-yielding processes. This means that an excessive supply of amino acids becomes an over-supply of energy, for which there are two possible outlets: the animal can increase its heat production (see Chapter 29), and the substrates can be used as precursors for the synthesis of fat which is deposited in adipose tissue. Once the maximum capability for protein synthesis is reached, any further increases in nutritional input lead to the production of a fatter carcass. This situation can arise very easily in pig production and has also been shown to occur in sheep.

30.3 MUSCLE GROWTH

The function of muscle and the structures of its major proteins are considered in Chapter 32, but the tissue's importance in agriculture is largely as the most saleable part of meat, and thus its growth is a primary consideration for the producer. Muscle is composed mainly of bundles of muscle cells surrounded by a collagen sheath which is continuous with the tendons that are attached to the skeleton. Muscle growth takes place by two processes, one involving an increase in the number of cells, and the other an increase in the amount of protein within each muscle cell. The DNA content in muscle remains relatively constant during juvenile growth, indicating that muscle protein content and numbers of cell nuclei are growing at about the same rate. However, in later life the level of protein increases at a far greater rate than the number of nuclei. Thus growth in the juvenile is a mixture of protein synthesis and nuclear proliferation. In the adult, the nuclei remain relatively constant but the amount of protein per nucleus increases.

30.3.1 CELLULAR GROWTH AS A COMPONENT OF MUSCULAR GROWTH

In addition to the contractile cells, muscle contains other cells such as those that surround the blood capillaries, fat cells, and a special type known as satellite cells. Muscle fibre cells are multinuclear, so their growth cannot take place by the simple processes of mitotic cell division (see Chapter 32).

Muscle cells begin their existence in foetal life as myoblasts, cells with a single nucleus and few of the characteristics of muscle. As foetal development progresses, the myoblasts start to fuse, forming multinuclear cells. They also begin to produce some of the proteins, such as actin and myosin, troponin and tropomyosin, characteristic of functional muscle. A collagen sheath of connective tissue forms around the developing cells and they take on the characteristic form of muscle cells. At the same time the developing muscle is invaded by nerves. The presence of nerve fibres seems to be essential for the further development of the muscle, and it has been suggested that the nerve cells secrete growth factors that change the pattern of muscle development.

In agricultural species, both mammals and birds, the number of muscle fibre cells is established at birth, and as the animal grows these fibres increase in length and width. This growth is accompanied by an increase in the number of nuclei per cell. Work with radioactively labelled compounds has shown that the new DNA does not originate inside the muscle fibre but is passed over from the accompanying satellite cells.

In younger animals satellite cells are able to undergo mitosis, and therefore growth and development of muscle fibre cells depends upon the division of this different type of cell. As an animal matures, the proliferation of satellite cells (and therefore any increase in overall muscle DNA content) comes to an end. At about the same time there is a tendency for animals to start to deposit lipid rather than muscle fibres.

30.3.2 PROTEIN ACCRETION AS A COMPONENT OF GROWTH

Muscle proteins are continually formed only to be broken down again as part of a large-scale turnover of body resources. The net deposition of protein is determined by the rates of protein synthesis and protein degradation. When considering how muscle protein is laid down, the ways in which synthesis and degradation are controlled must be considered separately. The overall process of protein synthesis is described in Chapter 21, and only the potential points of control are considered below. The degradation of protein is considered in Chapters 29 and 32.

30.3.3 CONTROL OF PROTEIN SYNTHESIS

The pathway for protein synthesis can be split into two events: transcription of DNA into mRNA, and the translation of the information contained in the mRNA into the form of polypeptides. In broad terms, controlling transcription alters the types of proteins formed, whereas the regulation of translation changes the rate at which proteins are synthesized.

Translation takes place through the ribosome cycle (Figure 30.3) and the main point of control is exerted at the initiation of peptide-chain synthesis, which takes place when the 40S subunit of the ribosome binds to various initiation factors (eukaryotic initiation factors, eIFs).The first amino acid incorporated into the nascent polypeptide is always methionine linked to a special tRNA, which differs from the tRNA which is used to carry methionine when it is to be incorporated into the middle of a polypeptide chain. The 40S ribosomal subunit also has to carry an active (phosphorylated) version of eIF-2. The eIF-2 complex is released in an inactive form when the 40S and

60S units combine to form the ribosome. The eIF-2 protein is reactivated by phosphorylation under the influence of a protein kinase enzyme. This allows a further cycle of polypeptide synthesis to be initiated. The involvement of an allosterically modified kinase provides a point at which protein synthesis becomes susceptible to control by a 'second messenger' system.

There are other possible control points in the regulation of translation which may involve other initiation factors, although the evidence is, as yet, rather uncertain.

30.4 GROWTH OF COLLAGEN

The synthesis of collagen is central to the growth of almost all animal tissues because it accounts for over a quarter of the protein in the body. Much of the collagen is in tissues such as skin and bone that have a low commercial value. Despite this, such tissues are important to the overall chemical economy of the animal because they are vital to survival. Collagen also is important because its synthesis diverts amino acids away from the formation of muscle fibre protein.

The term collagen relates to a family of at least 14 different proteins which are present in various tissues. The most widespread collagen, denoted as collagen I, is found in bone, tendons, skin and the cornea.

The importance of collagen lies in its physical properties, particularly its insolubility in water and its physical strength under tension. These unusual properties are matched by some extraordinary features in its chemical structure. The main distinguishing features are its atypical amino acid composition and its quaternary structure.

Collagens have a very high content of glycine which constitutes about 30% of the amino acid residues in the molecule. It also has very high concentrations of the hydroxy amino acids hydroxyproline and hydroxylysine. These are not incorporated into the polypeptide during elongation, rather they

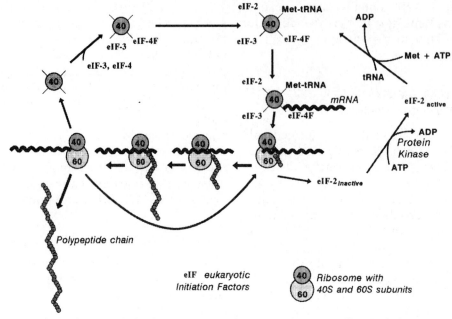

Figure 30.3 Initiation factors involved in protein synthesis in eukaryotic cells. eIF, Eukaryotic initiation factors.

are formed by the modification of the amino acids after the translation step of protein synthesis. Hydroxyproline is formed from proline under the influence of the enzyme prolyl hydroxylase. The enzyme has iron at its centre and requires molecular oxygen (O_2). α-Ketoglutarate is also required and is decarboxylated to succinate during the reaction. One interesting feature of this reaction is the need for a supply of ascorbic acid (vitamin C) to maintain the iron in the II oxidation state. It follows that a shortage of vitamin C will lead to a reduction in collagen synthesis, with consequences for those tissues such as skin that have high collagen contents. Many of the symptoms of vitamin C deficiency do indeed involve lesions of skin and other tissues with high collagen levels (Chapter 8).

Collagens are actually glycoproteins, although the number of sugar residues is small in comparison to the quantity of amino acids. The point of attachment for the carbohydrate (mainly mono- or disaccharides of glucose and galactose) is the hydroxyl group of hydroxylysine.

Collagen is formed from a group of polypeptide chains known as α-chains which intertwine to form a triple helix. This shape is reminiscent of a rope, a man-made structure of great tensile strength. Collagen I comprises two chains of a type called α1(I) and one of α2, each of which has in the region of 1000 amino acid residues.

The constituent polypeptide chains of collagen do not spontaneously form the triple helix, and thus the association must be one which is guided enzymically or by some form of chaperone protein. The process starts with the formation, in the endoplasmic reticulum, of a triple helix using three polypeptide chains which are considerably longer than those that will eventually be found in the collagen. This precursor collagen is known as procollagen. Once formed, it is secreted from the cell into the extracellular space, where the extra amino acid residues are trimmed off under the action of specific procollagen peptidases. The resulting molecules are termed tropocollagen: these then associate together to form the collagen fibres. The chains of collagen are held firmly together by hydrogen bonding in a very regular pattern, with individual tropocollagen molecules displaced from one another by about a quarter of their length (Figure 30.4).

Collagen is rarely sold as a primary agricultural product but it does form the raw material for a very important industrial commodity – gelatin. If collagen is boiled for an extended period it separates into individual tropocollagen fibres, and these then break into their individual polypeptide strands. Once the association of the polypeptide strands is lost, so too is the structure of the collagen, and the polypeptides form themselves into random units with a great deal of inter-molecular hydrogen bonding both between polypeptide chains and between the chains and water molecules, so as to form a continuous gel phase. Heating the gelatin leads to a loosening of the hydrogen bonding and the melting of the gel. Gelatin is the basis of a huge number of products in the food, confectionery and pharmaceutical industries.

30.5 GROWTH OF BONE

The final stature of an animal is linked to the size of its skeleton: if bone growth is reduced an animal's potential to deposit muscle is also reduced. Bone is not simply a mineral material, it is much more closely related to composite materials such as glass-reinforced plastic (fibreglass) where the mineral glass exists in a polymer matrix of resin. In bone, the mineral is mainly calcium phosphates and the polymer matrix is collagen. Bone does not just provide the rigid frame of the skeleton, it also functions as a shock absorber against mechanical damage and as a reservoir for the storage of minerals, particularly calcium and phosphorus. The matrix of protein gives to bone the necessary slight flexibility to allow it to absorb impact, and thus to flex under stress rather than to break. Calcium phosphate acts as the rigid component and also the mineral store. Bone development therefore requires both protein synthesis and mineral deposition. In Chapter 29, it was shown that tissues are dynamic and are continually being synthesized and broken down – the same processes occur in bone metabolism.

The size of the carcass of an animal is largely determined by the size of its long bones such as those of the legs, and these will be treated as the examples in this chapter. The mature size of an animal is fixed at or near puberty, and at this point there are big changes in the physiology of the bone.

A simplified form of the structure of bone is shown in Figure 30.5.

Bone formation is determined by two types of cells, osteoblasts that synthesize bone and osteoclasts that break it down. The processes start in embryonic life with the formation of cartilage (largely collagen) surrounded by a primary bone collar synthesized by the

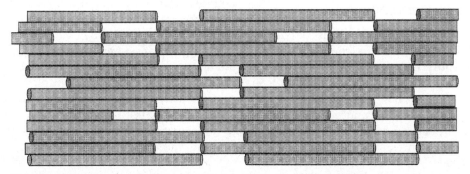

Figure 30.4 The arrangement of individual collagen chains in a collagen fibre.

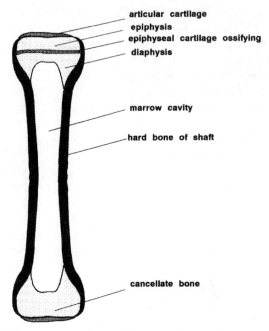

articular cartilage
epiphysis
epiphyseal cartilage ossifying
diaphysis

marrow cavity

hard bone of shaft

cancellate bone

Figure 30.5 Simplified diagram of the structure of a long bone.

osteoblasts. This solid structure is penetrated by other cells, amongst them osteoclasts, that hollow out an internal cavity where the cells of the bone marrow develop and grow. This remodelling process continues towards the ends of the bone. At each end (the epiphyses) an area of actively growing bone, known as the epiphyseal growth plate, is formed. Within the growth plate, cartilage-secreting cells (chondrocytes) proliferate and extend towards the shaft of the bone (diaphysis).

30.5.1 CALCIFICATION OF BONE

The mineral portion of bone consists largely of hydroxyapatite which is a complex calcium hydroxy phosphate. One commonly quoted formula is $Ca_{10}(OH)_2(PO_4)_6$, but this is only an approximation. A small proportion of the calcium may be replaced by another of the alkali earth metals, mainly magnesium, but under certain circumstances strontium may be laid down within the crystalline structure. Similarly, a small proportion of the hydroxyl ion may be replaced by fluoride. Some phosphate is replaced with carbonate ($CO_3 2-$). Pure hydroxyapatite with a Ca:P:OH ratio of 5:3:1 would be a solid with a hexagonal crystalline structure, with planes of weakness (cleavage) such that it would be more liable to fracture in one direction than another. A pure crystalline structure in bone would be disadvantageous and therefore much of the mineral that is laid down is as very small individual crystals, 0.02–0.03 mm in length. This lack of consolidated crystal structure is seen particularly in younger animals where the mineral component may be almost completely amorphous. As animals mature the degree of hydroxyapatite crystallisation increases. The first crystalline mineral to be formed may not be hydroxyapatite itself but simpler calcium phosphates (tricalcium phosphate and octacalcium phosphate), with transformation to hydroxyapatite taking place subsequently.

Bone mineralization takes place in the extracellular matrix of the bone, which exists as an aqueous gel phase due to the presence of collagen. Also present are smaller amounts of a large range of other proteins, many of which are capable of temporarily binding to (sequestering) calcium. The formation of the calcified material occurs by a process of precipitation which needs four components:

- a source of extracellular calcium;
- a source of extracellular phosphate;
- a solid phase that can act as a centre of crystal nucleation;
- a matrix which can form the shape of the growing crystals.

In growing bone, calcium and phosphate are released into the extracellular space from the osteoblasts of the epiphyseal growth plate. The nucleation of crystal formation is thought to be provided by a family of proteolipids that contain phospholipid–phosphate complexes. Once crystal growth has been initiated, its orientation and the size of the crystals appear to be shaped by the collagen matrix.

30.6 GROWTH OF ADIPOSE TISSUE

Adipose tissue growth is a very topical issue in the study of human biochemistry, because of the great interest in preventing the excessive development of lipid deposits which are considered both unhealthy and unsightly. Interest in the adipose tissue of livestock is of a different nature. A moderate amount of fat deposited within meat is regarded as a commercially valuable indicator of quality. Dietary energy provided as fat is deposited very efficiently as carcass fat with a relatively small energy cost in biochemical terms to the animal. This is because fatty acids may be absorbed and incorporated in fat depots with little modification. In contrast, the conversion of dietary protein to tissue protein has a much higher energy cost because of the high ATP consumption associated with protein synthesis.

Excessive fat deposited in the carcass is removed during butchering. This is wasteful in terms of the utilization of dietary energy. Because carcass fat has a very low water con-

tent compared to protein, the deposition of 1 kg fat is more expensive, in terms of the supply of dietary energy, than the deposition of 1 kg protein.

As observed in Chapter 9, adult animals tend to lay down more fat than young ones. Thus, farm animals are often taken to slaughter before they reach full maturity so as to restrict the amount of fat in the carcass.

Most of the adipose tissue in the carcasses of farm livestock is of the white variety. Amounts of brown adipose tissue are small in the carcasses of all but the new-born, thus this section concentrates almost entirely on the white adipose tissue. This is distributed between a number of sites which vary in quantitative importance during the animal's life. Intermuscular fat develops at an earlier age than perinephric or subcutaneous adipose tissue.

The growth of adipose tissue, as with most tissues, rests upon two events: the increase in the number of specific adipose tissue cells (the hyperplasia of adipocytes); and the quantity of fat that is deposited in each cell (hypertrophy). There has been much debate about the relative importance of these processes. It was previously believed that adipocyte number was a constant determined at birth, and that the growth of fat deposits was purely a process of hypertrophy. This now appears to have been a mistaken assumption.

30.6.1 GROWTH IN ADIPOSE TISSUE CELL NUMBER

The main source of adipocytes in white adipose tissue is a line of cells known as adipocyte precursor cells (preadipocytes). These can be extracted from the adipose tissue of a wide range of animals and cultured in laboratory media where they will divide, doubling their numbers approximately every 26 hours. After a few weeks in culture they acquire the characteristic biochemical features of adipocytes, particularly the presence of lipoprotein lipase. In addition, in humans it has been shown that mature adipocytes are capable of DNA replication, indicating the possibility of increasing adipocyte numbers by the simpler process of mitotic cell division.

A number of locally acting hormones (growth factors) have been shown to be necessary for the proliferation of adipocyte precursor cells. Furthermore, their activity can be stimulated by insulin and an insulin-like growth factor (IGF-2).

30.6.2 DEPOSITION OF FAT WITHIN ADIPOCYTES

The control of fat deposition depends upon a balance between the rate at which triacylglycerols are formed in adipocytes (lipogenesis) and the rate at which they are broken down (lipolysis). Figure 30.6 shows the principal routes of fat metabolism in the adipocyte. The precursors for fat synthesis, preformed fatty acids released from circulating lipoproteins by lipoprotein lipase, acetate and glucose, are taken up from blood. These fatty acids, and those synthesized within the cell from glucose or acetate, are incorporated into triacylglycerols. The glycerol-3-phosphate required for triacylglycerol synthesis is also derived from glucose. The breakdown of triacylglycerols to non-esterified fatty acids and glycerol occurs by the action of lipases. Glycerol is exported from the adipocyte because these cells lack the glycerol kinase which is required to convert it to glycerol-3-phosphate. Some of the fatty acids formed may be activated within the cell and reincorporated into triacylglycerols. This is an example of a futile cycle, but it is one which may be used as a source of thermal energy within the animal (Chapter 29). The non-esterified fatty acids which are not reincorporated into triacylglycerols pass to the blood for circulation to the liver and subsequent use in other tissues.

Control of triacylglycerol breakdown

The initial step in the hydrolysis of triacylglycerols is catalysed by triacylglycerol lipase, the activity of which is regulated by numerous hormonal factors. The β-adrenergic receptor-mediated activation of this enzyme is described in more detail in section 30.7.3, and is shown in Figure 30.11.

Triacylglycerol lipase exists in two forms: the active, phosphorylated enzyme; and the inactive, dephosphorylated enzyme. Conversion from the inactive to the active form occurs when the enzyme is phosphorylated by a protein kinase (sometime called the A-kinase) that is activated by an increase in the concentration of cyclic AMP (cAMP). The reverse process, inactivation of the lipase, is catalysed by a phosphatase enzyme. Adenosine, and the hormones glucagon and adrenaline, are known to stimulate the release of fatty acids from adipose tissue in response to some defined physiological state. Glucagon signals an overall energy deficit in the animal. Levels of adrenaline in blood plasma are increased by stress which brings with it a need for energy. Adenosine is thought to act as a local modifier of adipocyte activity.

Insulin has an antilipolytic effect probably caused by its activation of phosphodiesterase, the enzyme which breaks down cAMP (Figure 30.6).

The control of lipid synthesis

Lipid synthesis is regulated in several ways. One of these depends on the availability of the glucose for the synthesis of fatty acids and the glycerol-3-phosphate needed for triacylglycerol formation. These processes are primarily regulated by insulin, which stimulates the uptake of glucose by tissues. Control is also exerted via the enzymes involved in fatty acid synthesis. This process is regulated by its rate-limiting step which is the transformation of acetyl-CoA into malonyl-CoA under the influence of acetyl-CoA carboxylase (Chapter 19). Although the major changes in its activity are brought about by citrate via allosteric regulation, the activity of acetyl-CoA carboxylase can also be modified by phosphorylation through protein kinases. The effect of phosphorylation

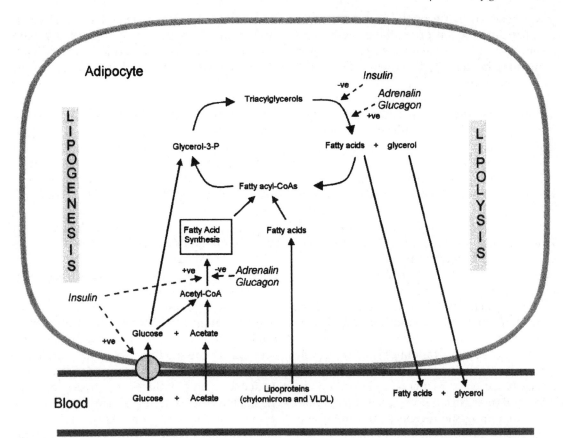

Figure 30.6 Metabolism of triacylglycerols on adipose tissue, showing the main sites of action of hormones on lipogenesis and lipolysis.

is to reduce the activity the enzyme, and therefore to decrease fatty acid synthesis. Thus the overall effect of phosphorylation of acetyl-CoA carboxylase and triacylglycerol lipase is an increase in the rate of lipolysis. Changes in the phosphorylation state of these enzymes are regulated hormonally.

30.7 MANIPULATION OF GROWTH

The last 30–40 years have seen major advances in our understanding of the biochemical basis of animal nutrition and the endocrinology of growth at a tissue level. This has led to the potential to manipulate animal performance via nutritional means, and by the use of naturally occurring and synthetic compounds such as natural and recombinant growth hormone,

anabolic oestrogens and androgens, β-adrenergic agonists, and gut-active growth promoters such as the ionophore and non-ionophore antibiotics.

30.7.1 GROWTH HORMONE AND INSULIN-LIKE GROWTH FACTORS

The effects on growth and carcass composition of growth hormone (GH), also known as somatotropin (ST), and of the insulin-like growth factors (IGFs), also known as the somatomedins, are interrelated and often difficult to distinguish. Growth hormone is a protein hormone (20–22 kDa) produced in an episodic manner by the anterior pituitary. A number of animal tissues have been demonstrated to have GH receptors. For many years it was believed that growth hormone was the primary determinant of animal growth. Until recently, studies on the effects of growth hormone have been limited by the need to extract the hormone from pituitary glands; however, the advent of recombinant DNA techniques has made it possible to produce growth hormone in much larger quantities using modified bacteria. It is now clear that some of the effects of growth hormone are mediated by the IGFs, small proteins with a molecular weight of around 7.5 kDa. Growth hormone acts on the liver to stimulate the production and secretion of the IGFs, particularly IGF-1, which then act on peripheral tissues by a classical endocrine mechanism. GH and probably other growth factors may also act on non-hepatic tissue to stimulate the local production of IGFs, which act within the tissue by local control mechanisms. In addition, GH may have direct effects on a number of tissues.

Chronic administration of GH generally increases plasma insulin concentration, which would be expected to increase glucose utilization by tissues. In adipose tissue this should result in a lipogenic response, and in skeletal muscle provide substrates and energy for protein synthesis. However, both *in vivo* and *in vitro*, GH tends to decrease lipogenesis, apparently by reducing glucose uptake and its incorporation into glycerol and fatty acids. Studies *in vitro* of adipose tissue from sheep, cattle and pigs suggest that, despite increases in plasma insulin concentration, GH actually antagonizes the effects of insulin on adipocytes, possibly by decreasing the sensitivity of the tissue to insulin. GH has also been shown to increase lipolysis in some species. This may be due to the decrease in insulin sensitivity and a reduction of the inhibitory effects of insulin on lipolysis. As lipid accretion in adipose tissue is determined by the balance between lipogenesis and lipolysis (see Figure 30.6), the net effect of GH is the mobilization of fatty acids from adipose tissue triacylglycerols.

Much less is known at the molecular level about the effects of GH on tissue protein metabolism. Studies of the whole animal clearly indicate increased protein accretion and improved utilization of feed for protein deposition. However, it is still not known how these gross changes are related to intracellular changes in protein metabolism. Increased protein accretion could result from an increase in protein synthesis, a decrease in protein degradation or a combination of both effects.

The direct effects of IGFs on fat and protein metabolism remain largely unknown. In plasma, virtually all IGFs are bound to carrier proteins, the insulin-like growth factor-binding proteins. Bound IGFs have little effect on growth and it is the free form which interacts with cell membrane receptors. Removal of the N-terminal amino acids from IGF-1 results in the production of (des 1–3)IGF-1 which is a more potent stimulator of protein synthesis than the original IGF-1. This has been attributed to a lower affinity of the plasma-binding proteins for the modified IGF, and therefore a higher concentration of free (des 1–3)IGF-1 available to bind with cell receptors. This suggests that the N-terminal amino acids are involved in the binding to plasma IGF-1 binding proteins, but not with IGF-1 cell membrane receptors.

At an intracellular level the mode of action of IGF-1 is poorly understood. It has been shown that the cell membrane contains distinct receptors for insulin and IGFs. The IGF receptors have been classified as types 1 and 2, but their structures and the mechanisms by which they initiate changes in cellular metabolism are unknown. They may be similar to the insulin receptor which is a glycoprotein with four subunits, two α and two β. The α chains are extracellular, attached to the outer surface of the cell membrane where they interact with insulin. The β chains are transmembrane proteins which contain tyrosine kinase activity on the intracellular domain of the protein. It is suggested that binding of insulin to the receptor activates the protein tyrosine kinase. The net effect is that an intracellular target protein is phosphorylated on a tyrosine residue. This protein may act directly within the cell or may be part of a second messenger system (Figure 30.7).

Assuming that one or both of the IGF receptors is similar in structure and mechanism to the insulin receptor, there remains the question of how the primary signal, binding of IGF to its receptor, is translated into a change in cellular metabolism. Measurements have shown that IGF-1 is somatogenic: it increases glucose uptake, DNA synthesis and proliferation of specific types of cell, e.g. chondrocytes, fibroblasts, epithelial cells and muscle satellite cells. At the level of protein metabolism it increases protein synthesis and decreases protein degradation. The links between these changes and events at the receptor have yet to be discovered. So far no second messenger mechanism has been identified.

30.7.2 OESTROGENS AND ANDROGENS

Natural sex steroids

Steroid hormones have a marked effect on behaviour, growth and carcass composition in animals. The oestrogens (17β-oestradiol and oestrone) are particularly associated with intact females, although they are produced (albeit in smaller amounts) in male testes, and may be responsible for the development of some male secondary sex characteristics. The androgens (chiefly testosterone) are thought of as being hormones of the intact male. It is, however, interesting to note that the blood plasma concentrations of androgens in females may actually exceed those of the oestrogens. The third group of steroids, the progestagens, do not have large effects on growth.

The outward effects of the sex steroids are fairly simple to demonstrate by looking at the appearance of farm livestock which have been castrated or left intact. For example, intact male animals differ in shape, with a preponderance of muscle growth around the shoulders when compared with females or castrates. The differences are not only seen in gross anatomy; in general entire males grow more rapidly than females or castrates and produce carcasses that, at the same weight, have lower levels of fat.

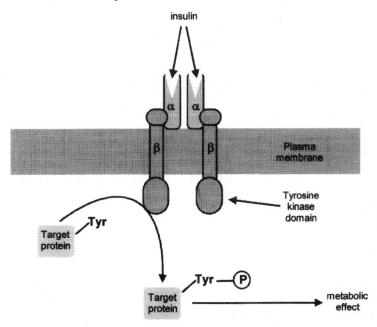

Figure 30.7 Mode of action of the insulin receptor.

These differences in carcass growth and conformation have stimulated a great deal of research into the use of steroid hormones in manipulating growth.

Synthetic androgens and oestrogens

A number of synthetic oestrogenic and androgenic compounds have also been produced which elicit growth responses in animals. These include the non-steroidal oestrogens, diethylstilbestrol (now banned from use because of its carcinogenic properties), hexoestrol and zeranol, and the steroidal androgen trenbolone acetate (TBA).

Numerous studies have indicated that administration of naturally occurring or synthetic oestrogens and androgens can increase growth responses and cause changes in carcass composition in farm animals. The mechanisms by which these changes are mediated are complex and still not fully understood. There are clear sex and species differences in the responses to treatment with these compounds, which suggests that there is no single mechanism of action which can explain the observed growth and carcass changes. The complex interrelationships between the component parts of the endocrine system, combined with differing responses due to sex, age and nutrition, have been difficult to unravel – there is still much to be learned about the mode of action of these compounds. The structures of natural and synthetic oestrogens and androgens are shown in Figure 30.8.

Typical responses to treatment with oestrogens and androgens

In castrate male ruminants, exogenous oestrogens generally increase lean tissue deposition and decrease subcutaneous and intramuscular fat. Similar, but smaller, responses are seen in females. In the entire male, body fat is often unchanged or increased by oestrogens. TBA, the androgen most commonly used in ruminants, increases lean tissue growth and decreases fat deposition in dairy and beef cows. Measurements in heifers have indicated increased nitrogen retention and protein deposition in response to treatment. Responses in entire males, which already have high circulating levels of natural androgens, are small. In ruminants, the optimum responses to anabolic steroid treatment have been shown to occur when oestrogens and androgens are given in combination where their effects appear to be additive.

Treatment with anabolic steroids is generally less effective in pigs and poultry. In boars, oestrogens increase growth rate and show little effect on, or may even increase, carcass fat. Androgens increase lean tissue deposition and nitrogen retention in barrows and gilts, and have variable effects on growth rates. In broiler chickens, oestrogens have the undesirable effect of increasing body fat content.

Mode of action

Oestrogens have effects similar to growth hormone; it is therefore possible that the effects of oestrogens may be mediated indirectly via increased secretion of GH. This idea is supported by evidence showing that oestrogen administration causes an increase in the size of the anterior pituitary and elevated plasma GH secretion in ruminants. However, oestrogens must also exert their effects via other mechanisms, as the response to the combined administration of oestrogens and growth hormone appears to be additive. Furthermore their individual effects on protein accretion differ. For example, GH increases accretion by increasing protein synthesis but seems to have no effect on protein degradation, whereas oestrogens have little effect on (or decrease) protein synthesis, but reduce protein degradation to a greater extent. In addition, the administration of oestrogens to rats increases plasma GH concentrations but does not produce a growth response, whereas direct administration of GH stimulates growth.

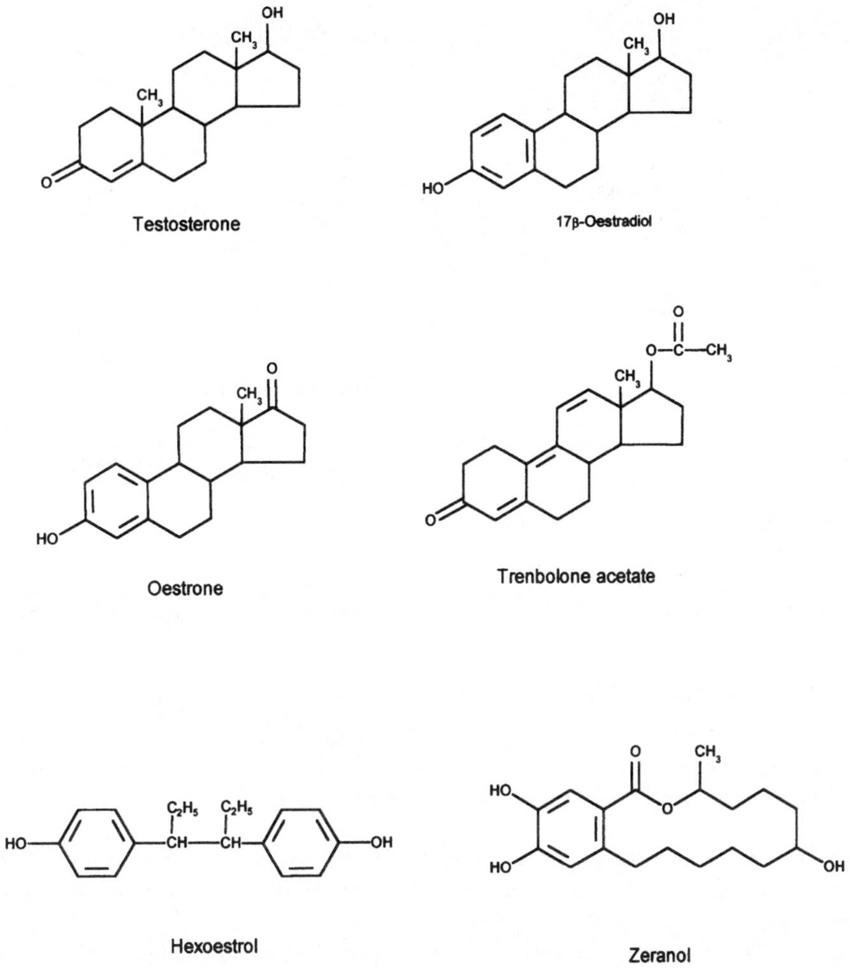

Figure 30.8 The structures of natural and synthetic anabolic steroidal and non-steroidal compounds.

An alternative hypothesis is that oestrogens may mimic growth hormone effects by increasing the production and/or sensitivity of tissues to the insulin-like growth factors (IGF-I and IGF-II). Thus it is possible that oestrogens modify the sensitivity and number of GH and/or IGF receptors in liver and other tissues.

Androgens and oestrogens also have direct effects in cells. These hormones are carried in the blood stream bound to carrier proteins. When they reach their target tissue they are released and, because they are lipophilic, pass through the plasma membrane. In the cytoplasm the steroids bind with specific receptor proteins. The steroid–receptor complex is translocated to the nucleus where it interacts with specific binding sites on the DNA called hormone-responsive elements (HREs). As a result of the binding to the HREs, the transcription of some genes is enhanced while for others it is suppressed. In this way androgens and oestrogens can influence protein metabolism (Figure 30.9).

30.7.3 THE β-AGONISTS

Adrenaline (epinephrine) and noradrenaline (norepinephrine) are examples of naturally

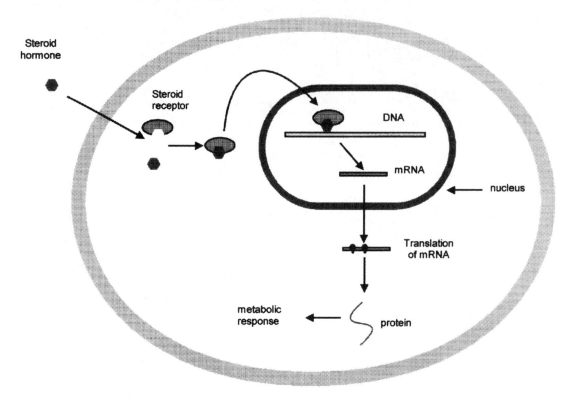

Figure 30.9 Mode of action of steroid hormones on protein synthesis.

occurring hormones, the catecholamines. Adrenaline is produced by the adrenal medulla and noradrenaline is a neurotransmitter produced by post-ganglionic sympathetic nerve endings. These hormones have many profound physiological and biochemical effects. For example, at a cellular level they stimulate lipolysis in adipose tissue, glycogenolysis and gluconeogenesis in liver, and glycolysis in muscle. At an organ level they increase heart rate and blood pressure and cause dilation of the respiratory passages. Many analogues of the natural catecholamines have been synthesized for medical and veterinary purposes. In general, catecholamines decrease the fat content in animal carcasses and may increase protein deposition, resulting in the production of leaner carcasses. Because they appear to shift the balance in the utilization of nutrients from adipose tissue to skeletal muscle, they are often referred to a repartitioning agents. The three most widely used compounds are clenbuterol, cimaterol and ractopamine, the structures of which are compared with adrenaline and noradrenaline in Figure 30.10.

The differing and apparently tissue-specific effects of catecholamines are, in part, due to the fact that they act via different plasma membrane adrenergic receptors classified as α_1, α_2, β_1 and β_2, which elicit different cellular responses. Analogues which produce similar or greater effects than the naturally occurring catecholamines are referred to as agonists. As clenbuterol, cimaterol and ractopamine act via the β-receptors they are also known by the generic term β-agonists.

Typical effects of β-agonists in meat-producing animals

In general, effects of β-agonists on carcass composition are a visible reduction of subcuta-

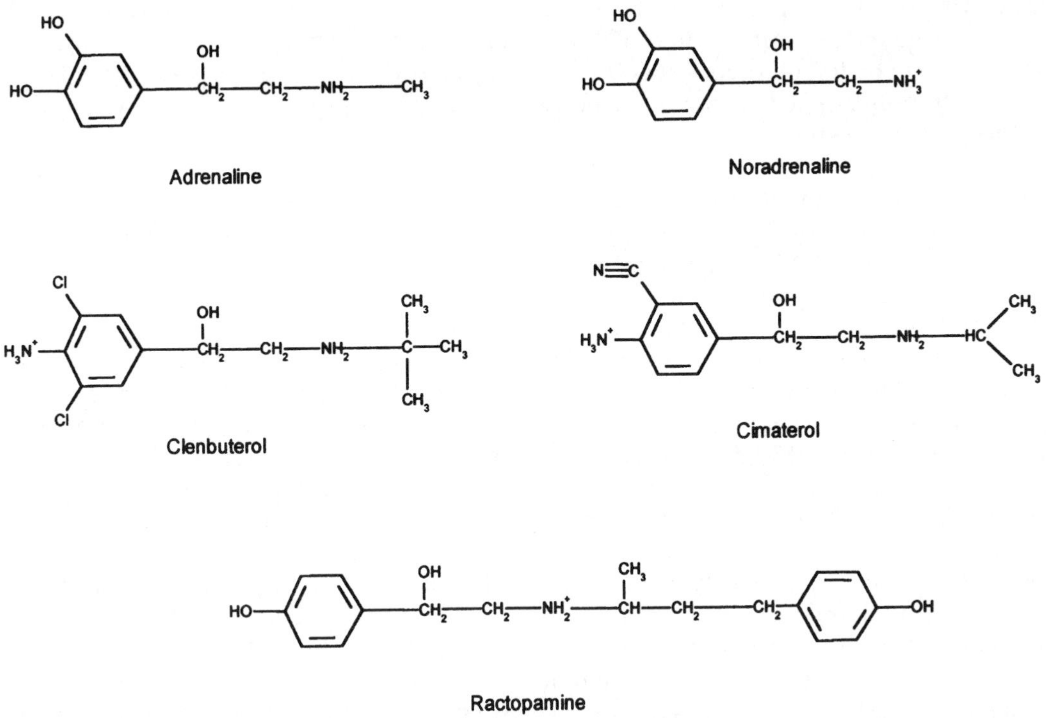

Figure 30.10 Structures of the catecholamines adrenaline and noradrenaline, and the β-agonists clenbuterol, cimaterol and ractopamine.

neous fat cover and of internal fat depots (e.g. perinephric, mesenteric, etc.). Intramuscular fat is also reduced. Typically, reductions of 25–30% in fat have been observed in beef animals and even greater reductions are seen in sheep. Muscle accretion is also increased by treatment, and in cattle, sheep and pigs the cross-sectional area of the *longissimus dorsi*, a measurement routinely used to assess muscle protein accretion in animals, is increased significantly. In poultry the main fat depot, the abdominal fat pad, is reduced in size by treatment and the protein content of the carcass increases.

Advantages of the use of β-agonists are that they can be added directly to the diet or the water supply; their effects are largely independent of sex; and they are relatively short-acting and so can be used at the most advantageous time during the growth period. The disadvan- tages are that these compounds are very potent pharmaceuticals and must be handled with great care. Typical dose levels are 2–10 mg kg^{-1} in cattle and sheep diets, 1–2 mg kg^{-1} in pig diets and 0.25–0.5 mg kg^{-1} in poultry diets. In addition, β-agonist treatment tends to increase meat toughness. It is thought that this occurs due to an increase in muscle connective tissue content and the cross-linking between adjacent collagen molecules. There is also a decrease in the activity of proteolytic enzymes which are involved in the *post mortem* transformation of muscle into meat.

Mode of action

Of all the hormones, the biochemical mode of action of the catecholamines is probably best understood. Receptors in the plasma membrane are coupled to signal transduction proteins which are called G-proteins because they bind to the guanosine nucleotides GDP and GTP. A number of different types of G-protein have now been identified. Those involved in the classical adrenergic responses in cells, resulting in a change in the intracellular cyclic AMP (cAMP) concentration, are designated G_s (s for stimulation of cAMP production) and G_i (i for inhibition of cAMP production). All G-proteins are subunit proteins composed of three non-identical, dissociable subunits, α, β and γ. In their inactive state these proteins exits in their trimeric form with GDP bound to the α subunit. The α subunit also contains a weak GTPase activity which is involved in the deactivation of the G-protein. The binding of a β-agonist to a β-adrenergic receptor sets in train a cascade of events which changes intracellular metabolic activity. Firstly, the binding of the hormone to the receptor causes a conformational change in the receptor. As a consequence, the GDP on the G-protein α subunit exchanges for GTP and the α subunit dissociates from the $\beta\gamma$ dimer. The GTP-α subunit is free to move in the plane of the membrane and interacts with another membrane-bound protein. In the case of the G_s-protein this is the enzyme adenyl cyclase. This catalyses the conversion of ATP to 3′,5′ cyclic AMP, one of the second messengers within the cell. The next stage in the cascade involves the enzyme cAMP-dependent protein kinase, which can exist in inactive or active forms. In the inactive form it consists of four subunits, two regulatory subunits and two catalytic subunits. The regulatory subunits of this tetrameric form mask the active sites on the catalytic subunits, preventing expression of enzyme activity. The regulatory subunits have binding sites for cAMP, and when these are occupied the enzyme dissociates to release two free active catalytic units. In the stimulation of lipolysis in adipose tissue, the target substrate for the active protein kinase is the inactive form of triacylglycerol lipase. The protein kinase-catalysed phosphorylation of triacylglycerol lipase causes a conformational change in the enzyme, converting it to its active form which catalyses the first stage in the hydrolysis of triacylglycerols to fatty acids and glycerol (Figure 30.11).

The effects of β-agonists on protein metabolism in skeletal muscle are less clear but may be mediated via a different type of G-protein, G_p. This has the same trimeric structure as G_s and G_i but when activated by GTP binding the α subunit interacts with membrane-bound phospholipase C. This enzyme acts upon a specific phospholipid component of the plasma membrane, phosphatidylinositol-4,5-bisphosphate, to release inositol-1,4,5-trisphosphate into the cytoplasm, leaving diacylglycerol in the membrane. Both of these products act as second messengers within the cell. Inositol-1,4,5-trisphosphate binds with specific receptors on the endoplasmic reticulum and causes Ca^{2+} channels to open, releasing sequestered Ca^{2+} into the cytoplasm. The released Ca^{2+} can act in two ways. Firstly, it can bind to the Ca^{2+} binding protein, calmodulin. The Ca^{2+}/calmodulin complex then acts on a Ca^{2+}/calmodulin-dependent protein kinase which modifies the activity of specific target enzymes. Some of these are involved in protein synthesis and degradation. Secondly, the released Ca^{2+} interacts with the membrane-bound diacylglycerol to activate yet another protein kinase, protein kinase C, which in turn regulates the activity of intracellular enzymes, some of which may be involved in protein turnover (Figure 30.12).

In addition to these β-adrenergic receptor-mediated effects, β-agonists may act in other ways. For example, it is possible that they decrease the sensitivity of adipocytes to insulin which would have an antilipogenic effect. It is not thought that changes in the circulating growth hormone or IGFs are involved in the mode of action of β-agonists, however they may change the sensitivity of cells to IGFs via their effects on insulin receptors or specific IGF receptors.

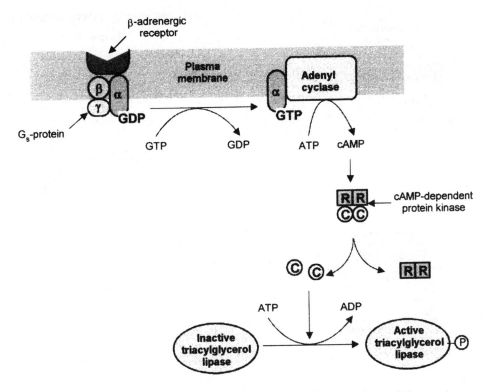

Figure 30.11 Activation of triacylglycerol lipase via the β-adrenergic receptor and G$_s$ protein system.

30.7.4 GLUCOCORTICOIDS

Elevated levels of glucocorticoid hormones are usually associated with chronic stress although they may have other roles, for instance associated with lactation. Raised levels of these hormones can induce muscle wasting. This may be due to an increase in the rate of protein degradation rather than a lack of synthesis. It is still not certain whether the glucocorticoids have a role in controlling protein turnover at the concentrations found in normal animals. Glucocorticoids may have long-term effects on the mobilization of adipose tissue reserves by increasing the number of β-adrenergic receptors. These hormones are not used commercially to manipulate carcass growth.

30.7.5 THYROID HORMONES

It has been known for many years that the administration of thyroid hormones, or of iodinated compounds capable of stimulating thyroid hormone production, can increase the metabolic rate of farm animals. At the tissue level, thyroid hormones increase the rate of protein degradation. This would be expected to decrease growth rate; however, animals with hyperactive thyroid function actually grow faster. It is possible that the overall increase in metabolic rate actually increases the availability of substrates for protein synthesis. This stimulation of synthesis may actually be greater than the activation of degradation.

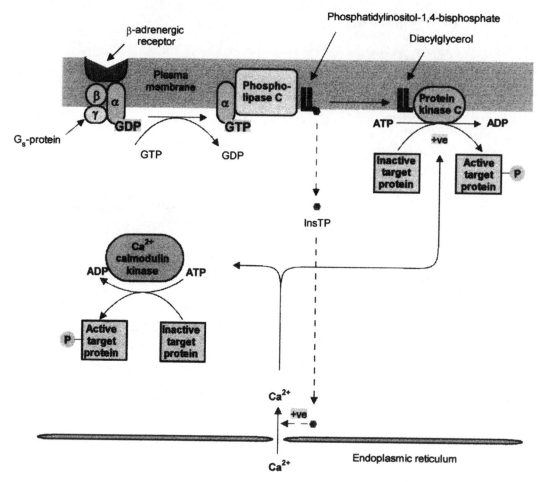

Figure 30.12 The effects of β-agonists in muscle tissue mediated via the β-adrenergic receptor/G_p protein complex.

30.7.6 ANTIBIOTICS

Antibiotics have been used for many years for veterinary purposes in farm animals. Some, particularly those used to control bacterial infection of the digestive tract, produce a growth response and this has led to the development of a number of compounds, now in widespread use, specifically designed as gut active growth promoters. Those commonly used in the livestock industry are listed in Table 30.1. They are divided into two main groups, ionophores and non-ionophores,

based principally on their mode of action. Ionophores are used almost exclusively in ruminant species and, indeed, can be fatal if given to equine species. Non-ionophores are used in both ruminant and non-ruminant species.

Ionophores

Cells have developed mechanisms for maintaining a relatively constant intracellular environment despite quite large changes in the chemical composition of the medium in which

Table 30.1 Commonly used ionophore and non-ionophore gut active growth promoters

Growth promoter	Alternative names
Ionophores	
Monensin	Rumensin
Lasalocid	Bovatec
Salinomycin	Coxistac
Tetranasin	
Narasin	Monteban
Non-ionophores	
Avoparcin	Avotan
Bacitracin	Altacin, penitracin
Virginiamycin	Staphylomycin, eskalin
Flavomycin	Bambermycin, moenomycin
Tylosin	Tylan, tylon

they grow. Considerable energy is expended by cells to transport inorganic ions in both directions across the plasma membrane to maintain the electrolyte composition of the intracellular fluid. The ability to maintain ion concentration gradients is usually linked to the consumption of ATP via the process of active transport (Chapter 22).

The ionophores disrupt the ion equilibrium in cells by facilitating the uncontrolled movement of ions across the cell membrane. Although they vary considerably in structure and specificity, ionophores all possess this common function, and have been defined as 'organic substances which bind to polar substances in cell membranes and act as ion transfer agents through lipid membranes'.

Gramicidin A and valinomycin are well characterized ionophores. Although they are not used in farm animals, they illustrate the principal mode of action and provide examples of the two main types: those that form ion channels in the membrane, and those that act as mobile carriers in the lipid bilayer. Gramicidin A is a linear peptide of 15 amino acids with hydrophobic side chains. Two molecules of gramicidin A link together, via their N-terminal amino acids, to form a channel in the plasma membrane. The hydrophobic side chains of the amino acids are located on the outside of the channel and stabilize the struc-

ture in the lipid bilayer. The hydrophilic interior of the channel has dimensions that allows the movement of protons and potassium ions down their electrochemical gradients. This dissipates the proton motive force generated by the membrane electron transport chain, disrupting ATP synthesis and proton-linked transport systems.

Valinomycin is an example of a mobile ion carrier. This peptide antibiotic is doughnut-shaped and contains valine residues which are configured in such a way that the oxygens of their carboxyl groups are arranged octahedrally around the centre of the molecule. This allows it to form a coordination complex with potassium ions, facilitating the movement of potassium across the membrane.

The ionophores used as gut active growth promoters in farm animals are lipid-soluble mobile carriers, but they are not peptides. Monensin, the best characterized of the ionophore gut active growth promoters, is a polyether ionophore. Its structure and that of another commonly used ionophore, lasalocid, are shown in Figure 30.13.

These molecules are flexible and the binding of a cation initiates the formation of a cyclic lipophilic cation–ionophore complex that can diffuse through the lipid bilayer. Monensin shows preference for the transport of sodium ions and as it can pass through the membrane only when bound to a cation or when protonated, the net movement of sodium depends on the relative concentrations of sodium and protons inside and outside the cell. Lasalocid acts in a similar way to monensin but exhibits highest affinity for potassium, and lower but equal affinity for sodium and calcium.

Monensin dissipates the proton gradient across the cell membrane, causing a decrease in the proton motive force and intracellular ATP concentration. In addition, its effect on the sodium ion flux across the membrane alters the ability of microorganisms to accumulate essential nutrients via sodium-dependent active transport mechanisms. A characteristic effect of ionophores on the end products of carbohy-

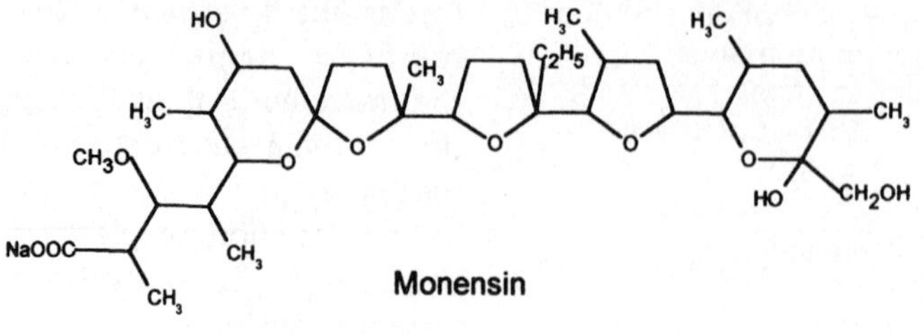

Monensin

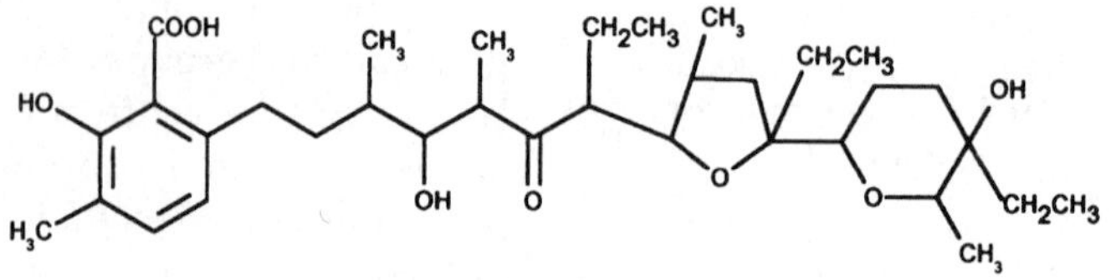

Lasalocid

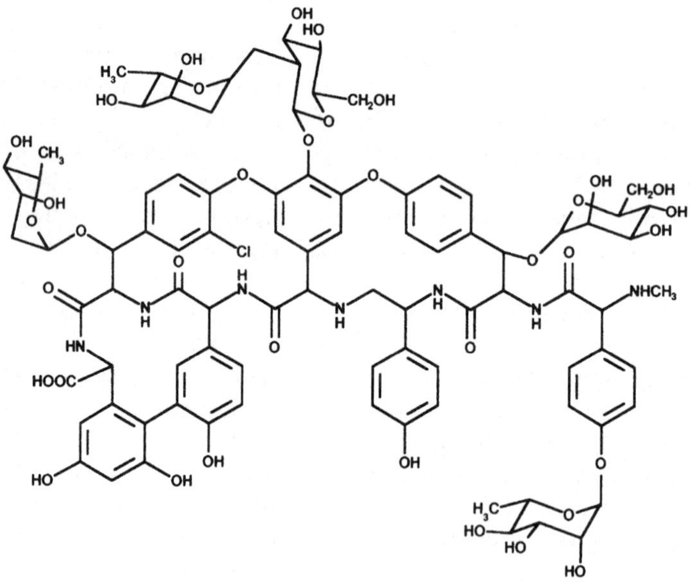

α-Avoparcin

Figure 30.13 The structure of the ionophore antibiotics monensin and lasalocid, and the non-ionophore antibiotic α-avoparcin.

drate digestion in the rumen is an increase in the relative proportion of propionic acid and a decrease in the proportion of acetate. This has been shown to be due to an adaptive change in the rumen microflora in which Gram-positive bacteria, the primary producers of acetate and butryrate, are reduced in numbers, whereas the Gram-negative species, which produce succinate and propionate, increase in numbers. Monensin also has direct effects on methane production in certain bacteria. This is due to the dissipation of the proton gradient across the cell membrane which is an essential factor in methane production. The shift in the microbial population from acetate producers to propionate producers also contributes to the decrease in rumen methane production, as propionate synthesis from pyruvate uses hydrogen which would otherwise be lost as methane.

Non-ionophores

The non-ionophore antibiotics used as gut active growth promoters in livestock production have more complex structures than the ionophores and contain sugar and/or peptide groups. In general, non-ionophores show more marked effects on Gram-positive microorganisms in the digestive tract than on the Gram-negative bacteria.

In Gram-positive organisms, the cell wall consists of peptidoglycans composed of layers of linear chains of polysaccharide made up from repeating disachcharide units of N-acetylglucosamine and N-acetylmuramic acid. The polysaccharide chains in each layer are linked at numerous points by a pentapeptide consisting of four amino acids, L-alanine, D-glutamic acid, L-lysine and D-alanine. To strengthen the peptidoglycan structure further these short peptides are also cross-linked by a glycine pentapeptide.

In Gram-negative organisms the cell-wall structure is simpler, consisting of a single layer of peptidoglycan sandwiched between the inner plasma membrane and an outer lipid membrane containing large lipoproteins, which extend through the peptidoglycan layer and are anchored in the plasma membrane.

A number of antibiotics, including some of the non-ionophore gut active growth promoters, inhibit bacterial growth by disrupting the assembly of the peptidoglycan layer. As Gram-positive microorganisms contain a more complex peptidoglycan structure they are more susceptible to the effects of some antibiotics than Gram-negative bacteria. The synthesis of peptidoglycan and the sites of action of non-ionophore antibiotics are shown in Figure 30.14.

The initial stage in peptidoglycan synthesis is the formation of a UDP-N-acetylmuramylpentapeptide. The second step in the synthesis involves the transfer of the N-acetylmuramylpentapeptide from UDP to the isoprenoid compound undecaprenol phosphate and the subsequent formation of the dissacharide repeating unit by the addition of N-acetylglucosamine. The final step before transport of this building block across the plasma membrane is the sequential addition of five glycine residues to the pentapeptide. Undecaprenol phosphate plays a crucial role in the translocation of this highly polar building block across the plasma membrane. On the outer surface of the plasma membrane, the peptidodisaccharide is transferred from the undecaprenol phosphate to the reducing end of an existing peptidoglycan chain.

Avoparcin (Figure 30.13) disrupts peptidoglycan synthesis at two stages. Firstly, it inhibits the extracellular transfer of the peptidodisaccharide to the existing peptidoglycan. Secondly, at higher concentrations avoparcin also inhibits the transfer of the N-acetylmuramylpentapeptide from UDP to undecaprenol phosphate. Avoparcin was banned as a feed additive in the countries of the European Union in 1997.

When released from the peptidodisaccharide the undecaprenol is in the pyrophosphate form, and must undergo dephosphorylation to the monophosphate before it can be recycled to the cytoplasm where it can react with another UDP-N-acetylmuramylpentapeptide. Bacitracin inhibits this recycling process. Bacitracin also inhibits other bacterial reactions which use undecaprenol phosphate. It also forms a complex with the eukaryotic equivalent of undecaprenol, dolichol, which is used in the synthesis of cell membrane glycoproteins. Much higher concentrations of bacitracin are required to affect eukaryotic cells compared with prokaryotic cells.

Flavomycin is a phosphorus-containing glycolipid with considerable structural similarity to undecaprenol phosphate which interferes with the synthesis of proteoglycans. It inhibits the glycosylation reaction which adds *N*-acetylglucosamine to the undecaprenol phos-

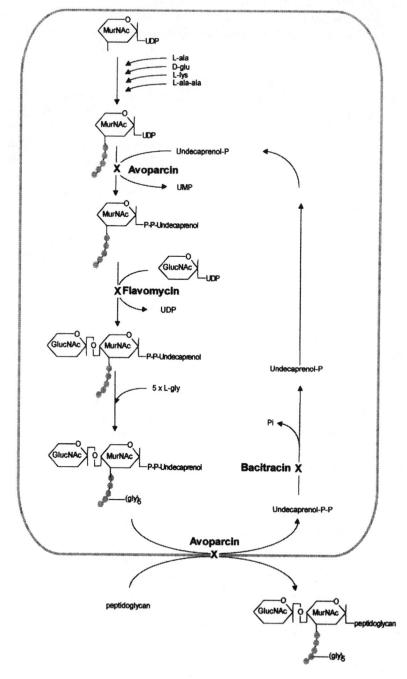

Figure 30.14 Site of action of the non-ionophore antibiotics on the synthesis of peptidoglycan.

phate-linked *N*-acetylmuramylpentapeptide, probably by substituting for undecaprenol phosphate and forming an unreactive analogue. The outer membranes of Gram-positive bacteria form an effective permeability barrier to flavomycin, and this type of bacterium is less susceptible to flavomycin action.

Virginiamycin and tylosin act via a different mechanism: they are both potent inhibitors of protein synthesis. Virginiamycin consists of two components, virginiamycin M and virginiamycin S. The M and S components bind tightly to the 50S subunit of prokaryotic ribosomes. The M unit appears to block the P and A binding sites on the ribosome to prevent binding of the aminoacyl-tRNAs and the subsequent peptidyltransferase reaction. The mode of action of the S unit is not fully understood but is appears to be different from that of the M unit. In addition, virginiamycin causes the breakdown of polyribosomes (see Chapter 21 for detailed coverage of protein synthesis). Tylosin has a similar mode of action. It binds to the 50S ribosomal subunit and is a potent inhibitor of ribosomal binding of aminoacyl-tRNA and the tRNA linked to the nascent peptide. It also inhibits the peptidyltransferase reaction.

The growth improvements seen in animals in response to the use of these antibiotics are due mainly to changes in nutrient availability in the gut, but the underlying cause of these changes is the altered microbial population in the digestive tract. In ruminants, the shift in fermentation pattern increases the availability of propionate for hepatic gluconeogenesis and thus increases the supply of glucose to the peripheral tissues. The increased supply of propionate and/or glucose may alter the profile of tissue metabolism possibly mediated via insulin and growth hormone. The decrease in methane production conserves dietary energy, effectively increasing the metabolizable energy content of the diet. Ionophores and non-ionophores usually reduce rumen ammonia concentration indicating a decrease in the degradation and deamination of dietary protein sources. Although the total amount of protein degraded in the small intestine does not increase, the composition of the protein may change due to reduced microbial protein synthesis in the rumen and the escape of dietary protein from rumen degradation. The resulting shift in the composition of amino acids available for absorption in the small intestine is considered to improve the overall efficiency of protein deposition.

The use of antibiotics as growth promoters in farm animals is a subject of some controversy. There is growing concern about the increased incidence of bacterial pathogens which are resistant to a broad spectrum of common antibiotics. Many of these microorganisms also occur in animal species and may have developed some antibiotic resistance due to the practice of adding antibiotics to animal feedstuffs.

31.1 INTRODUCTION

The majority of the world's supply of milk comes from cows, with water buffalo, goats, sheep and camels producing much smaller amounts. The mammary gland of a high-yielding animal has an extraordinary rate of metabolic activity, which may be greater than the rest of the whole of the animal's metabolism put together. This metabolic activity is initiated very rapidly: the cow is able to make the transition from a non-lactating state to a high milk yield within a very few days. High-yielding dairy cows are capable of producing well over 10 tonnes of milk within a period of 10 months. During this time the rate of production is not constant but builds up to a peak over the first few months and then declines steadily to less than a half of maximum. In early lactation, the cow's food intake is insufficient to meet the requirements for milk production. Thus, the cow must contribute metabolites from her own body reserves to enable milk secretion to continue. Food intake increases during lactation, so in the later months the cow is able both to lactate and to replace the lost body reserves (Figure 31.1).

The peak yield may exceed 50 kg per day which represents a daily secretion of about 1.65 kg of protein, 1.95 kg of fat and 2.25 kg of lactose. There is a correspondingly large demand for substrates and energy to fuel the synthetic processes. A comparison of these metabolic activities with those of meat animals may serve to put them into context. A very rapidly growing beef animal of perhaps

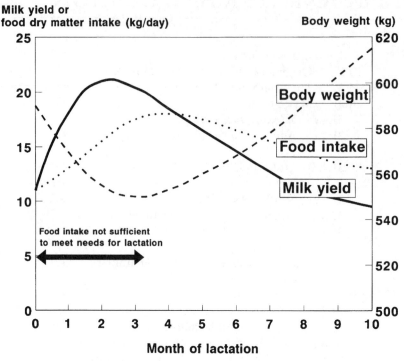

Figure 31.1 Lactation curve for a dairy cow. The milk yield increases during the first 2 months of lactation and then declines. In the early months of lactation, food intake is insufficient to meet the needs of lactation. Body weight initially declines but recovers in late lactation.

350 kg live weight may add a maximum of 0.3 kg of protein to its body mass each day. A sheep of 30 kg live weight may gain 50 g protein per day. Assuming that the animal's metabolic activity is related to its metabolic body weight ($W^{0.75}$), then the protein deposited by fast-growing ruminants such as cattle or sheep is in the order of 3.5 g protein $kg^{-0.75}$. By contrast, the high-yielding dairy cow has a body weight of about 600 kg which means that she is secreting 13.6 g protein $kg^{-0.75}$ – nearly four times as much.

There is a further advantage to protein production from milk; in a meat animal a large proportion of the protein is in the form of collagen or in tissues that are not normally eaten, and is thus not available to the eventual consumer. In dairy production it can be assumed that all of the protein is available for use.

31.2 ORIGINS OF THE COMPONENTS OF MILK.

The mammary glands of animals vary greatly in their outward size, shape, number and even anatomical location, but their internal structures are very similar. The functioning gland consists of a series of secretory cells connected by fine ducts to larger ducts which communicate with even larger ones until they meet the teat or nipple area. Figure 31.2 shows the secretory cells, organised in clumps (alveoli) surrounding a space (the lumen) into which the milk is initially secreted prior to transport through the duct system. Each alveolus is surrounded by fine blood vessels, through which the precursors required for the synthesis of milk are delivered. Also surrounding the alveolus is a network of contractile cells (myoepithelial cells) which, in response to hormonal signals, can squeeze the whole alveolus and thus push the milk out of the central lumen and towards the ducts.

The secretory cells (Figure 31.3) have a highly organized internal structure with a well developed endoplasmic reticulum, numerous mitochondria, ribosomes and large Golgi bodies. The cell membrane facing the lumen (the apical membrane), has numerous projections (villi) which increase the surface area. At the opposite end of the cell the basal membrane lies in contact with the walls of capillary blood vessels.

31.3 THE ORIGIN OF LACTOSE

The disaccharide lactose is almost unique to milk. For the individual steps of the synthetic pathway, see the section on disaccharide synthesis (Chapter 18). The enzyme that catalyses the final step, lactose synthase, is situated on the inner surface of the membrane of the Golgi bodies (Figure 31.4). The active form of the enzyme consists of galactosyl transferase whose specificity is modified by the presence of α-lactalbumin (see Figure 18.9). Lactose synthase has a requirement both for Mn^{2+} ion and another divalent cation (X^{2+} in Figure 31.4).

Glucose enters the Golgi body freely from the cytoplasm via pores, but the other precursor, UDP-galactose, has to be actively carried inwards. After the formation of lactose, the liberated UDP is hydrolysed within the Golgi body to UMP and P_i, the UMP is actively exported to the cytoplasm, and the P_i crosses the membrane through a pore. This completes a uridine phosphate cycle.

The lactose produced is stored in the lumen of the Golgi body ready for transport to the apical membrane and secretion into the alveolar lumen.

In comparison with the other constituents of milk, lactose has a relatively low molecular weight and is readily soluble in water. Therefore one of its effects is to raise the osmotic pressure to isotonic level. During synthesis, the local concentrations of lactose must be much higher than they are in the final product, which would render the contents of the secretory cell hypertonic. This unsatisfactory state is avoided by the fact that the high concentrations of lactose are retained within the structure of the Golgi bodies before transport to the apical membrane in vesicles. As lactose is the principal osmoregulator of milk, its secretion controls

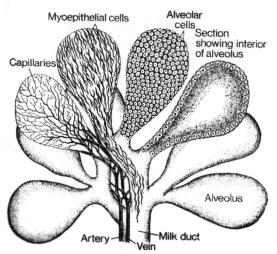

Figure 31.2 A cluster of alveoli in the mammary gland of a goat. (Reproduced with permission from Cowie, A.T. (1972) Lactation and its hormonal control, in *Reproduction in Mammals* (eds C.R. Austin and R.V. Short), Cambridge University Press, Cambridge, UK.)

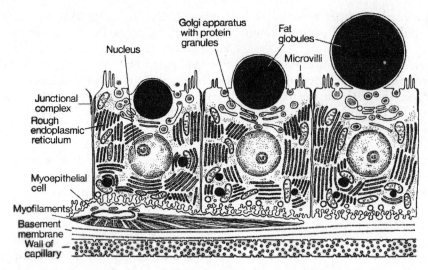

Figure 31.3 The ultrastructure of three alveolar cells and a myoepithelial cell. (Reproduced with permission from Cowie, A.T. (1972) Lactation and its hormonal control, in *Reproduction in Mammals* (eds C.R. Austin and R.V. Short), Cambridge University Press, Cambridge, UK.)

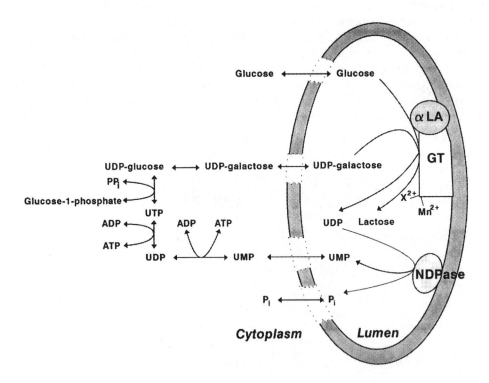

Figure 31.4 Role of the Golgi body in the synthesis of lactose. The lactose synthase complex [glycosyl transferase (GT) and α-lactalbumin (αLA)] and nucleotide diphosphatase (NDPase) are bound to the inner membrane of the Golgi body.

the amount of water secreted and hence the quantity of milk produced.

31.4 MILK PROTEINS

31.4.1 THE ORIGIN OF MILK PROTEINS

The proteins of milk are synthesized within the gland from amino acids taken up from blood. Small amounts of blood proteins, such as immunoglobulins and serum albumin, can be detected in milk. Excessive amounts of these, particularly of serum albumin, are usually associated with 'leakage' through a mammary gland that has been damaged in some way, possibly by the diseases known collectively as mastitis.

In most species the commonest milk proteins are the caseins, which are distinguished in two ways. Firstly, they have isoelectric points in the region of 4 and precipitate under mildly acidic conditions: this property forms the basis of the production of fermented milk products such as yoghurt and cheese. The second unusual feature is that they are highly phosphorylated. In milk, the negative charges on the phosphate groups are neutralized by the presence of calcium (Ca^{2+}) ions. This means that, in addition to supplying the young growing animal with a dietary supply of amino acids, casein also contributes much to its needs for calcium and phosphorus. The caseins of cows' milk can be separated by electrophoresis into at least five bands, denoted; α_{S2}-, α_{S2}-, β-, γ- and κ-. The α_{S1}- and β- fractions are present in the largest amounts (see Figure 31.5).

If the pH is reduced to approximately 4 then all of the caseins are precipitated. The proteins that remain in solution are known as the milk serum or whey proteins. The major protein of the whey fraction is β-lactoglobulin. This has an important role in the nutrition of the young due to its relatively high levels of sulphur amino acids, which help to balance the low and potentially limiting levels found in casein. There is also a suggestion that β-lac-

toglobulin may function as a carrier for some of the vitamins.

α-lactalbumin is the specifier protein for lactose synthesis, and thus in its absence no lactose and hence no milk can be produced. Milk fat globules form at the endoplasmic reticulum and migrate to the apical membrane. As they cross this membrane they acquire a coating of cell membrane material. In this way, membrane glycoproteins enter milk.

31.4.2 AMINO ACID SUPPLY TO THE MAMMARY GLAND

Work has shown that the proportions of amino acids absorbed from the blood are different from those found in milk proteins, so within the gland there must be a large degree of inter-conversion of amino acid. For instance, in goats the mammary uptake of arginine is many times the amount which is secreted in milk but more serine is passed to the milk proteins than can be accounted for by uptake from blood. Both of these are non-essential amino acids. The essential amino acids are either supplied in amounts equal to their secretion in milk, or are oversupplied. Amino acids are taken up by the secretory cells by active transport.

31.4.3 PROTEIN SYNTHESIS

The synthesis of the proteins of milk follows the normal, ribosomal pattern (see Chapter 21) –what is unusual about the mammary gland is that a very high proportion of the ribosomes are attached to the membranes of the endoplasmic reticulum. Milk proteins are synthesized with a 'signal' peptide of 15–21 amino acids at the amino-terminal end of the growing chain. This peptide is recognised by receptors on the membrane of the endoplasmic reticulum, allowing it to be transported through the membrane with the simultaneous removal of the signal sequence by a protease. It is at this point that any sugars are added to form the cell membrane glycoproteins.

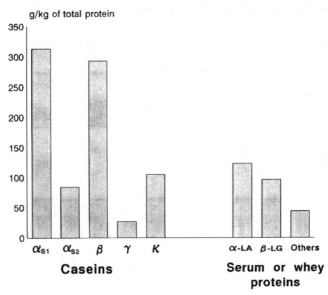

Figure 31.5 Quantities of individual protein fractions in cow's milk.

Caseins have a high concentration of phosphate groups which are attached to serine residues in the polypeptide chain, after its transfer to the Golgi body, by the activity of a casein kinase enzyme. κ-casein is also a glycoprotein and probably receives its carbohydrate moiety within the Golgi body.

31.5 THE FATS

Most of the fat in milk is in the form of triacylglycerols, together with far smaller amounts of phospholipid, mono- and diacylglycerols, and steroids. The fatty acids of milk have their origins in two distinct sources:

- synthesis of fatty acids within the mammary gland itself (sometimes called synthesis *de novo*);
- uptake from the blood of fatty acids that have been synthesized or stored in other tissues.

In general, about 40% of fatty acids are synthesized in the mammary gland and 60% transported in blood as pre-formed units.

31.5.1 SYNTHESIS *DE NOVO*

Fatty acid synthesis in the cow takes place by the elongation of a primer molecule of acetyl-CoA by the sequential addition of pairs of carbon atoms from malonyl-CoA (see Chapter 19). In non-ruminants the major part of the carbon for fatty acid synthesis is supplied by acetyl-CoA which has been formed from glucose by glycolysis. On the other hand, ruminants gain much of their nutrition in the form of the volatile fatty acids produced in the rumen, particularly acetate, propionate and butyrate. Only one of these, propionate, can be used as a supplier of glucose, but the other two are excellent sources of acetyl-CoA and can easily be used as precursors for fatty acid synthesis. Acetate is absorbed through the rumen wall and circulates to tissues such as the mammary gland, where it is activated by the formation of its CoA ester.

$$\text{acetate} + \text{CoASH} + \text{ATP} \rightarrow \text{acetyl-CoA} + \text{AMP} + \text{PP}_i$$

The reaction is catalysed by acetyl-CoA synthetase.

Butyrate is absorbed into the rumen wall but is there transformed to 3-hydroxybutyrate (one of the ketone bodies) before it is passed to the blood stream for circulation to other tissues. The activation of butyrate to butyryl-CoA and its transfer to the mitochondrial matrix are similar to those shown in Figure 13.1. In the matrix, two hydrogen atoms are removed to form a double bond and yield Δ^2 trans-butenoyl-CoA. The reaction uses FAD as the acceptor of the hydrogens. An enoyl hydratase adds water across the double bond to give L-3-hydroxybutyryl-CoA, which loses its coenzyme group to liberate the 3-hydroxybutyrate.

Within the mammary gland, 3-hydroxybutyrate can be oxidized to yield acetoacetyl-CoA which can be cleaved, with the addition of a further molecule of coenzyme A, to give two molecules of acetyl-CoA. This acetyl-CoA and that produced from acetate are precursors for fatty acid synthesis.

31.5.2 UPTAKE OF FATTY ACIDS FROM BLOOD

Despite the fact that there is a high level of non-esterified fatty acids (NEFA) in the blood of lactating cows, these are not used as a source of the fatty acids for milk. Rather it is triacylglycerols bound to proteins and circulating as very fine emulsions in the form of either chylomicrons or very low density lipoproteins (VLDLs) that supply the pre-formed fatty acids. The triacylglycerol components of these lipoproteins arrive at the capillaries of the mammary gland where they are hydrolysed and the products, glycerol and fatty acids, immediately taken up into the secretory cell. Once in the secretory cell, the fatty acids move to the endoplasmic reticulum where they are re-esterified to triacylglycerols.

31.5.3 MODIFICATIONS TO FATTY ACIDS IN THE MAMMARY GLAND

Fatty acids which are either synthesized *de novo* in the mammary gland or absorbed from blood may be modified before they are incorporated into the triacylglycerols of milk. In ruminants, the most important changes are the desaturation at the Δ^9 position of palmitate (C16:0) and stearate (C18:0) to yield palmitoleate (C16:1) and oleate (C18:1), respectively (see Chapter 19). In the mammary glands of non-ruminants such as the rat there is a large-scale elongation of palmitate to stearate, but this does not appear to operate to any great extent in the mammary gland from ruminants.

The distribution of fatty acids of milk fat

Milk fat is unusual in the very wide range of fatty acids that it contains. These can be divided into three main groups: short, medium and long, on the basis of chain length.

- Short- and medium-chain fatty acids (C4:0 – C6:0 and C8:0 – C12:0). These acids are not present in circulating lipids and must be synthesized in the gland. As under most circumstances fatty acid synthesis produces palmitate (C16:0), the premature release of short- and medium-chain acids from the fatty acid synthase demonstrates the presence of a specialized thioesterase in mammary tissue.
- Long chain fatty acids (C14 and above). The majority of these acids are supplied from the triacylglycerols of blood.

Lipoprotein lipase provides the mechanism for the uptake of circulating fatty acids of both dietary and endogenous (adipose tissue) origin for use in the synthesis of milk fat. It is interesting to note that in mammals the activities of lipoprotein lipase in adipose tissue and mammary glands change in a reciprocal manner. During lactation there is a marked increase in the activity of this enzyme in the mammary gland and a decrease in adipose tissue. This change is probably coordinated by changes in hormonal status to allow repartitioning of lipid to the mammary gland at a time when substrates for milk triacylglycerol synthesis are required.

31.6 THE SUPPLY OF ENERGY IN THE MAMMARY GLAND

In looking at the synthesis of the components of milk, it is tempting to think only of the pathways in which the carbon skeletons of the lactose, fats and protein are formed and to ignore the supply of energy, which is equally important. For lactose synthesis, ATP is required for the activation of galactose to form UDP-galactose. Similarly, in protein synthesis there are ATP requirements associated with the processes of translation and transcription. The source of this ATP is the TCA acid cycle and the electron transport chain. The synthesis of triacylglycerols also requires energy in the form of ATP to drive a number of reactions.

There is, however, another large energy requirement, this time for reducing power in the form of NADPH, for fatty acid synthesis.

Fatty acid synthetase is located in the cytoplasm which is where the NADPH is needed. Most of the reducing power produced during catabolism is in the form of NADH, and the greater part of this is formed during the operation of the TCA acid cycle in the mitochondrion. There are two ways in which the cytoplasm can gain NADPH. One is to transform reducing power in the form of mitochondrial NADH into cytoplasmic NADPH, and the other is to use a pathway such as the pentose phosphate pathway, which is capable of producing NADPH in the cytoplasm. Both of theses routes are used in the mammary gland although their relative importance probably differs between species. The pentose phosphate pathway requires glucose as its precursor, whereas mitochondrial pathways use acetyl-CoA which can be formed from acetate.

At the start of lactation there are large increases in the activity of fatty acyl synthetase which are paralleled by a rise in the activity of the dehydrogenases of the pentose phosphate pathway.

31.7 METABOLISM IN LACTATION.

The start of lactation requires some enormous changes in the metabolic activity of the dairy cow. During the early months of lactation the cow is unable to eat enough food to supply all of the nutrients that leave in her milk, resulting in a drop in body weight (Figure 31.1). This might be regarded as a prime feature of a dairy cow as compared with a beef cow which, in response to a shortage of food, will tend to reduce her milk yield so as to ensure a metabolic balance.

A cow has to partition nutrients between the needs for her own body metabolism and the requirements of lactation. This process of partition is carefully controlled in a dairy cow. As lactation starts the metabolism of tissues changes:

- adipose tissue – all aspects of lipogenesis are reduced, lipolysis is stimulated;
- liver – large increase in the production of glucose and 3-hydroxybutyrate;
- skeletal muscle – protein degradation increased, glucose utilization reduced.

All of the changes are designed to reduce the use of biochemical precursors in other tissues and to pass them to the mammary gland. The control of these processes must come from the activities of the animal's endocrine system.

31.7.1 ENDOCRINE CONTROL OF LACTATION

One point to note at the beginning of this discussion is that there are quite large differences in the ways in which hormones control lactation in ruminants and non-ruminants. Much of the research on endocrine control of lactation has been performed on rats but, as lactation in rodents is not a common economic activity, discussion here will be limited to ruminants. Hormones are involved in two main areas in the process of lactation:

- lactogenesis – the process by which a non-lactating gland is prepared for lactation and gains the ability to synthesize the specific products of milk, e.g. α-lactalbumin;

• galactopoesis – the process by which the body regulates the continuing lactation and passes precursors from the rest of the body to the gland.

Lactogenesis is a complex process and is beyond the scope of this book, the interested reader is referred to specialized texts on lactation. Galactopoesis, however, is mainly a process for the redirection of biochemical resources.

There are two parts to the activity of any hormonal control mechanism. The first is the action of the endocrine tissue in releasing the hormone into the bloodstream. Equally important is the responsiveness of the target tissue to the hormone in question, and both of these factors have to be considered together.

Metabolic hormones

Lactation is affected by the levels of many hormones, the principal ones are those normally regarded as metabolic hormones, although reproductive hormones may also exert some influence on milk synthesis. Insulin, glucagon, growth hormone (somatotropin) and the insulin-like growth factors (IGF-1 and IGF-2) are intimately associated with the control of milk synthesis. Other hormones such as those of the thyroid (triiodthyronine and thyroxin), the adrenal cortex (cortisol and corticosterone), and the adrenal medulla and sympathetic nervous system (adrenaline and noradrenaline) may be needed for the continuation of lactation and have effects within the general metabolic milieu. Prolactin is required for lactogenesis and might be expected to have an important role in galactopoeisis. It is needed for the maintenance of lactation in non-ruminants but appears to have a much reduced role in the ruminant.

Removal of growth hormone from blood leads to a large decrease in milk production in ruminants. Supplementation of GH leads to large increases in milk output. This hormone has no direct effect on the metabolic activity of isolated mammary tissue. Thus its action must be expressed through some other factor, probably IGF-1, which is released from many tissues (liver, muscle, bone) in response to increases in growth hormone level. The mammary gland in the ruminant has receptors for IGF-1, and infusing IGF-1 into the blood of goats leads to an increase in milk yield. The local effects of IGF-1 include:

• stimulation of lipolysis in adipose tissue leading to an increased availability of fatty acids for milk fat synthesis;
• stimulation of liver gluconeogenesis (mainly from propionate);
• stimulation of glycogen breakdown in muscle.

These factors all allow a greater part of the animal's resources to be used in the mammary gland.

Insulin has a major role in the partitioning of nutrients during lactation. However, in ruminants insulin has little or no direct effect on the metabolism of mammary gland tissue. During lactation in ruminants the level of insulin in blood is suppressed, leading to a reduction of its action on other tissues. Insulin has a central role in promoting the uptake of substrates by muscle and other peripheral tissues so that the decrease in insulin will lead to:

• reduced uptake of glucose in insulin-sensitive tissues and an increased availability to the insulin-insensitive mammary gland;
• decrease of lipogenesis in adipose tissues leaving more precursors available to the mammary gland.

31.8 MANIPULATION OF LACTATION

Milk production from cows is a commercial activity which usually has high costs (feed etc.) but results in a product of high value. The margin between these two figures may be very small, and it is essential that the producer is able to manipulate lactation in such a way as to optimize the returns. Apart from changing the genetic composition of the herd and the environmental conditions, there are two main mechanisms for control. The first of these is to modify the diet and the second is to adjust the metabolism of the cow by the use of exogenous hormones.

31.8.1 DIETARY MANIPULATION OF LACTATION

The quantity of milk produced by the cow is determined largely by the amount and quality of nutrients ingested in the diet. In early lactation the cow uses some of her body reserves as a source of nutrients, particularly amino acids and lipids, but the extent of this contribution is quite limited in comparison to the quantity of metabolites secreted by the high-yielding animal. Nutrient intake depends upon two factors: the amount of food dry matter eaten per day, and the concentration of nutrients within that dry matter (nutrient density).

Rumen fermentation

The food intake of a ruminant is determined, in large part, by the ability of the rumen to degrade the feed quickly, and this depends upon the rate of fermentation. Diets containing large amounts of soluble carbohydrate (either starch or sucrose) ferment rapidly, however they also lead to the development of acid conditions in the rumen and the production of lactate. These diets tend to have the highest nutrient densities (in terms of metabolizable energy per kg of food). The feeding of such diets produces a fermentation pattern in the rumen which is biased towards propionate production and away from acetate (see Chapter 16). Such conditions in the rumen lead to a marked decrease in the proportion of fat in milk. An increase in the molar percentage of propionate from 20–40% would be expected to lead to a halving of the quantity of fat in milk. In many countries there is a legal requirement for milk to contain a minimum content of fat (generally between 30 and 35 g fat per kg milk). When high rumen propionate levels are present, the milk may fail to meet this minimum standard.

High levels of dietary soluble carbohydrate also lead to the formation of a rumen microflora with a lowered ability to hydrolyse cellulose. The rate of degradation in the rumen is reduced, as is the intake of food. Accompanying these changes are reductions in rumen pH that may lead to the development of acidosis. Addition of extra buffering capacity to the rumen has been shown to alleviate some of the deleterious effects of feeding high-concentrate diets. A common technique is to add 1–2% of sodium bicarbonate to dairy rations, particularly in conditions where heat stress is also a potential problem.

Added ionophores (see Chapter 30) stimulate rumen function and have been widely used in meat animals. The high-propionate fermentations that result are advantageous to a meat animal because they lead to conditions under which tissue growth is favoured. Unfortunately, in dairy animals this leads to an unacceptable reduction in the fat content of milk and a diversion of nutrient resources towards body tissue rather than the mammary gland.

One approach which has proved valuable is the addition of live yeast (*Saccharomyces cerevisiae*) to the feed. This has been shown to stimulate the ruminal breakdown of cellulose, particularly in high-yielding animals that are receiving diets with a high content of digestible (chiefly α-linked) carbohydrate. The improvements have been linked to two effects:

- improved ability to buffer rumen pH so as to reduce the time during which it becomes very acid;
- lowered oxygen tension in the rumen consequent on the yeast's ability to 'scavenge' oxygen, resulting in conditions which favour the strictly anaerobic bacteria responsible for cellulolysis.

The low levels of fat in milk are caused by a reduction in the uptake by the mammary gland of preformed fatty acids from blood. There is a consequent increase in the uptake of these acids by adipose tissue, leading to a repartitioning of metabolites between milk secretion and tissue growth. The effect of propionate appears to be mediated by insulin through changes in the relative activities of the lipoprotein lipases in adipose and mammary tissues.

Protein supply

It is well accepted that the yield of milk in a dairy cow is greatly influenced by the quantities of amino acids arriving at the mammary gland through the blood. The first limiting amino acid is generally lysine, with methionine the second. In ruminants there are two possible sources of supply. One is the microbial protein synthesized in the rumen, and the other is the dietary protein which is undegraded in the rumen but which is digested in the small intestine (digestible undegraded protein, DUP, also known as digestible by-pass protein).

In ruminants at a low level of production, the quantity of microbial protein synthesized in the rumen will be sufficient to satisfy the needs of the animal. Microbial protein is normally regarded as being a high-quality protein source in terms of its spectrum of amino acids. As the level of animal production increases, so too does the need for a supply of essential amino acids from DUP. Where DUP represents only a small part of the animal's amino acid supply, its composition is not very critical. Any deficiencies or imbalances in amino acid supply are masked by the contribution of microbial protein. In the high-yielding dairy cow, DUP may contribute over half of the needs for amino acids, and at this point any deficiencies in amino acid in the protein start to exert a negative effect upon the composition of the supply to the mammary gland. Work in a number of centres has shown that increasing the supplies of lysine and methionine can lead to increases in milk yield.

31.8.2 MANIPULATION OF MILK PRODUCTION BY EXOGENOUS HORMONES

It has been known for many years that administration of bovine growth hormone leads to increases in milk yield (section 31.7.1). This was of theoretical interest when the only source of the hormone was its extraction and purification from pituitaries removed at slaughter. With the development of recombinant gene techniques, it has been possible to produce a synthetic product normally denoted bST (bovine somatotropin). The hormone is a protein and would therefore be degraded if given orally, for this reason it is administered daily by injection.

Typically, increases of between 25 and 40% in milk yield may be expected in cows treated daily with bST and fed on diets with a relatively high energy content (see Figure 31.6). Part of the effect of bST is due to increases in the food intake of treated cows, but there are also increases in the efficiency with which dietary energy is used for milk production. A further effect of bST is to change the partition of nutrients so that there is a diversion of resources away from tissue deposition into milk production.

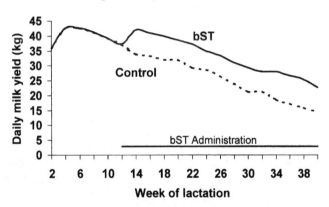

Figure 31.6 Typical effects of administering recombinant bovine somatotropin (bST) on milk yield in a cow.

MUSCLE AND MEAT 32

32.1 INTRODUCTION

Much of animal agriculture is devoted to the production of meat for which skeletal muscle provides the raw material. Although meat is derived from muscle there are important biochemical differences between the physiologically active tissue and the commercial commodity, and changes to the carcass after slaughter have to be carefully controlled to ensure the provision of a hygienic product of high eating quality.

32.2 BIOCHEMISTRY OF MUSCULAR CONTRACTION

The characteristic striated appearance of skeletal muscle is due to the arrangement of muscle fibres within the tissue. Within striated muscle it is rather difficult to define what is meant by an individual cell. The fibres of the muscle are arranged within a plasma membrane called the sarcolemma; each of these membranes may contain many flattened nuclei so that it is difficult to say whether each of these envelopes actually represents a separate cell or several cells; this distinction becomes important in looking at processes of growth (Chapter 30).

Most of the space within the sarcolemma is occupied by the actual fibres, the myofibrils, which are the distinctive operating mechanism of muscle fibres. Each of the fibres can be seen as a series of repeating units known as sarcomeres (Figure 32.1), at each end of which is a dark band known as the Z disk. Closer examination of the sarcomere shows that it is made up of two different types of filament: at each end, and attached to the Z disk, are thin filaments, whilst interdigitating between these are the thick filaments (Figure 32.1b). The process of contraction in the muscle is illustrated in Figure 32.1c, where it can be seen that the thick and thin filaments slide past one another so as to shorten the distance between the Z disks.

The thin filaments are predominantly made up of the protein actin, whereas the major part of the thick filaments is myosin. This protein is

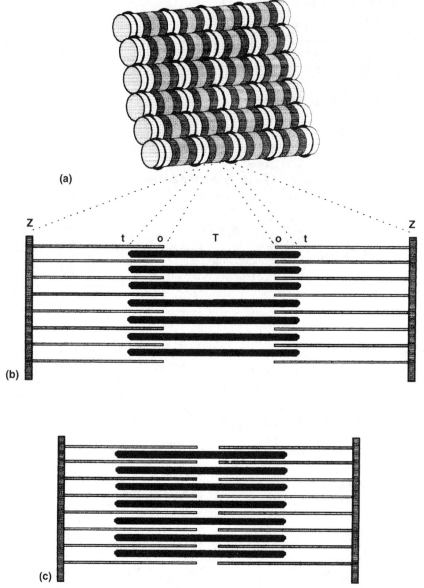

Figure 32.1 Structure of muscle. The striated structure of muscle (a) is composed of a series of repeating subunits (sarcomeres). Each sarcomere (b) has a Z disc at each end to which are attached the thin filaments, t. Interdigitated between the thin filaments are thick filaments, T. The region of overlap between the fibres is marked o; contraction of muscle fibres is achieved by an increase in the degree of overlap between fibres, (c).

known as myosin II, to distinguish it from myosin I which is found in some cells which are motile but not muscular. The process of muscle contraction is therefore to be found in the relationship between the individual molecules of actin and myosin.

32.2.1 STRUCTURE OF THICK FILAMENTS

Individual molecules of myosin are composed of two heavy subunits (molecular weight about 200 kDa) and four light ones (molecular weight about 20 kDa). The heavy subunits each have a globular 'head' and a long 'tail' region; the tertiary structure of the polypeptide chain is almost completely that of an α-helix. The tails of the heavy subunits are coiled together (Figure 32.2a). Many of these individual molecules polymerize to form a single heavy filament in such a way that all the globular regions are clustered at each end of the filament (Figure 32.2b).

Purified myosin can be shown to have ATPase activity associated with the globular region. Myosin also has the ability to bind to actin, the principal protein of the light filaments within the myofibril.

32.2.2 STRUCTURE OF THIN FILAMENTS

Actin is formed from a globular protein with a molecular mass of about 42 kDa which is known as 'G actin'. This simple subunit aggregates into long chains that come together as pairs (F actin) and comprise the 'backbone' of the light filaments (Figure 32.3), accounting for about 67% of its weight. Actin bears binding sites on its structure that attach to complementary sites on the globular part of myosin; the sites are all orientated in the same direction.

Attached along the length of the F actin chain is another protein, tropomyosin. At intervals along the fibre are the molecules of the troponin complex; troponin T, troponin I, and troponin C. Tropomyosin and troponin are integral parts of the mechanism that controls the action of the muscle.

32.2.3 MECHANISM OF MUSCLE MOVEMENT

The sliding action of muscle movement is achieved by changes in the way in which myosin binds to actin. There are two important requirements for the process of movement:

- the binding between actin and the head of myosin is reversible and depends upon the presence or absence of ATP and ADP;
- the shape of myosin, particularly of the 'head' region, can change depending on its chemical state.

The mechanism is illustrated in Figure 32.4. The process starts with the individual 'head' units of the myosin each lying close to a binding site on an actin molecule of a thin filament. At this time a molecule of ATP is bound to the globular region of the myosin. There is no binding between the myosin and actin. The movement starts with the hydrolysis of the ATP to ADP and orthophosphate (P_i) which both remain attached to the myosin. In this combination the myosin binds firmly to the corresponding site on actin. Once the bond is formed, the myosin molecule changes shape by tilting the 'head' part and pushes the actin in one direction and the myosin in the oppo-

a.

b.

Figure 32.2 Structure of the thick filaments in muscle. A molecule of myosin (a) is composed of two heavy subunits, each of which is composed of a long 'tail' region and a globular 'head'. The head units also carry the light subunits which include the ATPase activity. Myosin molecules cling together to form the thick filaments (b) with the tail units closely woven into a single strand and clusters of globular units at each end.

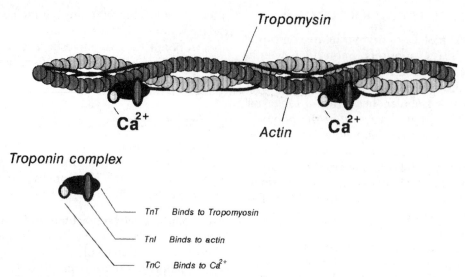

Tropomysin

Ca^{2+}

Actin   Ca^{2+}

Troponin complex

TnT   Binds to Tropomyosin

TnI   Binds to actin

TnC   Binds to Ca^{2+}

Figure 32.3 Structure of actin. The main part of the molecule consists of two twisted strands (F actin), each composed of a chain of globular (G actin) subunits. Associated with actin is the troponin complex which controls the action of muscle in response to the presence or absence of Ca^{2+}.

site one. After the movement, the bound ADP and P$_i$ are liberated and another molecule of ATP is added to the myosin. The myosin is now freed from the actin and the myosin 'head' finds itself opposite the next binding site on the actin molecule. The cycle is now ready to start again.

32.2.4 CONTROL OF MUSCLE MOVEMENT

Obviously, muscle activity must be one of the best coordinated and most closely regulated of all processes. The movement of muscle must therefore be under the control of a series of very precise mechanisms. Additionally, muscle tissue has an important role in heat production which is not related to coordinated movement.

The physiological trigger to muscle contraction is Ca^{2+} which changes the interaction of myosin and actin. The role of Ca^{2+} is mediated through the two other groups of proteins in the thin filament: tropomyosin, which is a long fibrous protein running alongside the double filament of actin; and an assembly of

three linked proteins known as the troponin complex. In the absence of Ca^{2+} the troponin complex and tropomyosin together block the binding sites on actin and prevent the interaction with myosin.

Within the muscle cell, the supply of Ca^{2+} is regulated by a series of internal membranes known as the sarcoplasmic reticulum, which concentrate and bind it in combination with the protein calsequestrin until required. The calcium concentration in the sarcoplasmic reticulum is greater than 10^{-3} M. The calcium is held within vesicles known as calciosomes. The trigger for muscle movement is the release of Ca^{2+} from this system via a protein channel, ryanodine (see section 32.4.5).

There are always two important parts to any control system – the part that turns it on and the part that turns it off. In the absence of any stimulus to muscle movement, the supply of Ca^{2+} in the cell is rapidly reduced to less than 10^{-6} M by Ca^{2+}-ATPase similar to Na$^+$-K$^+$ ATPase. The system moves two Ca^{2+} ions for every ATP molecule that is hydrolysed. This use of ATP in the removal of the muscle stim-

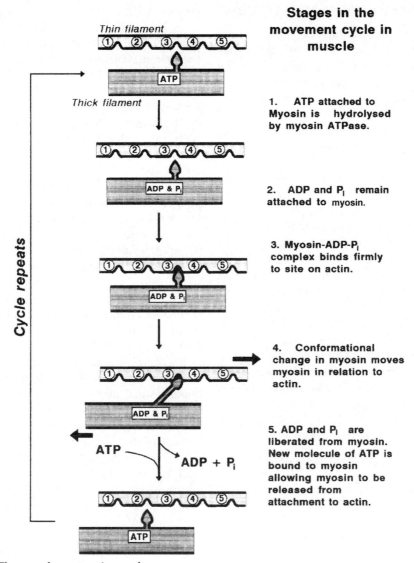

Stages in the movement cycle in muscle

1. ATP attached to Myosin is hydrolysed by myosin ATPase.

2. ADP and P_i remain attached to myosin.

3. Myosin-ADP-P_i complex binds firmly to site on actin.

4. Conformational change in myosin moves myosin in relation to actin.

5. ADP and P_i are liberated from myosin. New molecule of ATP is bound to myosin allowing myosin to be released from attachment to actin.

Figure 32.4 The muscle contraction cycle.

ulus adds further to the amount of energy that is required for muscular work.

The stimulus to Ca^{2+} release from the sarcoplasmic reticulum is transmitted by the depolarization of the outer cell membrane at the junction between nerve and muscle cells. The outer membrane (sarcolemma) of muscle cells has specialized zones, which run into and through the muscle cell as a series of tubules arranged at right angles (transverse) to the muscle fibres in the region of the Z disks. This gives the cell the ability to transmit the depolarization of the outer membrane, in response to incoming nervous signals in positions which are physically close to the operating areas of the fibres.

32.3 ENERGY PROVISION IN MUSCLE TISSUE

It is a common observation that all muscles are not of the same colour; this is particularly easy to see in the carcass of a chicken, where the muscle of the breast is pale pink whereas the meat of the legs is a much redder shade. On cooking these yield meats which are light and dark, respectively. The different colours are related to the functions of the different muscle areas and to their biochemical and physiological properties. Three different types of fibre are commonly recognised: two of these are white and one red. Within any particular muscle it is common for all three types of fibre to be found, although the predominant one will differ. Basically, the red fibres are found in muscles that have a high and persistent level of physical activity. The red fibres are denoted as type I whereas the white are type II, a category which is further split into IIA and IIB depending upon the type of oxidative metabolism that predominates. All types of muscle share a common contractile mechanism and a consequent need for ATP. Where they differ is in the fuel which supplies their ATP.

There are four potential fuels that can provide the ATP for muscular activity:

- phosphocreatine
- glycogen
- plasma glucose
- plasma lipids.

The importance of individual fuels varies in the different types of muscle fibre and may also change with circumstances.

The characteristics of the three type of muscle fibre are summarized in Table 32.1. The red, type I fibres have a slow twitch rate after stimulus (they are often referred to as 'slow-twitch' fibres). The motor neurones of this type of muscle have a very low threshold of activation which means that these fibres are the ones which will be activated first. These are predominantly fibres which gain their ATP from mitochondrial oxidation of substrates that are taken up from blood, in particular plasma non-

esterified fatty acids (NEFA) and plasma glucose. They are richly supplied with capillaries to supply the large amounts of substrates that they need. The pattern of metabolism is predominantly aerobic oxidation, leading to a high oxygen demand within the cytochrome system of the mitochondria. This need is met by the action of myoglobin, an oxygen carrier within the cytosol of the fibres. When resting, the major fuel source (about 80%) for the slow-twitch fibres is NEFA. As the physical demands on the muscle increase during exercise, the proportion of energy that can be derived from this source drops and more glucose is used. At the highest levels of muscle output, the principal energy substrate is glucose.

Phosphocreatine is used particularly in the white fibres as a reservoir of group transfer potential. Phosphocreatine and ADP have a reversible relationship with creatine and ATP under the influence of the enzyme creatine kinase (Figure 32.5).

Phosphocreatine brings advantage to muscle because, although it is a nearly instantaneous source of ATP, it does not use up the cell's supply of adenosine. If the muscle cell stored energy only in the form of ATP, very soon all of the adenosine would be in the form of ATP and none in ADP or AMP. This would lead to the inhibition of a wide range of allosteric enzymes (e.g. phosphofructokinase) and a halt to substrate use and respiration.

Muscle glycogen acts as a reserve of energy within muscles. The amount stored is quite small but is able to provide energy for muscles that are not in persistent activity. These tend to be the white, fast-twitch fibres and, not surprisingly, these have much higher capacity for gycogenolysis and glycolysis. There are two families of fast-twitch cells (A and B) which differ in the ways in which they are able to utilize their supplies of glycogen. Type B fibres have high glycolytic ability but a low potential for mitochondrial oxidation, they must therefore gain their energy by lactate production and recycling of glucose through the Cori cycle (see Chapter 16). Type A fibres can oxi-

Table 32.1 Differences between muscle fibre types

Property	Fibre type		
	Type I	*Type II A*	*Type II B*
Colour	Red	White	White
ATPase activity	Low	High	High
Creatine kinase	Low	High	High
Glycolytic enzymes	Low	High	High
Mitochondrial oxidation	High	High	Low
Myoglobin	High	Low	Low
Twitch speed	Slow	Fast	Fast
Fatigue	Resistant	Resistant	Fatiguable

dize glycogen completely to CO_2 and water, thereby increasing their ATP yield from each of the glucose units of glycogen by about 18 times. The type B fibres are easily fatigued because their energy supply is, in the short term, non-renewable.

32.3.1 MYOGLOBIN

Myoglobin shares the same haem structure (iron protoporphyrin IX) with haemoglobin, although unlike haemoglobin it exists as a monomeric protein structure with a single polypeptide chain of 153 amino acids. Unlike haemoglobin, myoglobin is purely a carrier of oxygen and does not transport CO_2 or H^+. The presence of the haem within the protein structure gives the characteristic red colour to muscle.

The iron atom is held into the structure of the porphyrin ring as a coordination compound. Iron has a coordination number of 6, and of these bonds, four are with the nitrogen atoms in the individual five-membered rings of the porphyrin. The fifth is with a nitrogen atom in the side chain of a histidine group (amino acid 92 out of 153) in the polypeptide.

The sixth is the one that can be occupied by a pair of electrons from a molecule of oxygen.

32.4 CHANGES IN MUSCLE AFTER DEATH

Immediately after death, the oxygen supply to the muscle is cut off and conditions become anaerobic, so that there is no longer a supply of ATP from oxidative phosphorylation. Even in those muscles that normally operate under anaerobic conditions through the action of the Cori cycle (Chapter 16), the ability to recycle lactate/glucose through the liver is lost due to the cessation of blood circulation. The immediately available energy sources in muscle, ATP and creatine phosphate, are used up in a very short time. Thereafter the supply of ATP must come from the anaerobic oxidation of the limited supplies of glycogen to lactate (Chapter 16). The demand for energy is not completely lost at death: the Ca^{2+} ATPase continues to maintain the Ca^{2+} gradient in the sarcoplasmic reticulum and uncoordinated muscle twitching can be observed.

The actin and myosin components of the muscle fibre are only able to separate when ATP is bound to myosin, and as supplies of

Figure 32.5 Creatine kinase catalyses the reversible transfer of phosphate groups from ATP to creatine.

ATP become exhausted the overlapping actin and myosin become strongly bonded together and are thus firmly cross-linked. This results in the muscle becoming very hard and the process known as *rigor mortis* takes hold. Maximum rigor occurs at about 24 hours after death, depending upon both temperature and the species of the animal.

Prior to rigor, muscle is quite extendable under load and acts like a spring, returning to its original length when the load is removed. After the cross-linking of the filaments this elasticity is lost. On cooking, meat taken from an animal before *rigor mortis* is relatively tender. However, it becomes extremely tough as the actin and myosin filaments link together.

In order for muscle to be converted into meat it has to go through a process of conditioning where it is allowed to mature for a period of days or even weeks before consumption. During this period the meat will gain greatly in tenderness and possibly in flavour. Perhaps surprisingly, the changes in tenderness are not due to a decrease in the amount or quality of collagen in the muscle, which tends to remain constant during conditioning. Muscle with a high level of collagen, particularly that from old animals, leads to the formation of meat with a high level of collagen. This gives one biochemical reason for the fact that meat from older animals tends to be tough.

The main changes in conditioning occur within the fibres of the sarcomeres. The area around the Z disk of each fibre (where the thin filaments are attached) starts to break down. The thin filaments are thus free to pull away. Meat that has been conditioned for about 3–4 days will be as extendable under load as muscle from a freshly killed animal; however, once the applied load is removed the fibres will not contract back to their original size. These changes imply a gradual breakdown of protein structure within the myofibrils. They occur under the influence of enzymes from within the muscle itself rather than from any bacterial fermentation.

32.4.1 ENZYMES LEADING TO THE DEGENERATION OF MYOFIBRILS

The changes in myofibril structure require the presence of Ca^{2+} ions; if these are removed by the use of a complexing agent (such as EDTA) then only limited degeneration takes place. It is interesting to see that this is a further role for calcium which is so closely involved in the process of muscle contraction. The enzyme responsible for myofibrilar degradation is known as calcium-activated factor (CAF) and has been prepared from a wide range of species. Its distribution is not limited to muscle and it can be prepared from many different tissues.

CAF specifically hydrolyses troponin T and, to a lesser extent, troponin I. It can also trigger the release of α-actinin (a protein which binds to actin within the Z disk). Another protein of the Z disk, desmin, is also degraded by CAF, further weakening the attachment of the thin fibres to the disk.

Within the living animal, most cellular components are constantly undergoing breakdown and resynthesis. One of the mechanisms for breakdown involves the activity of the cell's lysosomes which act very much as intracellular stores of a whole range of hydrolytic enzymes. Amongst those responsible for protein degradation are collagenase (breaking down collagens), peptidases (acting on smaller peptides) and a whole range of proteolytic enzymes known as the cathepsins. Although their levels are much higher in liver, two of the cathepsins, B and D, are found in striated muscle. Both of these enzymes are able to catalyse a limited breakdown of actin and myosin to smaller subunits.

Experiments on muscle from cattle have indicated that the activity of CAF is more important than that of the cathepsins in meat conditioning.

32.4.2 THE ROLE OF MYOGLOBIN

Transport of oxygen within the extended cell structure of the sarcomere requires that muscle has its own intracellular carrier system. Myoglobin can exist in three different forms dependent upon the structure of the haem group. At its heart, the haem molecule has an atom of iron attached to the centre of the porphyrin ring. The iron is usually in oxidation state II (ferrous), and is attached covalently to a nitrogen atom on the side chain of a histidine residue in myoglobin's polypeptide chain. Oxygen attaches directly to the iron atom to form oxymyoglobin; the configuration without oxygen is known as deoxymyoglobin. In a third and physiologically inactive variant the iron atom is oxidised to the III state (ferric).

After death the oxygen attached to myoglobin is used up very rapidly, and the molecule assumes the reduced form of deoxymyoglobin, which has a very dark, almost purple colour. On cutting meat, the myoglobin picks up oxygen, forming the bright red oxymyoglobin. This is the colour that consumers associate with fresh meat and one which retailers often try to accentuate with appropriately tinted lighting. If meat is left exposed to the atmosphere for extended periods the haem group oxidizes from iron (II) to iron (III), and this ferrimyoglobin is of a brown colour (also referred to as metmyoglobin). The colour development is not a simple event – at depth within the meat, the mitochondrial cytochrome system is able to reduce ferrimyoglobin back to oxymyoglobin. This may also permit limited aerobic metabolism to continue within the meat.

32.4.3 CONDITIONING

The process of conditioning is central to the production of meat from a newly killed carcass and can make a very great deal of difference to the eventual quality of the product. The temperature of conditioning is an important component of the process. If the carcass is allowed to achieve too high a temperature then the risk of bacterial contamination of the product becomes serious. The principal improvement during conditioning is in the tenderness, although in some species there may also be changes in flavour. Meats from the major

ruminants, cattle, sheep and goats, take similar lengths of time (approximately 7–10 days) to achieve maximum tenderness, whereas pig meat requires a shorter period and chicken may require only a few days.

32.4.4 COLD SHORTENING

A complete carcass, particularly of a beef animal, takes a long time to cool down after death. Some of this is due to the poor thermal conductivity of the thick pieces of flesh, but a further factor, heat production in the carcass is that, due to ongoing biochemical reactions, may continue for a considerable period. If the temperature of the carcass is reduced too quickly to assist in the maintenance of hygiene, the process known as cold shortening results.

The contraction of muscle is controlled by the levels of free Ca^{2+} within the fibre: normal control mechanisms ensure that levels of Ca^{2+} are maintained at low levels as long as the muscle does not receive neural stimuli. Rapid chilling of muscle in the early *post-mortem* period can lead to an increase in intracellular Ca^{2+} at a time before the cell's ATP resources have been depleted. Under these conditions the thin and thick filaments move together, increasing their overlap so that the muscles take on their most contracted form. In extreme cases their length may reduce to 40% of the resting state. With the onset of *rigor mortis*, cross-linking of thick and thin filaments becomes fixed. Meat in which the muscles have become shortened is tough, and even after extended conditioning does not become as tender as normal meat. The effect of chilling on Ca^{2+} levels is probably due to an impairment of the ability of the sarcoplasmic reticulum to bind the ion and possibly also to its being released during the *post-mortem* degeneration of mitochondria.

32.4.5 EFFECTS OF STRESS PRE-SLAUGHTER

One of the main effects of animal stress, whether caused by physical exhaustion, rough

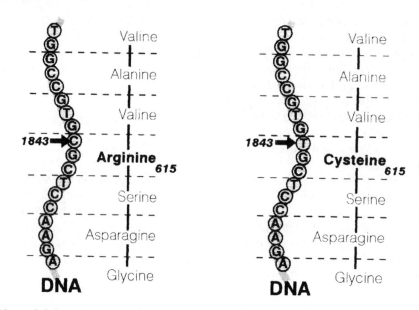

Figure 32.6 Normal (left) and mutated (right) gene responsible for malignant hypothermia (porcine stress syndrome). A single nucleotide substitution (T for C) leads to the incorporation of cysteine instead of arginine in the calcium channel protein, ryanodine.

handling, transport, fear, severe climatic conditions or hunger, is to increase the levels of catecholamine hormones produced in the adrenal medulla. In turn this leads to elevation of the activity of intracellular protein kinases and the activation of glycogen phosphorylase. Stressed animals therefore have reduced levels of glycogen in muscle, leading to a reduced capacity for lactate production (see Chapter 16) and a final pH of above 5.9 in the meat. The result is that in some muscles, but not all, the meat is darker and may be slightly firmer and drier (DFD meat: dark, firm, dry) than that from non-stressed animals. In beef animals the muscles most affected tend to be those of the hindquarter which are normally the most expensive cuts.

A further problem with a very definite biochemical basis is to be found in particularly stress-prone breeding lines of pigs. This problem is not confined to pigs but is also found in many species of livestock and even in man. It has perhaps been made more obvious in pigs through the intensive breeding and monitoring schemes introduced to produce animals that grow quickly and economically, and pro-

duce lean carcasses. It was found that in several breeds there were family trees of animals with a genetic abnormality such that quite mild stress may be fatal. In most species the problem is known as malignant hypothermia, but in pigs it is called porcine stress syndrome (PSS). The clinical findings associated with the condition included raised intracellular calcium levels which increase muscle glycogenolysis (and therefore heat production) and muscle tone. The lesion is a biochemical one at the level of the gene which codes for ryanodine, one of the proteins of the calcium channel that releases calcium from the sarcoplasmic reticulum. There is a simple substitution of a thymine residue for a cytosine at base 1843 in the DNA strand that codes for the protein, which changes the triplet codon from CGC to TGC (Figure 32.6). The amino acid that is coded by CGC is arginine, but TGC leads to cysteine incorporation into the polypeptide. There are other abnormalities in the base sequence of this gene in stress-susceptible pigs, but these do not lead to variations in the amino acid which is coded for. For instance, base 3942 codes for thymine in normal pigs and cytosine in susceptible ones. This leads to a change in triplet codon from GGT to GGC. Both of these codes are valid combinations for glycine so that there is no change in the eventual polypeptide formed.

The change from the normal arginine to cysteine leads to the formation of an abnormal variety of ryanodine, which is far more sensitive to the stimuli leading to Ca^{2+} release. Pigs which suffer from this problem were found to be extremely sensitive to anaesthetics, particularly halothane, and this has been used as a very simple way of distinguishing these animals. As the condition is due to a gene abnormality, much progress has been made in its elimination by selective breeding away from affected lines. After death, affected animals which have survived the stresses of transport and pre-slaughter handling produce carcasses with serious problems. Due to the elevated intracellular calcium levels, *rigor mortis* within major muscle groups sets in almost immediately, there is a very rapid production of lactate, and the pH drops to 5 or below. The muscles most affected are those yielding the high-value cuts of meat in the back and legs. The meat produced from these animals tends to be unusually pale and soft, and oozes fluid on cutting. Such meat is termed pale, soft, exudative (PSE) pork.

FURTHER READING

Alberts, B., Bray, D., Lewis, J., Raff, M., Roberts, K. and Watson, J.D. (1989) *Molecular Biology of the Cell*, 2nd edn, Garland Publishing Inc., New York.

Allen, J.C. and Hamilton, R.J. (eds) (1989) *Rancidity in Foods*, 2nd edn, Elsevier Applied Science, London and New York.

Anderson, J.W. and Beardall, J. (1991) *Molecular Activities in Plant Cells, An Introduction to Plant Biochemistry*, Blackwell Scientific Publications, Oxford.

Bender, D.A. (1992) *Nutritional Biochemistry of the Vitamins*, Cambridge University Press, Cambridge, UK.

Buttery, P.J., Boorman, K.N. and Lindsay, D.B. (eds) (1992) *The Control of Fat and Lean Deposition*, Butterworth–Heinemann, Oxford.

Campion, D.R., Hausman, G.J. and Martin, R.J. (eds) (1989) *Animal Growth Regulation*, Plenum Press, New York.

Christie W.W. (ed.) (1981) Digestion, Absorption and Transport of Lipids in Ruminant Animals, in *Lipid Metabolism in Ruminant Animals*, Pergamon Press, Oxford, pp. 57–94.

Christie W.W. (ed.) (1981) Lipid metabolism in the Rumen, in *Lipid Metabolism in Ruminant Animals*, Pergamon Press, Oxford, pp. 21–56.

Davis P.J. (ed) (1995) *Plant Hormones: Physiology, Biochemistry and Molecular Biology*, 2nd edn, Kluwer, Dordrecht.

Dennis, D.T. and Turpin, D.H. (eds) (1990) *Plant Physiology, Biochemistry and Molecular Biology*, Longman Scientific and Technical, Harlow, Essex.

Finean, J.B., Coleman, R. and Michell, R.H. (1984) *Membranes and their Cellular Functions*, 3rd edn, Blackwell Scientific Publications, Oxford.

Gale E.S. (ed.) (1981) *The Molecular Basis of Antibiotic Action*, 2nd edn, John Wiley, London.

Gunstone, F.D., Harwood, J.L. and Padley, F.B. (1994) *The Lipid Handbook*, 2nd edn, Chapman & Hall, London.

Gurr M.I. (1992) *Role of Fats in Food and Nutrition*, 2nd edn, Elsevier Applied Science, London.

Gurr, M.I. and Harwood, J.L. (1991) *Lipid Biochemistry, An Introduction*, 4th edn, Chapman & Hall, London.

Horton, R.H., Moran, L.A., Ochs, R.S., Rawn, J.D. and Scrimgeour, J.G. (1993) *Principles of Biochemistry*, Prentice Hall International, Englewood Cliffs, New Jersey.

Kaneko, J.J. (ed.) (1989) *Clinical Biochemistry of Domestic Animals*, 4th edn, Academic Press, London.

Kigel, J. and Galili, G. (1995) *Seed Development and Germination*, Marcel Dekker, Philadelphia.

Lawlor, D.W. (1993) *Photosynthesis: Molecular, Physiological and Environmental Processes*, 2nd edn, Longman Scientific and Technical, Harlow, Essex.

Lea, P.J. and Leegood, R.C. (1993) *Plant Biochemistry and Molecular Biology*, John Wiley and Sons Ltd, Chichester.

Lehninger, A.L., Nelson, D.L. and Cox, M.M. (1993) *Principles of Biochemistry*, 2nd edn, Worth, New York.

Machlin L.J. (ed) (1991) *Handbook of Vitamins*, 2nd edn, Marcel Dekker, New York.

Marschner, H. (1995) *Mineral Nutrition of Higher Plants*, 2nd edn, Academic Press, London.

Mathews, C.K. and Van Holde, K.E. (1996) *Biochemistry*, 2nd edn, Benjamin/Cummings, Menlo Park, California.

McDonald, P., Edwards, R.A., Greenhalgh, J.F.D. and Morgan, C.A. (1995) *Animal Nutrition*, 5th edn, Longman Scientific and Technical, Harlow, Essex.

McDowell, L.R. (1989) *Vitamins In Animal Nutrition, Comparative Aspects to Human Nutrition*, Academic Press, San Diego, California.

Mepham, T.B. (1987) *Physiology of Lactation*, Open University Press, Milton Keynes.

Mertz, W. (1987) *Trace Elements in Human and Animal Nutrition*, 5th edn, Academic Press, New York.

Pearson, A.M. and Dutson, T.R. (eds) (1991) *Growth Regulation in Farm Animals, Advances in Meat Research, Volume 7*, Elsevier Science, London.

Primrose, S.B. (1991) *Molecular Biotechnology*, Blackwell Scientific Publications, Oxford.

Roberts, J.A. and Hooley, R. (1988) *Plant Growth Regulators*, Blackie and Son Ltd, Glasgow and London.

Ruckebusch, Y., Phanent, L.P. and Dunlop, R. (eds) (1991) *Physiology of Small and Large Animals*, Marcel Dekker, Philadelphia.

Salisbury, F.B. and Ross, C.W. (1992) *Plant Physiology*, 4th edn, Wadsworth Publishing Co., Belmont, California.

Van Soest. P.J. (1994) *Nutritional Ecology of the Ruminant*, Cornell University Press, Ithaca, New York.

Yeagle, P. (1987) *The Membranes of Cells*, Academic Press, San Diego, California.

INDEX